Proceedings of the I-ESA Conferences

Volume 7

For further volumes:
http://www.springer.com/series/10534

Preface

INTEROP-VLab defined in 2007 Enterprise Interoperability (EI) as "the ability of an enterprise system or application to interact with others at a low cost in a flexible approach." This definition was the result of several analyses with the existing EI interface. Often the development of these interfaces was very costly and about all was not able to evolve when the IT applications of the both enterprises were changing.

The research on EI started in 2001 based on an action initiated by European Commission on the demand of the European Industry to reduce the cost of EI development.

Since this time, several research programs were launched including a Network of Excellence "Interoperability Research for Networked Enterprises Applications and Software" INTEROP-NoE (IST-508011, http://www.interop-noe.org) at the end of 2003 which gave, among other results, the creation of INTEROP-VLab, the Virtual European Laboratory for the development of Enterprise Interoperability (www.interop-vlab.eu).

One of the main results promoted by INTEROP-VLab is that the EI solutions must have a multidisciplinary vision by merging three research areas supporting the development of Enterprise Interoperability:

- ICT: the basic technology to develop EI solutions including Future Internet.
- Enterprise Modeling: defining requirements and supporting implementation solutions for the organization.
- Ontology: ensuring the semantically consistency of organizations and solutions.

The motto for I-ESA 2014 was "*interoperability for agility, resilience and plasticity of* collaborations." Actually, *agility* is a crucial concept in collaborative situations. Consequently, interoperability in an agile framework is definitely one of the next main steps in the management of interoperability. We suggest considering the concept of agility according to four main aspects:

- Being agile requires, first, being able to supervise the relevant environment to *detect* any elements requiring agility.
- Being agile requires, second, being able to control its own structure and its own behaviour to *adapt* to the faced situation.

- Being agile requires, third, being able to proceed detection and adaptation in a *reactive* manner (that is dynamically relevant with the evolution of context).
- Being agile requires, four, being able to proceed detection and adaptation in an *efficient* manner (that is functionally relevant with the evolution of context).

Consequently, agility in interoperability may be defined as: "*the ability of an interoperable system to detect its potential unsuitability to the current environment and to perform a relevant adaptation according to its composing systems, in a reactive and efficient manner.*" This definition may be simply and roughly formulated as:

$$\textbf{\textit{Agility}} = (\textbf{\textit{Detection}} + \textbf{\textit{Adaptation}}) \times (\textbf{\textit{Reactiveness}} + \textbf{\textit{Efficiency}}).$$

In an ecosystem more and more fluid, enterprises and organizations have to take part into collaborations and to perform interoperability. However, this interoperability should also be agile in order to ensure *resilience* and *plasticity* of collaborative systems.

Interoperability for Enterprise Systems and Applications (I-ESA 2014) is the seventh Conference after the six previous successful experiences: Genève (2005), Bordeaux (2006), Madeira (2007), Berlin (2008), Coventry (2010), and Valencia (2012), and a special edition in Beijing (2009). This time the motto is "Interoperability for Agility, Resilience and Plasticity of Collaborations." The I-ESA'14 Conference was organized by Ecole des Mines Albi-Carmaux, on behalf PGSO ("Grand Sud-Ouest" French Pole of INTEROP-VLab), and the European Virtual Laboratory for Enterprise Interoperability (INTEROP-VLab) and sponsored by the International Federation for Information Processing (IFIP).

The program proposed several keynotes presented by high-level renowned experts from industry, government, and academia:

- Mr. Gerald Santucci, European Commission, EU
- Pr. Wil Van der Aalst, University of Technology of Eindhoven, NL
- Pr. Luis Camarinha-Matos, New University of Lisbon, PR
- Dr. Bartel Van de Walle, Tilburg School of Economics and Management, NL
- Mr. Sergio Gusmeroli, TXT e-Solutions SpA, IT.

World's leading researches and practitioners in the area of Enterprise Integration from government, industry, and academia contributed to this book. As a result, Enterprise Interoperability VI is a unique anthology presenting visions, ideas, research results, industrial experiences, and problems on business interoperability.

This book is organized into 11 parts addressing the major research in the scope of Interoperability for Enterprise Systems and Applications:

I. Introduction
II. Business Interoperability
III. Enterprise Modeling for Enterprise Interoperability
IV. Semantics for Enterprise Interoperability
V. Architectures and Frameworks for Interoperability
VI. Future Internet and Enterprise Systems
VII. Platforms for Enterprise Interoperability
VIII. Services for Enterprise Interoperability
IX. Enterprise Interoperability Science-Based
X. Standards for Interoperability
XI. Interoperability Scenarios and Case Studies

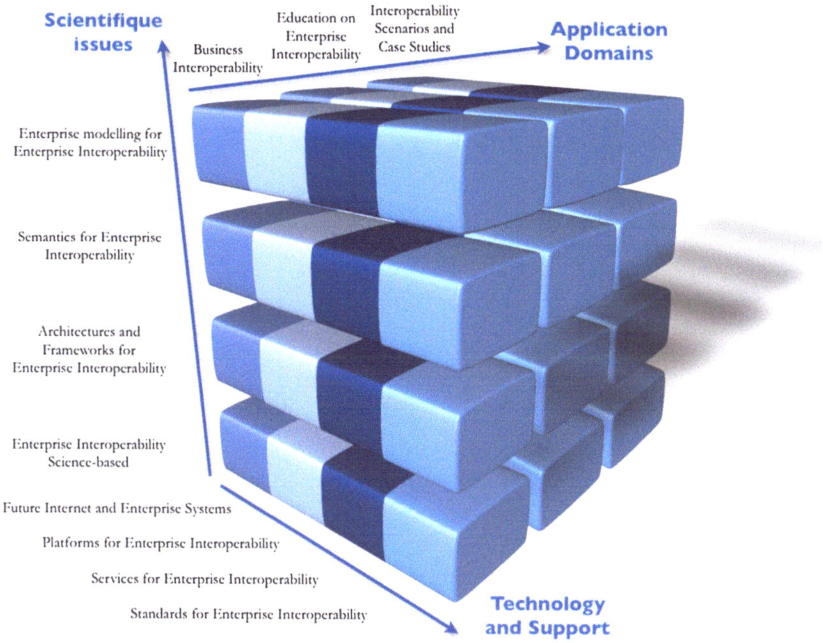

Berlin, March 2014 Kai Mertins
Albi Frédérick Bénaben
Alcoy Raúl Poler
Bordeaux Jean-Paul Bourrières

Acknowledgments

École des Mines d'Albi-Carmaux and INTEROP-VLab would like to thank all the authors, invited speakers, International Programme Committee members, International Senior Programme Committee members, International Honorary Committee members, Steering Committee members, and participants of the conference that made this book a reality and the I-ESA'14 Conference a success.

We express our gratitude to all organizations that supported the I-ESA'14 Conference preparation, especially PGSO, the Pole Grand Sud-Ouest of INTEROP-VLab, the International Federation for Information Processing (IFIP) and all the Poles members of INTEROP-VLab.

We are deeply thankful to the local organization support notably Paul Gaborit, Isabelle Fournier, Nelly Delmas, Jacques Lamothe, Matthieu Lauras, and Raquel Sanchis for their excellent work for the preparation and the management of the conference and our special thankful to Cathy Lieu for her strong involvement in the dissemination activities of I-ESA'14 Conference.

Contents

Part X Standards for Interoperability

Part XI Interoperability Scenarios and Case Studies

Part I
Introduction

Collaborative Networks: A Mechanism for Enterprise Agility and Resilience

Luis M. Camarinha-Matos

Abstract The collaborative networks paradigm, particularly when focusing the rapid formation of consortia, represents an important mechanism to support enterprises' agility and resilience in turbulent business environments. By dynamically combining the best fitting set of competencies and resources, communities of enterprises can be reshaped in different organizational forms, in order to cope with unexpected changes and disruptions, while also seeking to take advantage of new business opportunities. In this context, this paper provides a brief survey of the area, summarizing the main classes of collaborative networks, current state of developments, and challenges ahead.

Keywords Collaborative networks · Agility · Resilience · Business sustainability

1 Introduction

There seems to be a wide consensus that enterprises are nowadays under a big pressure, having to cope with rapidly and continuously changing market conditions and related business environments [1, 2]. The accumulated effects of a number of factors such as the acceleration of the globalization, changes in regulations for environmental protection and working conditions, more demanding quality standards, economical crisis in some regions, demographic shifts, and fast technological evolution, led to what is often called market turbulence. Under these conditions, the threats to business sustainability lead to higher levels of risk; furthermore, trends show that unexpected disruptive events are increasing in

L. M. Camarinha-Matos (✉)
Faculty of Sciences and Technology, New University of Lisbon,
2829–516 Monte Caparica, Portugal
e-mail: cam@uninova.pt

K. Mertins et al. (eds.), *Enterprise Interoperability VI*,
Proceedings of the I-ESA Conferences 7, DOI: 10.1007/978-3-319-04948-9_1,

frequency and in their effects [3]. In this environment, generally accepted rules and business norms become uncertain and volatile, thus inherently unstable [4].

As such, agility and resilience appear as relevant qualities for enterprises' survival and business sustainability. The notion of agility represents the ability to quickly and effectively cope with unexpected changes in the environment [2, 5]. Some authors also add the capability to take advantage of the changes [1]. Resilience, although also related to changes, has a more specific focus, representing the ability of a system to cope with severe disturbances or disruptions and return to its original or desired state [1, 6, 7]. In other words, the ability to repair or reconstitute lost capability or performance after damaging perturbations in the environment. In both cases, there is the underlying desire of a fast and effective reaction to unexpected and even disruptive changes.

Collaboration has been pointed out as a mechanism to facilitate agility and resilience, and thus a way to mitigate the effects of disruptions [6, 8, 9]. For instance, in supply chains, an increased level of visibility along the chain, which is achieved through collaboration, can help enterprises to quickly adjust to demand fluctuations and disruptions [6, 8]. On the other hand, advances in ICT, and particularly the progress on Internet-related technologies, have induced or enabled new organizational forms such as the extended enterprise, virtual enterprise, virtual organization, business ecosystem, and many others, materializing different cases of collaborative networks and constituting highly interconnected and dynamic value chains. Associated to these organizational forms, and also led, to some extent, by a technology push, new business models emerged. In this way, technology represents an important enabler for the implementation of agility and resilience [9].

However, global interconnectivity and effective and transparent information flows, although facilitating interactions and timely feedback, also bring increased complexity and dynamics, which contribute to more uncertainty, emergent and not well-understood behaviours, new risks, including cyber risks, and increased time pressure [2]. Furthermore, in globally interconnected environments, problems that used to remain confined, have now far-reaching impacts [10].

In this context, this brief *position paper* discusses the role of collaborative networks as a facilitator for agility and resilience, pointing out the main achievements in this domain, but also identifying critical research challenges ahead.

2 Motivation for Collaborative Networks

Collaborative networks (CNs) have long been associated with agility and business sustainability in turbulent markets [9, 11]. The reason being the flexibility of the corresponding organizational structures. Instead of pursuing an effort to increase in-house competencies and resources to address each new challenge, what takes time even when feasible and tends to create rigid structures, the idea is to focus on a small number of core competencies and then seek complementarities by dynamically joining efforts with other enterprises according to the needs of each business opportunity. In this way, different collaborative networks, with different "shapes", can be formed more quickly, according to the needs. At least in theory, rapid reconfiguration of consortia allows to rapidly adjust to market demand and environment constraints.

This idea has led to many works on (optimal) consortia formation, including requirements analysis and consortia planning, partners' search and selection, negotiation, and consortia launching [12–14], which is perhaps one of the most addressed topics in CNs. Nevertheless, practical implementation of the idea faces some challenges, as discussed Sect. 3.

There are two main perspectives of collaboration—the enterprise-centric view and the network-centric view. The first one focuses on management of the relationships with clients and suppliers, as reflected in the areas of Client Relationships Management (CRM) and Multiple Relationships Management (XRM). This perspective puts the enterprise in the center ("egocentric view") and, as such, is more easily assimilated by the traditional enterprise culture. However, it is biased by the client-supplier and subcontracting notions, focusing on one-to-one relationships, not really capturing the potential of agile collaborative structures.

The second perspective, represented by the area of Collaborative Networks (CNs), focuses on the network as a whole ("holistic view"), emphasizing global performance, group governance, collective/emerging behaviors, etc., thus embedding the notion of "business community" or "business ecosystem". Agility and resilience can more effectively be supported under this perspective, which pursues global (community) optimization and not only individual benefits.

It shall be noted that collaboration implies opening or diluting organizational borders, which by itself brings new risks, especially when dealing with non-trustable parties. There is, therefore, an issue of finding the right balance between competition and collaboration, which goes hand-in-hand with trust building.

3 Classes of Collaborative Networks

As mentioned above, velocity, i.e. rapid adaptation to unexpected changes or disruptions is a key pillar of agility and resilience. The dynamic formation of a consortium, combining the most adequate set of competencies and resources to satisfy the needs of each new situation sounds indeed as a very appealing approach. However, reaching a rapid "alignment" among a diverse group of heterogeneous entities is not that simple and may require considerable time to achieve. Besides technical issues such as interoperability, establishment of proper sharing mechanisms, and setting up a collaboration platform, a number of other difficult issues including trust building, establishment of proper business agreements and intellectual property management rules, governance structures and principles, alignment of value systems and business cultures, among others, require considerable time, especially if involved participants do not have experience of working with each other. Rapid formation of an effective consortium in fact requires that the involved entities are prepared to work together.

On the other hand, reaching some "universal preparedness" for collaboration is currently not realistic. This situation led to the emergence of the concept of virtual organizations breeding environment (VBE) [15], aimed at the creation of long-term communities whose members invested in being prepared to collaborate with each other and thus be in conditions, after the initial preparation effort, to rapidly respond to new business opportunities or drastic changes in the business environment. Dynamic virtual enterprises or virtual organizations can then rapidly emerge in a VBE context, seeking to find the best consortium configuration for each situation. This concept emerged in the sequence of other more primitive organizational structures such as industry clusters, or industry districts [16]. Another variant of this concept is the business ecosystem, which is inspired in the biological ecosystems and represents an alliance of stakeholders, often from the same geographical region, that aims to preserve and leverage local specificities, tradition, and business culture. Besides the aim of preparedness for collaboration, this organizational structure more clearly embeds the idea of community and the objectives of collective optimization and sustainability.

Therefore, two main classes of CNs can be identified: the mentioned breeding environments or strategic alliances focused on preparedness for collaboration, and goal-oriented networks, comprising well-focused consortia which combine competencies and resources in order to achieve a common goal or a set of compatible goals [16]. Among the goal-oriented networks, we can find dynamic and often short-term organizations such as virtual enterprises and virtual organizations, which are formed within the context of a VBE, and dissolve once the triggering business opportunity is achieved. But under this category, we can also find long-term networks, reflecting some form of continuity in production or servicing, such as supply chains, collaborative transportation networks, distributed manufacturing systems, collaborative smart grids, etc. In these cases, given the long life cycle, it is affordable to invest some time in the initial preparation phase and thus the existence of a VBE is not a pre-requisite for them.

Similarly to networks of organizations, two classes of networks of professional individuals have emerged - the Professional Virtual Community (PVC), a kind of breeding environment for dynamic goal-oriented Virtual Teams formation.

4 Trends and Further Research Challenges

The area of collaborative networks has shown considerable progress along the last decades. Figure 1 briefly illustrates the main milestones of this evolution.

In the last few decades, substantial efforts have been put at various levels, as illustrated in Fig. 2. Some overviews of achievements in these areas can be found in [17–19]. An aspect that is less studied is the dissolution phase of the CN's life

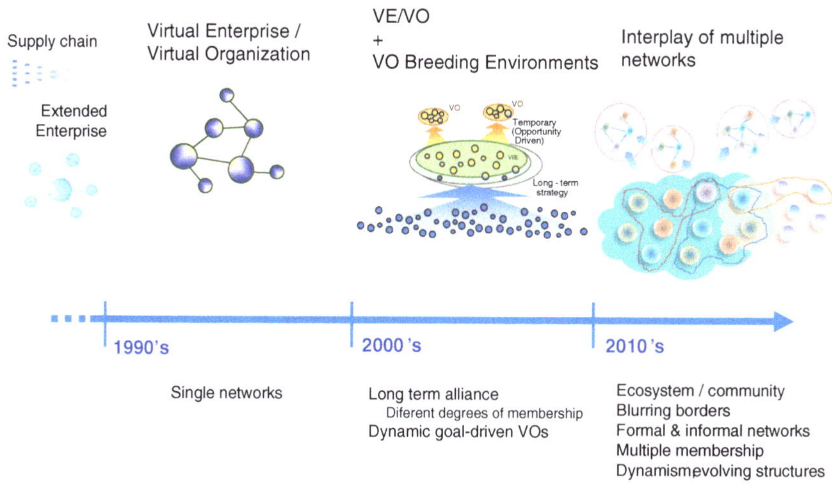

Fig. 1 Brief historic perspective of collaborative networks evolution

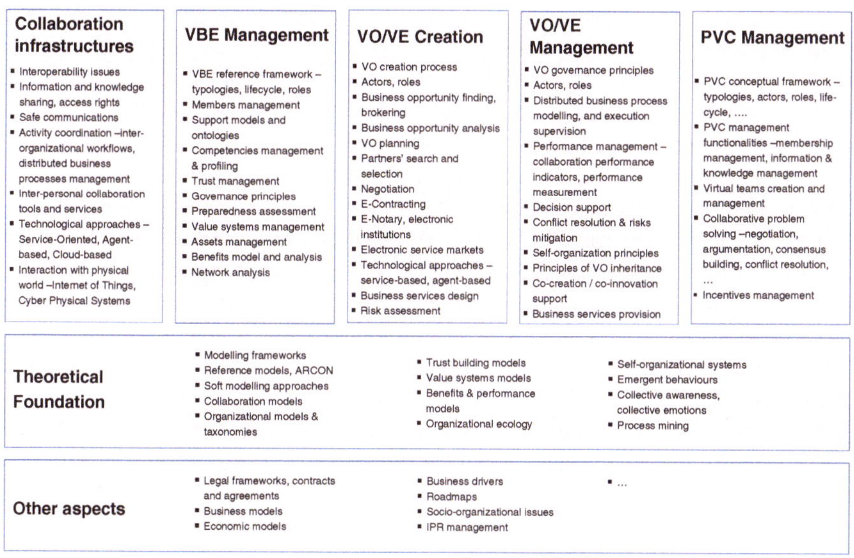

Fig. 2 Examples of key research areas in CNs in recent decades

cycle, which is becoming more relevant as dynamism and interplay among multiple networks increase.

In addition to manufacturing and supply chains, diverse forms of collaborative networks are being established in different application domains, although sometimes using different terminologies. Examples include: collaborative logistics networks, intelligent transportation systems, product-service systems, elderly care networks, collaborative sensor networks, smart environments, collaborative learning networks, etc. This growing scope of applications, combined with the possibilities offered by new technologies (e.g. cloud computing, smart mobile devices, natural user interfaces, etc.), induce new organizational forms and new business models, all pointing to a strong and dynamic interconnectivity, which in turn raises new research challenges. Examples of current challenging areas include:

- *Behavioural aspects*—The success and sustainability of collaboration requires better understanding of the involved behavioural aspects, which will provide a basis for the development of sounder governance principles and support tools. Examples of relevant research questions include [20]: How to cope with the evolution of business ecosystems (or VBEs) and emerging endogenous behavioural patterns? Can self-organizing and emergence (including co-evolution, bounded instability, recombination, etc.) play a role in changing behaviours? How to support and promote collective emotional health at the ecosystem level? Which mechanisms are adequate to induce collaborative behaviour? Which negotiation and mediation mechanisms can support conflict resolution? How can trust be promoted? How to facilitate alignment in case of disruptions? etc. In order to properly address these questions, a multi-disciplinary approach is needed (socio-technical systems).
- *Multiplex networks*—More and more, complex applications require the involvement and interplay of multiple networks. For instance, in the area of service-enhanced products (or product-service systems), various collaborative networks are involved, namely for product manufacturing, creation or co-creation of business services that enhance the product, service provision along the life cycle of the product, involvement of the customer and other local stakeholders close to the customer in the process of co-creation/co-innovation, etc. [21]. Additional challenges come from the fact that enterprises can be involved in multiple business communities, with different degrees of membership. Furthermore, it is also necessary to consider the co-existence of formal and informal networks.
- *Risks and complexity*—Although expected, risks in collaborative networks are, surprisingly, one of the least developed areas [6, 22]. Particularly in turbulent environments, it is necessary to deal not only with endogenous risks (due to misalignments), but also with exogenous ones (terrorism, natural disasters and occurrences, acceleration of globalization, demographic shift, etc.). Cyberspace risks are becoming an increasingly relevant subject of concern. As complexity of business environments increase, namely with the multiplicity and volatility of the involved organizational structures, also the risks and vulnerabilities increase, calling for urgent research actions.
- *Interconnected worlds*—Fast progress towards smart environments, i.e. context sensitive systems in which the physical and the cyber worlds are interwoven through seamless integration of sensors, actuators and other everyday objects, progressively enriched with computational and decision making power, and

interconnected through networks. This trend is reflected in a number of contemporary terms that focus on partial perspectives of the larger notion of "interconnected worlds" where collaboration can play a significant role: Internet of Things, Internet of Objects, Cyber-Physical Systems, Ambient Intelligence, Smart Environments, Collective Awareness Systems, Sensor Networks and Big data, and Sensing Enterprise.

In addition to these areas, which are only given as examples, further developments will be needed in the various sections shown in Fig. 2. A transversal challenge in all these areas is the validation issue. Since most of the effects of any change on the organizational structures and business models can only be observed much later than the normal duration of a research project, the validation process becomes a challenge on its own. A substantial part of the literature in the area remains at the "position paper" level with only minimal validation. Nevertheless, in recent years, the community is more aware of the need to find adequate validation methods for this new discipline. Attempts to combine quantitative and qualitative strategies are resorting to a mix of experimental research (when data can be available), simulation, ethnographic approaches, case studies, etc., but this will remain as a critical issue for the next years.

5 Concluding Remarks

The collaborative networks paradigm can facilitate enterprises' agility and resilience in turbulent business environments. The potential benefits of collaboration are not limited to providing a survival capability, but can also facilitate the identification and exploitation of new opportunities in such contexts.

In line with an increased set of possibilities offered by ICT and especially the so-called Future Internet, collaborative networks are "spreading" over many application domains. As a result, also new research challenges and even new research and validation approaches are emerging. In fact, the nature and wide scope of issues addressed in CNs require a multi-disciplinary approach and the involvement of different communities with distinct research cultures.

In terms of practical implementations, it shall be noted that the addressed challenges are not only a matter of technology. The introduction of this paradigm in existing business environments often requires a cultural change and a new mindset (e.g. going from a sub-contracting/outsourcing model to a collaborative culture), which is not a trivial task.

Acknowledgments This work has been partly funded by the European Commission through the Project GloNet: *Glocal enterprise network focusing on customer-centric collaboration* (Grant Agreement No. 285273). The author wishes to acknowledge the Commission for their support. He also wishes to acknowledge with gratitude and appreciation to all the GloNet Project partners for their contribution during the development of various ideas and concepts presented in this paper.

References

1. WEF. (2013). *Building resilience in supply chains.* World Economic Forum, Retrieved November 25, 2013, http://www3.weforum.org/docs/WEF_RRN_MO_BuildingResilienceSupplyChains_Report_2013.pdf
2. Alberts, D. S. (2011). *The agility advantage—a survival guide for complex enterprises and endeavors.* DoD, ISBN 978-1-893723-23-8, Retrieved November 25, 2013, http://www.dodccrp.org/files/agility_advantage/Agility_Advantage_Book.pdf
3. CDS. (2007). *The agile and resilient enterprise—a thought leadership roundtable on digital strategies.* Report from Center for Digital Strategies at the Tuck School of Business. Retrieved November 25, 2013, http://digitalstrategies.tuck.dartmouth.edu/cds-uploads/publications/pdf/Round_Overview_AgileandResilient.pdf
4. AMA. (2006). *Agility and resilience in the face of continuous change—a global study of current trends and future possibilities 2006–2016.* Report of American Management Association, Retrieved November 25, 2013, http://www.amanet.org/images/hri-agility06.pdf
5. Mansouri, M., Ganguly, A., & Mostashari, A. (2011). Evaluating agility in extended enterprise systems: A transportation network case. *American Journal of Engineering and Applied Sciences 4* (1), 142–152.
6. CU. (2003). *Creating resilient supply chains: A practical guide.* Cranfield University report, Retrieved November 25, 2013, www.som.cranfield.ac.uk/som/dinamic-content/research/lscm/downloads/57081_Report_AW.pdf
7. Lenort, R., & Wicher, P. (2011). Agile versus resilient supply chains: Commonalities and differences. Carpathian Logistics Congress, 27–30. Podbanské, Slovakia, http://konsys2.tanger.cz/files/proceedings/09/reports/1023.pdf
8. Peters, C. (2010). *Improving supply chain resilience with network centric manufacturing, DSN innovations. White Paper,* Retreived November 25, 2013, http://thelucrumgroup.com/documents/ImprovingSupplyChainResliencewithNCM_FINAL_cpeters.pdf
9. Camarinha-Matos, L. M., Afsarmanesh, H., Galeano, N., & Molina, A. (2009). Collaborative networked organizations—concepts and practice in manufacturing enterprises. *Journal of Computers and Industrial Engineering, 57,* 46–60.
10. Deloitte. (2012). Supply chain resilience: A risk intelligent approach to managing global supply chains. White paper, Retrieved November 25, 2013, http://www.deloitte.com/assets/Dcom-UnitedStates/Local%20Assets/Documents/us_consulting_supplychainresilience_052312.pdf
11. Camarinha-Matos, L. M., Afsarmanesh, H., & Boucher, X. (2010). The role of collaborative networks in sustainability. *Collaborative Networks for a Sustainable World,* IFIP AICT Series, *336,* Springer, pp. 1–16.
12. Camarinha-Matos, L. M., Oliveira, A. I., Sesana, M., Galeano, N., Demsar, D., Baldo, F., et al. (2009). A framework for computer-assisted creation of dynamic virtual organizations. *International Journal of Production Research, 47*(17), 4661–4690.
13. Baldo, F., Rabelo, R. J., & Vallejos, R. (2009). A framework for selecting performance indicators for virtual organisation partners' search and selection. *International Journal of Production Research, 47*(17), 4737–4755.
14. Crispim, J. A., & de Sousa, J. P. (2010). Partner selection in virtual enterprises. *International Journal of Production Research, 48*(3), 683–707.
15. Afsarmanesh, H., Camarinha-Matos, L. M., & Msanjila, S. S. (2011). Models, methodologies, and tools supporting establishment and management of second-generation VBEs. *IEEE Transactions on Systems, Man and Cybernetics—C, 41*(5), 692–710.
16. Camarinha-Matos, L. M., & Afsarmanesh, H. (2005). Collaborative networks: A new scientific discipline. *Journal of Intelligent Manufacturing, 16*(4–5), 439–452.
17. Camarinha-Matos, L. M., Afsarmanesh, H., & Ollus, M. (2005). Virtual organizations: Systems and practices. Springer, ISBN 0-387-23755-0.

18. Camarinha-Matos, L. M., Afsarmanesh, H., & Ollus, M. (2008). *Methods and tools for collaborative networked organizations*. New York: Springer.
19. Kühnle, H., & Dekkers, R. (2012). Some thoughts on interdisciplinarity in collaborative networks' research and manufacturing sciences. *Journal of Manufacturing Technology Management, 23*(8), 961–975.
20. Camarinha-Matos, L. M., & Afsarmanesh, H. (2011). *Behavioral aspects in collaborative enterprise networks. Proceedings of INDIN 2011—IEEE 9th International Conference on Industrial Informatics* (pp. 26–29). Caparica, Lisbon, Portugal.
21. Camarinha-Matos, L. M., Ferrada, F., & Oliveira, A. I. (2013). Interplay of collaborative networks in product servicing. *Collaborative Systems for Reindustrialization, IFIP Series, 408*, 51–60.
22. Oliveira, A. I., & Camarinha-Matos, L. M. (2013). Negotiation support and risk reduction in collaborative networks. *Technological Innovation for the Internet of Things,* IFIP AICT Series, *394*, Springer, pp. 15–24.

Data Scientist: The Engineer of the Future

Wil M. P. van der Aalst

Abstract Although our capabilities to store and process data have been increasing exponentially since the 1960s, suddenly many organizations realize that survival is not possible without exploiting available data intelligently. Out of the blue, "Big Data" has become a topic in board-level discussions. The abundance of data will change many jobs across all industries. Moreover, also scientific research is becoming more data-driven. Therefore, we reflect on the emerging *data science* discipline. Just like computer science emerged as a new discipline from mathematics when computers became abundantly available, we now see the birth of data science as a new discipline driven by the torrents of data available today. We believe that the *data scientist* will be the engineer of the future. Therefore, Eindhoven University of Technology (TU/e) established the Data Science Center Eindhoven (DSC/e). This article discusses the data science discipline and motivates its importance.

Keywords Data science · Big data · Process mining · Data mining · Visual analytics · Internet of things

1 Always on: Anything, Anytime, Anywhere

As described in [9], society shifted from being predominantly "analog" to "digital" in just a few years. This has had an incredible impact on the way we do business and communicate [12]. Society, organizations, and people are "Always On". Data is collected *about anything*, *at any time*, and *at any place*. Gartner uses the phrase "The Nexus of Forces" to refer to the convergence and mutual

W. M. P. van der Aalst (✉)
Department of Mathematics and Computer Science (MF 7.103), Eindhoven University of Technology, PO Box 513 ML-5600 MB Eindhoven, The Netherlands
e-mail: W.M.P.v.d.Aalst@tue.nl

K. Mertins et al. (eds.), *Enterprise Interoperability VI*,
Proceedings of the I-ESA Conferences 7, DOI: 10.1007/978-3-319-04948-9_2,

reinforcement of four interdependent trends: social, mobile, cloud, and information [10]. The term "Big Data" is often used to refer to the incredible growth of data in recent years. However, the ultimate goal is not to collect more data, but to turn data into real value. This means that data should be used to improve existing products, processes and services, or enable new ones. *Event data* are the most important source of information. Events may take place inside a machine (e.g., an X-ray machine or baggage handling system), inside an enterprise information system (e.g., a order placed by a customer), inside a hospital (e.g., the analysis of a blood sample), inside a social network (e.g., exchanging e-mails or twitter messages), inside a transportation system (e.g., checking in, buying a ticket, or passing through a toll booth), etc. Events may be "life events", "machine events", or both. We use the term the *Internet of Events* (IoE) to refer to all event data available. The IoE is composed of:

- The *Internet of Content* (IoC): all information created by humans to increase knowledge on particular subjects. The IoC includes traditional web pages, articles, encyclopedia like Wikipedia, YouTube, e-books, newsfeeds, etc.
- The *Internet of People* (IoP): all data related to social interaction. The IoP includes e-mail, facebook, twitter, forums, LinkedIn, etc.
- The *Internet of Things* (IoT): all physical objects connected to the network. The IoT includes all things that have a unique id and a presence in an internet-like structure. Things may have an internet connection or tagged using Radio-Frequency Identification (RFID), Near Field Communication (NFC), etc.
- The *Internet of Locations* (IoL): refers to all data that have a spatial dimension. With the uptake of mobile devices (e.g., smartphones) more and more events have geospatial attributes.

Note that the IoC, the IoP, the IoT, and the IoL are partially overlapping. For example, a place name on a webpage or the location from which a tweet was sent. See also Foursquare as a mixture of the IoP and the IoL. Content, people, things, and locations together form the IoE as shown in Fig. 1.

Data science aims to use the different data sources described in Fig. 1 to answer questions grouped into the following four categories:

- Reporting: *What happened?*
- Diagnosis: *Why did it happen?*
- Prediction: *What will happen?*
- Recommendation: *What is the best that can happen?*

The above questions are highly generic and can be applied in very different domains. Wikipedia states that "Data science incorporates varying elements and builds on techniques and theories from many fields, including mathematics, statistics, data engineering, pattern recognition and learning, advanced computing, visualization, uncertainty modeling, data warehousing, and high performance computing with the goal of extracting meaning from data and creating data

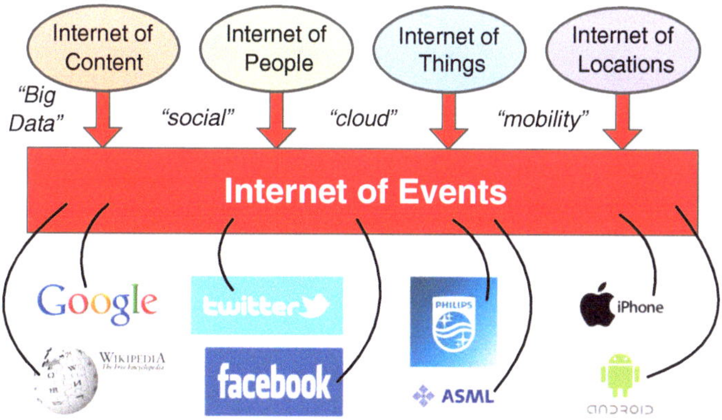

Fig. 1 The Internet of Events (IoE) is based on the Internet of Content (IoC), the Internet of People (IoP), the Internet of Things (IoT), and the Internet of Locations (IoL)

products" [21]. Many alternative definitions of data science have been suggested. For a short overview of the history of data science, we refer to [17].

The remainder is organized as follows. In Sect. 2 we discuss the unprecedented growth of (event) data and put it in a historical perspective. Section 3 compares data with oil, followed by Sect. 4 which discusses the value of this new oil. Section 5 describes the required capabilities of the data scientist. Section 6 lists some of the core technologies available to transform data into results. Finally, Sect. 7 describes the recently established Data Science Center Eindhoven (DSC/e).

2 Our Growing Capabilities to Store, Process and Exchange Data

Figure 1 describes the different sources of data contributing to the Internet of Events (IoE). As an example, take a modern smartphone like the iPhone 5S. As illustrated by Fig. 2 such phones have many sensors. These may be used to collect data on a variety of topics ranging from location (based on GPS) to usage.

It is difficult to estimate the growth of data accurately. Some people claim that *humanity created 5 exabytes (i.e., 5 billion gigabytes) of data from the Stone Age until 2003, that in 2011 that amount of data was created every 2 days, and that now (2013) it takes about 10 min to generate 5 exabytes* [18]. The expanding capabilities of information systems and other systems that depend on computing, are well characterized by Moore's law. Gordon Moore, the co-founder of Intel, predicted in 1965 that the number of components in integrated circuits would double every year. During the last fifty years the growth has indeed been exponential, albeit at a slightly slower pace. For example, as shown in Fig. 3, the number of transistors on integrated circuits has been doubling every two years.

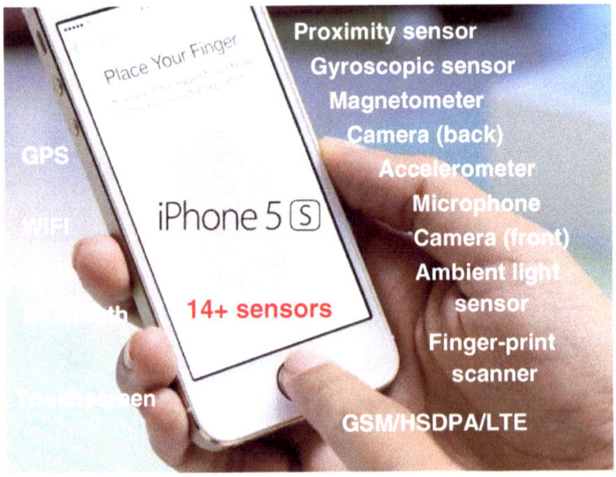

Fig. 2 Modern smartphones have many sensors that can be used to collect data

Disk capacity, performance of computers per unit cost, the number of pixels per dollar, etc. have been growing at a similar pace.

Note that Fig. 3 uses a logarithmic scale: the number of transistors on a chip increased by a factor $2^{40/2} = 1048576$ over a 40-year period. To truly grasp this development, let us illustrate this using a few comparisons. If trains would have developed like computer chips, we could now travel by train from Eindhoven to Amsterdam in approximately 5 ms (1.5 h divided by $2^{40/2}$). Airplanes could fly from Amsterdam to New York in 24 ms (7 h divided by $2^{40/2}$), and we could drive around the world using only 38 milliliters of petrol. These examples illustrate the spectacular developments associated to Moore's law.

3 Big Data as the New Oil

Data science aims to answer questions such as "What happened?", "Why did it happen?", "What will happen?", and "What is the best that can happen?". To do this, a variety of analysis techniques have been developed. However, such techniques can only be applied if the right input data is available. *Fancy analytics without suitable data are like sports-cars without petrol.* In fact, already in 2006 Clive Humby (co-founder of Dunnhumby) declared: *"Data is the new oil"*. However, only recently it became evident that data indeed represents incredible economic and societal value.

Using the metaphor "data = oil" we can definitely see similarities:

- *Exploration*: just like we need to find oil, we need to locate relevant data before we can extract it.
- *Extraction*: after locating the data, we need to extract it.
- *Transform*: clean, filter, and aggregate data.
- *Storage*: the data needs to be stored and this may be challenging if it is huge.

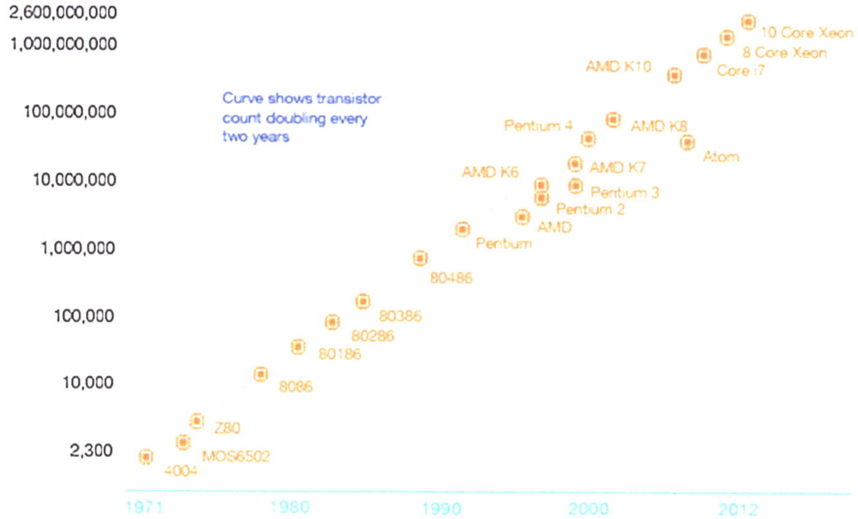

Fig. 3 Moore's law applies not only to the exponential growth of transistors on a chip: it also applies to processor speeds, communication speeds, storage space on hard disks, and pixels on a screen

- *Transport*: getting the data to the right person, organization or software tool.
- *Usage*: while driving a car one consumes oil. Similarly, providing analysis results requires data.

So the different stages from exploring crude oil to using it to drive a car also apply to data science. However, there are also important differences between data and oil:

- Copying data is relatively easy and cheap. *It is impossible to simply copy a product like oil.* (Otherwise gas prices would not be so high.)
- Data is *specific*, i.e., it relates to a specific event, object, and/or period. Different data elements are *not exchangeable*. When going to a petrol station, this is very different; drops of oil are not preallocated to a specific car on a specific day. Production to stock of data is seldom possible. Typically, data elements are unique; therefore it is difficult to produce them in advance.
- Typically, *data storage and transport are cheap* (unless the data is really "Big Data"). In a communication network data may travel (almost) at the speed of light and storage costs are much lower than the storage costs of oil.

As pointed out before, Moore's law does not apply to classical means of transport by car, trans, or plane (cf. speed, fuel consumption, etc.). The end of Moore's law has been wrongly predicted several times. However, it is clear that the ultimate limits of the law come in sight. At some stage transistors cannot be made any smaller and clock speeds cannot be further increased. Therefore, the only way to keep up with the growing demands for storage and communication is to increase the number of computing entities. See the increasing numbers of cores in processors and the trend to use large clusters of commodity hardware in the context of Hadoop. Consider for example Google. Instead of relying on expensive proprietary hardware to store and process data, Google uses industry-standard servers that both store and process the data, and can scale without limits by using distributed parallel processing. Such massive parallelization results in a huge energy consumption. This is the reason why Google invests in renewable energy and decides on the location of its data centers based on the availability of energy sources.

Energy costs and the costs of hardware are also influencing the infrastructure most suitable for large-scale data science applications. Figure 4 shows the evolution of the costs of storage. The lower line refers to the decreasing costs of disk storage. However, as shown in Fig. 4, the costs of in-memory storage are decreasing at a similar pace. Hence, the current prices of in-memory storage are comparable to the prices of disk storage of a few years ago. This explains the growing interest in in-memory databases and in-memory analytics. It now becomes affordable to load entire databases in main memory. The SAP HANA in-memory computing platform [16] is an illustration of this trend.

To understand the importance of storing data at the right place, consider the characteristics of the Xeon Intel chip shown in Fig. 5. If the CPU requires a data element and it is available in it's L1 cache, then this takes only 1.5 ns. Assume that this corresponds to a distance of 90 cm. If the data is not in the L1 cache, but in main memory, then this takes 60 ns. This corresponds to a distance of 36 meters (using our earlier assumption that 90 cm equals 1.5 ns). If the data is not in main memory, but on a Solid-State Drive (SSD) then this takes 200.000 ns. This corresponds to a distance of 120 km. To get the data from a regular hard disk takes 10.000.000 ns and corresponds to a distance of 6000 km. Hence, shifting data from hard disk to main memory may result in incredible speed-ups.

Having the right "oil infrastructure" is crucial for data science. Moreover, innovations in hardware and software infrastructures (e.g., Hadoop) allow for types of analysis previously intractable. When using MapReduce techniques and distributed computing infrastructures like Hadoop, we are trying to optimize the alignment between data and computation (e.g., bringing computation to data rather than bringing data to computation).

4 On the Value of Data

In [4] the value per user was computed by dividing the market capitalization by the number of users for all main internet companies (Google, Facebook, Twitter, etc.). This study (conducted in 2012) illustrates the potential value of data. Most user accounts have a value of more than \$100. Via the website www.twalue.com one can even compute the value of a particular twitter account, e.g., the author's twitter account (@*wvdaalst*) was estimated to have a value of \$321. Adding up the

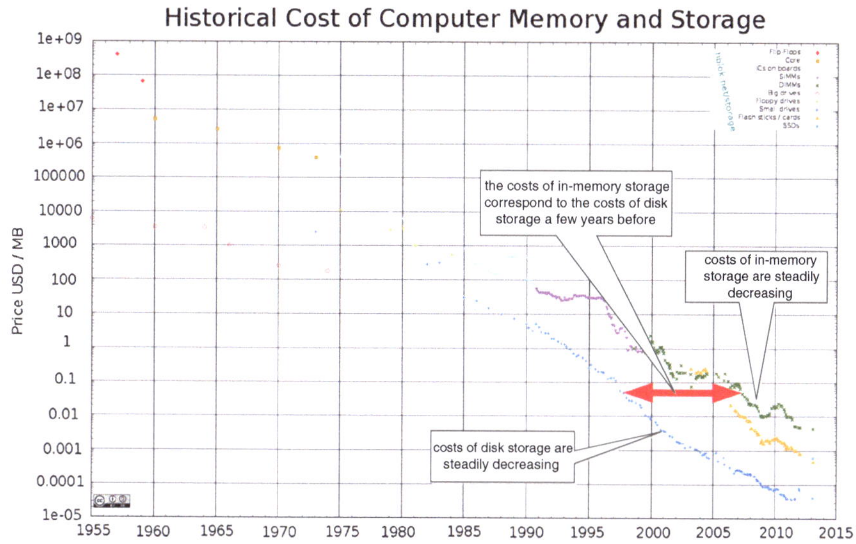

Fig. 4 Comparing the costs of different types of storage over time (taken from [13])

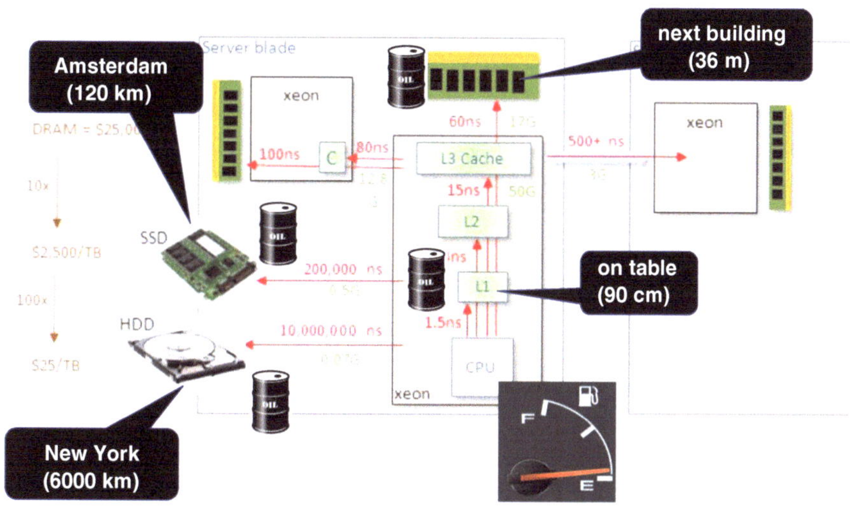

Fig. 5 How to get new oil? The power of in-memory computing becomes obvious by relating travel distances to the time required to fetch data in a computer

different social media accounts of a typical teenager may yield a value of over \$1000. Such numbers should not be taken very serious, but they nicely illustrate that one should not underestimate the value of data. Often the phrase “If you’re not paying for the product, you are the product!” is used to make internet users aware of the value of information. Organizations like Google, Facebook, and Twitter are spending enormous amounts of money on maintaining an infrastructure. Yet, end-users are not directly paying for it. Instead they are providing content and are subjected to advertisements. This means that other organizations are paying for the costs of maintaining the infrastructure in exchange for end-user data.

The internet is enabling new business models relying on data science. Some examples:

- PatientsLikeMe.com connects patients having similar medical problems and sells this information to professionals. The community platform is based on the sharing of information that is resold to a third party.
- Groupon.com provides a broker platform where customers can get a discount by buying as a group. If the deal takes place, Groupon gets parts of the revenue.
- AirBnb.com connects people so that they can rent out spare rooms to one another. AirBnb gets commission.

In all cases data is used to connect people and organizations so that information, products, or services can be exchanged.

Besides enabling new business models, data science can be used to do things more efficient of faster. Moreover, data science plays a pivotal role in Customer Relationship Management (CRM). For example, data originating from different information sources (websites, sales, support, after sales, and social media) can be used to map and analyze the so-called *customer journey*. Organizations may use analytics to maximize the opportunities that come from every interaction customers have with them. Loyal customers are more cost effective to retain than acquiring new ones, since they are likely to purchase more products and services, are less likely to leave, and may help to promote the brand.

Optimizing the customer journey is one of the many ways in which organizations benefit from data science and extract value from data. Increased competition makes data science a key differentiator. Organizations that do not use data intelligently, will not survive. This is illustrated by various studies. See for example the results of a Bain & Company study [15] shown in Fig. 6. We believe that in the future organizations will compete on analytics.

5 Data Scientist: The Sexiest Job of the Twenty-first Century

Hal Varian, the chief economist at Google said in 2009: "The sexy job in the next 10 years will be statisticians. People think I'm joking, but who would've guessed that computer engineers would've been the sexy job of the 1990s?". Later the article "Data Scientist: The Sexiest Job of the Twenty-first Century" [7] triggered a discussion on the emerging need for data scientists. This was picked up by several media and when analyzing job vacancies, one can indeed see the rapidly growing demand for data scientists (see Fig. 7).

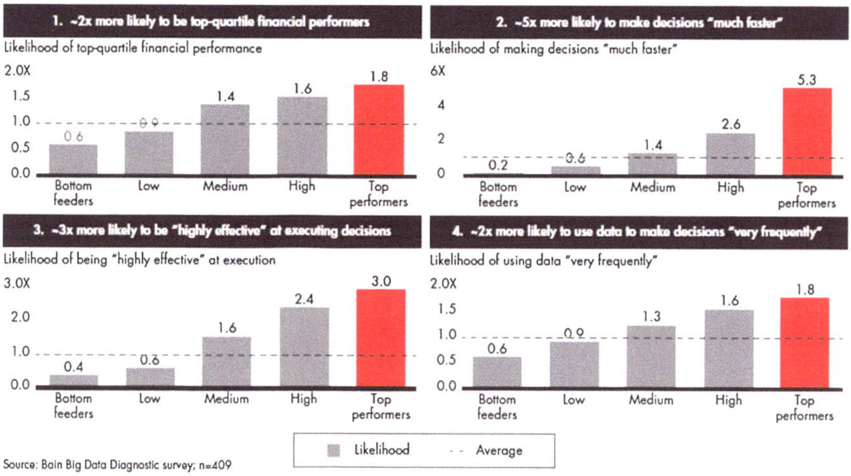

Fig. 6 Survival of the fittest: results of a Bain & Company study suggesting that companies with the best data science capabilities outperform the competition [15]

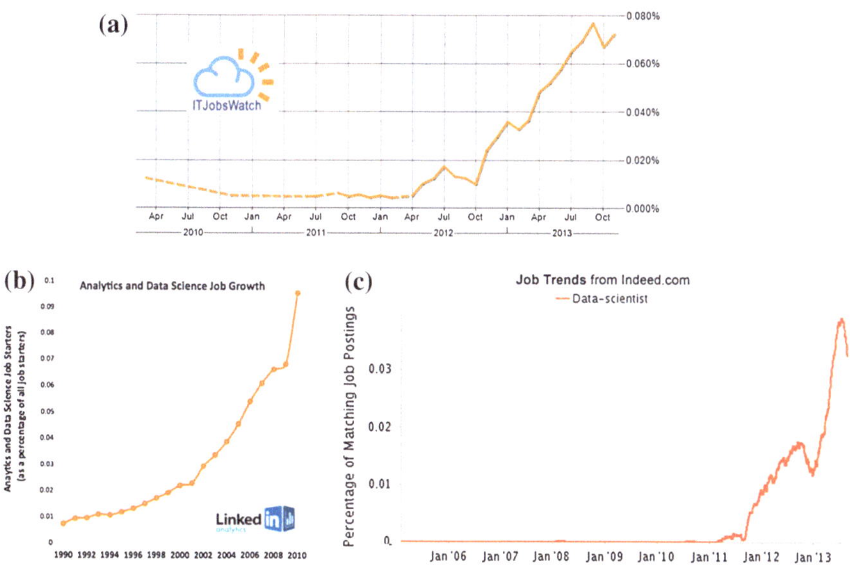

Fig. 7 The demand for data scientists is growing. **a** ITjobsWatch. **b** LinkedIn jobs. **c** Indeed jobs

So, what is a data scientist? Many definitions have been suggested. For example, [7] states “Data scientists are the people who understand how to fish out answers to important business questions from today’s tsunami of unstructured information”. Figure 8 describes the ideal profile of a data scientist. As shown, data science is multidisciplinary. Moreover, Fig. 8 clearly shows that data science

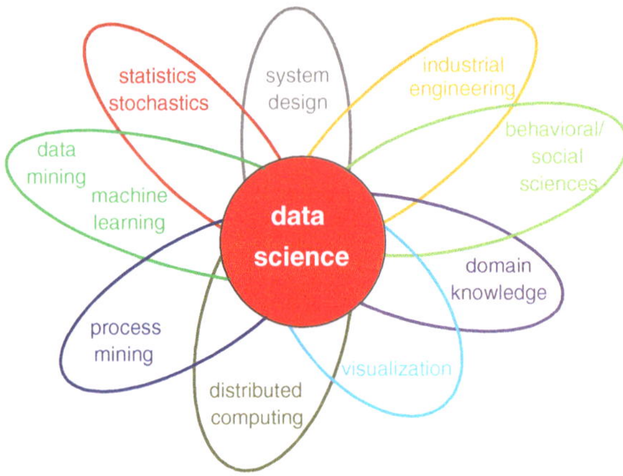

Fig. 8 Profile of the data scientist: different subdisciplines are combined to render an engineer that has quantitative and technical skills, is creative and communicative, and is able to realize end-to-end solutions

is more than analytics/statistics. It also involves behavioral/social sciences (e.g., for ethics and understanding human behavior), industrial engineering (e.g., to value data and know about new business models), and visualization. Just like Big Data is more than MapReduce, data science is more than mining. Besides having theoretical knowledge of analysis methods, the data scientist should be creative and able to realize solutions using IT. Moreover, the data scientist should have domain knowledge and able to convey the message well.

It is important to realize that data science is indeed a new *discipline*. Just like computer science emerged from mathematics when computers became abundantly available in the 1980-ties, we can now see that today's data tsunami is creating the need for data scientists. Figure 9 shows that data science is emerging from several more traditional disciplines like mathematics and computer science.

6 Turning Data into Value: From Mining to Visualization

Although data science is much broader (cf. Figure 8) we would now like to briefly describe three "data science ingredients": data mining, process mining, and visualization.

In [8] *data mining* is defined as "the analysis of (often large) data sets to find unsuspected relationships and to summarize the data in novel ways that are both understandable and useful to the data owner". The input data is typically given as a table and the output may be rules, clusters, tree structures, graphs, equations, patterns, etc. Initially, the term "data mining" had a negative connotation especially among statisticians. Terms like "data snooping", "fishing", and "data dredging" refer to ad-hoc techniques to extract conclusions from data without a sound statistical basis. However, over time the data mining discipline has become mature as characterized by solid scientific methods and many practical applications [2, 5, 8, 14, 22]. Typical data mining tasks are *classification* (e.g., constructing a decision tree), *clustering*, *regression*, *summarization*, and *association rule learning*.

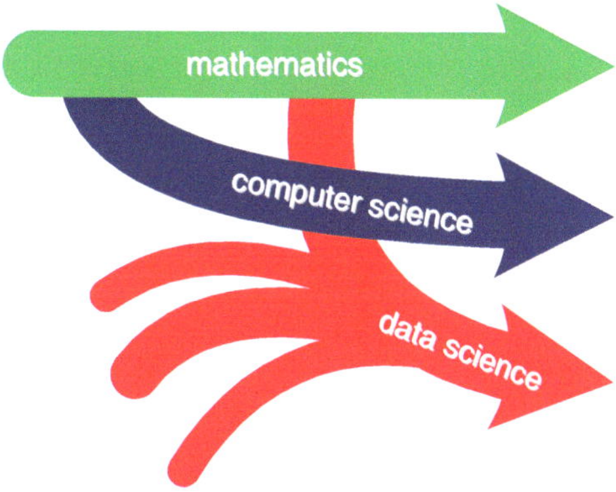

Fig. 9 Just like computer science emerged as a discipline when computers became widely available, data science is emerging as organizations are struggling to make sense of torrents of data

All of these are based on simple tabular data where the rows correspond to instances and the columns correspond to variables.

Process mining aims to *discover, monitor and improve real processes by extracting knowledge from event logs* readily available in today's information systems [1]. Starting point for process mining is an *event log*. Each event in such a log refers to an *activity* (i.e., a well-defined step in some process) and is related to a particular *case* (i.e., a *process instance*). The events belonging to a case are *ordered* and can be seen as one "run" of the process. Event logs may store additional information about events. In fact, whenever possible, process mining techniques use extra information such as the *resource* (i.e., person or device) executing or initiating the activity, the *timestamp* of the event, or *data elements* recorded with the event (e.g., the size of an order).

Event logs can be used to conduct three types of process mining [1]. The first type of process mining is *discovery*. A discovery technique takes an event log and produces a model without using any a priori information. Process discovery is the most prominent process mining technique. For many organizations it is surprising to see that existing techniques are indeed able to discover real processes merely based on example behaviors stored in event logs. The second type of process mining is *conformance*. Here, an existing process model is compared with an event log of the same process. Conformance checking can be used to check if reality, as recorded in the log, conforms to the model and vice versa. The third type of process mining is *enhancement*. Here, the idea is to extend or improve an existing process model thereby using information about the actual process recorded in some event log. Whereas conformance checking measures the alignment between model and reality, this third type of process mining aims at changing or extending the a priori model. For instance, by using timestamps in the event log one can extend the model to show bottlenecks, service levels, and throughput times.

Data and process mining techniques can be used to extract knowledge from data. However, if there are many "unknown unknowns" (things we do not know we don't know), analysis heavily relies on human judgment and direct interaction with the data. *Visualizations* may reveal patterns that would otherwise remain unnoticed.

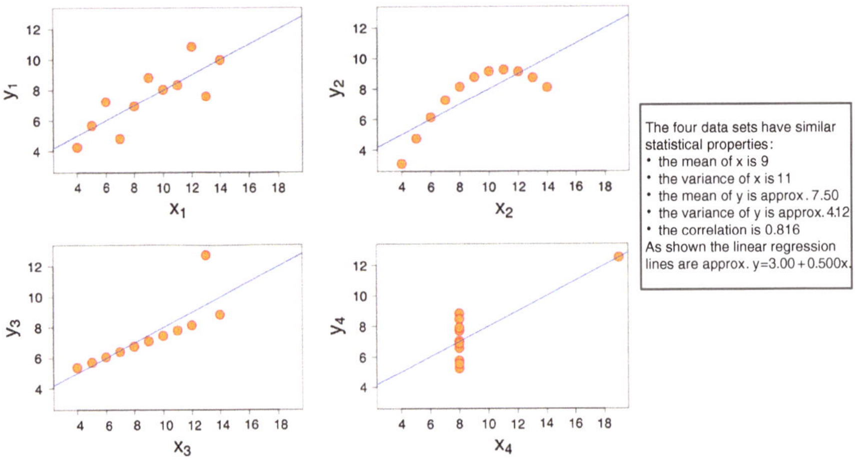

Fig. 10 Anscombe's Quartet [3]: Although the four data sets are similar in terms of mean, variance, and correlation, a basic visualization shows that the data sets have very different characteristics

A classical example is Anscombe's Quartet [3] shown in Fig. 10. The four data sets have nearly identical statistical properties (e.g., mean, variance, and correlation), yet the differences are striking when looking at the simple visualizations in Fig. 10.

The perception capabilities of the human cognitive system can be exploited by using the right visualizations [20]. Information visualization amplifies human cognitive capabilities in six basic ways: (1) by increasing cognitive resources, such as by using a visual resource to expand human working memory, (2) by reducing search, such as by representing a large amount of data in a small space, (3) by enhancing the recognition of patterns, such as when information is organized in space by its time relationships, (4) by supporting the easy perceptual inference of relationships that are otherwise more difficult to induce, (5) by perceptual monitoring of a large number of potential events, and (6) by providing a manipulable medium that, unlike static diagrams, enables the exploration of a space of parameter values [6, 19].

The term *visual analytics* was coined by Jim Thomas to advocate a tight integration between automatic techniques and visualization. Visual analytics combines automated analysis techniques with interactive visualizations for an effective understanding, reasoning and decision making on the basis of very large and complex data sets [11]. For example, data and process mining can be used in conjunction with interactive visualization.

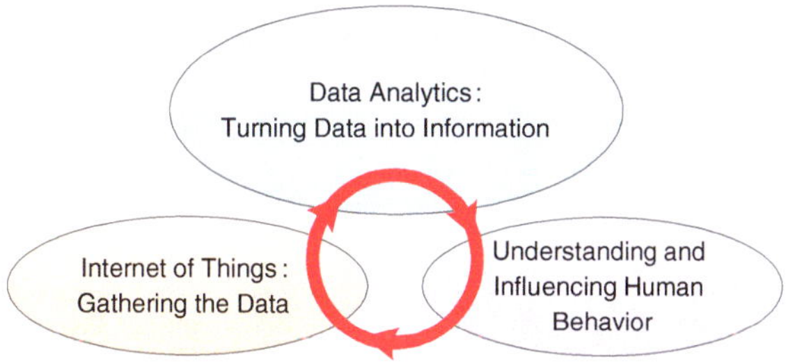

Fig. 11 The three main research lines of DSC/e

7 Data Science Center Eindhoven (DSC/e)

In 2013, the *Data Science Center Eindhoven* (DSC/e) was established as Eindhoven University of Technology's (TU/e) response to the growing volume and importance of data. About 20 research groups of the Department of Mathematics and Computer Science, the Department of Electrical Engineering, the Department of Industrial Engineering and Innovation Sciences, and the Department of Industrial design of TU/e are involved in this center.

In line with the TU/e policy, DSC/e's research contributes to the challenges of the TU/e Thematic Research Areas: Health, Energy, and Smart Mobility. Each of these areas witnesses a rapid growing volume of data triggering a variety of scientific challenges. Data science is also highly relevant for the high-tech industry in the Brainport region ("the smartest region in the world"). However, DSC/e is not limited to the TU/e's thematic research areas or the Brainport region. In fact, industries such as the financial industry and the creative industry heavily depend on data science.

TU/e has strong research groups in areas related to data science: computer science, mathematics, electrical engineering, industrial engineering, innovation sciences, and industrial design. In subdisciplines such as process mining, which are at the very heart of data science, TU/e is globally leading. The DSC/e aims to further strengthen research in three broader areas (Fig. 11).

- *Internet of Things: Gathering the Data*
- *Data Analytics: Turning Data into Information*
- *Understanding and Influencing Human Behavior.*

DSC/e's research focuses on developing new insights (models, theories, tools) to be able to add and extract value from real sets of heterogeneous data. On the one hand, the groups involved will continue to conduct focused research in particular areas relevant for data science. On the other hand, the DSC/e initiative will fuel multidisciplinary research combining expertise in the different DSC/e research groups contributing to DSC/e.

Given the empirical nature of data science, DSC/e collaborates with a wide range of organizations. Collaborations include larger joint research projects, PhD projects, master projects, and contract research. Examples of organizations collaborating within DSC/e are Philips, Adversitement, Perceptive Software, Magnaview, Synerscope, and Fluxicon.

References

1. van der Aalst, W. M. P. (2011). *Process mining: Discovery, conformance and enhancement of business processes*. Berlin: Springer-Verlag.
2. Alpaydin, E. (2010). *Introduction to machine learning*. Cambridge: MIT press.
3. Anscombe, F. J. (1973). Graphs in statistical analysis. *American Statistician, 27*(1), 17–21.
4. Bergstein, B., & Orcutt, M. (2012). Is Facebook worth it? Estimates of the historical value of a user put the IPO hype in perspective. *MIT Technology Review*, http://www.technologyreview.com/graphiti/427964/is-facebook-worth-it/
5. Bramer, M. (2007). *Principles of data mining*. Berlin: Springer-Verlag.
6. Card, S. K., Mackinlay, J. D., & Shneiderman, B. (1999). *Readings in information visualization: Using vision to think*. San Francisco: Morgan Kaufmann Publishers.
7. Davenport, T. H., & Patil, D. J. (2012, October). Data scientist: The sexiest Job of the 21st century. *Harvard Business Review*, 70-76.
8. Hand, D., Mannila, H., & Smyth, P. (2001). *Principles of data mining*. Cambridge: MIT press.
9. Hilbert, M., & Lopez, P. (2011). The world's technological capacity to store, communicate, and compute information. *Science, 332*(6025), 60–65.
10. Howard, C., Plummer, D. C., Genovese, Y., Mann, J., Willis, D. A., & Smith, D. M. (2012). *The nexus of forces: Social, mobile, cloud and information*. http://www.gartner.com
11. Keim, D., Kohlhammer, J., Ellis, G., & Mansmann, F. (Ed.). (2010). *Mastering the information age: Solving problems with visual analytics*. VisMaster. http://www.vismaster.eu/book/
12. Manyika, J., Chui, M., Brown, B., Bughin, J., Dobbs, R., Roxburgh, C., & Byers, A. (2011). *Big data: The next frontier for innovation, competition, and productivity*. McKinsey Global Institute. http://www.mckinsey.com/insights/business_technology/big_data_the_next_frontier_for_innovation.
13. McCallum, J. C. (2013). *Historical costs of memory and storage*. http://hblok.net/blog/storage/
14. Mitchell, T. M. (1997). *Machine learning*. New York: McGraw-Hill.
15. Pearson, T., & Wegener, R. (2013). *Big data: The organizational challenge. bain and company*. San Francisco: Bain & Company. http://www.bain.com/publications/articles/big_data_the_organizational_challenge.aspx/
16. Plattner, H., & Zeier, A. (2012). *In-Memory data management: Technology and applications*. Berlin: Springer-Verlag.
17. Press, G. (2013). *A very short history of data science*. Forbes Technology. http://www.forbes.com/sites/gilpress/2013/05/28/a-very-short-history-of-data-science/
18. Smolan, R., & Erwitt, J. (2012). *The human face of big data*. Against All Odds Productions. New York.
19. Thomas, J. J., & Cook, K. A. (Ed.). (2005). *Illuminating the path: The research and development agenda for visual analytics*. IEEE CS Press. Los Alamitos, CA.
20. van Wijk, J. J. (2005). The value of visualization. In C. Silva, H. Rushmeier & E. Groller (Eds.) *Visualization 2005* (pp. 79-86). IEEE CS Press. Los Alamitos, CA.
21. Wikipedia. (2013). *Data science*. http://en.wikipedia.org/wiki/data_science
22. Witten, I. H., & Frank, E. (2005). *Data mining: Practical machine learning tools and techniques (second edition)*. San Francisco: Morgan Kaufmann.

Part II
Business Interoperability

Computing the Strategies Alignment in Collaborative Networks

Beatriz Andrés and Raul Poler

Abstract Research in collaborative networks has increased due to the opportunities associated with collaboration across the networked partners. This paper focuses on the development of an approach to generally model the network, considering a set of five objects: networks, enterprises, objectives, strategies and key performance indicators (KPI). The model relates the objectives and strategies of the networked partners through KPIs defined to measure the objectives. The strategies influence, on improving the objectives, is modelled through computing the increase of the KPIs when a certain strategy is activated in an enterprise of the network. The model proposed allows researchers to formally identify the aligned strategies, in order to improve the enterprises' objectives and, consequently, the network performance.

Keywords Collaborative networks · Modelling · Objectives · Strategies alignment · KPIs

1 Introduction

Collaboration in a network involves enterprises to jointly work in order to achieve common objectives [1], often beyond their capabilities, without negatively influence the individual ones defined by the enterprises [2]. The increasing research interest in collaboration has resulted in the emergence of a wide variety of network topologies [3].

B. Andrés (✉) · R. Poler
Research Centre on Production Management and Engineering (CIGIP), Universitat Politècnica de València (UPV), Plaza Ferràndiz y Carbonell 2, 03801 Alcoy, Spain
e-mail: beaanna@cigip.upv.es

R. Poler
e-mail: rpoler@cigip.upv.es

K. Mertins et al. (eds.), *Enterprise Interoperability VI*,
Proceedings of the I-ESA Conferences 7, DOI: 10.1007/978-3-319-04948-9_3,

Making the enterprise information systems interoperable is a key factor to establish collaborative relationships within partners of the same network [4]. However, interoperability is not an only process whereby collaboration can be supported. The alignment of objectives and strategies set down is a process to be also considered to deal with the barriers appearing in collaborative networks. An idealistic situation would be one that (i) the objectives of one enterprise promote the achievement of the objectives defined by other companies in the network and (ii) the strategies carried out by one network partner are aligned with the strategies of other networked partners [5]. Nevertheless, it must be considered that individual enterprises take part in several networks and, it is very likely that some of these networks have contradictory objectives. Therefore, for enterprises belonging to a collaborative network the defined objectives and the strategies formulated by one enterprise, to achieve those objectives, could favour, or not, the objectives and strategies of other enterprises. In order to achieve the ideal situation, an enterprise belonging to a collaborative network should be able to identify those strategies that are aligned; this is, those strategies whose activation promotes the improvement of both the objectives defined by the own enterprise and the objectives defined by the rest of enterprises.

Considering how important collaboration is within the networks, and how important is to identify aligned strategies to collaborate (i.e. align interoperability strategies); this paper is focused on modelling the network objectives and strategies and the key performance indicators (KPIs) associated in order to improve the collaborative relationships among the network partners. That, will give experts a useful tool to give a shape on the strategies alignment and to identify how this alignment improves the objectives and network performance.

2 Background

A set of relevant processes have been defined and analysed, through the literature review carried out by [5], in order to characterise the collaborative networks. Amongst the identified processes, this paper focuses on the treatment of the strategy alignment process. The strategy alignment analysis allows researchers and practitioners to assess the alignment, amongst others, of interoperability strategies raised between enterprises belonging to a collaborative network.

In this paper we focus on the modelling approach, to characterise the network objectives and strategies that will allow to determine if two strategies are aligned.

Going beyond the literature review given by Andrés and Poler [5], it has been found that the models proposed in the literature are based on modelling specific domains of collaboration and specific collaborative processes, such as interoperability [6], collaborative product development [7], knowledge management [8] and planning [9] amongst other processes. The aforementioned approaches do not consider high levels of generalisation, what gives us the motivation to (i) provide a formal model to generally characterise all the objects participating at the global level of the collaborative networks, such as objectives and strategies—Sect. 3—and, taking into account the model defined at the global level, (ii) focus on a specific level in order to model the particular process of strategies alignment—Sect. 4.

Considering the above said, a formal model for networks is proposed, to allow researchers to identify the objects participating within the collaborative networks, taking into account *general* criteria to model the network as whole and *particular* criteria to model the strategies alignment process.

3 Collaborative Network Model

The main aim is to provide a general model mathematically constructed in order to (i) identify the objects that characterise the collaborative networks—Sect. 3.1, and (ii) identify the connections among these objects—Sect. 3.2. The model will allow researchers to determine how the network system is (AS-IS) and formally conceptualise the network objectives and strategies through using KPIs; giving researchers an insight of how to analyse the strategies influences within the networked partners and how to identify those that are aligned.

3.1 Network Objects

The formal model contributes to identify all the objects that influence and take part in a collaborative network.

For defining the objects of the network we have inspired on different documents: Enterprise Ontology (EO) documents [10], papers modelling specific process [6–9, 11, 12] and papers considering the relations among the network objects [13]. Furthermore, a set of definitions are also taken into account in order to clearly determine the network objects.

Lazzarini et al. [14] introduce the concept of a network chain through defining the *netchain* as "a set of networks with horizontal ties between firms, which are sequentially arranged based on vertical ties between firms in different layers". For defining the network, different perspectives are considered; the perspective provided by Christopher [15] defines a *network* as "a set of enterprises related to each other through upstream or downstream links of the different processes and activities that generate value by delivering products and services to end customers". Considering that a network is a set of enterprises, an *enterprise* is defined as a system that interacts with its environment materialising an idea, in a planned way, by satisfying the demands and desires of customers. The enterprise define the *objectives* to be achieved, and the definition of *key performance indicators* (KPI) provides the necessary information to monitor the accomplishment of these objectives [13]. Finally, the *strategies* are formulated in order to define how to reach the defined objectives. In the light of this, the KPIs can be used for measuring either objectives or strategies [13]. According to these definitions, the network model involves a set of objects defined as a 5-tuple **{N, E, O, KPI, S}**, where:

- **N** denotes the network, set of networks.
- **E** denotes the enterprises belonging to the network, set of enterprises.
- **O** denotes the objectives defined by each enterprise, set of objectives.
- **KPI** denotes the key performance indicators to measure the defined objectives, set of key performance indicators.
- **S** denotes the strategy needed to achieve the objectives set of strategies.

A minimum number of objects are identified in order to characterise the network. The objects identified allows, in next section, to model the influences (i) between the strategies activated at each enterprise belonging to the network and (ii) between strategies activated and the objectives improved.

3.2 Model

A model to characterise the objectives and strategies within the enterprises belonging to a network is proposed. This characterisation is done throughout the definition of KPIs for each of the objectives; and the strategies defined that can be activated (or not), in order to improve or achieve an enterprise objective. The modelling process is top–down performed, and starts with the set of networks (**N**) definition that consist of a group of networks, networks involve a set of enterprises (**E**), the enterprises define a set of objectives (**O**) that are measured by a set of KPIs (**KPI**), and finally, in order to achieve the objectives a set of strategies (**S**) are described. Therefore:

There is a set of networks $\boldsymbol{N}$ consisting of $\boldsymbol{n}$ networks, $N_n \in N$.

There is a set of enterprises $\boldsymbol{E}$ consisting of $\boldsymbol{i}$ enterprises, $E_i \in E$. One enterprise E_i can belong to one or more networks, $E_{ni}|E_i \in N_n$ (Eq. 1).

$$\alpha_i = \sum_n \gamma_n, \quad \begin{cases} \gamma_n = 1 & E_i \in N_n \\ \gamma_n = 0 & E_i \notin N_n \end{cases}, \tag{1}$$

where α_i is the number of networks to which the enterprise E_i belongs to and γ_n can be *1* or *0* depending on if the enterprise E_i belongs to the network N_n, or not.

There is a set of objectives $\boldsymbol{O}$ consisting of $\boldsymbol{x}$ objectives, $O_{ix} \in O$. The objectives are defined by each enterprise, $O_{ix}|O_x \in E_i$.

There is a set of key performance indicators $\boldsymbol{KPI}$ consisting of $\boldsymbol{k}$ KPIs, $KPI_{ixk} \in KPI$ where $KPI_{ixk}|KPI_k \in E_i \wedge KPI_k \in O_{ix}$.

There is a set of strategies $\boldsymbol{S}$ consisting of $\boldsymbol{s}$ strategies, $S_{is} \in S$. The strategies are defined by each enterprise, $S_{is}|S_s \in E_i$.

Rodriguez et al. [13] review the KPIs treatment and state that many authors coincide in defining some sort of KPIs to monitor the accomplishment of the objectives. As the strategies are determined by the enterprises in order to fulfill the defined objectives, the KPIs can be also related with the strategies [13]. Therefore, the KPIs can be used to measure how the strategies influence on the achievement of the objectives. Furthermore, establishing relations among the KPIs and the strategies also allows to identify the relationships between the strategies set down.

Considering the properties *active* and *not active* for the strategies. The state *active* defines that a strategy is started and carried out by an enterprise, otherwise, the state *not active* defines that a strategy is not put into practice.

$$S_{is}^{z} = \begin{cases} z = 1 & \textit{the strategy } \mathbf{s} \textit{ of the enterprise } \mathbf{i}\,(E_i) \textit{ is ACTIVE} \quad S_{is}^{1} \\ z = 0 & \textit{the strategy } \mathbf{s} \textit{ of the enterprise } \mathbf{i}\,(E_i) \textit{ is NOT ACTIVE} \quad S_{is}^{0} \end{cases} \tag{2}$$

The influence between the strategies and objectives is computed through the KPIs increase (∇KPI_{ixk}). This increase quantitatively analyses the improvement of an objective comparing the results when a strategy is active and not active. Two scenarios are considered when modeling the ∇KPI_{ixk} (KPI defined by the enterprise E_i for measuring the objective O_{ix}), when a strategy S_{is}/S_{js} is activated:

- The active strategy $\left(S_{\mathrm{is}}^{1}\right)$ is carried out in the same enterprise (E_i) in which the objective (O_{ix}) is defined. In this case, the increase of the ∇KPI_{ixk} when a strategy S_{is} is activated is defined by the Eq. (3); that models the influence of an objective when a strategy activated at the same enterprise.

$$\nabla KPI_{ixk}^{is} = \frac{KPI_{ixk}\left|S_{is}^{1} - KPI_{ixk}\right|S_{is}^{0}}{KPI_{ixk}\left|S_{is}^{0}\right.} \; | i \in E_i, x \in O_{ix}, k \in KPI_{ixk}, s \in S_s \tag{3}$$

- The active strategy $\left(S_{\mathrm{js}}^{1}\right)$ is carried out in a different enterprise (E_j) of the network in which the objective (O_{ix}) is defined. In this case, the increase of the ∇KPI_{ixk} (that measures the O_{ix} in the E_i) when a strategy S_{js} is activated is defined by the Eq. (4), that models the influence of an objective, defined at enterprise E_i, when a strategy activated at a different enterprise E_j.

$$\nabla KPI_{ixk}^{js} = \frac{KPI_{ixk}\left|S_{js}^{1} - KPI_{ixk}\right|S_{js}^{0}}{KPI_{ixk}\left|S_{js}^{0}\right.} \; | i,j \in E_i, i \neq j, x \in O_{ix}, k \in KPI_{ixk}, s \in S_s \tag{4}$$

If the $KPI_{ixk}|S_{\mathrm{js}}^{1} < KPI_{ixk}|S_{\mathrm{js}}^{0}$ implies a *good increase* (∇^{+}) therefore the Eqs. (3) and (4), defined to compute $\nabla KPI_{ixk}^{is}/\nabla KPI_{ixk}^{js}$ are reformulated as:

$$\nabla KPI_{ixk}^{is} = -\left(\frac{KPI_{ixk}\left|S_{is}^{1} - KPI_{ixk}\right|S_{is}^{0}}{KPI_{ixk}\left|S_{is}^{0}\right.}\right) \tag{3'}$$

$$\nabla KPI_{ixk}^{js} = -\left(\frac{KPI_{ixk}\Big|S_{js}^{1} - KPI_{ixk}\Big|S_{js}^{0}}{KPI_{ixk}\Big|S_{js}^{0}} \right) \tag{4'}$$

An example can be found when the KPI defined is related with a *costs measure*. If the KPI decreases due to the costs decrease when a strategy is activated, $KPI_{ixk}\Big|S_{js}^{1} < KPI_{ixk}\Big|S_{js}^{0}$, this implies a *good increase* (∇^{+}), therefore we have to consider the negative sign before the formula.

In order to model how good or bad is a strategy S_{is}/S_{js} as regards the KPI_{ixk}, defined within the enterprise E_i in order to measure the objective O_{ix}, the *KPIs increase rate* is provided through the Eqs. (5) and (6).

$$\nabla KPI_{ix}^{is} = \sum_{k} \nabla KPI_{ixk}^{is}\ \forall k, i, s|k \in KPI, i \in E, s \in S \tag{5}$$

$$\nabla KPI_{ix}^{js} = \sum_{k} \nabla KPI_{ixk}^{js}\ \forall k, j, s|k \in KPI, i, j \in E\, such\, that\, i \neq j, s \in S, \tag{6}$$

∇KPI_{ix}^{is} models the ***KPIs increase rate*** when the strategy S_{is} (defined at the same enterprise E_i) is activated,

∇KPI_{ix}^{is} models the ***KPIs increase rate*** when the strategy S_{js} (defined at a different enterprise E_j) is activated.

Once we have already defined the objects of the model and their relations, as regards the network. Now we are going to proceed to identify the alignment of the strategies established within partners belonging to the same collaborative network.

4 Strategies Alignment

According to the formulated model (Sect. 3.2), in this section an alignment function is provided in order to identify those pairs of strategies that are aligned.

The strategy alignment involves that the strategies activation benefits in the same way the objectives defined in different enterprises (Fig. 1). That is, the strategies activation implies a positive increase (∇^{+}) in the KPIs (KPI_{ixk}, KPI_{jyl}) associated to the defined objectives (O_{ix}, O_{jy}).

According to the Fig. 1, *four* different *scenarios* are identified:

SCENARIO 1/SCENARIO 2. The enterprise E_i/E_j defines an objective O_{ix}/O_{jy}. The objective O_{ix}/O_{jy} has associated a key performance indicator KPI_{ixk}/KPI_{jyl} that allows to measure if the objective O_{ix}/O_{jy} is improved or is worsen. In order to achieve the objective O_{ix}/O_{jy} the strategy $S_{is}^{1}\Big/S_{jr}^{1}$ is activated. Therefore,

- a strategy S_{is}/S_{jr} has a positive relation with the objective O_{ix}/O_{jy} if the KPI_{ixk}/KPI_{jyl} associated has a positive increase ($\nabla^{+}KPI_{ixk}/\nabla^{+}KPI_{jyl}$).

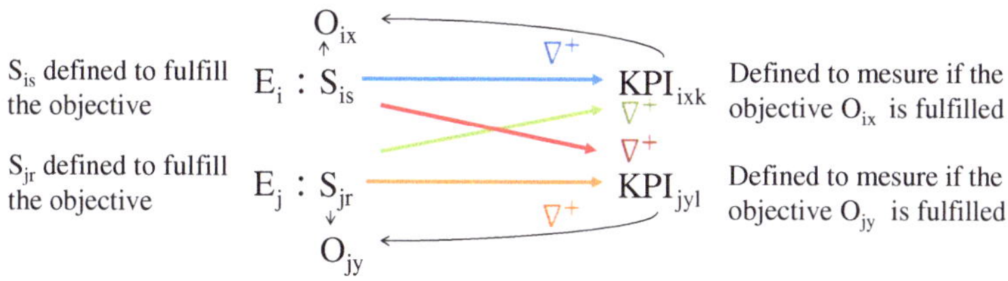

Fig. 1 Strategies alignment

- a strategy S_{is}/S_{jr} has a negative relation with the objective O_{ix}/O_{jy} if the KPI_{ixk}/KPI_{jyl} associated has a negative increase ($\nabla^{-}KPI_{ixk}/\nabla^{-}KPI_{\mathrm{jyl}}$).

SCENARIO 3. The enterprise E_i defines an objective O_{ix}. The objective O_{ix} has associated a key performance indicator KPI_{ixk} that allows to measure if the objective O_{ix} is improved or is worsen. If we want to identify into which extent the strategy activated in another enterprise $E_j\left(S^1_{jr}\right)$ positively or negatively affects the O_{ix}, the key performance indicator (KPI_{ixk}) to measure the O_{ix} is to be analysed. Two situations are possible:

- The KPI_{ixk} positively increases (∇^{+}) when the S^1_{jr} is activated. Therefore it can be stated that a strategy S^1_{jr} has a positive relation with the objective O_{ix} due to the KPI_{ixk} associated has a positive increase $(\nabla^{+}KPI_{ixk})$.
- The KPI_{ixk} negatively increases (∇^{-}) when the S^1_{jr} is activated. Therefore it can be stated that a strategy S_{jr} has a negative relation with the objective O_{ix} due to the KPI_{ixk} associated has a negative increase $(\nabla^{-}KPI_{ixk})$.

SCENARIO 4. The enterprise E_j defines an objective O_{jy}. The objective O_{jy} has associated a key performance indicator KPI_{jyl} that allows to measure if the objective O_{jy} is improved or is worsen. If we want to identify into which extent (positive or negative) the strategy activated in another enterprise E_i $\left(S^1_{is}\right)$ affects positively or negatively to the O_{jy}, the key performance indicator (KPI_{jyl}) to measure the O_{jy} is to be analysed. Two situations are possible:

- The KPI_{jyl} positively increases (∇^{+}) when the $\left(S^1_{is}\right)$ is activated. Therefore it can be stated that a strategy S_{is} has a positive relation with the objective O_{jy} due to the KPI_{jyl} associated has a positive increase $\left(\nabla^{+}KPI_{\mathrm{jyl}}\right)$.
- The KPI_{jyl} negatively increases (∇^{-}) when the S^1_{is} is activated. Therefore it can be stated that a strategy S_{is} has a negative relation with the objective O_{jy} due to the KPI_{jyl} associated has a negative increase $\left(\nabla^{-}KPI_{\mathrm{jyl}}\right)$.

Considering the four scenarios, a *function of alignment* $\boldsymbol{a(S_{is}, S_{jr})}$ is proposed to quantitatively analyse if two strategies are aligned (Eq. 7). In the light of this, two strategies S_{is} and S_{jr} are aligned if and only if being both strategies active S^1_{is} and S^1_{jr} the increases of the KPIs associated are positive $\left(\nabla^+ KPI_{ixk} / \nabla^+ KPI_{jyl}\right)$.

$$a\left(S_{is}, S_{jr}\right) = \begin{cases} a\left(s_{is}, S_{jr}\right) = 1 \leftrightarrow \begin{cases} S^1_{is} \rightarrow \nabla^+ KPI_{ixk} \wedge \\ S^1_{is} \rightarrow \nabla^+ KPI_{jyl} \wedge \\ S^1_{jr} \rightarrow \nabla^+ KPI_{ixk} \wedge \\ S^1_{jr} \rightarrow \nabla^+ KPI_{jyl} \wedge \end{cases} \\ a\left(s_{is}, S_{jr}\right) = 0 \end{cases} \tag{7}$$

Equation (7) allows to define pairs of strategies (S_{is} and S_{jr}) aligned when $\boldsymbol{a(S_{is}, S_{jr}) = 1}$. If $\boldsymbol{a(S_{is}, S_{jr}) = 0}$ it is concluded that the strategies S_{is} and S_{jr} are not aligned.

Through the proposed model, a *KPIs IMPROVEMENT INDEX* (Eq. 8) is defined as a result of the sum of the **KPIs increase rate** when aligned strategies ($a(S_{is}, S_{jr}) = 1$) are activated. This allows researchers to determine the extent to which the objectives are improved when aligned strategies are activated.

$$\textit{KPIs IMPROVEMENT INDEX} = \sum_k \nabla KPI^{is}_{ixk} + \sum_k \nabla KPI^{js}_{ixk} \left| a\left(S_{is}, S_{jr}\right) = 1 \right. \tag{8}$$

The provided model allows researchers to identify if pairs of strategies are aligned. A numerical example is proposed to better understand the proposed model. Considering a network with $\boldsymbol{i}$ enterprises, as shown in Fig. 2, amongst them two are chosen, E_1 and E_2. Each enterprise defines a set of objectives, among which is the O_{11} defined by the enterprise E_1 and objective O_{21} defined by the enterprise E_2. In order to measure the objectives, two key performance indicators are defined KPI_{111} (to measure the O_{11} defined in E_1) and KPI_{211} (to measure O_{21} defined in E_2). To meet these objectives a set of strategies have been proposed, S_{11} and S_{21} proposed by E_1 and S_{12} and S_{22} proposed by E_2. It should be noted that the model can also be applied within a company in order to determine if two strategies defined in a same company are aligned. Nevertheless, in order have a general insight, the numerical example studies the alignment of strategies formulated in two different network companies. In the numerical example, the strategies that are going to be assessed as regards their alignment are:

- *Scenario 1*: S_{11} (raised to achieve the objective O_{11} of enterprise E_1) and S_{21} (raised to achieve the objective O_{21} of enterprise E_2).
- *Scenario 2*: S_{12} (raised to achieve the objective O_{11} of enterprise E_1) and S_{22} (raised to achieve the objective O_{21} of enterprise E_2).

Depending on the strategies used, the key performance indicators KPI_{111}, KPI_{211} (defined to measure O_{11} and O_{21} respectively) will vary (Table 1).

Applying the Eqs. (3) and (4) we compute the KPIs increase when the strategies are active. The results can be seen in Fig. 3:

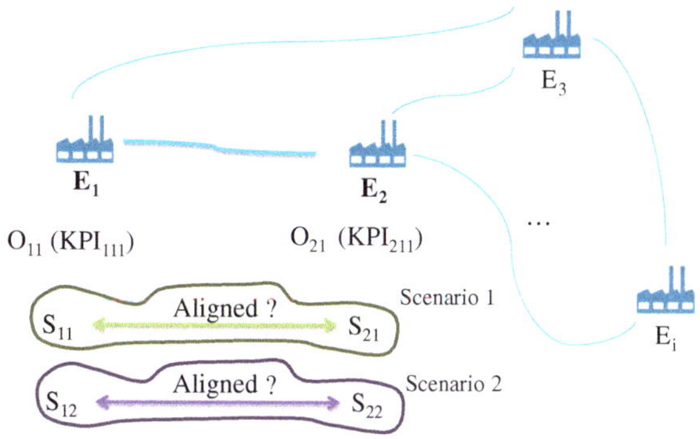

Fig. 2 Network diagram for the numerical example

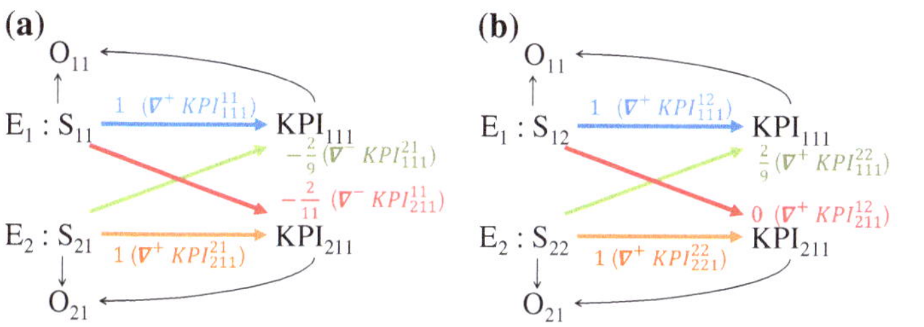

Fig. 3 Results from the numerical example **a** scenario 1, **b** scenario 2

Table 1 KPI values in scenario 1 and scenario 2

(a) Scenario 1			(b) Scenario 2		
	KPI_{111}	KPI_{211}		KPI_{111}	KPI_{211}
S_{11}^{1}	10	9	S_{12}^{1}	10	11
S_{21}^{1}	7	10	S_{22}^{1}	11	11
S_{11}^{0}	5	11	S_{12}^{0}	5	11
S_{21}^{0}	9	5	S_{22}^{0}	9	5

The obtained results from the Eq. (4) show that S_{11} and S_{21} are not aligned due to $\boldsymbol{a}\ (\boldsymbol{S_{is}}, \boldsymbol{S_{jr}}) = \boldsymbol{0}$ (see Eq. 7). On the contrary, S_{12} and S_{22} are aligned due to all the increase are positive resulting $\boldsymbol{a}\ (\boldsymbol{S_{is}}, \boldsymbol{S_{jr}}) = \boldsymbol{1}$. Being E_1 and E_2 collaborative, the strategies to be activated will be S_{12}^{1} and S_{22}^{1} due to positively affects the accomplishment of the objectives defined by each of the enterprises (O_{11} and O_{21}).

In order to consider the applicability of the model within the interoperability area of research, the interoperability assessment through KPIs is considered by Camara et al. [16] therefore, this model perfectly fits when interoperability strategies are raised.

5 Conclusions

This paper proposes a formal model mathematically constructed to define and represent. The objects identified (network, enterprises, objectives, KPIs and strategies) are related to each other in order to have a clear representation of the network. The interaction between the objects is also modelled.

The model is considered as an appropriate tool to (i) determine the, positive or negative, relations between the strategies activation and the objectives improvement, (ii) identify pairs of strategies aligned, and (iii) compute the extent into which the activation of aligned strategies improves the objectives and network performance (*KPIs IMPROVEMENT INDEX*). Generally speaking, the formal model, help researchers to increase their understanding on the relations within the objectives and strategies defined by the enterprises, and compare different scenarios in order to improve the collaboration within a network.

Despite the advantages of the model application, two limitations must be taken into account. First, when computing the KPIs depending on if a strategy is *active* or *not active* $\left(KPI_{ixk}|\mathrm{S}_{\mathrm{is}}^{1}/KPI_{ixk}|\mathrm{S}_{\mathrm{is}}^{0}\right)$, the enterprise needs to know what are the KPI values when a certain strategy is active/not active. In this case it may happen that:

- If the enterprise has not already activated the strategy, it does not know the value $KPI_{ixk}|\mathrm{S}_{\mathrm{is}}^{1}$. According to this, the enterprise can (i) estimate the value $KPI_{ixk}|\mathrm{S}_{\mathrm{is}}^{1}$ or (ii) wait until activate the $\mathrm{S}_{\mathrm{is}}^{1}$ and measure the $KPI_{ixk}|\mathrm{S}_{\mathrm{is}}^{1}$.
- If the enterprise has stored the KPI_{ixk} in both cases, $\left(KPI_{ixk}|\mathrm{S}_{\mathrm{is}}^{1}/KPI_{ixk}|\mathrm{S}_{\mathrm{is}}^{0}\right)$, the enterprise can objectively compute ∇KPI_{ixk}^{is}, for strategies activated in the same enterprise, and ∇KPI_{ixk}^{is}, for strategies active in different enterprises.

Secondly, when computing either ∇KPI_{ixk}^{is} or ∇KPI_{ixk}^{js}, the $KPI_{ixk}|\mathrm{S}_{\mathrm{is}}^{1}$ and $KPI_{ixk}|\mathrm{S}_{\mathrm{js}}^{1}$, respectively, only measures the KPI_{ixk}, *when only one strategy is active* (S_{is} or S_{js}). The formula raised to compute the increase of a $KPI_{ixk}\left(\nabla KPI_{ixk}^{is}/\nabla KPI_{ixk}^{js}\right)$ does not consider if the KPI_{ixk} measured is influenced by more than one strategy. Nevertheless, the KPI_{ixk} can be influenced by the activation of various strategies and not as the result of the activation of a specific strategy. In order to overcome this limitation, the KPI_{ixk} is obtained as a marginal value. Accordingly, if we want to measure the KPI_{ixk} when two or more strategies are activated, we will compute the sum of the $KPI_{ixk}\left(\sum_{k,s} KPI^{s}_{ixk}\right)$ when a set of strategies are activated; the marginal value of KPI_{ixk} permits this aggregation.

Future research lines are lead to propose a mathematical model to support collaborative enterprises on the decision of strategies that should activate or deactivate, to be aligned in order to obtain a *max (KPIs IMPROVEMENT INDEX)*. The model will serve both to the strategies defined for a network and the strategies defined for a set of networks. The model allows to identify the strategies that must be collaboratively activated in order to improve the achievement of the objectives and thus improve network performance. Statistically we could estimate the results as regards the ∇KPI^{is}_{ixk} (when the strategies are active for the same enterprise) and/or ∇KPI^{js}_{ixk} (when the strategies are active for different enterprises). This will identify the set of strategies that should be activated to maximize the increase of KPIs $\left(\nabla KPI^{is}_{ixk}/\nabla KPI^{js}_{ixk}\right)$ and, thus maximize the number of improved objectives.

References

1. Beyerlein, M. M., Freedman, S., McGee, C., & Moran, L. (2003). *Beyond teams: Building the collaborative organization*. San Francisco: Jossey-Bass/Pfeiffer.
2. Durugbo, C., Hutabarat, W., Tiwari, A., & Alcock, J. R. (2011). Modelling collaboration using complex networks. *Information Sciences, 181*(15), 3143–3161.
3. Camarinha-Matos, L. M., Afsarmanesh, H., Galeano, N., & Molina, A. (2009). Collaborative networked organizations—concepts and practice in manufacturing enterprises. *Computers and Industrial Engineering, 57*(1), 46–60.
4. IDEAS. (2002). IDEAS Consortium. *Thematic Network, IDEAS: Interoperability Development for Enterprise Application and Software—Roadmaps, Annex1—Description of work*, May 2012.
5. Andrés, B., & Poler, R. (2012). Relevant problems in collaborative processes of non-hierarchical manufacturing networks. In J. C. Prado, J. García, J. A. Comesaña, & A. J. Fernández (Eds.), *6th International Conference on Industrial Engineering and Industrial Management, Vigo, Spain* (pp. 90–97), July 18–20, 2012.
6. Coutinho, C., Cretan, A., & Jardim-Goncalves, R. (2012). Negotiations framework for monitoring the sustainability of interoperability solutions. In *Enterprise Interoperability* (pp. 172–184). Berlin, Heidelberg: Springer.
7. Kim, K. Y., Manley, D. G., & Yang, H. (2006). Ontology-based assembly design and information sharing for collaborative product development. *Computer-Aided Design, 38*(12), 1233–1250.
8. Jiang, Y., Peng, G., & Liu, W. (2010). Research on ontology-based integration of product knowledge for collaborative manufacturing. *The International Journal of Advanced Manufacturing Technology, 49*(9–12), 1209–1221.
9. Rajpathak, D., & Motta, E. (2004). An ontological formalization of the planning task. In *International Conference on Formal Ontology in Information Systems (FOIS 2004)* (pp. 305–316).
10. Uschold, M., King, M., Moralee, S., & Zorgios, Y. (1998). The enterprise ontology. *The Knowledge Engineering Review, 13*(1), 31–89.

11. Andres, B., & Poler, R. (2013). Method to quantify the power distribution in collaborative non-hierarchical networks. In: L. M. Camarinha-Matos & R. J. Scherer (Eds.), *Collaborative Systems for Reindustrialisation. 14th IFIP WG 5.5 Working Conference on Virtual Enterprises, PRO-VE 2013. IFIP AICT* (Vol. 408). (pp. 660–669). Heidelberg, New York: Springer
12. Andres, B., & Poler, R. (2013). Methodology to identify SMEs needs of internationalised and collaborative networks. In *Advances in production management systems. Competitive manufacturing for innovative products and services* (pp. 463–470). Berlin, Heidelberg: Springer.
13. Rodriguez, R. R., Saiz, J. J. A., & Ortiz, A. (2009). Quantitative relationships between key performance indicators for supporting decision-making processes. *Computers in Industry, 60*(2), 104–113.
14. Lazzarini, S. G., Chaddad, F. R., & Cook, M. L. (2001). Integrating supply chain and network analyses: The study of netchains. *Journal on Chain and Network Science, 1*(1), 7–22.
15. Christopher, M. (1992). *Logistics and supply chain management*. London: Pitman Publishing.
16. Camara, M., Ducq, Y., & Dupas, R. (2010). Methodology for prior evaluation of interoperability. In L. M. Camarinha-Matos, X. Boucher, & H. Afsarmanesh (Eds.), *Collaborative networks for a sustainable world. 11th IFIP WG 5.5 Working Conference on Virtual Enterprises, PRO-VE 2010; Saint-Etienne* (pp. 597–704). France: Springer.

Business Process Model Alignment: An Approach to Support Fast Discovering Complex Matches

Jimin Ling, Li Zhang and Qi Feng

Abstract It is common for large organizations to maintain repositories of business process models and model comparison happens when organizations merge or measure the gap between their own processes and industry-wide standards. Any comparison between process models relies on a construction of relationship between the elements of one model and the elements in the other model. To resolve this automatic construction issue, a three-step approach is proposed to align business process models based on lexical and structural matching to support discovering complex matches especially. The potential node matches, which are first identified by lexical and context similarity, are further grouped to potential complex matches according to the rules we defined. Then an extended graph structure based algorithm is used to select the optimum mapping in the potential matches. Finally, an experiment based on real-world process models from BPM AI is conducted to evaluate the effectiveness and efficiency of our approach.

Keywords Business process modelling · Business process alignment · Process similarity · Complex matches

J. Ling (✉) · L. Zhang · Q. Feng
Department of Computer Science and Engineering, Beihang University, Xueyuan Road 37#, Haidian District, Beijing, China
e-mail: lingjimin@cse.buaa.edu.cn

L. Zhang
e-mail: lily@buaa.edu.cn

Q. Feng
e-mail: miffy_feng513@cse.buaa.edu.cn

K. Mertins et al. (eds.), *Enterprise Interoperability VI*,
Proceedings of the I-ESA Conferences 7, DOI: 10.1007/978-3-319-04948-9_4,

1 Introduction

Process model is a useful technique to capture system requirements at the early stage of project development and it can be easily understood by both software engineers and business people. In the scenario of enterprise interoperability, process model comparison happens when organizations merge or measure the gap between their own processes and industry-wide standards. The foundation to do such comparison and measurement is always to determine which elements in one business process model correspond to which elements in another, which is called business process model alignment [1].

Facing the challenge of the alignment issue, a same activity in different models may have totally different labels or the same functionality can be realized as various structures of model elements. Thus, the following problems probably exist in the current techniques of process model alignment: (1) Most lexical matching methods only take activity name labels into account which results in an unsatisfied outcome. (2) Most matching techniques only support 1-1 mapping, which called elementary correspondence, while some other methods which can identify complex matches only restricted to 1-n matches commonly.

To resolve the aforementioned problems, we proposed a three-step approach to align business process models based on lexical and structural matching, which supports discovering n-m complex matches especially. First, in order to identify potential elementary correspondence precisely, we make full use of the attributes of name, input/output data and resource used as well as node context to calculate similarity. Then we present a bottom-up grouping approach utilizing the concept of process structure tree to detect potential complex correspondence. At last, we provide a greedy-based method to identify the optimal alignment using the extended graph-edit distance evaluation. Besides, an experiment is performed to evaluate the approach in this paper. The result shows a desirable precision and recall, and the efficiency of the method is in an acceptable range.

The rest of the paper is organized as follows. Section 2 provides preliminary about the process model graph which we describe process models. In Sect. 3 we illustrate our three-step method to align process model in detail. Section 4 provides an experiment evaluation of the correspondence discovering technique. Section 5 discusses related work before Sect. 6 concludes the article.

2 Preliminaries

Numerous notations have been developed to capture business processes, including Business Process Modelling Notation (BPMN), Event-driven Process Chains (EPCs) and UML Activity Diagrams. In this paper, our aim is to identify matches that can be applied to all these different notations, so we will illustrate our method based on process model graph (PMG) rather than a specific modelling language.

A PMG is simply a graph that captures nodes and edges whose properties such as names, types or resources used are treated as attributes of them.

Definition 1 *(Process Model Graph).* A process model graph (PMG) is a tuple (N, E, T, Ω, α), in which:

- N is a set of nodes, including semantic nodes and structural nodes;
- $E \subseteq N \times N$ is a set of directed edges;
- T is a set of attribute names, e.g. TYPE, LABEL, RESOURCE, INPUT, OUTPUT, etc.;
- Ω is a set of text string values;
- $\alpha : (N \cup E) \rightarrow (T \rightarrow \Omega)$ is a function that maps nodes or edges to attributes, where an attribute is a mapping from an attribute name to a text string value.

The concept of process alignment is composed of a series of correspondences. Each correspondence is measured by a match score. Two types of correspondence are defined, i.e. the elementary correspondence and the complex correspondence. The former one relates two element set with exactly one model element while the latter one, in turn, relates at least one set composes more than one element, and the complex correspondence probably includes several elementary correspondences.

3 Business Process Model Matching Technique

Our approach can be divided into three steps. First, the attribute and context similarity of node pairs is measured to identify potential elementary matches precisely. Then, an efficient grouping method and rules based on refined process structure tree is presented to support complex matches discovering. Finally, based on the potential correspondence set, a greedy-based algorithm to identify the optimal alignment is proposed using the extended graph-edit distance evaluation.

3.1 Identify Potential Elementary Correspondences

It is unrealistic to assume that two nodes are related only if they have exactly the same name, thus, as the basis of process model alignment, we should definite the similarity metrics between their nodes at first.

The attribute *TYPE* of node is an important factor to measure the similarity. In this article, we only consider the potential similarity of nodes in case they are of the same *TYPE*. Thus, the *type similarity* is defined as follows.

Definition 2 *(Type similarity)* Let $(N_1, E_1, T_1, \Omega_1, \alpha_1)$ and $(N_2, E_2, T_2, \Omega_2, \alpha_2)$ be two PMGs. For the node $n_1 \in N_1$ and $n_2 \in N_2$, the *type similarity* of them is marked as:

$$\mathrm{SimType}(\mathrm{n}_1, \mathrm{n}_2) = \begin{cases} 1 & \text{if } \alpha_1(\mathrm{n}_1) \cdot \mathrm{TYPE} = \alpha_2(\mathrm{n}_2) \cdot \mathrm{TYPE} \\ 0 & \text{otherwise} \end{cases} \tag{1}$$

For two semantic nodes with the same *TYPE*, we can further evaluate their similarity by their attribute values. The *attribute similarity* is defined as the weighting average of the same attribute name.

Definition 3 *(Attribute similarity).* Let (N_1, E_1, T_1, Ω_1, α_1) and (N_2, E_2, T_2, Ω_2, α_2) be two PMGs. For the node $n_1 \in N_1$ and $n_2 \in N_2$, assume that we define the *attribute similarity* of them as follows:

$$\mathrm{SimAttribute}(\mathrm{n}_1, \mathrm{n}_2) = \sum_{\substack{(t_{i1}, l_{i1}) \in \alpha_1(n_1), \\ (t_{i2}, l_{i2}) \in \alpha_2(n_2), \\ 1 \le i \le x,\, t_{i1} = t_{i2}}} \omega_i \cdot \mathrm{SimText}(l_{i1}, l_{i2}) \tag{2}$$

where x means the number of common attribute that the node pair shares, and ω_i is the weight of each attribute and $\sum_{1 \le i \le x} \omega_i = 1, 0 \le \omega_i \le 1$.

The SimText is the textual similarity of two text string values. To achieve a better result, we use the *semantic similarity* defined in [2] which based on the equivalence between the words they consist of to calculate their similarity rather than the syntactic way.

For some structural nodes like gateways, they probably don't have any attribute except for *TYPE*. Obviously, it is unrealistic to measure them by *attribute similarity*. Instead, we use a notion of *context similarity* which means the similarity degree of their presets and postsets. Here we first introduce the notion of *transitive input* and *transitive output* which means a variant of pre-set and post-set that ignore some certain types of node, such as gateways.

Definition 4 *(Transitive input, Transitive output).* Let (N, E, T, Ω, α) be a PMG. There is a path p between two nodes $n \in N$ and $m \in N$, denoted $p = n \Rightarrow m$, iff there exists a sequence of nodes $n_1, \ldots, n_k \in N$ with $n = n_1$ and $m = n_k$ and for all $i \in 1, \ldots, k-1$ holds $(n_i, n_{i+1}) \in E$. Let $ti \sqsubseteq \{\alpha(n).\mathrm{TYPE} \mid n \in N\}$ be a set of node types to be ignored. A *typed chain* is a path that for any node $n' \in \{n_2, \ldots, n_{k-1}\}$ holds $\alpha(n')$. TYPE $\in$ ti, denoted $n \overset{ti}{\Rightarrow} m$. We define the *transitive input* of node n as $n^{in} = \left\{ n' \in N \mid n' \overset{ti}{\Rightarrow} n \right\}$ and *transitive output* of node n as $n^{out} = \left\{ n' \in N \mid n \overset{ti}{\Rightarrow} n' \right\}$.

Definition 5 *(Context similarity).* Let (N_1, E_1, T_1, Ω_1, α_1) and (N_2, E_2, T_2, Ω_2, α_2) be two PMGs and let $n_1 \in N_1$ and $n_2 \in N_2$ be two nodes from them. The *context similarity* is defined as:

$$\mathrm{SimContext}(\mathrm{n}_1, \mathrm{n}_2) = \frac{\sum_{(n_i, n_j) \in M^{opt}_{Simi}(n_1^{in}, n_2^{in})} \mathrm{Simi}(n_i, n_j) + \sum_{(n_p, n_q) \in M^{opt}_{Simi}(n_1^{out}, n_2^{out})} \mathrm{Simi}(n_p, n_q)}{\max\left(\left|n_1^{in}\right|, \left|n_2^{in}\right|\right) + \max\left(\left|n_1^{out}\right|, \left|n_2^{out}\right|\right)} \tag{3}$$

where the $M^{opt}_{Simi}(\mathcal{N}_1, \mathcal{N}_2)$ means the *optimal node match set* between the node sets $\mathcal{N}_1$ and $\mathcal{N}_2$ based on the similarity function of *Simi*, which can be defined as follows.

Definition 6 *(Optimal node match set).* Let $(N_1, E_1, T_1, \Omega_1, \alpha_1)$ and $(N_2, E_2, T_2, \Omega_2, \alpha_2)$ be two PMGs and let $\mathcal{N}_1 \subseteq N_1$ and $\mathcal{N}_2 \subseteq N_2$ be two sets from them. Let $\mathrm{Sim} : \mathcal{N}_1 \times \mathcal{N}_2 \rightarrow [0..1]$ be a similarity function (*attribute similarity* is used). A partial injective mapping $M_{Sim} : \mathcal{N}_1 \rightarrowtail \mathcal{N}_2$ is called a *node match set*, and an *optimal node match set* $M^{opt}_{Sim} : \mathcal{N}_1 \rightarrowtail \mathcal{N}_2$ is a *node match set* for which all other *node match sets* M'_{Sim} holds that:

$$\sum\nolimits_{(n_1,n_2) \in M^{opt}_{Sim}(\mathcal{N}_1,\mathcal{N}_2)} \mathrm{Sim}(n_1, n_2) \geq \sum\nolimits_{(n_1,n_2) \in M'_{Sim}(\mathcal{N}_1,\mathcal{N}_2)} \mathrm{Sim}(n_1, n_2), n_1 \in \mathcal{N}_1, n_2 \in \mathcal{N}_2 \tag{4}$$

Based on the similarity measuring method mentioned above, we can make a conclusion about how to evaluate the similarity of two nodes from different process models. We call the evaluation as *node similarity* which can be defined as follows.

Definition 7 *(Node similarity).* Let $(N_1, E_1, T_1, \Omega_1, \alpha_1)$ and $(N_2, E_2, T_2, \Omega_2, \alpha_2)$ be two PMGs and let $n_1 \in N_1$ and $n_2 \in N_2$ be two nodes from them. Let $ts \sqsubseteq \{\alpha(n) \cdot \mathrm{TYPE} | n \in N\}$ be a set of node types that have not enough attributes (structural node), and the nodes of these types should be evaluated by *context similarity*. Therefore, the *node similarity* of them is marked as:

$$\mathrm{SimN}(n_1, n_2) = \begin{cases} \mathrm{SimType}(n_1, n_2) \cdot \mathrm{SimAttribute}(n_1, n_2) & \text{if } \begin{array}{l} \alpha(n_1).\ \mathrm{TYPE} \notin ts, \\ \alpha(n_2).\ \mathrm{TYPE} \notin ts \end{array} \\ \mathrm{SimType}(n_1, n_2) \cdot \mathrm{SimContext}(n_1, n_2) & \text{otherwise} \end{cases} \tag{5}$$

The final output of our first step, named p*otential elementary correspondences set* (PECS), includes all node pairs whose *node similarity* is higher than a pre-defined ***cutoff* value**.

3.2 Discover Potential Complex Correspondences

It is clear that not all compositions have a sense, because the element group usually expresses relatively complete business logic. Thus, to reduce the searching space to a tolerable scale, we propose the prerequisite of grouping elements based on the concept of single-entry-single-exit (SESE) process fragments [3].

Definition 8 *(SESE process fragment).* A node x is said to *dominate* node y in a directed graph if every path from start to y includes x. A node x is said to *postdominate* a node y if every path from y to end includes x. A *SESE process* *fragment* (*process fragment* for short) in a process graph G is an ordered edge pair (a, b) of distinct control flow edges a and b where:

- a dominates b,
- b postdominates a, and
- every cycle containing a and contains b and vice versa.

We can get the process fragments from process model in linear time using the cyclic equivalent algorithm [4]. The parent of a fragment F (or a node n) is the smallest fragment F′ that contains F (or node n), denoted as pa(F) or pa(n), and we also say that F is a child fragment of F' (n is a child node of F′), and the set of all child nodes and child fragments of F is denoted as ch(F). The process in Fig. 1. gives an example of the dividing result of process fragments.

It has been proved that two process fragments are either nested or disjointed [3], thus process fragments of a process model can constitute a tree structure which is called refined process structure tree (RPST) [5]. The corresponding RPST of the process in Fig. 1 is illustrated in Fig. 2.

Here we can present the prerequisite of grouping process nodes based on the above concepts. Process fragments and nodes can be assembled to a set $\mathcal{P}$ which is called a *feasible group*, if they satisfy the following conditions:

- The fragments and nodes of $\mathcal{P}$ have the same parent S′, and
  - if the relationship of elements in ch(S′) is sequence, the elements of P should be consecutive, i.e. for any process fragment R $\in \mathcal{P}$, if the entry edge (exit edge) of R is also an exit edge (entry edge) of another process fragment T and pa(T) = S′, then T $\in \mathcal{P}$ as well.
  - otherwise, $\mathcal{P}$ is a complete process fragment, i.e. ch(S′) $\subseteq \mathcal{P}$.

Considering the process structure tree in Fig. 2, the following groups are feasible groups: {J}, {K}, {L}, {M}, {J, K}, {g5, L, M, g6}, {g3, P, Q, g4}, {I}, {N}, {I, R}, {R, N}, {I, R, N}.

Based on the optional set obtained by the above condition, we need a criterion of which two pairs of group could be considered as a potential correspondence. Given two feasible groups $\mathcal{P}_1$ and $\mathcal{P}_2$ from two related process models, there is a *potential complex correspondence* between them if:

- $\mathcal{P}_1$ includes more than one node or $\mathcal{P}_2$ includes more than one node,
- and there are two process fragments $F_1 \in \mathcal{P}_1$ and $F_2 \in \mathcal{P}_2$ that a potential correspondence exists between F_1 and F_2,
- and the similarity of $\mathcal{P}_1$ and $\mathcal{P}_2$ is higher than the matching value of any potential correspondence F_1, F_2 where $F_1 \in \mathcal{P}_1$ and $F_2 \in \mathcal{P}_2$.

The criterion requires a definition of similarity measurement between two feasible groups. Referring to node similarity, the similarity between groups based on similarity of their attributes and context of adjacent nodes can be defined similarly. The major challenge of attribute similarity part is that different attributes

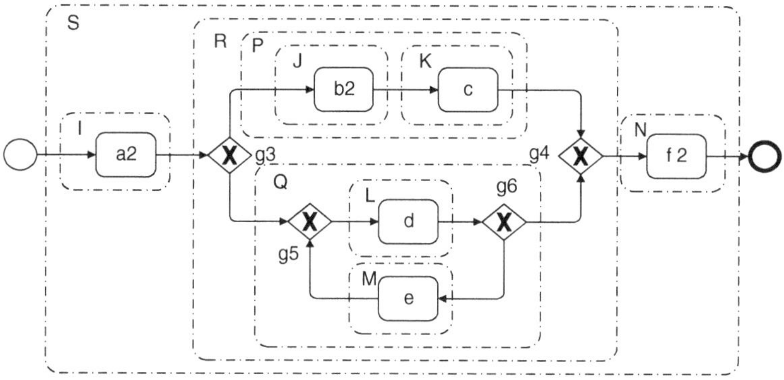

Fig. 1 An example of process fragment

Fig. 2 The RPST of the process in Fig. 1

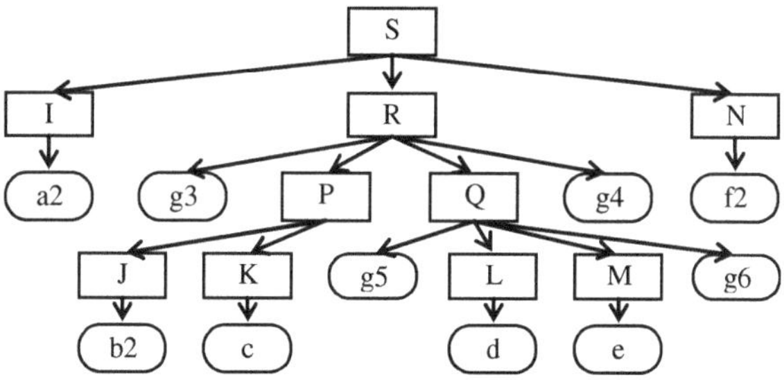

may have different composing policy, e.g. the attribute LABEL of a group can be simply merged by the name label of the containing nodes while the merging of attribute INPUT or OUTPUT should ignore the input or output data produced or consumed inside the group itself. Therefore, we can select specific attributes to be taken into consideration, and different merging policy for these frequently-used attributes is defined. Finally, the elements in two feasible group sets are checked one by one according to the bottom-up order in process structure tree, and if the group pair satisfies the condition of building correspondence, the group pair with their match value is added to the complex correspondence set.

3.3 Select the Optimal Mapping Set

So far one node element may appear in more than one correspondence, thus correspondences should be selected to construct an optimal non-overlapping mapping set. To support complex matches and improve the efficiency, we extend the concept of graph-edit distance [2] to evaluate the global matching score of two process models based on the greedy algorithm.

Note that a greedy strategy may lead to a suboptimal mapping set, but experiments in a similar context proved that results obtained by a greedy strategy are close to those obtained with an exhaustive strategy, and there are obvious advantages in executing efficiency [6]. For convenience, in the rest of this section, we consider that the mapping only includes group matching relationships, i.e. a node is regarded as a group with one element.

Definition 9 *(Extended graph-edit distance similarity).* $G_1 = (N_1, E_1, T_1, \Omega_1, \alpha_1)$ and $G_2 = (N_2, E_2, T_2, \Omega_2, \alpha_2)$ are two PMGs and let M be their mapping function including complex correspondences. Let $n_1 \in N_1$ be a node and $P_1 \subseteq N_1$ be a node group in G_1 which $n_1 \in P_1$, n_1 is a *substituted node* iff $\exists P_2 \subseteq N_2$, $M(P_1) = P_2$. A node is a *skipped node* if it's not a *substituted node*. Let $P_{11}, P_{12} \subseteq N_1$, $P_{21}, P_{22} \subseteq N_2$ and $e_1 \in E_1$ is the link between P_{11} and P_{12}, i.e. be the exit edge of P_{11} and also the entry edge of P_{12}, denoted as $\mathbb{L}(P_{11}, P_{12})$. e_1 is a *substituted edge* iff $\exists \mathbb{L}(P_{21}, P_{22}) \in E_2, M(P_{11}) = P_{21} \wedge M(P_{12}) = P_{22}$. For $n_{11}, n_{12} \in N_1$ and $n_{11}, n_{12} \in P_1$, $(n_{11}, n_{12}) \in E_1$ is a *inside edge* iff $\exists P_2 \subseteq N_2$, $M(P_1) = P_2$, and the *inside edge* set of group P is denoted as E^P. Other edges except for *substituted edges* and *inside edges* are called *skipped edge*. Let subn, skipn, inse and skipe be the sets and ω_{subn}, ω_{skipn}, ω_{inse} and ω_{skipe} be the weights of *substituted nodes, skipped nodes, inside edges* and *skipped edges* respectively. The *extended graph-edit distance similarity* of G_1 and G_2 induced by the mapping M is:

$$EGSim(G_1, G_2, M) = 1.0 - \frac{\omega_{subn} \cdot fsubn + \omega_{skipn} \cdot fskipn + \omega_{inse} \cdot finse + \omega_{skipe} \cdot fskipe}{\omega_{subn} + \omega_{skipn} + \omega_{inse} + E_{skipe}} \tag{6}$$

$$fsubn = \frac{\sum_{(P_1,P_2) \in M}[1.0 - Sim(P_1, P_2)] \cdot (|P_1| + |P_2|)}{|subn|}, fskipn = \frac{|skipn|}{|N_1| + |N_2|} \tag{7}$$

$$fskipe = \frac{|skipe|}{|E_1| + |E_2|}, finse = \frac{\sum_{(P_1,P_2) \in M}[1.0 - Sim(P_1, P_2)] \cdot (|E^{P_1}| + |E^{P_2}|)}{|inse|} \tag{8}$$

Here we illustrate the procedure to select the optimal mapping set. Initially, all potential correspondences are added to the CanPairs and the optimal mapping M is empty. In each iteration, a new EGSim is computed for adding every pair in CanPairs into M to see which pair increases the EGSim the most. This pair is added to the optimal mapping M and any other pair overlapping with this optimal pair is removed from the CanPairs. The procedure terminates when there is no group pair in CanPairs that can increase the matching score EGSim any more.

4 Experimental Evaluation

We evaluated our approach experimentally based on real-world process models from BPM Academic Initiative (BPM AI) which is a joint effort of academic and industry partners that offers a process modeling platform for teaching and research purposes. 20 pairs of related BPMN process models are selected. These models

Table 1 Result of various cutoff values

Cutoff	Precision	Precision_c	Recall	Recall_c	F-score	Time cost (ms)
0.5	0.67	0.60	0.74	0.62	0.70	15227
0.6	0.71	0.64	0.72	0.61	0.71	9830
0.7	**0.78**	**0.73**	**0.68**	**0.57**	**0.73**	**6124**
0.8	0.81	0.75	0.54	0.40	0.65	4437
0.9	0.77	0.73	0.42	0.28	0.54	3549

have 30.4 nodes on average while a minimum of 12 nodes and a maximum of 62 nodes for a single process model. The average number of edges pointing into or out of a single node is 1.22 and the average number of node label words is 3.08. Two experienced process modelers manually built the mapping respectively. After discussion, 383 matched activity pairs were finally determined and the complex matches reached the proportion of 37.9 %.

We evaluate our approach by calculating the precision, recall and F-score. The *precision* (or *precision_complex*) is the proportion of correct matches (or complex matches) in all discovered matches (or complex matches). The *recall* (or *recall_complex*) is the proportion of correct matches (or complex matches) in all human-detected matches (or complex matches). The F-Score combines precision and recall in one value.

As an elementary experiment, the experimental models in this paper only include two node attributes of TYPE and LABEL and few edges which aren't control-flow are ignored, thus our approach is evaluated in a simplified way. Our technique depends on the following parameters: (1) the *cutoff* value and (2) the weights of ω_{subn}, ω_{skipn}, ω_{inse} and ω_{skipe}. The human-determined matching results are pre-stored into computer to support automatic judgment between experiment results and human decision.

We consider an appropriate *cutoff* value to be 0.6 experientially at first. All combinations of the four EGSIM parameters are tested between 0 and 1 in the interval of 0.1. Note that the potential correspondences of each model pair can be recorded statically, thus only the optimal mapping building algorithm need to be run repeatedly. The result shows that the combination of (0.2, 0.2, 0.3, 0.7) can reach the highest F-Score by 0.71.

Table 1 summarizes the metric values under various *cutoff* values. From the result we can see that, in the low *cutoff* value, the lower precision means more false matches are found. However, the precision doesn't decline too much with the decreasing *cutoff* value and this probably benefit from our optimal mapping selecting algorithm. In the case of higher *cutoff* value, the precision does not increase obviously while the recall decreases rapidly, especially for the recall of complex matches which means much missing of true matches. The result shows that the *cutoff* value of 0.7 may lead to a desirable outcome of precision and recall, especially for the *recall_complex* metric comparing with [7]. The right column of Table 1 presents the total time cost of matching 20 process model pairs. With the increasing of *cutoff* value, the smaller set of potential elementary correspondences leads to a faster detection of complex correspondences and results in a smaller set of potential complex matches, thus the time cost of our approach is also reduced. By the optimal *cutoff* value of 0.7, the average time cost of one model pair alignment is about 306 ms which is an acceptable efficiency.

5 Related Work

Various research works try to construct correspondences of model elements by automated techniques based on textual or structure matching. The sub-graph isomorphism and the graph-edit distance have been used with syntactic or semantic similarity techniques [2]. Three approaches based on lexical or graph matching are compared in [6], and the result showed that a greedy graph matching technique produces the best effectiveness. The ICoP framework [7] used an architecture to enable the optional creation of matchers from the reusable components which can support complex correspondence detecting, but detailed algorithms of the components have not been illustrated clearly. The alignment method of two semantic annotated process models is illustrated in [8] which take into account activities' attributes (inputs/outputs), but this research only supports 1-n complex matching and doesn't include experimental verification. Besides, the techniques of process model merging [9] and change propagation between process models [10] are directly benefit from process model alignment techniques.

Business process similarity analysis is another closely related topic. A previous paper [11] of our research group utilized the idea of similarity propagation to measure the similarity of two process models, but the method is restricted to VPML process modeling language and only elementary matches are supported. Remco Dijkman etc. presented and evaluated three similarity metrics and the result showed that the structural similarity slightly outperforming the other two metrics of node matching and behavioral similarity [2]. A fast business process similarity estimating based on characteristic model features presented three techniques to improve the efficiency of current graph-based algorithm [12]. Four major graph matching algorithms for business process model similarity search are evaluated in [13]. For all these above similarity analysis techniques, a pre-step of node matching is usually required, however, complex correspondences and data or resource aspects are seldom taken into consideration.

6 Conclusion

In this paper, we proposed a three-step approach to align business process models based on lexical and structural matching to support discovering n-m complex matches especially. Data and resource aspects and similarity measurement idea were considered in elementary correspondence discovering. Then an efficient grouping method based on refined process structure tree was presented to support complex matches. Finally, a greedy-based algorithm to identify the optimal alignment was proposed using the extended graph-edit distance evaluation. The optimum value of the parameters in our approach was achieved by an experiment based on 20 pairs of real-world process models. The experiment results showed that our approach achieved a desirable effectiveness with an acceptable efficiency. In future research, we aim to improve the accuracy and efficiency of our approach and conduct a more comprehensive experiment evaluation by huge amount of process models in various model repositories.

Acknowledgements This work is supported by the National Natural Science Foundation (No. 61170087) and the Fundamental Research Funds for the Central Universities of China.

References

1. Brockmans, S. et al. (2006). *Semantic Alignment of Business Processes*. Paper presented at ICEIS (3), Paphos, Cyprus.
2. Dijkman, R. et al. (2011) Similarity of business process models: Metrics and evaluation. *Information Systems, 36*(2), 498–516.
3. Vanhatalo, J., Völzer, H., Leymann, F. (2007). Faster and more focused control-flow analysis for business process models through SESE decomposition. In *Proceedings of the Service-Oriented Computing–ICSOC 2007* (pp. 43–55). Heidelberg: Springer.
4. Polyvyanyy, A., Vanhatalo, J., Völzer, H. (2011). *Simplified computation and generalization of the refined process structure tree: Proceedings of the Web Services and Formal Methods* (pp. 25–41). Heidelberg: Springer.
5. Vanhatalo, J., Völzer, H., Koehler, J. (2008). *The refined process structure tree: Proceedings of the Business Process Management* (pp. 100–115). Heidelberg: Springer.
6. Dijkman, R., et al. (2009). *Aligning business process models: Proceedings of the IEEE International Enterprise Distributed Object Computing Conference EDOC'09*, Enschede: IEEE.
7. Weidlich, M., Dijkman, R., Mendling, J. (2010). *The ICoP framework: Identification of correspondences between process models*: *Proceedings of the Advanced Information Systems Engineering*. Heidelberg: Springer.
8. Gater, A., Grigori, D., Bouzeghoub, M. (2010). *Complex mapping discovery for semantic process model alignment: Proceedings of the 12th International Conference on Information Integration and Web-based Applications & Services*. New York: ACM.
9. La Rosa, M., et al. (2013). *Business process model merging: an approach to business process consolidation*. Paper presented at ACM Transactions on Software Engineering and Methodology (TOSEM) (vol. 22, 2), New York.
10. Weidlich, M., Mendling, J., & Weske, M. (2012). Propagating changes between aligned process models. *Journal of Systems and Software, 85*(8), 1885–1898.
11. Gao, J., Zhang L. (2009). On Measuring Semantic Similarity of Business Process Models. In *Proceedings of the International Conference on Interoperability for Enterprise Software and Applications IESA'09*. Beijing, China: IEEE.
12. Yan, Z., Dijkman, R., & Grefen, P. (2012). Fast business process similarity search. *Distributed and Parallel Databases, 30*(2), 105–144.
13. Dijkman, R., Dumas, M., García-Bañuelos, L. (2009). *Graph matching algorithms for business process model similarity search: Proceedings of the Business Process Management* (pp. 48–63). Heidelberg: Springer.

An Approach to BPMN 2.0 Serialization Constraints Specification

Marija Jankovic and Zoran Marjanovic

Abstract Correct and standard compliant serialization of BPMN process models is decisive for model execution and interoperability between tools. The official BPMN 2.0 specification document does not offer an immense set of all constraints concerning the correctness of model serialization. This paper fills this gap by presenting an approach to elaborate a generic list of technology independent constraints stated by the standard. The issues in BPMN model serialization, which are the consequence of complexity and inconsistency of BPMN 2.0 specification document, are being analyzed herein. In addition, results and future research directions are also discussed.

Keywords BPMN 2.0 · Interoperability · XML serialization · Standard compliance

1 Introduction

Business Process Model and Notation (BPMN) is an Object Management Group (OMG) standard for business process modeling that is widely adopted in practice and academia today [1]. BPMN provides a standardized set of graphical shapes, which is easy to use for modeling purposes such as visualization and documentation [2]. Version 2.0 takes BPMN to a new level. It introduces standardized meta-model and XML-based serialization format for BPMN, which allows users to

M. Jankovic (✉) · Z. Marjanovic
Faculty of Organizational Sciences, University of Belgrade, Jove Ilica 154, 11000 Belgrade, Serbia
e-mail: marija.jankovic@fon.rs

Z. Marjanovic
e-mail: zoran.marjanovic@fon.rs

K. Mertins et al. (eds.), *Enterprise Interoperability VI*,
Proceedings of the I-ESA Conferences 7, DOI: 10.1007/978-3-319-04948-9_5,

interchange business process models between tools of different vendors. Two serialization formats have been defined: XML Metadata Interchange (XMI) and XML Schema Definitions (XSDs) [1, 3–5].

The main problem of the former BPMN versions lies in the lacking standardization of the serialization format [2, 6, 7]. This led to a plethora of different proposals to serialize BPMN models. Modeling tools used their own proprietary file formats, but also popular were mappings and transformation to various serialization formats, e.g. Business Process Execution Language (BPEL) and XML Process Definition Language (XPDL) [8–11]. Without side agreements between tool A and tool B, their model interoperability was impossible [12].

In theory, the standardized model interchange based on BPMN 2.0 serialization should support collaboration and improve interoperability between different organizations within and across different enterprises. In practice, both standardized serialization formats face a serious problem. They do not guarantee correct and standard compliant BPMN models. BPMN serialization may be realized in different ways, but all variants must respect the constraints stated in the standard document. Standard document [1] does not provide a complete list of all constraints to validate compliance of BPMN models. To the best of our knowledge, there are no tools or tests to check if model conforms to all restrictions stated in [1].

The aim of our work is to provide extensive and complete list of all serialization constraints defined in the standard document [1]. The constraints should be generic and independent from any serialization format in order to ensure the correctness of all BPMN models. In this paper, the term correctness regards to BPMN 2.0 specification compliance and not to semantic correctness. The complete list of serialization constraints is a necessary step in developing compliance checking mechanism for BPMN process models. We focus on process model serialization, and therefore the list does not contain other BPMN aspects such as execution semantic rules or graphical constraints.

In this context, problem statement and background is summarized in Sect. 2. Related work is discussed in Sect. 3. Description of our approach to detecting and extracting constraints is presented in Sect. 4. Section 5 elaborates on the results achieved and lessons learned. Finally, Sect. 6 concludes the paper and gives an outlook on the future work.

2 Problem Statement and Background

Before presenting our approach and results, we have to clarify the problem statement and limitations of our work. The main problem of the BPMN1.x versions is, as mentioned above, the missing standardization of the serialization format [2, 6, 7]. Most of the effort in creating BPMN 2.0 involved development of the formal meta-model and its corresponding XML representation. The meta-model is published in two alternative XML formats, XML Metadata Interchange (XMI) and XML Schema Definition (XSD) [1, 3–5]. Most BPMN tool vendors use

XSD format. Therefore, in this paper we will focus on XSD representation of the BPMN meta-model. A full explanation of XSD based BPMN 2.0 serialization is beyond the scope of this paper. Basic serialization concepts relevant for the problem statement are introduced.

2.1 BPMN 2.0 XML Serialization

A XML serialization for BPMN models consists of a sequence of *definitions*. There are two kinds of definitions, *process definitions* and *graphical definitions*. The process definition describes the elements of the process diagram and their relations. The graphical definition, called BPMN Diagram Interchange (BPMNDI), complements the process definition for presentation purposes, e.g. giving their exact location as (x, y) coordinates. A valid BPMN model may omit BPMNDI entirely, but may not omit the semantic model. In a particular serialization of BPMN model, graphical definitions are optional or may only describe a partial view of a process definition [1].

The BPMN 2.0 schema is distributed as a set of five XSDs files: BPMN20.xsd, Semantic.xsd, BPMNDI.xsd, DI.xsd, and DC.xsd [13]. BPMN20.xsd is the top level. It includes Semantic.xsd and imports BPMNDI.xsd, which in turn imports DI.xsd and DC.xsd. For the semantic and graphical data, BPMN 2.0 standard defines a XML schema file: Semantic.xsd and BPMNDI.xsd, respectively. BPMNDI.xsd uses type definitions from DI.xsd (the XML schema file for Diagram Interchange (DI)) and DC.xsd (the XML schema file for Diagram Common (DC)). DI and DC are parts of Diagram Definition standard (DD) that is developed by the Object Management Group (OMG) to describe various types of object elements, e.g., shapes, edges, and labels. BPMNDI.xsd extends the data structures defined in DI and DC with additional attributes specific to BPMN [1].

Every XML schema file is associated with a *target namespace*. The purpose of the target namespace is to give a globally unique name to all elements defined in the schema. The BPMN 2.0 serialization uses several namespaces. For example, all types and elements defined in Semantic.xsd are part of the BPMN 2.0 Model namespace: http://www.omg.org/spec/BPMN/20100524/MODEL. Figure 1 shows the XML schema files of the BPMN 2.0 standard with their dependences and associated namespaces.

A dashed line from one file to another, for example from BPMN20.xsd to Semantic.xsd means that BPMN20.xsd uses objects defined in Semantic.xsd. The simple process model is depicted and serialized in Fig. 2. The serialization does not include BPMNDI.

Several things are worth nothing about the serialization, generated by Process Modeler for Visio [14]. *Ids* for all elements are tool-generated globally unique values. The *targetNamespace* is common for all models serialized by this tool. The *default namespace* is declared to be BPMN20.xsd namespace. Two other namespaces are declared in the *definition* element. The prefix *xsi* references the

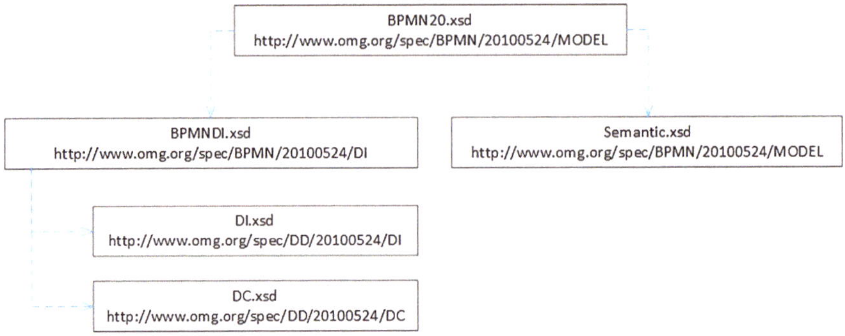

Fig. 1 BPMN 2.0 schema files and associated namespaces

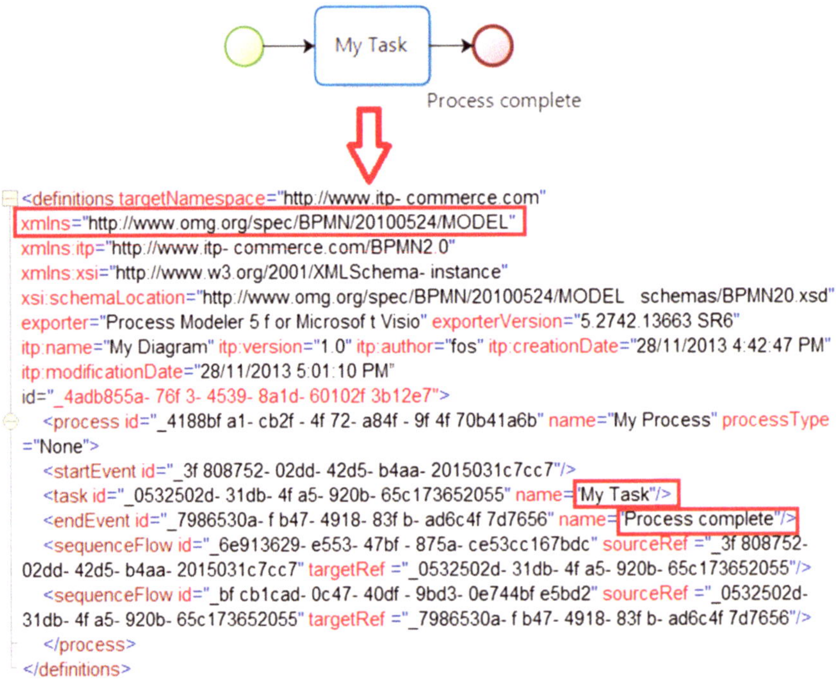

Fig. 2 Serialization of a simple process model

schemaLocation attribute which indicates that this instance document is to be validated against the BPMN2.0 schema found at the relative file location *schemas/BPMN20.xsd*. The prefix *itp* references tool vendor proprietary elements and attributes. The *name* attribute of the *task* and *endEvent* elements match their labels in the diagram.

2.2 Assumptions and Limitations

The main contribution of our work is the establishment of an extensive, a complete, and a generic list of serialization constraints that should be verified in order to ensure the correctness of BPMN models. The constraint list is not limited to XSD based serialization of BPMN models. Therefore, the set of rules is technology independent and can be used to analyze all types of BPMN serialization formats. The main restriction of our approach is that we strictly focus on the specification of constraints regarding semantic serialization of BPMN process models. The serialization of BPMN diagrams and all graphical constraints is out the scope of this work. Other aspects which are not covered are all constraints regarding the execution semantics of BPMN.

3 Related Work

Academic research focuses mainly on semantic correctness of BPMN models and assumes that they are already standard compliant [15, 16]. This paper targets issues regarding process correctness from a different perspective, model serialization. There are already various tools available which perform a consistency check to a certain extent, but hardly any vendors or authors have published their findings.

An exception is Bruce Silver, who published a list of 39 mandatory process modeling rules directly derived from the standard [1, 13]. He introduces 28 additional style rules in order to improve comprehensibility of BPMN process models. The rules are implemented in Process Modeler for Visio and Signavio modeling tools [14, 17]. Silver does not provide the complete list of all serialization constraints defined in the standard document. He leaves out of the analysis all structural and cardinality aspects that are already covered with the XSDs. Moreover, Silver divided the modeling using BPMN into levels which refer to the definition of Process Modeling Conformance Subclasses in [1, pp. 2–9]. His rules cover level 2 which is equal to Analytic Process Modeling Conformance Subclass [13].

According to Silver [13], schema validation is not enough to guarantee BPMN 2.0 model correctness. The model must be created in accordance with the rules that are defined in the specification. A model can be valid per BPMN 2.0 XSD, but the rules might be violated. For example, a sequence flow connecting elements in separate pools is valid with respect to the schema, but illegal based on the rule that the sequence flow cannot cross the boundaries of the pool [1, p. 112].

An interesting proposal that aims to identify the reasons underlying the missing interoperability between BPMN 2.0 tools and highlights serialization issues has appeared in [15]. The serialization issues are further discussed in [18]. Authors have identified more than 600 different constraints. In comparison to Silver [13], their constraint set is far more extensive. They take into account all basic structural, value and reference restrictions. Rules regarding the common executable conformance subclass [1, p. 2], which are left out in [13] are listed in their constraint list.

OMG has recognized model serialization and interchange between BPMN tools as an important issue. They initiated the BPMN Model Interchange Working Group (BPMN MIWG) with the goal to provide support for modeling tool developers and the standard-related issues which inhibit interchange of models between tools [19]. Planned outputs are test cases, feature tests, a set of BPMN 2.0 issues and interchange guidelines. The group is still working in an initial project phase, therefore no official outcomes have been released yet.

4 Details of the Approach

Considering model serialization constraints BPMN specification document consists of four important sources: the running text, class diagrams, tables specifying attributes of BPMN elements and XSD normative files. In order to perceive all constraints, an overview of all sources in [1] is crucial. Unfortunately, BPMN 2.0 specification document does not provide such an overview. In following, the main steps of the proposed approach are presented, as illustrated in Fig. 3.

In the first step, two main constraint categories have been identified: *basic* and *advanced*. *Basic* category consists of all constraints regarding cardinality (CARD), default values of attributes/elements (DEF), and references (REF) to other BPMN elements. All other constraints that could not be assigned to the *basic* are categorized in *advance* category.

Second, depending on the category the extraction approach is defined:

- *Basic* constraints can be specified by analyzing the tables in [1], which include all needed information such as attribute names, type definitions, value and cardinality restrictions. The same tables provide information for all reference constraints. References can be determined as attributes with the suffix *Ref*. In the normative XSD files, they can be recognized through the usage of the datatypes *xsd:IDREF* and *xsd:QName*.
- Some *advanced* constraints can be revealed by analyzing the column *Description/Usage* of the tables just mentioned. The running text often states further restrictions and rules, which have to be respected. Extracting the rules from text usually requires additional interpretation. Although some rules are not explicitly stated, they are necessary. Besides, the running text, class diagrams and XSD files are often inconsistent. In such cases, additional sources (e.g. vendors implementation in their modeling tools) were used to evaluate which constraint was implied by the standard authors.

In the third step, serialization constraints are determined. For both categories a short introduction is given and all corresponding constraints are documented as stated in the extraction approach defined in the previous step. The *Serialization*

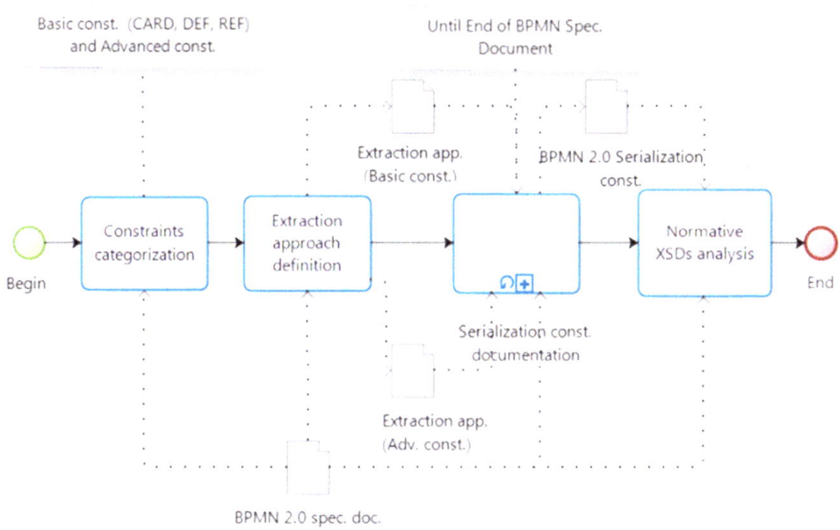

Fig. 3 Main steps to apply BPMN 2.0 serialization constraints specification approach

Fig. 4 Decomposition of serialization const. Specification subprocess

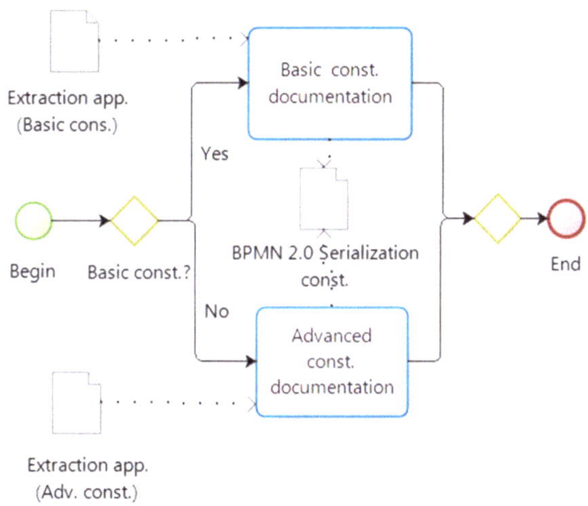

const. documentation subprocess is defined as a standard loop with loop condition, *Until End of BPMN 2.0 Spec. document*, indicated in the text annotation (Fig. 3). The subprocess is further decomposed in the Fig. 4.

Finally, the analysis of normative XSDs is performed in order to see which constraints are covered and implemented correctly. The analysis is useful as most of the BPMN tools and engines use XSD-based serialization format.

5 Results

We worked through the standard document and in the first iteration we derived more than 350 basic and 100 advanced constraints. Only the main findings are described in this section. An overview of all extracted constraints will be available soon in the form of a detailed technical report.

The most elementary rules are the basic attribute and element cardinality constraints. They declare attributes and model associations applicable to each BPMN element. For each identified attribute, a name, datatype and a cardinality have been extracted. The main problem was inconsistency between the text, class diagrams and XSD normative files. Tables and class diagrams often provide different information for minimum and maximum occurrence of the relations between BPMN model elements. Frequently some attributes are defined as mandatory in the text and the class diagrams, but the schema definition marks the same attribute as optional. For example, the attributes *innerConversationNodeRef* and *outerConversationNodeRef* are defined in the Table 9.14 in [1, p. 139] as optional, but in the class diagram in Fig. 9.31 [1, p. 139] and in the normative XSD file the attributes are marked as mandatory.

For attributes, which refer to basic datatypes (e.g. boolean or integer), the BPMN standard defines value restrictions and default values. For each attribute a name, datatype, default value and value restriction have been identified. Most of them are correctly implemented in normative XSD files, but there still exist some inconsistencies. For example, the boolean attribute *cancelActivity* defined in the Table 10.91 [1, p. 266] is mandatory, but no default value is fixed. In normative XSD, the attribute is optional, but the default value is true (see line 102).

The BPMN standard document comprises more than 100 different associations between elements that are implemented as references. For the technology agnostic view of the references, it is sufficient to know which attribute is a reference and what type of BPMN elements is referenced. However, for the purpose of the analysis in the third step of the proposed approach XSD-specific attributes *naming* and *implementation* are extracted. Important sources for reference usage are sequence and message flow definitions.

In distinction to the earlier introduced constraints from the basic category which can be represented with a few attributes as they all follow the same pattern, the advanced constraints need to be specified in more detail. Aforementioned is mainly because the rules in this category cover complex requirements. Advanced constraints cannot be further classified as they focus on different aspects, but some descriptive attributes are relevant for all of them. The common structure, which is used to ensure a clear representation, is shown in Table 1.

Id represents the unique number for each constraint. Most rules concern a single BPMN element, but some complex constraints affecting various elements have been identified. If no single *element/attribute* is affected, the corresponding values in the table might be empty (denoted as "–"). In order to describe the constraint, a textual description is always needed. *Pre-condition* is the condition under which

Table 1 Common structure for advance constraints specification

Id: ADV065	
Name: IfEndEventExistsStartEventIsMandatory	
Element	Start Event
Attribute/sub-El.	–
Constraint	A Start Event must be present when an End Event belongs to the same process level
Pre-condition	End Event used in the process
Source	If there is an End Event, then there must be at least one Start Event
Chapter-pg.	10.4.2-246

the rule applies. In order to reproduce and verify the extracted condition, the standard document is always quoted and referenced.

After the constraints have been extracted, the analysis of their coverage in normative XSD is performed. Although most of the basic (structure, value and reference) constraints can be expressed using XML schema restrictions, implementation of the advanced constraints is hard. Those facts evidently show that the usage of XML schema validation alone is not sufficient to enforce the standard compliance of models.

6 Conclusion and Outlook

In this paper, we have presented an approach to specify an extensive and technology independent set of BPMN serialization constraints. We have identified more than 450 constraints categorized in two different categories. It is hard to prove that our constraint set is complete and free of errors, particularly in cases where rules are stated only implicitly.

Future work is planned in several directions. First, we plan to ensure the quality and correctness of our constraint set by comparing the results with other collections of BPMN constraints and consistency checks integrated in various tools [13, 18].

Next challenge is to develop compliance checking mechanism for BPMN process models. The presented set of constraints in the paper is a required substep for further analysis and research. The analysis of the standard XSD-based serialization format of BPMN models shows that XSD schema validation alone is not sufficient to enforce the correctness of the models. A manual review of the standard compliance of concrete models is not feasible, due to the numerous and extensive set of specified constraints. We are currently developing a tool to check all extracted constraints which are not covered with schema validation yet.

Another planned activity is an in-depth analysis of the compliance checking mechanism of current BPMN modeling tools. A lot of tools already perform some internal checks, but the first tentative evaluation of some significant editors showed that not all of our detected constraints are covered [14, 17, 20–22].

We expect that our research effort will contribute to establish the basis for standard compliance of serialized BPMN models.

References

1. Object Management Group (OMG). (2011, January). *Business process model and notation (BPMN) version 2.0.*
2. Business Process Management Initiative (BPMI). (2006, May). *Business process modeling notation (BPMN) version 1.0.*
3. Object Management Group (OMG). (2013, June). *OMG MOF 2 XMI mapping specification version 2.4.1.*
4. World Wide Web Consortium (W3C). (2004, October). *XML Schema part 1: Structures* (2nd ed.), (http://www.w3.org/TR/xmlschema-1/).
5. World Wide Web Consortium (W3C). (2004, October). *XML Schema part 2: Datatypes* (2nd ed.), (http://www.w3.org/TR/xmlschema-2/)
6. Object Management Group (OMG). (2008, January). *Business process modeling notation, version 1.1.*
7. Object Management Group (OMG). (2009, January). *Business process modeling notation (BPMN) version 1.2.*
8. OASIS. (2007, April). *Web services business process execution language, version 2.0.*
9. Workflow Management Coalition (WfMC). (2008, April). *XML process definition language (XPDL) version 2.1.*
10. Workflow Management Coalition (WfMC). (2012, April). *XML process definition language (XPDL) version 2.2.*
11. Ouyang, C., Dumas, M., HM, A., & Van Der Aalst, W. M. (2008). Pattern-based translation of BPMN process models to BPEL web services. *International Journal of Web Services Research, 5*(1), 42–62.
12. Jankovic, M., Ivezic, N., Marjanovic, Z., Knothe, T., & Snack, P. (2007). *A case study in enterprise modeling for interoperable cross-enterprise data exchange, I-ESA'07.* Funchal (Madeira Island), Portugal Monograph: R. J. Goncalves, J. P. Muller, K. Mertins, & M. Zelm (Eds.). (2007). *Enterprise interoperability II: New challenges and approaches* (pp. 541–552). London: Springer. ISBN-13:9781846288579, ISBN 978-1-84628-857-9, e-ISBN 978-1-84626-858-6.
13. Silver, B. (2011). *BPMN method and style: With BPMN implementer's guide* (2nd ed.). Aptos, CA: Cody-Cassidy Press.
14. Retrieved October 15, 2013 from http://www.itp-commerce.com/
15. Matthias Geiger, M., & Wirtz, G. (2013, February). Detecting interoperability and correctness issues in BPMN 2.0 process models. In *Proceedings of the 5th Central-European Workshop on Services and their Composition (ZEUS), Rostock, Germany, CEUR, Workshop Proceedings* (pp. 39–42). CEUR-WS.org.
16. Dijkman, R. M., Dumas, M., & Ouyang, C. (2008). Semantics and analysis of business process models in BPMN. *Information and Software Technology, 50*(12), 1281–1294.
17. Retrieved October 15, 2013 from http://www.signavio.com/
18. Geiger, M. (2013, May). *BPMN 2.0 Process model serialization constraints.* Bamberger Beiträge zur Wirtschaftsinformatik und Angewandten Informatik, no. 92, OttoFriedrich-Universität Bamberg.
19. http://www.omgwiki.org/bpmn-miwg/doku.php (accessed 15 October 2013)
20. Retrieved October 15, 2013 from http://www.bizagi.com/index.php/products/bizagi-process-modeler#freeware
21. Retrieved October 10, 2013 from http://www.bpmnvisiomodeler.com/
22. Retrieved October 10, 2013 from http://www.bonitasoft.com/

Ontology-Based Approach for Semantic Service Selection in Business Process Re-Engineering

Sophea Chhun, Néjib Moalla and Yacine Ouzrout

Abstract This research aims to provide the possibility to the business analysts to be able to know whether their design business processes are feasible or not. In order to solve this problem, we proposed a model called BPMNSemAuto that makes use of the existing services stored in the service registry UDDI (Universal Description Discovery and Integration). From the data extracted from the UDDI, the WSDL files and the tracking data of service execution on the server, a Web Service Ontology (WSOnto) is generated to store all the existing services. The BPMNSemAuto model takes an input of business process design specifications, and it generates an executable business process as an output. It provides an interface for business analysts to specify the description of each service task of the design business process. For each service task, the business analysts specify the task objective (keywords), inputs, outputs and weights of the Quality of Service (QoS) properties. From the design business process with the service task specifications, a Business Process Ontology (BPOnto) is generated. A service selection algorithm performs the mapping between the instances of the WSOnto and the BPOnto to obtain possible mappings between these two ontologies. The obtained mappings help the model to acquire web services to execute the desired service tasks. Moreover, the consistency checking of the inputs of the proposed model is performed before executing the service selection algorithm. WordNet is used to solve the synonym problems and at the same time a keyword extraction method is presented in this paper.

S. Chhun (✉) · N. Moalla · Y. Ouzrout
Université Lumière Lyon 2, 86 rue Pasteur, 69635 Lyon Cedex 07, Bron, France
e-mail: sophea.chhun@univ-lyon2.fr

N. Moalla
e-mail: nejib.moalla@univ-lyon2.fr

Y. Ouzrout
e-mail: yacine.ouzrout@univ-lyon2.fr

K. Mertins et al. (eds.), *Enterprise Interoperability VI*,
Proceedings of the I-ESA Conferences 7, DOI: 10.1007/978-3-319-04948-9_6,

Keywords Business process · Keyword extraction · Ontology · Ontology matching · QoS · Semantic web service · Service selection

1 Introduction

The Workflow Management Coalition provides a definition of Business Process, saying that "it is a set of one or more linked procedures or activities which collectively realize a business objective or policy goal, normally within the context of an organizational structure defining functional roles and relationships" [1]. Rummler and Brache defined business process as "the series of steps that a business executes to produce a product or service" [2]. The business process applications can be modeled with different modelling specifications such as Business Process Model and Notation (BPMN) [3], Petri Net [4, 5], Workflow and Unified Modeling Language (UML). This research study focuses on the modelling of business processes with BPMN specifications (BPMN2.0). Some examples of business process modeled with BPMN are presented in [6], such as "handling and invoicing process application", "taxi reservation application" and "online purchasing application". Correia and Abreu [7] state that "A BPMN2 process model diagram has around 100 different modeling constructs, including 51 event types, 8 gateway types, 7 data types, 4 types of activities, 6 activity markers, 7 task types, 4 flow types, pools, lanes, etc." However, this research work considers only the automatic implementation of the service task (one kind of task type) by using the existing services stored in the service registry UDDI (Universal Description Discovery and Integration).

A service task can be performed by a web service or a composite service. The web service is a software module created to perform a specific business task. It is described by the service description languages such as Web Service Description Language (WSDL) and Web Service Description Language-Semantic (WSDL-S). These languages provide different capabilities, for example, WSDL cannot store the pre-condition and post-condition of a service but WSDL-S and OWL-S do.

Researchers use the ontology to represent the semantic meaning of services and as a knowledge base. The Ontology is an explicit specification and hierarchy of different concepts. It defines properties, characteristics and behaviors of objects in the same domain; and it expresses the relationships between concepts [8]. An ontology consists of three elements: vocabularies, specifications and constraints. The vocabularies describe the domain of ontology and the constraints are used to capture knowledge about the ontology's domain. The specifications define the relationships between different concepts of the ontology. Moreover, some ontology languages have been proposed such as RDF, DAML-OIL, WSMO, OWL and OWL-2. This research study builds a Web Service Ontology for the semantic representation of the services stored in the service registry (UDDI).

Globally, this research study aims to provide an automatic implementation of business processes by re-using the existing web services stored in the service registry. A model called BPMNSemAuto is proposed and takes the input of business process design specifications by the users; and it generates an output of an executable business process. After designing the business processes, the business analyst provides the specifications of each service task through a user interface such as context, inputs, outputs and weight. After that the BPMNSemAuto model performs the service selection and composition to choose the most suitable services to execute every service task. The service selection is done by comparison between the ontology represents the user's requirements and Web Service Ontology; and it uses the Quality of Service (QoS) values to rank the matched services.

This work targets some research problems such as: (i) Semantic representation of the existing web services and the users' requirements. It is because the service registry UDDI (Universal Description, Discovery and Integration) supports only keywords matching and does not store the non-functional properties of web services. However, the non-functional properties of services are the important criteria of the service selection algorithm. (ii) A service selection and composition algorithm. (iii) A solution to solve the problems of synonyms because organizations usually use their own specific terms to name business elements and web services.

The rest of this paper is organized as follows. Section 2, presents the current existing solutions related to this research study. Section 3, introduces the proposed model architecture, the web service ontology structure and a keywords extraction method. Finally, this paper is finished by a conclusion and future work.

2 Related Works

2.1 Business Process Modeling and Modeling Languages

Business Process Modelling Notation (BPMN) is a standard notation for modeling the business processes. It bridges the gap between the design and the implementation of the business processes. The primary goal of BPMN is "to provide a notation that is readily understandable by all business users, from the business analysts that create the initial drafts of the processes, to the technical developers responsible for implementing the technology that will perform those processes, and finally, to the business people who will manage and monitor those processes" [3]. BPMN represents the business process as a Business Process Diagram (BPD) [9]. It divides the elements of business process into four categories: (i) Flow objects that define the behaviors of a business process. A flow object can be an event, activity and gateway; (ii) Connecting objects that connect between two flow objects or between a flow object with other resources. Three types of connecting objects exist: sequence flow, message flow and association; (iii) Swimlanes that group the primary modeling elements. There are two kinds of swimlanes, pool and lane; (iv) Artefacts allow to provide additional information about the process. Artefacts are sub-categorized into data object, group and annotation. In addition, BPMN permits the automatic translation of the graphical business process into BPEL (Business Process Execution Language).

sBPMN (Semantic Business Process Modelling Notation) ontology provides the semantic meaning of each element of the business processes, allows machine readable, and allows reasoning on the process description [10]. In [9], the authors concluded that a modeling language is chosen based on some criteria such as modeling approaches (graph based or rule based) and capabilities of the language (expressibility, flexibility, adaptability, dynamism and complexity).

In summary, the graphical process modeling approach is more used than the rule based approach because it provides a graphical interface that allows business users to be able to model their business processes.

2.2 *Ontology Representation*

In the literature reviews, experts generally define the hierarchy of ontologies and design ontologies for a specific domain of applications. This manual hierarchy of ontology structure is supposed to provide a better accurate and comprehensive representation of the domain information; because experts understand the domain of applications very well. The ontology building for a specific domain is easier than building a generic one that shares amount many application domains. The advantage of the generic domain ontology is the independent aspect of applications. It is rich in axioms, but a heavyweight ontology. The generic ontologies must complete some constraints such as modelling expressiveness, clear semantics and rich formalization, adaptability, harmonization with other standards and usability [11].

In addition, ontology is used to improve the semantic representation of web services in the service oriented architecture. It supports the service selection and composition process, and provides the ability to determine different matching degree between two concepts such as exact, plugin, subsume, intersection and disjoint [12].

Different ontology languages are proposed and they must verify a number of requirements in order to be useful for the business system modelling such as: well defined semantics, sufficient expressive power, powerful and understandable reasoning mechanisms, easy to use with reasonable compact syntax [13]. Each ontology language is different from each others by their expressiveness, supported data type of concept's properties, syntax, constraint checking, top level elements and its ability to support reasoning. Azleny et al. [14], state that an ontology language is selected based on four criteria: intended use, expressiveness, automated reasoning and user perception. In addition, the continuous evolution of ontology languages is also a main criterion for choosing an ontology language.

2.3 *Quality of Service*

In order to improve the accuracy of the result of the service selection and composition algorithms, the non functional properties of services has to be considered and not only their functional properties. The attributes of QoS are mainly defined for a specific domain and are categorized into groups based on their characteristic such as performance, security and context [15, 16]. It is hard for the users to define the value of each attribute of QoS, therefore it is easier for them if they just provide the weight value. The weight defines the importance level of each attribute of QoS [17].

However, it requires addition work to do when working with the QoS because the limitations of the current web service technologies. For example, the representation of the services with WSDL and OWL-S does not allow to express the QoS values. The service registry UDDI does not support the storing of QoS values. There is no standard structure of the QoS ontology and how to calculate them.

3 Proposed Solution

This research aims to provide a solution for the automatic implementation of the business processes from business process design specifications by reusing the existing services stored in the service registry. A model called BPMNSemAuto is proposed in order to solve this problem. The input of BPMNSemAuto model is the business process design specifications (bpmn file format) that can be designed with any supporting editor of BPMN specifications (ex. Jdeveloper and Eclipse). The output of BPMNSemAuto model is the executable business process corresponding to the business process designed by the users. In addition, the BPMNSemAuto model performs only the automatic implementation of the service task and not the other business tasks specified by the users.

3.1 *BPMNSemAuto Model Architecture*

The proposed model architecture called BPMNSemAuto (see Fig. 1) is described in detail step by step as follows:

- First, the uses design the business process with any supporting editors of BPMN specifications, and it is used as the input for BPMNSemAuto model.
- After that the BPMNSemAuto model provides an interface to the users for specifying the description of each service task including context, inputs, outputs and weights to identify the importance of the QoS properties.

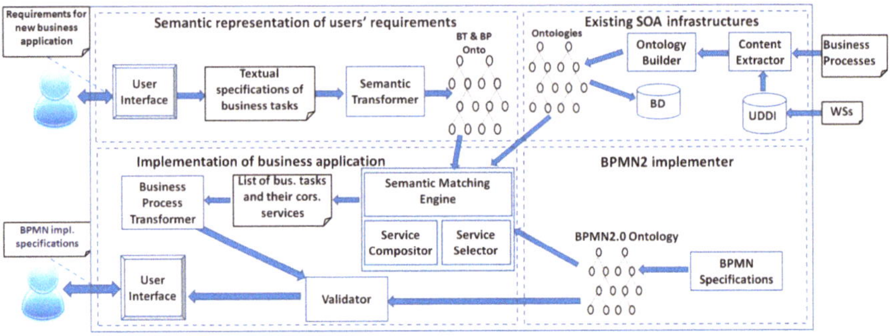

Fig. 1 BPMNSemAuto model architecture

- The BPMNSemAuto checks the data type consistency of the input and output specifications from the users. If any inconsistency is detected, it generates an alert to the users.
- The BPMNSemAuto model builds two different ontologies, a BPOnto ontology to represent the specifications of business process designed by the users; and a WSOnto ontology to represent all the existing services stored in the service registry UDDI.
- Then, the Semantic Matching Engine performs the instance matching between BPOnto and WSOnto ontology to obtain the possible matched services to perform the requested service tasks. It ranks the matched services based on the QoS values. Moreover, it performs the service composition algorithm to create the composite services if the existing atomic services cannot reply to the requirements of users.
- After that, the Business Process Transformer generates an executable business process corresponding to the designed business process of the users.
- Next, the Validator validates the generated business process to check the syntax inconsistency with the support of the BPMN 2.0 ontology defined in [18]. It corrects the syntax error if possible, if not it alerts to the users.

3.2 Web Service Ontology

The Web Service Ontology (WSOnto) is proposed to store all the existing services grouped in categories. This WSOnto provides enough information about the services for the service selection algorithm and for the implementation of services in the business process generation process. The service category is defined by the values of tModel of the UDDI. The Machine Learning Techniques such as Support Vector Machines (SVM) [19] and Nearest Semantic Similarity [20] require the training data in order to define the category patterns. However in this research study, the existing services are already published in the specific categories by publishers. In the future, the machine learning techniques might be needed for

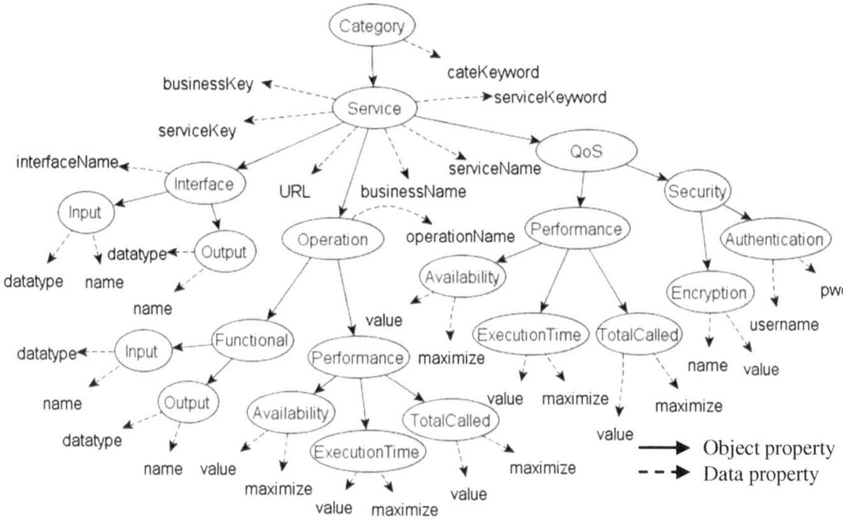

Fig. 2 Web service ontology

suggesting the category of service to the service publishers when they publish their services. Each service of the WSOnto is described by its functional and non-functional properties. The functional properties of services are specified by their service interface (for calling the service) and operations. The non-functional properties of services are specified by the QoS and some additional information such as service-key, business-entity-key and WSDL file location. The complete information about services is useful when inquiring, matching and ranking the services. The QoS values are used to rank the matched services because they can improve the re-usability of services and reduce the development cost. The content of the WSOnto ontology is extracted from the UDDI registry (serviceKey, businessKey, WSDL file location, businessName and service's security information), WSDL files (operations and interface of the services) and the tracking data of service execution on the server (performance value of the service and service's operations). The WSOnto ontology (see Fig. 2) contains of twenty one classes, twenty object properties and thirty three datatype properties.

3.3 Text Extraction

In the proposed WSOnto ontology (Fig. 2), each service and service category are linked to a list of keywords. The keywords of category are extracted from the description of tModel, description of service and name of service of all services in the same category. The keywords of service are extracted from the information of service (description, name, name of the service's operation, description of the service's operation, name of the inputs and name of the outputs).

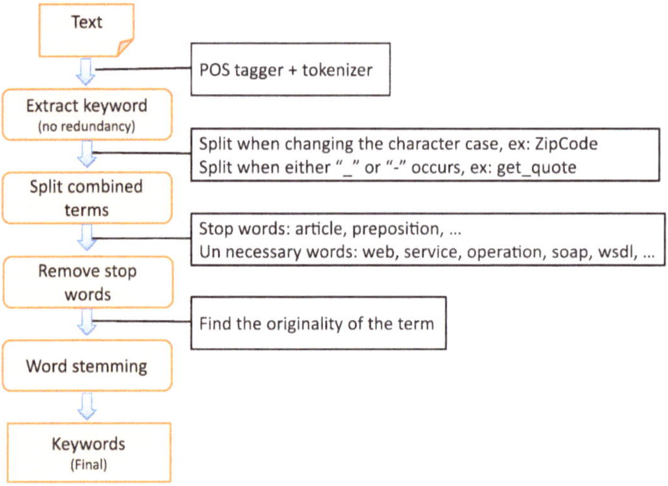

Fig. 3 Keyword extraction method

By adopting the method presented in [19, 20], the keywords are extracted by the following steps (Fig. 3):

1. **Extract Keyword**: From the text description, Part Of Speech (POS) tagger and Tokenizer are used to extract the words that have part of speech as noun, compound noun and verb.
2. **Split combined terms**: From the keywords generated in step 1, the keyword extraction module splits the keyword if the keyword is a combined word.
3. **Remove stop words**: This sub module removes the stop words, those stop words can be an article, proposition and some useless words such as service, operation, WSDL and soap.
4. **Word stemming**: It is a process to find the originality of a word. In English language, the nouns and verbs can be in singular and plural form, but they mean the same thing. Therefore, the best solution is just to store the infinitive form of the word.

3.4 WordNet

WordNet is an English online lexical reference system which provides synonym, hypernyms (generalization) and hyponyms (specialization) sets consisting of nouns, verbs, adjectives and adverbs. At the same time, it provides APIs (Application Programming Interface) in different programming languages that allow us to query to the server to obtain a list of synonym words. The proposed BPMN-SemAuto model uses WordNet to obtain the synonym terms. Using WordNet can solve the synonym problems cause by the use of specific terms to name the concepts specified by the companies.

3.5 Consistency Checking

Through the provided interface, the users are responsible for designing the business process and specifying the description of each service task based on a specific format provided. They have to provide the objective of the service tasks (in the form of keywords), inputs, outputs and the weights of the QoS attributes. Before performing the service selection algorithm, the BPMNSemAuto model checks the consistency of the design business process by comparing between the data type of the outputs of a service task with the data type of inputs of another service task in the sequence. The data type can be a simple type (integer, float, string) or a complex data type (object data type). The object data type is the composition of many simple data types. Therefore, the checking of data type consistency is the comparison of simple data types. For example, if an output is a string but the input is a float, then it shows an inconsistency. However, if the output is an integer and the input is float, then it is acceptable. This consistency checking process is really important, if it is not considered, maybe the generated business process cannot be executed.

3.6 Service Selection

From the user specifications of each service task, the correspondent Service Task Ontology (STOnto) is generated, or a Business Process Ontology (BPOnto) is generated to represent the design business process of the user. After that the BPMNSemAuto model makes the comparison between the STOnto or BPOnto and the WSOnto ontology to obtain possible mappings between these two ontologies. The obtained mappings help the model to acquire web service to execute the desired service task.

The ontology matching methods focus on two things, schema matching and instance matching. A survey of ontology matching tools and techniques is presented in [21]. However in this research study, the instance matching is focused. The OWL API is used to traverse the two ontologies in order to compare the corresponding individuals. The service selection algorithm matches first the keywords, then the inputs and the outputs (name and data type) with the support from WordNet dictionary. After obtaining the list of matched services, an existing service ranking algorithm presented in [22] is adopted to rank them. The authors of [22] presented a Multi-dimensional Multi choice 0–1 Knapsack Problem (MMKP) to choose the best solution out of the K groups of items. However, the algorithm is reduced to just apply with only one group of services that provide the same functionality. Therefore, the value of the utility function of each service can be calculated with the Eq. (1).

$$F = \sum_{i=1}^{\propto} w_i * \left(\frac{q_{ai} - \mu_{ai}}{\sigma_{ai}} \right) + \sum_{j=1}^{\beta} w_j * \left(1 - \frac{q_{bj}(K) - \mu_{bj}}{\sigma_{bj}} \right) \tag{1}$$

where α: the number of QoS properties that are required to maximize their values; β: the number of QoS properties that are required to minimize their values; w: weight of each QoS parameter that is set by users ($0 < w_i, w_j < 1$); μ and σ are the average value and the standard deviation of QoS attributes for all candidates in a service class; $\sum_{i=1}^{\alpha} w_i + \sum_{j=1}^{\beta} w_j = 1$; $\alpha + \beta =$ total number of QoS attributes; q: QoS value.

Finally, the service with the maximum value of the utility function F is selected. From the original business process design specifications with a list of services corresponding to the service tasks, an executable business process is generated. The BPMN2.0 ontology [18] is planned to use to validate the generated business process.

4 Conclusion and Future Work

A BPMNSemAuto model is defined to perform an automatic implementation of business processes from their design specifications and the existing services. A WSOnto Ontology is proposed to store all the necessary information of the existing services stored in the service registry in different categories. Moreover in order to check the usability of the WSOnto, the OWL API is used to traverse the WSOnto and STOnto or BPOnto ontology in order to compare the corresponding individuals to obtain the matched services; and an existing service ranking algorithm is adopted to rank the matched services. From different keyword extraction literature reviews, a synthesis of a keyword extraction method is presented.

Currently, the data description of the services is extracted from the WSDL files and UDDI data; therefore, it does not contain the value of the pre-condition and the post-condition. In the future work, OWL-S might be used to represent the services in order to have the value of the pre-condition and the post-condition.

Acknowledgments This project has been funded with support from the European Commission (EMA2-2010- 2359 Project). This publication reflects the views only of the authors, and the Commission cannot be held responsible for any use which may be made of the information contained therein.

References

1. Workflow Management Coalition. (1999). *Workflow management coalition terminology & glossary*.
2. Rummler, G. A., & Brache, A. P. (1995). *Improving performance: How to manage the white space on the organization chart* (2nd ed.). San Francisco: Jossey-Bass.
3. OMG. (2011). *Business process model and notation (BPMN) version 2.0*. Object Management Group.
4. Lohmann, N., Verbeek, E., & Dijkman, R. (2009). Petri net transformations for business processes–a survey. In *Transactions on petri nets and other models of concurrency II* (pp. 46–63). Berlin Heidelberg: Springer.

5. Genrich, H. J., & Lautenbach, K. (1981). System modelling with high-level Petri nets. *Theoretical Computer Science, 13*(1), 109–135.
6. Retrieved September 7, 2013 from http://www.conceptdraw.com/samples/business-process-diagrams-business-process-model-notation
7. Correia, A., & Abreu, F. B. e. (2012, January). Adding preciseness to BPMN models. *Procedia Technology 5*, 407–17.
8. Chandrasekaran, B., Josephson, J. R., & Benjamins, V. R. (1999). What are ontologies, and why do we need them? *Intelligent Systems and Their Applications, IEEE., 14*(1), 20–26.
9. Lu, R., & Sadiq, S. (2007). *A survey of comparative business process modeling approaches. Business information systems* (pp. 82–94).
10. Abramowicz, W., Filipowska, A., Kaczmarek, M., & Kaczmarek, T. (2007). Semantically enhanced business process modelling notation. In *Proceedings of the Workshop on Semantic Business Process and Product Lifecycle Management (SBPM-2007).*
11. Sabou, M. (2006). *Building Web service ontologies.*
12. Jaeger, M., Rojec-Goldmann, G., Liebetruth, C., Mühl, G., & Geihs, K. (2005). *Ranked matching for service descriptions using owl-s. Kommunikation in Verteilten Systemen (KiVS)* (pp. 91–102).
13. Kalibatiene, D., & Vasilecas, O. (2011). *Survey on ontology languages. Perspectives in Business Informatics Research* (pp. 124–141).
14. Azleny Kamaruddin, L., Shen, J., & Beydoun, G. (2012). Evaluating usage of WSMO and OWL-S in semantic web services. In *Eighth Asia-Pacific Conference on Conceptual Modelling (APCCM 2012)* (pp. 53–58). Melbourne: Australian Computer Society.
15. Neelavathi, S., & Vivekanandan, K. (2011). An innovative quality of service (QOS) based service selection for service orchrestration in SOA. *International Journal of Scientific and Engineering Research, 2*(4), 1–8.
16. Pradnya Khutade, A., & Rashmi Phalnikar, B. (2012). QOS based web service discovery using oo concepts. *International Journal of Advanced Technology & Engineering Research (IJATER), 2*(6), 81–86
17. Chaari, S., Badr, Y., Biennier, F., BenAmar, C., & Favrel, J. (2008). Framework for web service selection based on non-functional properties. *International Journal of Web Services Practices, 3*(2), 94–109.
18. Natschläger, C. (2011). Towards a BPMN 2.0 ontology (pp. 1–15).
19. Bruno, M., Canfora, G., Di Penta, M., & Scognamiglio, R. (2004). *Using SVM and concept analysis to support web service classification and annotation.*
20. Batra, S., & Bawa, S. (2010). Web service categorization using normalized similarity score. *International Journal of Computer Theory and Engineering, 2*(1), 1793–8201.
21. Castano, S., Ferrara, A., Montanelli, S., & Varese, G. (2011). Ontology and instance matching. In *Knowledge-driven multimedia information extraction and ontology evolution* (pp. 167–195). Berlin Heidelberg: Springer
22. Yu, T., & Lin, K. J. (2005). Service selection algorithms for composing complex services with multiple QoS constraints. *Service-Oriented Computing-ICSOC, 2005*, 130–143.

Part III
Enterprise Modelling for Enterprise Interoperability

Modeling Frameworks, Methods and Languages for Computerizing Small and Medium-Sized Enterprises: Review and Proposal

Nawel Amokrane, Vincent Chapurlat, Anne-Lise Courbis, Thomas Lambolais and Mohssine Rahhou

Abstract Small and Medium-sized Enterprises (SMEs) often lack of time, means and competences to define methodologically their requirements for a new software development. This creates a gap between the real needs and the system requirements identified by analysists and designers. It is thereby important to provide SMEs' stakeholders means to autonomously build and share knowledge about their organization. In this work, we present an analysis of a set of modeling frameworks, methods and modeling languages to identify these requirements in the case of such organizations. Taking into consideration unexpected limitations given by this analysis, we propose first an enrichment of the ISO19439 and ISO19440 standards to be applicable to SME typed organisations and second a requirements elicitation and validation process that is compliant with this new framework.

Keywords Enterprise modeling · Requirements engineering · Design verification · Software system for SMEs

N. Amokrane (✉) · V. Chapurlat · A.-L. Courbis · T. Lambolais
Laboratoire de Génie Informatique et d'Ingénierie de Production - LGI2P - site de l'Ecole des Mines d'Alès, Parc Scientifique Georges Besse, 30035 Nîmes cedex 5, France
e-mail: Nawel.Amokrane@mines-ales.fr

V. Chapurlat
e-mail: Vincent.Chapurlat@mines-ales.fr

A.-L. Courbis
e-mail: Anne-Lise.Courbis@mines-ales.fr

T. Lambolais
e-mail: Thomas.Lambolais@mines-ales.fr

N. Amokrane · M. Rahhou
RESULIS, 1 Rue de la Bergerie 30100 Alès, France
e-mail: mohssine.rahhou@resulis.com

K. Mertins et al. (eds.), *Enterprise Interoperability VI*,
Proceedings of the I-ESA Conferences 7, DOI: 10.1007/978-3-319-04948-9_7,

1 Introduction

When compared with large enterprises, Small and Medium-sized Enterprises (SMEs) usually have a simpler structure with limited sets of human, financial, and physical resources. Various Information System (IS) studies have focused on SMEs since the introduction of micro-computers in the early 1980s [1] and showed their lack of IS competences and strategic vision. Moreover, SMEs still find difficulties in selecting application packages, or specifying their needs and wishes in the development of new software to support their activity. This is due to their low IS expertise, and their lack of time and means to get works owner support. This induces difficulty for the analysts and engineers when performing software requirements elicitation and validation activities.

Requirements engineering (RE) constitutes the earliest, most crucial phase of any engineering project and by evidence a software development project. It aims at bridging the gap between SMEs' stakeholders (business experts, decision makers and end users) and the designers and developers team in charge of building here a (set of) software application(s). More precisely, RE is the branch of system engineering concerned with the elicitation, documentation and assessment of: (i) Stakeholder requirements that represent as simply and as formally as possible the needs, wishes and conditions expressed by the project stakeholders regarding the new system and reflecting the problem world; (ii) System requirements that represent the properties of the system, which constrain the conception of the solution. They are gathered in a well-written, verified and validated set of requirement specifications and are written by analysts and requirements engineers to meet the stakeholder requirements. The set of requirements is then considered as the foundation for a contractual agreement. The quality of the RE activities in the SMEs' software development project is often mitigated owing to: (i) The lack of well-established business procedures and practices, which induces difficulties in formulating and justifying the needs; (ii) The lack of formalization as SMEs' stakeholders express ambiguous and informal requirements, thus hard to verify and validate. These issues have raised a crucial need for providing SMEs' stakeholders assistance in formalizing their business and formulating their needs in a simple yet formal way. This work is therefore concerned with providing SMEs' stakeholders the means to: (i) autonomously specify their needs, build and share a common understanding about the structure and the behavior of the SME; (ii) check the consistency of the produced models and descriptions with a set of verification means and finally (iii) transfer them to the developers of the software vender side without loss of information and respecting model and data interoperability expectations.

This paper is structured as follows: in Sect. 2, we introduce the motivation behind this work. Next, we present in the state of the art section relevant enterprise modeling frameworks and reference models and a set of requirements engineering languages (Sect. 3). Then, we present our proposal for SMEs requirements elicitation and validation in Sect. 4. Finally, we conclude the paper and address future works in Sect. 5.

2 Statement

Once established and facing activity growth, many SMEs need to dispose of suitable and adapted software, designed around their business [1]. SMEs must then be able to define appropriate business requirements to perform their computerization. But in the current RE practices, most SMEs' stakeholders do not have the skills to use the requirements elicitation tools and the requirements are gathered informally by developers [2, 3]. This is error prone, because developers lack of knowledge regarding the business domain. Besides, SMEs' stakeholders are directly concerned with the description, argumentation and final decision about their business and needs. We consider that the SMEs' stakeholders have to be involved as active actors, as soon as possible during the requirements engineering activities. To do so, they should be provided with simple means that require no special knowledge in modeling notations or IT skills, to autonomously specify their requirements. It is also important for SMEs' stakeholders to communicate their vision of the organization to the designers and developers. In fact most SMEs have not defined their enterprise model i.e.: the way they manage their activities, the distribution the responsibilities in the organization and the stakeholders' roles, the processes in which they are involved and the information and resources they use. Many concerns have to be addressed, formalization with reference frameworks and methods is then needed to build the enterprise model. All stakeholders can then dispose of a common and non-ambiguous understanding of the SME and its environment.

In addition, the requirements and enterprise models defined by SMEs' stakeholders have to be formal enough to allow verifying their consistency and if they are well-formed. There is hence a trade-off between the necessity of the SMEs stakeholders' autonomy that requires a simple formalism and the level of formalization that allows verification and simulation. The enterprise models along with the stakeholder requirements constitute the base upon which the works owner support analysts specify the system requirements that represent properties of the system and requirements that guide and constrain the construction of the solution. The gaps that may appear between the stakeholder and system requirements, due to absences and inconsistencies have to be reduced by the application of a systematic engineering process and a tooled support.

3 State of the Art

Our work is about understanding and analyzing the structure and the behavior of SMEs to be able to define, verify and validate system requirements. It uses the business view of SMEs' stakeholders as an entry point for the construction of the enterprise specific model. In this section we first present enterprise modeling methods and frameworks that guide the construction of the enterprise model. Then we assess a set of requirements modeling languages according to certain criteria such as their accessibility and the verification techniques that are used. Because of size limits we do not address in this article model transformation methods.

3.1 Enterprise Modeling Methods and Frameworks

In order to manage the inherent complexity of enterprise systems due to their sociotechnical structural and behavioral characteristics, enterprise modeling methods, architectures and tools were developed and used in support of the life cycle engineering of complex and changing systems. Enterprise Modeling (EM) allows the construction of business models that formalize all or part of the business in order to understand or explain an existing situation or to achieve and validate a designed project [4]. A business model is thereby an inescapable and very useful means of communication among stakeholders. Several EM methods, languages, reference models, architectures and tools have been defined since the late 1970s such as the CIM Open System Architecture (CIM-OSA) that has been developed for integration in manufacturing enterprises [5]. It uses well defined modeling constructs structured into four enterprise modeling views: function, information, resources and organization. In the same way the standard ISO/DIS 19440 proposes constructs providing common semantics and enables the unification of models developed by different stakeholders [6]. The enterprise model view dimension enables the modelers to filter their observations of the real world by emphasizing on aspects relevant to their particular interests and context. It also guaranties enterprise modeling principals [7] such as the separation enterprise behavior and enterprise functionalities which provides a flexible and scalable representation of the enterprise.

The Generalized Enterprise Reference Architecture and Methodology (GERAM) [8] takes into account the CIM-OSA four modeling views to organize and define the generic concepts that are required to enable the creation of enterprise models. The framework for enterprise modeling (ISO/DIS 19439) standard [9] took back part of the GERA modeling framework. It provides a unified conceptual basis for model-based enterprise engineering that enables consistency and interoperability of the various modeling methodologies and supporting tools. The framework structures the entities under consideration in terms of three dimensions: the enterprise model view, the enterprise model phase and levels of genericity.

Along with the constructs for enterprise modeling standard ISO19440, the framework for enterprise modeling standard ISO19439 can be considered as an operational state of the art framework to manage the modeling activities of an enterprise or an information system [10]. However, even if requirements definition is part of the enterprise model life-cycle phases, the standards do not offer any constructs to support, share and save information about stakeholders or system requirements. Hence, the descriptions of the desired system are expressed in terms of solution; this makes the real problem not clearly understood and mitigates the validation of the solution.

3.2 Requirements Modeling Languages

Considering the (ISO/DIS 19439) modeling framework model phase dimension, RE activities cover the three first phases: domain identification, concept definition and requirements definition. The business functionality in terms of business processes, enterprise activities, their inputs and outputs and the resources needed for their fulfillment are then described. Therefore, RE encompasses requirements in term of functionality, information, organization and resources. It is highly related to EM activities, especially when RE is oriented to software development and IS definition. However, RE is considered as an independent engineering process and discipline that relies on an iterative process [11] that comprises several activities: (i) domain analysis and elicitation where the existing system in which the software to be built is studied and stakeholder requirements are gathered, (ii) negotiation and agreement about the possible conflicts among stakeholder and alternative situations, (iii) specification and documentation where the requirements are formalized and (iv) verification and validation of the quality of the requirements.

RE considers the matching level between the identified system requirements and the needs that motivated the project as the main indicator of success for software development projects. It is also important to check the quality of single or sets of requirements and enterprise models. Verification and validation techniques are hence used in both RE and EM fields. The main means used by enterprise modeling methods to perform verification and validation [12] are: (i) the use of standards, reference architectures and models to reduce the errors with guided modeling and as a support for consistency analysis, (ii) human expertise to review the models according to predefined criteria, (iii) tests, simulation and process execution techniques and (iv) formal methods used with languages that have a mathematical basis. They are used for property expression and verification and formal reasoning to prove correctness, completeness and consistency of the specifications. A study about RE practices in different industrial sectors [13] identifies the use of advanced techniques to ensure correct requirements writing such as: the use of Natural Language Processing (NLP) to verify the lexical correctness of requirements and the use of requirements boilerplates [14] for guiding requirements writing and checking. The use of ontologies or controlled natural languages coupled with NLP is a promising way to verify requirement consistency and detect missing and ambiguous requirements.

RE provides several methods and languages to guide and support the requirements elicitation, documentation and validation activities. We studied a set of languages and approaches that can be conducted upstream of detailed conception, namely: RDM (Requirement Definition Model) part of the CIM-OSA modeling process [15], goal oriented requirements languages: KAOS [16] and GRL (Goal-Oriented Requirement Language) [17], scenario oriented language Use Case Maps (UCM) [17] and Unified Requirements Modeling language (URML) [18] a high-level quality and functional system requirements language. We are not trying to make an inventory of the all languages proposed in the literature, but rather identify the information that should be addressed while collecting and modeling requirements, and assess their accessibility to SMEs' stakeholders. They can be classified according to:

The orientation and basic concept: this defines the way requirements are modeled and described, and aspects through which they are considered. Two main orientations can be distinguished in the literature [19] (i) scenario based requirements languages like UCM that describe pragmatic, behavioral interactions that end users have with the system to be able to achieve their tasks, and (ii) goal oriented requirements languages that focus on the intended properties of the system expressed by the stakeholders such as KAOS and GRL. Reasoning with goals for functional requirement elicitation is not an easy task for SMEs' stakeholders. They are more likely to describe their daily activities rather than thinking about the motivation behind them.

The accessibility of the syntax which directly influences the target users of the language: the more detailed a requirement language is the less accessible it gets. Considering their syntax, all the studied languages are meant to be used by experts (works owner support analysts, requirements engineers, designers, etc.).The languages use artefacts that require modeling competence; they so cannot be handled by SMEs' stakeholders without prior training. As we are aiming to allow SMEs' stakeholders autonomously express their needs, languages using very light modeling artefacts and Natural language (NL) would be more appropriate. Languages that use NL do not cover all the modeling views. For instance SBVR [20] is mainly used for business rules specification and data description using a controlled NL; it is accordingly only suitable for informational modeling.

The verification and validation techniques that are used: the languages that we studied mainly use non-formal or semi-formal techniques for verification and validation. RDM and UCM use simulation to detect potentially undesirable interactions and behaviors among processes or scenarios. Only KAOS proposes formal reasoning to check the consistency and completeness of models by rewriting the specifications into a verifiable model based on a mathematical formalism.

The modeling views covered by the language and the kind of information represented: we consider here the four modeling views recommended by the ISO19440 standard. We gather in a stakeholder and system requirement view, specific concepts that allow stakeholder and system requirements and goals to be defined, chosen and argued about; such as: goal, obstacle, system requirement and expectation. We distinguish for each language the concepts that are likely to represent the same information as the concepts from the ISO19440 standard to identify the modeling views that the language covers.

The studied languages propose concepts to collect knowledge about the enterprise (*activity, resource…*) necessary to build the information system that will be supported by the desired system, the latter is also addressed by concepts that specify its properties and the way it interacts with the environment (*requirements, expectation…*). However, these concepts are defined and modeled differently from one language to another. For instance, an activity in RDM details business processes and is defined by its inputs and outputs; whereas in KAOS, an operation expresses the realization of a requirement derived from the refinement of goals. Also the assessed concepts are not supported by all the languages. The concept of system requirement, as an independent and clearly defined concept, is only supported by KAOS and URML. Moreover, the concept of stakeholder requirements is represented by all the primary information that we can collect about the enterprise and its business. However there is no clearly defined construct to represent it.

As said before, when RE is oriented to software development and IS definition, it is highly related to enterprise modeling activities. All the modeling views are therefore necessary to gather the needed knowledge about the enterprise. But not all the studied languages cover these modeling views. Thereby a single language is not sufficient to model all the views, a combination of languages is necessary to tackle the complexity of the information system to be modeled. Also, we aim at using the business view of SMEs' stakeholders as an entry point for the construction of the enterprise specific model and the system requirements, but all the studied languages require a specific knowledge in modeling notations. Adequate formalisms have therefore to be chosen to foster the autonomy of end users.

From our base of knowledge, there is no method that guarantees the autonomy of end users and that covers the modeling views to build a common understanding of the enterprise during the requirements elicitation and validation activities.

4 Proposition

It is here important to help SMEs' stakeholders build and share knowledge about their organization so that designers and developers would be aware of the environment for which they must provide solutions. So, we propose to guide software requirements engineering activities considering EM principals. In addition to the specification of the users' needs and system requirements, requirements engineering stakeholders will be provided with means to build the SME's enterprise model which will be used as a reference in the downstream software development activities.

We rely on the combination of the framework for enterprise modeling standard ISO19439 and the constructs for enterprise modeling standard ISO19440, which we extend with a requirements modeling view. Indeed, even if stakeholder or system requirements use concepts from the other views (informational objects and their states, e.g.: a vehicle can have more than one owner, activities and system functions e.g.: the car pound agent assigns the vehicle to a removal agent, etc.) requirements represent independent concepts necessary to gather clients' needs and justify design decisions. They are also important in all the project phases; a requirement document can for instance be used for project planning, cost estimation, scenario generation, etc. Even after decommission of an enterprise entity, it is important to know what motivated its creation in the first place. However, the

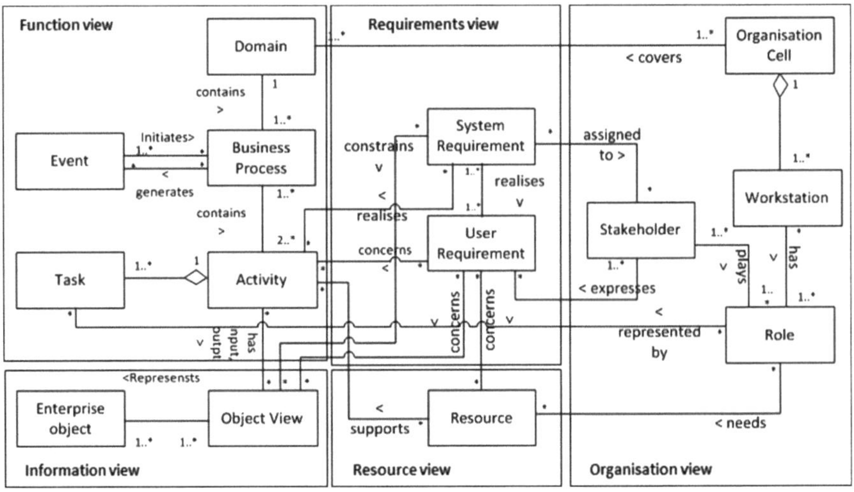

Fig. 1 Proposed constructs for the function, information, resource, organization and requirements modeling views

standards do not offer any constructs to represent or distinguish between stakeholder requirements and system requirements. Such constructs allow modeling the way requirements relate to the other modeling views, assessing the matching level between stakeholder and system requirements and the traceability to design constructs. In the framework for enterprise modeling standard ISO19439 we take into account:

(i) all three levels of genericity, the generic level will be populated by a subset of the constructs for enterprise modeling standard ISO19440 enriched by requirements related concepts, the partial level will constitute typical generic patterns for a category of enterprises to be instantiated for a specific client. The particular level will represent the SME's enterprise particular model,
(ii) the enterprise model phases from the domain identification to the implementation description are considered. As we are concerned with sharing the knowledge collected in the requirement elicitation phase to the downstream conception and development activities, we gather the domain definition and concept definition phases into a stakeholder requirements definition phase,
(iii) to the four modeling views (function, information, resource and organization) that represent the aspects to be considered and the knowledge that has to be captured and stored during modeling activities. We propose to add a complementary requirements view (Fig. 1) to represent concepts addressed in the requirements definition phase and used in the downstream design and implementation phases. We distinguish stakeholder requirements from system requirements.

This standard framework is a support for consistency analysis and verification between the levels of genericity and the modeling views. The verification process allows assessing the correctness of the models and their compliance to meta-models. It is carried out through the definition of modeling rules, consistency rules and completeness criteria.

We are concerned with the definition of requirements that will be derived from the enterprise model and the stakeholder requirements definition. The modeling activities will be achieved by SMEs' stakeholders to whom we will provide a simple notation. We consider reasoning about daily practices and the way tasks are carried out to be an intuitive way for each business stakeholder to describe his role in the enterprise. Indeed they usually do not have a full, common image of business processes in which they are involved especially when these processes are cross-cutting different enterprise departments or business domains. In the same way, business stakeholders who may have this knowledge (head of the enterprise, head of departments, decision makers, etc.) do not know the detailed tasks performed by other stakeholders according to their roles. We propose a role based function oriented modeling process that comprises the four following paradigms:

Organization modeling and role definition: it is necessary to define the stakeholders' responsibilities. According to stakeholders' *workstation*—which are grouped into organizational cells, single roles are defined where a *role* is the function that a *stakeholder* plays while intervening in business processes.

Function and behavior modeling: where the scope of the enterprise part to be modeled is defined and gathered into domains, a *domain* is a functional area representing a business of the enterprise, for each domain single roles are detailed into different *activities*, an activity being the locus of use and creation of information and resources, it can be defined by a number of elementary *tasks* and it has pre and post conditions represented by the state of the *objects* it uses. The activities will be gathered into *business process* models which are derived after the definition of business rules and control flow in the form of *stakeholder requirements*.

Information and resource modeling: *objects* and *resource*s models are derived from the functional models, the activities description and the definition of business rules.

Stakeholder requirements definition: will be performed all along the modeling process and *stakeholder requirements* will be collected to represent business rules and detailed descriptions. Other needs not always related to the ISO19440 modeling views can be collected, such as: non-functional requirements and expected behaviors of the SME' environment.

A tool will support the modeling process. It will be endowed with verification mechanisms in order to: (i) asses the correct utilization of the language by SMEs' stakeholders and the conformance of the produced models to their meta-models; (ii) detect contradictory behaviors among roles and processes definition. For instance situations where stakeholders intervening in the same business processes provide conflicting descriptions regarding the inputs, outputs or the order of the activities; and (iii) discern non-exhaustive descriptions where for instance the output of an activity (that does not represent the purpose of the business process) is not used by any other activity. Highlighting these errors, lacks and inconsistencies while producing enterprise models and stakeholder requirements will encourage SMEs' stakeholders to provide more information and organize their business processes.

Our SMEs computerization approach is based on an iterative requirements elicitation and validation process, which is conducted upstream in the model transformation process. It starts by involving the SMEs' stakeholders (end users and business experts who represent typical roles in the enterprise) to express with a simple formalism their needs and their knowledge about the enterprise organization and functionalities. Then, the stakeholder requirements will be refined by requirements engineers and will serve to formulate a well-written set of system requirements to be validated by the SME stakeholders. They will also serve to create a stable base of the enterprise model. Next, to speed up the process of understanding the business domain, we extract and represent the user's knowledge in some intermediary notation (standard UML diagrams, for example) through model generation. The generated models will then be completed and perfected by requirements engineers. Then, instead of only using requirements models and documents, the generated models serve as a basis for deriving code with model transformation techniques. This will accelerate the production of mock ups that will be validated and refined by the SMEs' stakeholders. The model generation and transformation process has the advantage to ensure conceptual and technical interoperability between the users' models and development languages.

5 Conclusion

Fostering the collaboration between the involved stakeholders during the software development's requirements elicitation and validation activities involves the construction, by the SMEs' stakeholders, of a common understanding about the structure and the behavior of the enterprise. We propose a requirements elicitation and validation process that is compliant with enterprise modeling reference frameworks and favorable to model transformations. The proposal is a work in progress. The notations, the modeling languages and the detailed concepts to be used have to be fixed. We will likely use a simple graphical formalism, coupled with natural language specifications. NLP techniques are under investigation to foster the autonomy of SMEs' stakeholders and provide a means for writing well-formed requirements with controlled vocabulary, rather than verifying requirements afterwards. To ensure the enhancement of the interoperability between the tools used in the computerization project, model transformation mechanisms from natural language to model based specification to code have to be settled.

References

1. Cragg, P., Caldeira, M., & Ward, J. (2011). Organizational information systems competences in small and medium-sized enterprises. *Information and Management, 48*(8), 353–363.
2. Cao, L., & Ramesh, B. (2008, January/February). *Agile requirements engineering practices: An empirical study. IEEE Software* (pp. 60–67).
3. Nikula, U., Sajaniemi, J., & Kälviäinen, H. (2000). *Management view on current requirements engineering practices in small and medium enterprises.* The Fifth Australian Workshop on Requirements Engineering (pp. 81–89). Queensland University of Technology, Brisbane, Australia.
4. Braesch, C., Hauratm, A., & Beving, J. (1995). L'entreprise-système. In C. Braesch, & A. Haurat (Eds.), La modélisation systémique en entreprise (pp. 83–88). Hermès, Paris.
5. Berio, G., & Vernadat, F. (1999). New developments in enterprise modelling using CIMOSA. *Computers in Industry, 40,* 99–114.
6. ISO/DIS Enterprise integration - Constructs for enterprise modelling. (2004). ISO/DIS 19440, ISO/TC 184/SC 5.
7. Vernadat, F. (2001). UEML: towards a unified enterprise modelling language. In *3eme Conference Francophone de Modelisation et simulation: Conception, Analyse et Gestion des systèmes Industriels (MOSIM'01).*
8. Bernus, P., & Nemes, L. (1997). The contribution of the generalised enterprise reference architecture to consensus in the area of enterprise integration. In *Proceedings of ICEIMT97.*
9. ISO/DIS CIM Systems Architecture—Framework for enterprise modelling. (2002). ISO/DIS 19439, ISO/TC 184/SC 5.
10. Millet, P.A. (2008). *Une étude de l'intégration organisationnelle et informationnelle: application aux systèmes d'informations de type ERP.* Thèse INSA de Lyon.
11. Van Lamsweerde, A. (2001). Goal-oriented requirements engineering: A guided tour. In *5th International Symposium of RE* (pp. 249–261). IEEE CS Press.
12. Chapurlat, V. (2007). *Vérification et validation de modèles de systèmes complexes: application à la Modélisation d'Entrepise.* HDR Université de Montpellier 2.
13. Fanmuy, G., Fraga, A., & Llorens, J. (2011). Requirements Verification in the Industry. In O. Hammami, D. Krob, & J.-L. Voirin (Eds.), *Complex Systems Design & Management (CSD&M), Paris,* (pp. 145–160). Berlin Heidelberg: Springer.
14. Farfeleder S., Moser T., Krall A., Stålhane T., Zojer H. & Panis C. (2011). DODT: Increasing requirements formalism using domain Ontologies for improved embedded systems development. In *Poster Session of the 14th IEEE Symposium on Design and Diagnostics of Electronic Circuits Systems (DDECS),* p. 271–274. Cottbus.
15. Zelm, M., Vernadat, F., & Kosanke, K. (1995). *The CIMOSA business modelling process. Computers in industry* (pp. 123–142). Amsterdam: Elsevier.
16. Darimont, R., Delor, E., Massonet, P., & Van Lamsweerde, A. (1998). GRAIL/KAOS: An environment for goal-driven requirements engineering. In *ICSE'9R—20th International Conference on Software Engineering.*
17. ITU–T Z.151: User requirements notation (URN)—language definition (2008). Series Z: languages and general software aspects for telecommunication systems, Formal description techniques (FDT), Geneva.
18. Schneider, F., Naughton, H., & Berenbach, B. (2012). A modeling language to support early lifecycle requirements modeling for SE. *Procedia Computer Science, 8,* 201–206.
19. Rolland, C., Grosz, G., & Kla, R. (1999). Experience with goal-scenario coupling in requirements engineering. In *Proceedings of the IEEE International Symposium on RE.*
20. Semantics of Business Vocabulary and Business Rules (SBVR), v1.0. (2008). Specification formal-08-01-02, Object Management Group, (2008, January).

Identification of Interface Information for a Virtual Data Integration

Konstantin Klein, Marco Franke, Karl A. Hribernik and Klaus-Dieter Thoben

Abstract Nowadays, a production and logistics chain consists of many companies. The establishment of a robust information flow consists of the exchange of diverse information between the companies and its corresponding heterogenous IT-systems. By changing suppliers and logistic partners, the interfaces between their IT-systems have to be adapted. The adaption process is a complex and a time consuming process and it is a significant disturbance variable in the establishment of dynamic production and logistics chains. The time reduction to bind the relevant systems to one's systems becomes more and more important. This gain of time benefits companies in relation of theirs competitors. But, the binding of heterogenous systems is not trivial. To bring data sources together, different data integration approaches have to be considered and challenging data integration problems have to be resolved. This includes e.g. the data sources have different meaning of the information, their structure and other context sensitive information. These facts leads to the important question: Which information about a data source is required and how it can be represented to enable an automated binding process of data sources. This paper explains why an exchange of interface information as a context information is important and how this exchange could look.

K. Klein · M. Franke · K. A. Hribernik
BIBA - Bremer Institut für Produktion und Logistik GmbH, Hochschulring 20, 28359 Bremen, Germany
e-mail: kle@biba.uni-bremen.de

M. Franke
e-mail: fma@biba.uni-bremen.de

K. A. Hribernik
e-mail: hri@biba.uni-bremen.de

K.-D. Thoben
University of Bremen, Bibliothekstrasse 1, 28359 Bremen, Germany
e-mail: tho@biba.uni-bremen.de

K. Mertins et al. (eds.), *Enterprise Interoperability VI*,
Proceedings of the I-ESA Conferences 7, DOI: 10.1007/978-3-319-04948-9_8,

Keywords Virtual data integration · Cyber physical systems · Supply chains · Logistics · Context driven data source binding

1 Introduction

In the today's business environment, production and logistics chains rely on e.g. suppliers, outsourced partners and alliances. Hribernik mentioned that "…traditional supply chains are evolving into complex networks with numerous stakeholders…" [1]. All stakeholders are interconnected due to realize the material, financial, information and decisional flows. To handle this complexity, Fiala mentioned that the supply chain management "… benefits from a variety of concepts that were developed in several different disciplines as marketing, information systems, economics, system dynamics, logistics, operations management, and operations research…" [2].

To achieve this kind of robustness different standards were developed. These standards enabled a cross-cutting process communication between stakeholders. The information which can be exchanged over a standard is only a subset of the information exchanged today. For information which isn't covered by the mentioned standards additional standards and proprietary formats were developed. The usages of different standards and proprietary formats in business interoperability have led to a heterogeneous IT-infrastructure. Hannus [3] has already mentioned this process in 1996, which is illustrated in Fig. 1.

Since 1996, the IT-architectures aim a paradigm shift, from monolithic and central oriented systems to decentralized, loosely coupled and autonomous systems. These new approaches had been explored successfully for logistic processes for example in the Collaborative Research Centre 637 [4] "Autonomous Cooperating Logistic Processes—A Paradigm Shift and its Limitations". The developed approaches could be applied in the future by the usage of the Future Internet technologies like Internet of Things (IoT). One of the main objectives of IoT is to upgrade objects to interact with the environment, to offer services and to work autonomously. The relevance of the above mentioned approaches is still high and analogical approaches are intended in the research initiative "Industrie 4.0" of the high-tech strategy announced by the German government [5]. From the technical perspective, the capabilities of IoT will be extended to Cyber-physical-Systems (CPS) [6].

In conclusion, the mentioned trends lead to an increasing number of autonomous systems in the application field of production and logistics. This evolution will challenge the interoperability of enterprises and the resulting technical implementation of supply chains. To enable a robust information flow between services and systems, high dynamic data integration approaches are necessary. This requirement can be satisfied by the application of virtual data integration approach into the world of interoperability. This concept gives up the idea to storage data from each data source into one data warehouse. It queries the necessary data of data sources on demand. To resolve a query, a binding procedure for each data source is required and implemented by a wrapper, [7] depicted in Fig. 2.

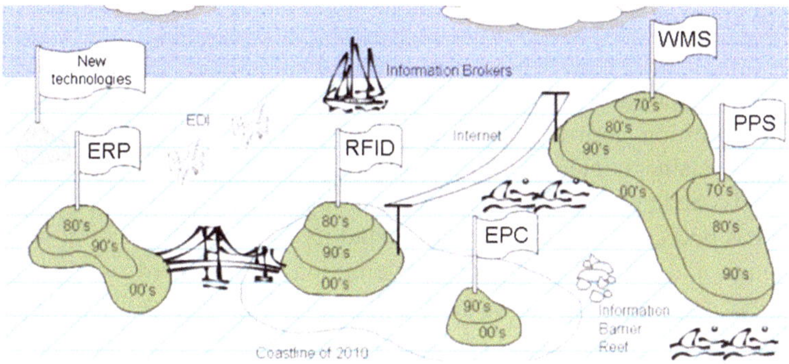

Fig. 1 IT-Infrastructure in the logistic (Hannus 1996)

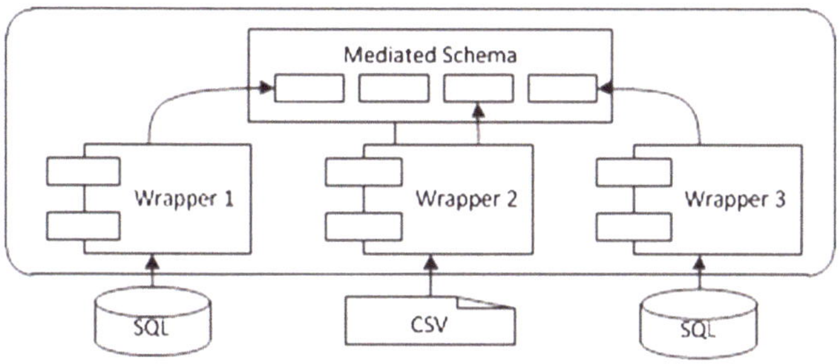

Fig. 2 Concept of a virtual data integration solution

A wrapper implements the communication interface between a data source and a virtual data integration approach. The implementation of wrappers is going fast and can be adapted for coming data sources. The core of a virtual data integration module remains untouched. The virtual data integration approach is applicable for the future trends, which are sketched in the following. Trends like cyber-physical systems make the kind and number of heterogeneous data sources more diverse.

The challenge to achieve an automated process is strong due to bind unknown, heterogeneous data sources. The data source integration challenges have especially a logical character [8]. This means that the schemata of data sources differ regarding its structural design. The following shows an overview of typical data integration problems: each data source uses different names for attributes with the same meaning; data representation can be different; each data source can use a different data model; each data source can cover different attributes. A complete overview is given in [9]. There are many more examples for problems, which must be handled to enable an automated binding process for data sources. The binding process requires for communication establishment and solving of data integration conflicts a better decision-making basis. After the State-of-the-Art is presented, the exact problem description and a proposed solution will be presented. The conclusion paper rounds out the article.

2 State of the Art

This paper describes the related work for the development of the following problem definition and a corresponding approach. The following State-of-Art focuses on business interoperability in the application field of production and logistics. To achieve a business interoperability on the technical perspective an overview is necessary about data sources and its context. The binding of relevant data sources and corresponding context definition is considered so that it is based on a virtual data integration approach which was motivated in the Introduction.

2.1 Data Sources

SQL and CSV files were introduced as examples for data sources in the introduction. To describe both, the structure of data sources and to explain the differences between data sources a working definition of a data source will be presented as follows [10]. A data source is a system which contains and offers data. The data is represented in a specific format. A data source offers an interface to handle data requests by others. For this purpose, an interface implements both the mechanism for data interpretation and the functionality to enable a data exchange process. In conclusion, a data source is characterized by the following criteria:

1. The information which are contained
2. The representation and interpretation mechanism (which language, which interpretation method)
3. The access method.

The second category is the basis to enable the exchange of information between different data sources. Two representations in those category are EDIFACT and ebXML [11].

Apart from the data perspective, the connection of at least one data source requires more knowledge as only the transformation mechanism. The above mentioned criterion "the access method" has to be considered. This includes both the *structural access information* like the address or the authentication data and the *proceeding information*. The *proceeding information* describes the steps to establish a connection and a continuous information exchange. The *structural access information, proceeding information* and the information above the representation and interpretation mechanism are required to enable the establishment of an information flow. This set of information is required and have to be exchanged.

In the following paper the term context is presented. Afterwards, it will be evaluated whether the required set of information can be summarized under the term of context.

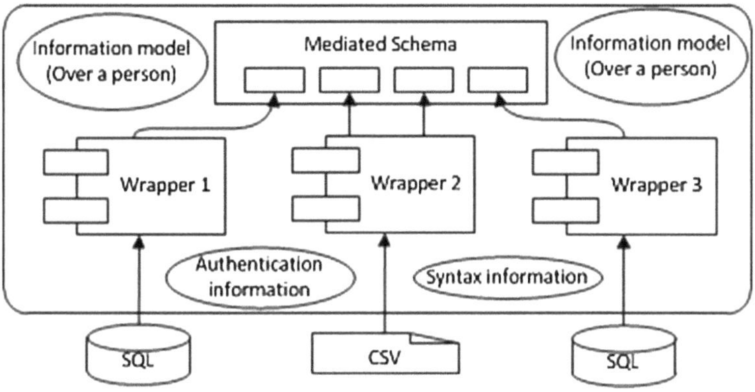

Fig. 3 Binding information

2.2 *Context Awareness as the Basis of a Virtual Data Integration Approach*

As it mentioned before, *structural access information, proceeding information* and the information above the representation and interpretation mechanism are required. By the application of a virtual and mediator based data integration approach, the *proceeding information* contains the following main steps.

1. Identifying whether the data source offers such kind of information
2. Identifying how the communication have to be proceed
3. Initiate the communication
4. Request information.

Within the first step, the mediator requires for each data source the information scheme or information model to determine whether the data source can deliver something to an information request (criterion: *The information which are contained*). If a data source has relevant information, a mediator has to connect to the data source. For this purpose, the mediator has to consider within the step 2 and 3 the data source specific constraints (*structural access information)* and proceedings to release a connection establishment. An example is given in Fig. 2.

In Fig. 3, the exchanged information is represented as ellipses. As you can see, a SQL based data source requires authentication information in contrast to a local file based data source. The set of information which is exchanged between a data source and a mediator to cover the four steps for establishing a connection will be defined as context in this paper. For example, the ellipse "Information model (Over a person)" and the ellipse "Authentication information" in Fig. 3 are summarized as the context of mediator and the SQL data source. In the context definition will be presented.

2.2.1 Machine to Machine

Nowadays, a broad variety of systems can communicate over different network topologies, whereby each network topology can be realized by different hardware and protocols. An analogical diversity results from the variety of systems which range from an intelligent product, over cyber-physical systems to monolithic systems. In consequence, an uncountable number of different communication scenarios can be listed which concern each kind of system over each possible network topology and with each possible protocol. This diversity prevents the identification of a unique set of information to carry out all relevant information for one context which is generally valid.

2.2.2 Context in Data Integration

Holger Wache describes in [10] a context as an element to describe a content of a data source. This definition neglects information about technical approaches that is needed for automated establishment of data exchange. He explains only the possibility to match the data structure and the meaning of data of data sources.

The above presented general definitions of context and the given specialized definitions show a broad spectrum of possible interpretations of the term concept regarding an application domain or specific information flow. To set up a connection establishment, each communication partner offers his context information and all contexts have to be considered. An overview of possible interpretation conflicts which is also known as data integration conflicts is given by [12].

2.3 Summary

The existing variety heterogeneous data sources, corresponding context definitions and corresponding implementation challenge the process of merging the context in such a way that there is no general valid and applicable context definition. On the technical perspective, there is no language to model and to exchange the necessary information as context automatically. In the future, the variety of context definitions will be increased strongly by the technologies of Internet of Things (IoT) [13] and the application of cyber physical systems in the production and logistics environment, which will increase the challenge sustainable.

3 Problem Description

In the previous paper, the existing definition of the term context was described. This paper explains the problem description for the concept of virtual data integration in the relation to the definition of the term context, and cyber physical systems driven logistics and digital supply chains.

The performance in the integration of logistic providers is a very important factor [14]. The increase of the decentralized control of production and logistic systems leads to a huge amount of heterogeneous systems in this field. This development leads to extreme high information interchange formats, demands and behaviours. The increase in fluctuations of partners in logistics networks leads to accelerated growth in integration of stakeholders in the data environment. This trend will be underlined through more becoming small systems in the relation of the cyber physical systems.

The important actual problem in this field is to know, what kind of the information must be exchanged to automate an establishment of a connection between a data source and the mediator in the logistic and supply chains. In the previous chapter of this paper, variables were summarized within the term context. This term has an enormous importance for the problem description in this field and describes the information for the mentioned purpose. The definition described in the state of the art is not meeting the core of the problem of the virtual data integration in logistics. To meet the core of the problem in the current environment a comprehensive term must be developed.

Moreover, the meaning of the new definition of the context for the technical development is non-existent. This means that the term context explains which scope the context information aims, but not how this information is represented and what it means. More exact, the concrete information for the generous exchange and how to model it are very vague.

Recapitulating, in this section the definition of three problems were established to be solved to reach the automated binding of different data sources:

- Insufficient definition of the term context
- Technical issues are uncertain
- Technologies are uncertain.

These problem description leads to the holistic problem description in logistics and digital supply chains. Nowadays, the binding procedures of systems in logistics and supply chains are made by hand. This leads to the procedure in which only the most significant systems are bound together. Smaller systems like cargo control systems or customs services are neglected because of the technical problem description above. In the following paper will be described an approach to deal with the current problem definition.

4 Approach

This section describes the approach to oppose the above described problem definitions. The approach describes first the definition of the term context in relation to the logistics, cyber physical system and internet of things. Second, this paper describes measures deduced from the established definition. Based on the deduced measures, the technical approaches to deal with the current problem definitions will be described.

4.1 Definition of the Term Context

As mentioned above, the definition of the term context related to the logistics, cyber physical systems and supply chains will be described below.

Generally, the term context describes the whole information that is required to establish a data exchange between at least two data sources [15]. In case of logistics and supply chains this means a data exchange between manufacturer and third or fourth party logistic contractor. As in the state of the art mentioned the context information includes four features of a data source. In most of the current network structures in logistic and supply chains, an authentification information is essential and is resolved e.g. via user name and password. This point is described in most significant standards like EDI, mentioned in the state of the art section or SAP systems. Furthermore is the information over a transport protocol TCP or UDP an essential point [16]. The two examples suggest that there is minimal information set which has to be covered in each context definition. This assumption includes in the case of logistics and cyber physical systems the following information to establish a data exchange.

The structure and the process of the authentification—security is nowadays a very important point in logistics and supply chains. The systems are containing very sensitive customer and company regarding information. This information are indicators for volume of money transactions, sales information and product sensitive information. Because of these factors, the systems are secured with strong name and password combinations and more complicated procedures like secureID.

The used protocols and the necessary information—affecting not only the networking transportation protocols. The protocols concerns security algorithms and other proprietary protocols applied in logistics and supply chains.

Information structure and semantic meaning—this point means the data structure and format in systems of a stakeholder in supply chains and logistics. The data transmitted by systems in logistics can be structured, unstructured or be transported as a stream. The semantic meaning is the information behind the value. The meaning of degree can be different. This value can sty for an angle or temperature.

The derived minimal definition of the term context in logistic and supply chain from the point above comes to: Minimal context in case of data integration describes the structure and the process of the authentication, the used protocols and their necessary information, information structure and meaning.

Based on this definition, the following section describes the deduced measures that contain the general approach to establish the data exchange in the CPS based logistics.

4.2 Deduced Measures

As described in the problem definition, the definition of the term concept related to logistics, cyber physical systems and supply chains is very important to deduce the general approach for automated binding procedures of deferent data sources.

As mentioned, a context describes the structure and the process of an authentification process. The structure means the data structure and the technical language needed to authenticate a data source. The languages in supply chains and logistics are usually XML for a SOAP service or a proprietary stream information [17]. The process explains how many stages are needed and which information are to exchange in the authentification process. The detailed information in this case is that the authentification has three steps and needs a session identification and a SecurID token. So far the structure and the process in the authentification in logistic and supply chain are closely combined. The wrapper implemented in the mentioned virtual data integration approach may apply the processing to handly the explained protocols and structures.

After the authentification, the communication process must be established. Data sources uses a protocol to fetch data. The protocols applied in logistics are FTP (File Transfer Protocol) to fetch structured CSV files, SQL or unstructured EDI stream data to retrieve data from target logistic provider system. To explained the spread in the used techniques, the following snippets shows the differences the virtual data integration approach have to deal with. Both snippets represent the sender and the recipient information of an order:

EDIFACT

```
UNA:+.? '
UNB+UNOC:3+Senderkennung+Empfaengerkennung+060620:0931+
1++1234567'
```

ebXML

```
<BinaryCollaboration name="OrderSystem" nameID="RefID" >
      <Role name="Buyer" nameID="ACP"/>
      <Role name="Seller" nameID="TechData"/>
<Start toBusinessState="Request Order"/>
```

To fetch the data from the data source additional information is needed. This includes a structure of the data, which data represents which kind of the key, which is a component of which table and which data represents which meaning.

If the above mentioned information is well-known, the data exchange process can proceed. The next section describes how the exchange of the context information can look.

4.3 Technical Approach

In the sections above the term context and the relevant information to this term were described. Based on this definition, a unique modelling approach for this information must be defined to ensure that two or more data sources have the same meaning of the context.

In the future research work will be evaluated how the combination of modelling languages like UML, SysML Ontologies can describe mentioned context information. An establishment of an exchange of this models between data sources and through a semantic mediator shall automate the binding process. To realize this approach a data source discovery similar to a service discovery must be established. By this service a data source can directly report its being, its entity and its modelled context information to a semantic mediator. New systems in logistics and supply chain can establish a robust and automated data flow by this process. A semantic mediator can report this information to already bound data sources to establish an information exchange between them. Moreover, the technical implementation of this approach needs additional research on how the information of a data source can be stored, exchanged and interpreted. This approach shall grant an automated semantic binding of data sources.

5 Conclusion

The present paper described a state of the art, a problem definition and an approach to automate the binding process in the CPS driven logistics and supply chains realised by the virtual data integration and implemented as a semantic mediator. The research work indicated demand to define an extended definition of a context, because the literature research of the term context hasn't shown a unique definition. In the future research a more comprehensive definition is needed to cover more specialised use cases. The shown approach to establish a communication between to or least data sources have a need of a comprehensive inclusion of use cases. By these use cases a modelling approach shall be developed to cover the possible context information of the structure and behaviour of each data source. After the exchange of the context information, the research work lies in the transformation of the data between sources.

Acknowledgments This article was written within the research project "Cyber-Physische Produktionssysteme—Produktivitäts- und Flexibilitätssteigerung durch die Vernetzung intelligenter Systeme in der Fabrik (CyProS)" and was sponsored by the Federal Ministry of Education and Research (BMBF) under support code 02PJ2460 to 02PJ2480.

References

1. Hribernik, K., Warden, T., Thoben, K.-D., & Herzog, O. (2010). An internet of things for transport logistics—an approach to connecting the information and material flows in autonomous cooperating logistics processes. In: *Proceedings of the 12th international MITIP conference on information technology & innovation processes of the enterprises*, pp. 54-67 (*subprojects: C2, B4*)
2. Fiala, P. (2005). Information sharing in supply chains, Omega, Bd. 33, Nr. 5, S. 419–423.
3. Hannus, M. (1996). Islands of automation in construction. In Ž. Turk (Ed.), *Construction on the information highway, number 198 in CIB publication*, p. 20. University of Ljubljana.
4. SFB637. (2013). Sonderforschungsbereich 637: Motivation. Verfügbar unter: http://www.sfb637.uni-bremen.de/forschungsprogramm.html?&L=2 [Zugegriffen: 20-Sep-2013].
5. BMBF. (2013). Industrie 4.0—Hightech-Strategie der Bundesregierung. Verfügbar unter: http://www.hightech-strategie.de/de/59.php [Zugegriffen: 20-Sep-2013].
6. Broy, M. (2010). Cyber-Physical Systems—Wissenschaftliche Herausforderungen bei der Entwicklung, Cyber-Physical Systems, pp. 17–31.
7. Laender, A. H., Ribeiro-Neto, B. A., da Silva, A. S., & Teixeira, J. S. (2002). A brief survey of web data extraction tools. *ACM Sigmod Record, 31*(2), 84–93.
8. Levy, A. Y. (2000). Logic-based techniques in data integration. In *Logic-based artificial intelligence* (pp. 575–595). Berlin, Heidelberg: Springer.
9. Wache, H. (2003). Semantische mediation für heterogene informationsquellen. KI, Bd. 17, Nr. 4, S. 56.
10. Doan, A., Halevy, A., & Ives, Z. (2012). Principles of data integration. Access online via Elsevier.
11. Rodon, J., Ramis-Pujol, J., & Christiaanse, E. (2007). A process-stakeholder analysis of B2B industry standardisation. *Journal of Enterprise Information Management, 20*(1), 83–95.
12. Goh, C. H. (1996). Representing and reasoning about semantic conflicts in heterogeneous information systems, Citeseer.
13. Atzori, L., Iera, A., & Morabito, G. (2010). The internet of things: A survey. *Computer Networks, 54*(15), 2787–2805.
14. Jayaram, J., & Tan, K.-C. (2010). Supply chain integration with third-party logistics providers. *International Journal of Production Economics, 125*(2), 262–271.
15. Goh, C. H., Bressan, S., Madnick, S., & Siegel, M. (1999). Context interchange: New features and formalisms for the intelligent integration of information. *ACM Transactions on Information Systems (TOIS), 17*(3), 270–293.
16. Ahmed, N., & Jensen, C. D. (2009). A mechanism for identity delegation at authentication level. In *Identity and privacy in the internet age* (pp. 148–162). New York: Springer.
17. Mortimore, C. (2013). JSON Web Token (JWT) Profile for OAuth 2.0 Client Authentication and Authorization Grants draft-ietf-oauth-jwt-bearer-06.

SLMToolBox: An Implementation of MDSEA for Servitisation and Enterprise Interoperability

Hassan Bazoun, Gregory Zacharewicz, Yves Ducq and Hadrien Boyé

Abstract Evolution of service concepts and competition standards in business domain has resulted in a new kind of collaboration between enterprises. As a result of this collaboration virtual enterprises and ecosystems were created. Due to this cooperation, several issues had arisen regarding interoperability concerns for data exchange, and service modeling of the new formed service systems. This paper presents the Model Driven Service Engineering Architecture (MDSEA) as a model driven approach targeting service development in collaborative environments, also it introduces the SLMToolBox, a software tool developed for this purpose.

Keywords Interoperability · Service modeling · Service systems · MDSEA · Service life cycle management · SLMToolBox

1 Introduction

Current world market and customer habits tendency show that traditional manufacturing enterprises, in Europe and around the world, will need to progressively migrate from a product-centric business to a service-oriented one due to

H. Bazoun (✉) · H. Boyé
Hardis/Hardis Conseil, 3 rue Guglielmo Marconi 44800 Saint Herblain, France
e-mail: hassan.bazoun@hardis.fr

H. Boyé
e-mail: hadrien.boye@hardis.fr

H. Bazoun · G. Zacharewicz · Y. Ducq
IMS, Univ Bordeaux, UMR 5218, F-33400 Talence, France
e-mail: gregory.zacharewicz@ims-bordeaux.fr

Y. Ducq
e-mail: yves.ducq@ims-bordeaux.fr

K. Mertins et al. (eds.), *Enterprise Interoperability VI*, 101
Proceedings of the I-ESA Conferences 7, DOI: 10.1007/978-3-319-04948-9_9,

competition standards. In order to manage this transition, companies will have to collaborate in one or several virtual enterprises [1] forming service systems [2]. Building this kind of systems is accompanied with interoperability issues arising between different partners.

This paper presents a model driven approach (MDSEA) as a solution for modeling, designing, and implemeting service systems taking into consideration interoperability concerns. In addition it introduces a software tool (SLMToolBox) which supports the realization of this approach covering design and requirement phases of service lifecycle management. The research work was developed in the frame of the FP7 MSEE Integrated Project [3].

2 State of the Art

2.1 MDA

The most known model driven methodology is Model Driven Architecture (MDA) [4]. This methodology is defined and adopted by the Object Management Group (OMG) in 2001. IT is designed to promote the use of models and their transformations to consider and implement different systems. It is based on an architecture which defines four levels, and goes from general considerations to specific ones:

- CIM Level (Computation Independent Model) is focusing on the whole system and its environment. It is also named "domain model", it describes all work field models (functional, organizational, decisional, process…) of the system with an independent vision from implementation.
- PIM Level (Platform Independent Model): models the sub-set of the system that will be implemented.
- PSM Level (Platform Specific Model): takes into account the specificities related to the development platform.
- Coding Level: the last level that consists in coding or more generally enterprise applications (ESA: Enterprise Software Application).

To complete this description, a Platform Description Model used for the transformation between PIM level and PSM level is added to these four kinds of models corresponding to four abstraction levels. MDA is a buzz word at the moment, its use for service implementation is important in research works. In most of the cases the final stage of the MDA or other MD Development method is the generation of a Service Oriented Architecture SOA [5] or an execution workflow language like BPEL [6]. Nevertheless the development is mainly IT directed and models already start with this IT goal without focusing on the user's point of view and requirements. To overpass this, MDI (see Sect. 2.2) tried to integrate different point of views starting from specification, while SDA tried to adapt MD method to service (Sect. 2.3) and more recently MDSEA (Sect. 3).

2.2 MDI

In standalone enterprises product/service development starts from business model creation at conceptual levels, and later reaches implementation level. In the context of enterprises working in collaborative networks (CN) [7], service modeling and development face interoperability issues at every modeling level. The "Model Driven Interoperability" (MDI) was an attempt to solve this issue.

The MDI works were realised in the frame of the Task Group 2 (TG2) of INTEROP-NoE [8] dedicated to define an approach inspired from OMG MDA. The goal is to tackle the interoperability problem at each abstraction level defined in MDA. In addition, it uses model transformation techniques to link vertically the different levels of abstraction or horizontally to ensure interoperability of models at each level. The main goal of this methodology, based on model transformation, is to allow a complete follow-up from expressing requirements to coding of a solution and also a greater flexibility thanks to the automation of these transformations.

2.3 SDA (Service Driven Architectures)

Several works have been proposed to give a guideline for service definition and implementation. The authors Camarinha-Matos et al. introduced in [9] the notion of "Transparent inter-enterprise plug-and-play infrastructure". They proposed the ARCON reference methodology framework that distinguishes different modeling steps from business to IT. The authors Lin et al. in [10] introduced the concept of "Service Driven Architecture" (SDA). This approach is more dedicated to the implementation of SOA considering mostly the technical constraints at the implementation step. In general, this research area is investigated by several activities, but no clear consensus on the definition of the concepts, research direction or action to be done has already emerged.

According to [9], most of the "Model Driven" mature initiatives have addressed only partial aspects of concrete service settings between enterprises, not properly supporting the various business entities and their interrelationships in complex and fast evolving business ecosystems. The ECOLEAD project [9], as a large international initiative, has experienced a more holistic approach considering both the long-term and temporary organization alliances and collaboration among organizations and individuals. They considered the solution is coming from an interface for networked organizations to be able to rapidly define and set-up relations with other organizations, which requires a plug-&-play-&-do-business infrastructure. Nevertheless this project admits that the modeling interoperability is still solved by proposing one model for all virtual organizations and the model transformation from one abstraction level to another is not automated or assisted by a tool.

2.4 Conclusion on "Model or Service" Driven Approaches

In conclusion, Model Driven approach is essential to allow the implementation of services in coherence with its definition at the business level using enterprise models. MDA defines the modeling levels and specifies the goals to reach at each level but without mentioning how to model or which modeling language to be used. The SDA approach is more detailed but mostly focuses on IT aspects. In addition, interoperability barriers represent a key issue for the development of collaborative networks and for the exchange of data between networked organizations but they are only tackled at all abstraction level by MDI. Therefore, it is necessary to develop a dedicated model driven approach defining accurately each modeling level, proposing modeling languages, interoperability and the transformation mechanisms from one level to another. This will be the objective of the parts hereafter.

3 MDSEA

As introduced earlier, model driven approaches separates the business view of a product-service system from the technical view, however these approaches lacks the ability to model and develop services in CN. As a result, there was a need for an engineering architecture that specifies a framework (i.e. a conceptual structure) for engineering activities, to provide a set of guidelines for structuring service's specifications not only through different vertical abstraction levels but also with horizontal dimensions to deal with interoperability questions at each abstraction level. The Model Driven Service Engineering Architecture (MDSEA) is inspired from MDA/MDI. This methodology is proposed in the frame of the MSEE project [3] that defines its first Grand Challenge as making SSME (Service Science, Management and Engineering) evolving towards Manufacturing Systems and Factories of the Future. MDSEA provides an integrated methodology dealing with modeling languages at various abstraction levels to support Service models and Service System design and implementation. The relationship between the MDSEA modeling levels (BSM, TIM, and TSM) and the Service System lifecycle phases (user-requirements, design and implementation) is established. One of the important innovations in MDSEA is to define the integration between domain components (IT, Organization/Human and Physical Means) at the BSM level in order to ensure that these integration aspects will be spread out at other levels. In this sense, this is therefore considered as an adaptation and an extension of MDA/MDI approaches to the engineering context of product related services in virtual enterprise environment. On the basis of MDA/MDI, the proposed MDSEA defines a framework for service system modeling around three abstraction levels: BSM (Business Service Model), TIM (Technology Independent Model) and TSM (Technology Specific Model).

3.1 Business Service Model (BSM)

BSM specifies models at a global level, describing the service running inside a single enterprise or inside a set of enterprises as well as the links between these enterprises. The models at the BSM level must be independent from future technologies that will be used for the various resources and must reflect the business perspective of the service system. In this sense, it's useful not only as an aid to understand a problem, but also it plays an important role in bridging the gap between domain experts and development experts. The BSM level allows also defining the link between Products' production and Services' production.

3.2 Technology Independent Model (TIM)

TIM delivers models at a second level of abstraction independent from the technology used to implement the system. It provides detailed specifications of the structure and functionality of the service system without including technological details. More concretely, it focuses on the operational details while hiding specific details of particular technology in order to stay technologically independent. At TIM level, the detailed specification of a service system's components are elaborated with respect to IT, Organization/Human and Physical means involved within the production of the service. This is important to mention that in comparison to MDA or MDI or SOMA (Service Oriented Modeling and Architecture) [11], the objective of MDSEA is not only IT oriented and this requires enabling the representation of human and physical resources from the BSM level. At TIM level, these representations must add some information in comparison to BSM models.

3.3 Technology Specific Model (TSM)

TSM enhances the specifications of the TIM model with details that specify how the implementation of the system uses a particular type of technology (such as, for example IT applications, Machine technology or a specific person). At TSM level, the models must provide sufficient details to allow developing or buying suitable software applications, hardware components, recruiting human operators/managers or establishing internal training plans, buying and realizing machine devices. For instance for IT applications, a TSM model enhance a TIM model with technological details and implementation constructs that are available in a specific implementation platform including middleware, operating systems and programming languages (e.g. Java, C++, EJB, CORBA, XML, Web Services, etc.). Based on the technical specifications given at TSM level, the next step consists of implementing the designed service system in terms of IT components

(Applications and Services), Physical Means (machine or device components or material handling), and human resources and organization.

3.4 Proposed Modeling Languages

Based on the described modeling levels, MDSEA proposes to associate relevant modeling languages at each level in order to represent confidently the existing system, future service product and future service system. For choosing modeling languages, the required abstraction level is important.

It is obvious to say that the first specification step of a service to be established between two partners is crucial. At the BSM level, the modeling language must be simple to use, powerful and understandable by business oriented users. Moreover, this (or these) language(s) must cover process and decision with coherent models. The choice is affected by the capacity of the language to propose a hierarchical decomposition (global view to detailed ones). Indeed, business decision-makers often have a global view of the running system and need languages allowing this global representation with few high level activities (process or decisions). This global view must be completed by more detailed activities models elaborated by enterprise sector responsible. These models are connected to top level models in a hierarchical and inclusive way. These are the principles of systemic and system theory which must be taken into account in the choice of the languages. But it is also obvious that the choice of modeling languages is also subjective, depending on the experience of the languages' practitioners and on the wide dissemination of these languages within enterprises.

As for process modeling at business level, several languages exist. Extended Actigrams Star (EA*), extended from GRAI extended Actigram [12], that was itself derived from IDEF0 [13], was chosen to model processes at BSM level due to its independence regarding IT consideration, its hierarchical decomposition and the fact it can model three supported resources: material, human and IT. It has been developed as an answer to previous issues encountered with GRAI extended actigram language regarding its interoperability. It intends to capture business process models at a high semantic level, independently from any technological or detailed specifications. Service Oriented Modeling and Architecture principles [14] developed by IBM were also considered, but these languages are more IT oriented and thus were far away from our requirements. Moreover, GRAI Grid [15] was selected for modeling governance in a service system. GRAI Grid aims at proposing a cartography of company's decisions which controls business processes, as proposed for instance in the ISO 9000-2008 standard [16]. The interest of GRAI Grid is to represent all decisions and their coordination, from the strategic to the operational levels. This representation is very important for business users because the results of decision making are also at the origin of performance evolution and achievement.

At the TIM level, BPMN 2.0 [17] was chosen in particular because this language offers a large set of detailed modeling construct, including IT aspects and benefits from the interoperability of many BPM IT platforms allowing the deployment and automated transformation to execution of BPMN processes. Moreover, BPMN enables also to represent human and technical resources which are required in the MDSEA principles of representation. BPMN has also the advantage to provide a meta-model developed by OMG which facilitates the implementation of the language. GRAI nets are proposed in order to detail the decision processes in coherence with the decisions identified in the GRAI Grid but with adding technical and organization information as the decision rules, the decision makers, and the decision support modules.

4 SLMToolBox

SLMToolBox is a software tool developed by Hardis [18] in the frame of MSEE project. The SLMToolBox will be used by enterprises willing to develop a new service or improve an existing one, within a single enterprise or a virtual manufacturing enterprise [1]. The tool will be used at the stage of "requirement" and "design" of the service engineering process.

4.1 Motivation and Requirements

The basic motivation for SLMToolBox development is the lack of reference tools for designing and managing service innovation projects. This fact is affecting European Manufacturers willing to invest in service innovation as they currently have to rely on various generic tools, mostly oriented on « business process management » and « software engineering » domains. Key requirements are based on needs related to service engineers and IT teams. A service engineer needs:

- To specify, evaluate, communicate and design the system supporting the service and its lifecycle.
- Appropriate formalisms (domain specific and easy to read).
- Interoperable data formats.

Also an IT team needs to:

- Design a solution which is aligned with business requirements.
- Focus on technical activities (e.g.: technical design, implementation, integration…).

To address the above requirements, an integrated modeling tool (SLMToolBox) is proposed. This software is dedicated to manufacturing services lifecycle management and permits to:

- Take benefit of a model based architecture (e.g.: syntactic validation; transformation; execution; …).
- Maintain the coherence through the whole engineering process—from Business requirements to IT implementation (modeling).
- Anticipate/simulate the result of the service (engineering).
- Design the governance of the service (monitoring and control).

4.2 Conceptual Architecture

The SLMToolBox is regarded to be an integration of several scientific concepts related to services into one tool. These concepts can be summarized into service modeling, engineering, simulation, monitoring and control.

4.2.1 Service Modeling

MDSEA Metamodels and Languages

MDSEA defines a set of constructs and relationships (described with "templates") which are specific to the domain of service system modeling, at 3 modeling levels: BSM/TIM/TSM. For each abstraction level, MDSEA suggest a set of references to standard or former graphical modeling languages (which are domain agnostic), in order to extend and complete the representation of the system to be modeled, under different perspectives (e.g.: decision structure; process; use cases; …).

This type of modeling architecture is based on a "view model" pattern (or "viewpoints framework") [19] as it defines a coherent set of views to be used, in the construction of a manufacturing service. The purpose of views and viewpoints is to enable humans to comprehend very complex systems, to organize the elements of the problem and the solution around domains of expertise and to separate concerns. In the engineering of physically intensive systems, viewpoints often correspond to capabilities and responsibilities within the engineering organization.

Both BSM (Business Service Models) and TIM (Technology Independent Models) are structured in the same manner. A "core" model gathers a set of generic (meta-) data in order to qualify the service to be modeled (specified/designed); this "core" model refers to external graphical modeling languages (e.g. : UML [20]) so that certain aspects of the service model can be elaborated in more details with the help of graphical languages.

This structure allows to map "view specific" modeling languages (e.g.: GraiGrid, UML Class Diagram) with "domain specific" constructs (i.e.: MDSEA BSM) without introducing modifications or restrictions to the MDSEA metamodel.

From the user point of view, it allows the possibility to edit core information, independent from any specific modeling language, and to retrieve and reuse this data under different views, accomplished with the help of several graphical diagrams. With this approach, MDSEA Core Constructs remain agnostic from any representation formalism. Their implementation is realized by a core model, which acts as domain specific (Service System Modeling) "glue" between several modeling languages. Thus, we can reuse standard modeling languages without introducing modifications to their metamodels (e.g.: BPMN, UML…). Graphical languages such as "Extended Actigram Star" or "GraiGrid" can continue to evolve, with (almost) no impact on MDSEA Core metamodels (i.e.: BSM and TIM).

Modeling Editors

The modeling environment will support service system modeling activities by providing editors for domain specific models (BSM, TIM) and related modeling languages to enhance the description of the BSM and TIM models. A set of language specific modeling editors is provided for each modeling language. These editors are either the result of a Hardis's specific development (BSM templates, EA*, GraiGrid, and TIM templates editors) or open source plugins integrated within the same environement (UML and BPMN editors).

Model Transformation

SLMToolBox supports specific model transformations, mostly to support the continuity between the service concepts and requirements phase to the service design phase. In addition model transformation aims to save effort and reduce errors by automating the development of models when possible. Information and requirements collected at BSM level are reused at TIM level. SLMToolBox supports the transformation of BSM data models into TIM data models, and the transformation of EA* model into BPMN process and collaboration diagrams [21].

4.2.2 Service Engineering and Simulation

In MDSEA service systems are modeled from different views: static view (formalism: GraiGrid) and dynamic view (formalism: Extended Actigram Star). Static model is more structural than behavioral, helps in depicting static constituents of the system, rigid as it is time independent view of a system, and can't be changed in real time. On the other hand, a Dynamic model is a representation of the behavior of the static components of the system, and consists of a sequence of operations, state changes, activities, and interactions. Dynamic model is flexible as it can change with time as it shows what an object does with many possibilities that might arise in time. As a result, in order to simulate a service system, simulations will be built on the basis of "Business Process" models and in specific on the dynamic Extended Actigram Star diagrams. In addition the service's quality is assessed through the evaluation of time and cost criteria from one side and service's objectives on the other side.

4.2.3 Monitoring and Control

SLMToolBox offers service's owners the ability to monitor their service's performance through GraiGrid diagrams. The functionalities proposed by the SLMToolBox consist in defining decision variables, objectives and primary indicators for each decision center of the system. In addition, the tool proposes a reference list of primary indicators, categorized by domain and aggregation level (i.e. enterprise or virtual enterprise) according to the service governance method defined in the MSEE project.

5 Conclusion

In this paper we covered the problem of interoperability in collaborative enterprises by proposing new methodology (MDSEA) as an extension of MDA/MDI approaches directed towards service systems. In addition we introduced the SLMToolBox which is currently used as a support tool for the four industrial pilot cases (IBAMIA, TP-VISION, INDESIT, and BIVOLINO) of the MSEE project. A final prototype has been released in October 2013, including the foundations which are necessary to build an advanced toolset, dedicated to service lifecycle management on top of the Eclipse Platform. In its current version, the prototype can be used to build models of "As-Is" and "To-Be" situations, in order to support the "requirement" and "design" phases of the service lifecycle—at both BSM and TIM levels.

Acknowledgments This work has been partly funded by the European Commission through the Project MSEE: Manufacturing Service Enterprise (Grant Agreement No. 284860). The authors wish to acknowledge the Commission for their support.

References

1. Thoben, K., Jagdev, H., Eschenbächer, J. (2001). *Extended products: Evolving traditional product concepts.* Paper presented at: The 7th international conference on concurrent enterprising, engineering the knowledge economy through co-operation, June 27–29, 2001. Bremen, Germany.
2. Cardoso, J., Pedrinaci, C., Leidig, T., Rupino, P., De Leenheer, P. (2012). *Open semantic service networks.* Paper presented at: The international symposium on service science (ISSS).
3. FP7—FoF-ICT-2011.7.3. (2011). Manufacturing service ecosystem project- Annex 1 description of work, July 29, 2011.

4. OMG. (2013). *MDA Guide Version 1.0.* Retrieved October 18, 2013, from http://www.omg.org/cgi-bin/doc?omg/03-06-01
5. Johnson, S. K., Brown, A. W. (2006). A model-driven development approach to creating service-oriented solutions. *Service-Oriented Computing—ICSOC 2006 Lecture Notes in Computer Science* vol. 4294, pp. 624–636. Springer (2006).
6. Orriëns, B., Yang, J., Papazoglou, M. P. (2003). Model driven service composition. *Service-Oriented Computing—ICSOC 2003 Lecture Notes in Computer Science.* vol. 2910, pp. 75–90.
7. Camarinha-Matos, L. M., Afsarmanesh, H. (2006). Collaborative networks value creation in a knowledge society. *Knowledge Enterprise: Intelligent Strategies in Product Design, Manufacturing, and Management—IFIP International Federation for Information Processing* (vol. 207, pp. 26–40). doi:10.1007/0-387-34403-9_4.
8. Bourey, J. P., Grangel, R., Doumeingts, G., Berre, A. (2007). Deliverable DTG2.3 from the INTEROP project. *Report on Model Driven Interoperability.* Retrieved May 15, 2013, from http://interop-vlab.eu
9. Camarinha-Matos, L. M., Afsarmanesh, H., Ollus, M. (2008). ECOLEAD and CNO base concepts. *Methods and Tools for Collaborative Networked Organizations.* pp. 3–32. Springer, US. doi:10.1007/978-0-387-79424-2_1.
10. Lin, H. L., Hung, S. S., Tsaih, D. (2009). Modeling service-driven architecture for distributed enterprise portal for maintaining path consistency, In *SEPADS'09 Proceedings of the 8th WSEAS International Conference on Software engineering, parallel and distributed systems.* pp. 135–143.
11. Arsanjani, A., Ghosh, S., Allam, A., Abdollah, T., Ganapathy, S., Holley, K. (2008). SOMA: A method for developing service-oriented solutions. IBM Systems Journal, 47(3).
12. Grangel Seguer, R., Cutting-Decelle, A. F., Bourey, J. P. (2008). A UML profile for transforming GRAI Extended Actigrams into UML. In *Proceedings of the 17th IFAC World Congress, Session: Architectures and Software Tools for Enterprise Integration and Networking in Manufacturing*, Paper ThB24.3, Séoul, Korea; 2008.
13. National Institute of Standards and Technology (NIST). (2013). *IDEF, Integrated Definition Methods.* Retrieved October 18, 2013, from http://www.idef.com/IDEF0.htm
14. Bell, M. (2008). Service-oriented modeling: Service analysis, design, and architecture. Hoboken, NJ: Wiley publishing.
15. Doumeingts, G., Vallespir, B., Chen, D. (1998). GRAI grid, decisional modeling. In *International Handbook on Architecture of Information System* (pp. 313–337). Berlin: Springer.
16. Goult, R. (2013). *Introduction to ISO 9001.2008 Handbook.* Retrieved October 18, 2013, from http://fr.scribd.com/doc/171647046/ISO-9001-2008-Handbook
17. OMG. (2013). *Business Process Model and Notation (BPMN) Version 2.0.* Retrieved October 18, 2013, from http://www.omg.org/spec/BPMN/2.0/PDF/
18. Hardis. (2013). Hardis is a software company with specialist expertise in management computing. Retrieved October 18, 2013, from http://www.hardis.fr/eng/jsp/site/Portal.jsp
19. ISO/IEC/IEEE 42010:2011. (2011). Systems and software engineering—Architecture description.
20. OMG. (2013). *OMG Unified Modeling Language (OMG UML), Infrastructure Version 2.4.1.* Retrieved October 18, 2013, from http://www.omg.org/spec/UML/2.4.1/Infrastructure/PDF/
21. Bazoun, H., Zacharewicz, G., Ducq, Y., boye, H. (2013). *Transformation of Extended Actigram Star to BPMN2.0 and simulation model in the frame of MDSEA.* Paper appeared In Proceedings of the symposium on theory of modeling & simulation.

Use of Service Patterns as an Approach to Modelling of Electronic Government Services

Wannessa R. Fonseca and Pedro L. P. Corrêa

Abstract The government appears to be a high potential scenario for the deployment of service-oriented applications. The business processes in the service-oriented computational model are modelled and implemented as services. Interoperable service is a major goal of service orientation. Government organizations are adopting service use in order to achieve interoperability of government systems, but there is still a lack of technical support to reuse already conceived service concepts, as well as the efforts and experience of the experts who conceive services. Accordingly, this paper proposes a Service Specification Method for Electronic Government (SSMe-Gov) to support the development of systems of government. The method supports the specification of e-government services from service patterns. A lifecycle of services is also proposed for the specification of new services from service patterns. The lifecycle of services includes the activity of finding patterns of candidate services. The conception of services combined with the concept of patterns can help software architects to identify recurrent functional elements and reduce redundant efforts in the conception of services with the same purposes. Previous case studies show that it is feasible to set service patterns from the analysis of existing services in government.

Keywords Service · Electronic government · Service-oriented computing · Reuse · Service patterns

W. R. Fonseca (✉) · P. L. P. Corrêa
University of São Paulo, Luciano Gualberto, travessa 3, 158. Prédio da Engenharia Elétrica, São Paulo, Brazil
e-mail: wannessafonseca@usp.br

P. L. P. Corrêa
e-mail: pedro.correa@usp.br

W. R. Fonseca
Cepromat, Cuiabá, Brazil

K. Mertins et al. (eds.), *Enterprise Interoperability VI*,
Proceedings of the I-ESA Conferences 7, DOI: 10.1007/978-3-319-04948-9_10,

1 Introduction

The business processes in the service-oriented computational model are modelled and implemented from the perspective of services. Services are autonomous and platform-independent computational entities. Services can be described, published, discovered, and dynamically assembled to develop distributed, interoperable and evolvable systems. Service-oriented computing (SOC) utilizes services as basic constructs to support the rapid development of low cost and easy composition of distributed applications even in heterogeneous computing environments. SOC is intended to make services available, where applications are assembled with little effort in a network of loosely coupled services, to create flexible business processes and agile applications that can span different organizations and computing platforms [1].

The government, in general, appears as a high potential scenario for the service-oriented solutions deployment, especially due to the large number of existing applications, technological diversity, the need for interaction between these applications and the need for service quality management. However, the scenario described does not exempt the government from the challenges related to the paradigm of service-oriented development.

In the electronic government (e-government) context when the term **service** is used, it easily relates to the term *electronic service* directly provided to the citizen, through an end-user interface. In this paper, the term *service* is used to represent a software interface, provided by a government, to be consumed by applications from governmental institutions or non-governmental institutions.

Currently a large number of information systems are created and developed in the three Brazilian government levels (Federal, State and Municipal) and its various sectors (Executive, Legislative and Judiciary), but the services are generally created as from basic principles, without considering the reuse of service-oriented solutions by other public entities. Thus, efforts to devise solutions through the paradigm of services are not taken advantage of. Although there are electronic means to disseminate services, such as the Federal Government interoperable service catalog, these are not disclosed in order to reuse the solution, but in order to consume those services. What is observed is the lack of support to encourage the reuse of service-oriented solutions designed for business processes related to electronic government.

The reuse of solutions that have already been devised and that worked in the past is a good practice in the development of systems, regardless of the paradigm this implies. Although Gamma et al. [2] address the object-oriented paradigm, they point out that the best designers know they should not solve a problem based on basic principles or from scratch.

Research related to e-government is mostly multidisciplinary [3], involving several areas of study, such as politics, management and information technology. In that sense, this paper presents a research related to SOC in the e-government scenario.

The goal is to define a method to support the reuse of the service concept in the setting of e-government. To achieve this goal, the use of service patterns for government scenario is suggested.

2 Interoperability and Service-Oriented Architecture

The government to government (G2G) interaction is classified as back office while government to citizen (G2C) and government to business (G2B) interactions are front office. The back office interactions are considered problematic due to different difficulties of interoperability [4].

It is often difficult to achieve interoperability in government organizations. In some situations, government agencies are reluctant to change existing work processes, make their data and services available to external partners and renegotiate their transactions with external parties [5].

Aguair et al. [6] emphasize that the definition of patterns, norms and methods facilitates and improvement the interaction among the various sectors and levels of government, as well as with society in general. It is essential that there is communication and integration between management and technological aspects to proceed with the actions of e-government.

Several interoperability standards for e-government have been defined, e.g., e-Government Interoperability Framework (e-GIF) defined by the United Kingdom [7]. The e-GIF has become a reference standard for interoperability of e-government.

The Brazilian Federal Government developed the Interoperability Standards of Electronic Government (e-PING) which defines a minimum set of premises, policies and technical specifications governing the use of Information and Communication Technology in the federal Government, establishing the conditions of interaction with the other sectors and levels of government and society in general [8]. The areas covered by e-PING are: Interconnection, Security, Means of Access, Information Organization and Exchange, and Integration of Electronic Government. Specifically in the area of integration for Electronic Government, the goal is to approach and to explore the boundaries between the technological, semantic and organizational guidelines and policies seeking public management improvement from the view of an interoperable technology platform. Some products available are: Interoperability Guide and Catalog of Interoperability composed of Services Catalog (Web Services) and Data Standard Catalog.

According to the e-PING [8] reference document *the clearly defined policies and specifications for interoperability and information management are essential to facilitate the connection of government, both internally and in its contact with society, and at a higher level scope, with the rest of the world—other governments and companies in the global market.*

The e-Government Interoperability Guide provided by the United Nations Development Program [9] has suggested that a *Service-Oriented Architecture*

(SOA) is the best underlying paradigm with which to begin to roll out e-government services that can be used in cross-agency and cross-border situations. The use of SOA is also recommended by e-PING [8], as a technical guideline for the integration of information systems.

The term SOC is often confused with the term Service-Oriented Architecture and sometimes used interchangeably, although they are distinct. SOC is a paradigm of distributed computing. This computing model is composed of several elements, including the service-oriented design paradigm, the proper services, service composition, services inventory and SOA. SOA is characterized by the introduction of new technologies and platforms to support the creation, implementation and evolution of service-oriented solutions. A SOA implementation is unique to each company and may have different combinations of technologies, products, APIs and supporting infrastructure [10].

According to Sommerville [11], the principle of service-oriented software engineering is building a program by composing independent services that include reusable functionalities.

The most recommended technology for interoperability is SOA, implemented using Web Services [12].

The interoperability of services is a key goal of service orientation and it lays the foundation to achieve the strategic benefits of SOC [10].

Challenges of developing services for the purpose of implementing G2G interoperability are dealt with by Klischewski and Askar [13]. These challenges are related to several factors, including the definition of service scope and the definition of the service interface granularity.

3 Service Patterns

Generally speaking, patterns are reusable solutions for recurrent design problems [14]. The concept of service patterns used in this study is similar to that defined by Fki et al. [15] an abstract service representing a generic and reusable description. Besides this definition, service patterns must contemplate the description of atomic services and compound services, as well as the interactions between services. Thus, the service patterns will be able to meet a government task or business process.

Table 1 illustrates some differences in the terms services and service patterns used in this study.

In the service-oriented development process, the unit of software to be reused is the service. When inserting the concept of service patterns in the process of service-oriented development, the goal is to reuse the service concept and service logic represented by the pattern.

The use of the design principles of service orientation (e.g. Standardized Service Contract, Low Coupling, Service Reuse Capability) assists in defining how the logic should be decomposed and modeled into services. These principles

Table 1 Services and service patterns

	Goal	Related artefacts	Reuse
Services	Be part of an inventory of highly reusable services	Service specification Source Web Services Description Language (WSDL) Code of deployment	Through service composition
Service patterns	Be part of a highly referenced service pattern catalog	Service patterns specification (as described in Sect. 4.1)	Reuse of the concept and logic abstracted from a service

support or contribute to the interoperability of services. The goal is to produce inventories of highly reusable services to meet new business demands. In this sense, the organization must adopt methods that help identifying services to build their inventories.

The goal of using service patterns is to assist the specification of new services from existing services in the government, aiming at the reuse of already devised solutions. The goal of SOC is to have service inventories that can meet business processes. Therefore, based on the modelled business process the service patterns catalog should be consulted in order to locate service patterns that may be suitable for the business process.

The service pattern should serve as a reference for the specification of a service. Figure 1 shows a conceptual view of service patterns and related elements. The catalog consists of service patterns, each of which contains a description, a service interface, and each service capability contains an activity diagram that represents the behavior of that capability.

4 Proposed Method

Considering the redundancy of functional characteristics in the government scenario, this research aims to: (1) provide mechanisms to support the reuse of e-government service concepts; (2) add value to service specification activity in the life cycle of services in e-government scenario and (3) reduce redundancy of efforts in the design of new services for the same purposes in other levels and sectors of the government.

The Service Specification Method for e-Government (SSMe-Gov) was proposed in order to assist the specification of e-government services. In this method, the service patterns are defined from existing services in the government. The service patterns should be cataloged in a repository to serve as a reference for the creation of new services. Figure 2 illustrates an abstract view of SSMe-Gov.

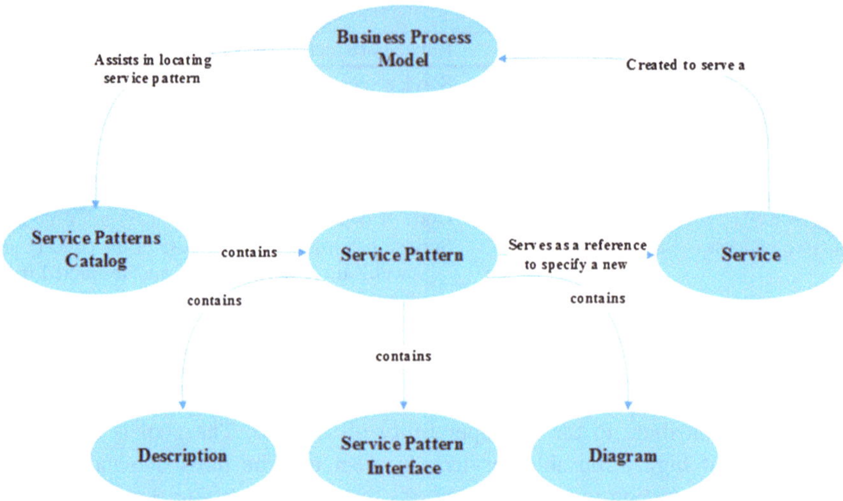

Fig. 1 Elements related to service pattern

(a) Definition of Service Patterns: in order to define service patterns, the concept linked to a service should be abstracted and represented as a service pattern. The service patterns are documented using the *Service Pattern Description* template and service patterns are cataloged in the service pattern repository.
(b) Creation of new services: in order to create new services cataloged service patterns should be used as reference.

4.1 Definition of Service Patterns

This activity consists in abstracting concepts linked to a service and representing it as a service pattern. For this activity some government services were identified and analyzed, and from this analysis it was possible to generalize the concept and represent it as a service pattern allowing the concept to be reused.

Among the principles of service-oriented design, the capability of service reuse is strongly bonded to the research described in this paper.

The service pattern description structure should contain a set of aspects that supports the understanding of the pattern. According to Li et al. [14], the main advantage of a pattern description structure is that it introduces a structured way to understand, explain and reason out patterns. Also, it allows for more effective communication between software engineers and domain experts, because a pattern description structure in particular provides a common language to talk about patterns.

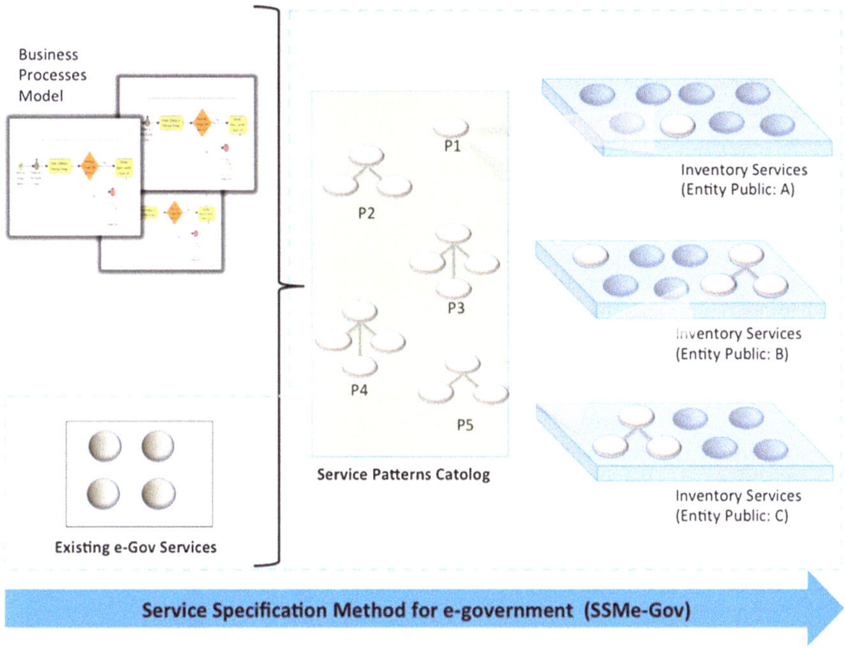

Fig. 2 Proposed method

There is no established rule on the format used to describe a service pattern, therefore this work defined a template named *Service Pattern Description*, as shown in Table 2. This template was defined from the analysis of the works, such as Tchuta and Chunhua [16], Li et al. [14] and Fki et al. [15].

The aim of the catalog is to bring together service patterns defined and documented in order to expedite the location of service patterns. Service patterns will be classified in the catalog according to government areas. Examples of government areas are education, health, finance, social security and work.

Services portfolio aims to ensure that services can be reused and shared. On the other hand, catalogs of service patterns, in this work, aim to support the reuse of the service concepts.

Obtaining reuse depends not only on technical issues; it is strongly related to the issues of management and organizational culture. While creating the service patterns catalog itself does not ensure the reuse, it can support reuse and service designing. The creation of service patterns can support the reuse of the concepts associated with the services.

Table 2 Service pattern description template

Service pattern name	Name that identifies the service pattern
Service pattern description	Meaningful description of the service pattern
Version	Version number
Authors	The authors that contributed to the service pattern
Problem	Describes the recurring problem for which the solution was defined
Solution	Describes the fundamental principle of the solution proposed by service pattern
Participants elements	Presents which parts are involved in the services, for example, the systems and service layers
Pattern type	Specifies the type of service rendered by the pattern, whether it is atomic or compound
Related patterns	Name of the patterns related to the pattern described
Catalog	Describes the name of the catalog the pattern belongs to
Service interface	Representation of the service pattern interface
Capability name	Service capacity name
Capability description	Service capacity description
Diagram	Graphical representation of the service pattern capacity

4.2 Creation of New Government Services

Cataloged service patterns form the basis for the design of new services. Those interested can consult the catalog of service patterns and from the cataloged service patterns create a new service, using the selected service patterns as reference.

A service pattern can be used as a basis for designing services for multiple inventories or it can also be the basis for creating multiple services for a same inventory. The decision whether a service pattern will result in a service in an inventory or multiple services on the same inventory, will depend on the level of granularity required for the service to be created.

Although the discussion on the granularity of a service created from a service pattern is outside the scope of this research, it is worth noting the following considerations:

- whereas public service inventories are to meet the business process of the government, such inventories should theoretically have a similar level of granularity;
- there are currently no government guidelines regarding the granularity of services. Government entities are most frequently responsible for designing services and conducting SOA governance, it is their role to decide on the granularity.

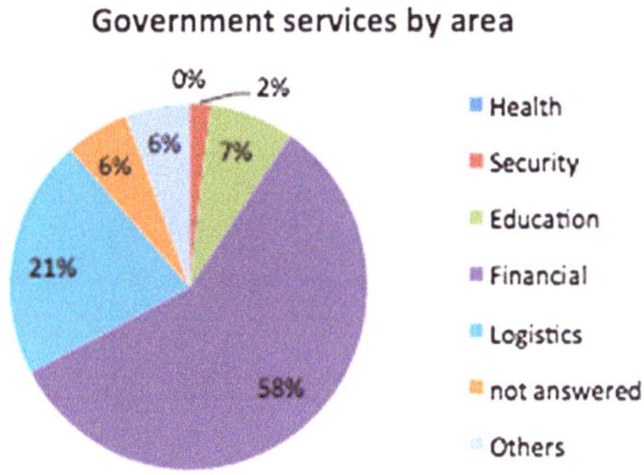

Fig. 3 Government services by area

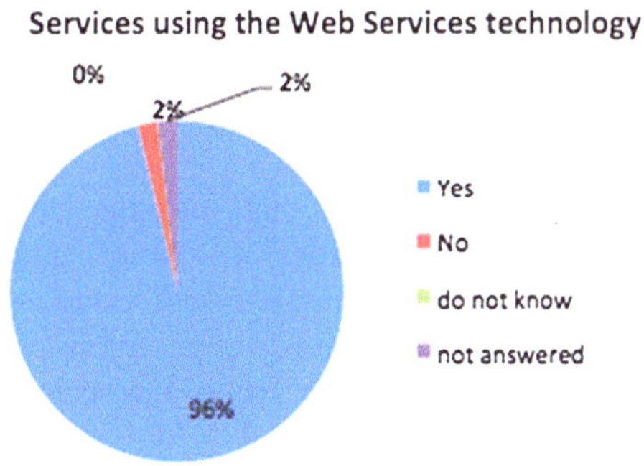

Fig. 4 Services using the Web Services technology

4.3 Evaluation

In order to evaluate the proposed method, survey and experimental research approaches were used.

A survey of services was initially carried out with the aim of collecting information about the use of services (Web Services) and SOA within the Brazilian government.

Data collection was performed by means of an electronic questionnaire developed with the Emailmeform [19] tool. The questionnaire was sent to 125 organizations, out of which, 14 organizations responded to the questionnaire. These 14 organizations have described a total of 57 services and 97 capabilities. Several issues about the services have been raised, such as: service objective, system name of service supplier, government level that meets the service, if the service is composed, the service implementation technology (Fig. 4) and the government area where the service is used (Fig. 3).

Following the data collection, the experiment, divided into 2 stages, was conducted to apply the method in SSMe-Gov e-government scenario:

1. Definition of service patterns—The services listed in the survey were analyzed and subsequently service patterns were generated. Examples of service patterns created in this step are: Financial Entry and Electronic Tax Document Batch Reception.

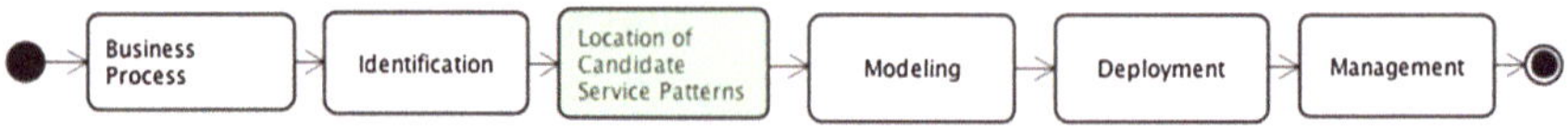

Fig. 5 Proposed lifecycle

2. Creation of new services—services were created from the patterns set in the previous step.

This experiment shows that it is possible to define service patterns from the analysis of existing services in the government and benefit from the experience of experts represented in the service patterns.

4.4 Service Lifecycle

Marks and Bell [17] state that the lack of interoperability can result from the differential applications of policies, standards, and process. The way to achieve interoperable services is to enforce a body of SOA policies across the service lifecycle: identification, design and implementation.

Several service life cycles have been defined, e.g. those defined by Marks and Bell [17], Arsanjani et al. [18], and Gu and Lago [19]. In general, all of these cycles include the following activities: business process analysis, identification, modeling (analysis and design), development and service management.

Thus the proposed use of service patterns presented in this paper aims to add an activity in the lifecycle of services, called "Location of Candidate Service Patterns", as illustrated in Fig. 5.

The activity "Location of Candidate Service Patterns" must be performed by consulting the catalog of e-government service patterns. The aim is to find patterns of services that meet the business process in question. The service patterns will be used as reference to create new services.

5 Conclusions

The SSMe-Gov method based on service pattern has been proposed in order to specify e-government services. The proposal of service designing allied to the concept of service patterns aims to reduce redundancy of efforts in the design and project of new services for the same purposes in other levels and sectors of the government.

Besides the SSMe-Gov, other items have been proposed: a conceptual view of the elements related to the service patterns and the use of service patterns in the lifecycle of services.

This solution was designed considering the lack of subsidies to encourage the reuse of oriented services solutions designed to meet the business processes related to e-government. To demonstrate the relevance of the proposal described in this work, the SSMe-Gov method was applied to a government scenario. From government services, service patterns have been defined and from these patterns, new services have been created.

Future studies may consider aspects of the creation and location of service patterns in the catalog of services, the level of granularity of the service to be created from the service patterns and other sources for the setting of service patterns, such as the business processes.

Acknowledgments The authors thank CAPES, the Brazilian government entity dedicated to the training of human resources and FAPEMAT, Foundation for Research Support of the State of Mato Grosso, for providing support towards the viability of the EPUSP/UFMT/CEPROMAT agreement for the completion of the PhD on which this work is based.

References

1. Papazoglou, M. P., Traverso, P., Dustdar, S., & Leymann, F. (2008). Service-oriented computing: A research roadmap. *International Journal of Cooperative Information Systems, 17*(2), 223–255.
2. Gamma, E., Helm, R., Johnson, R., & Vlissides, J. (2000). *Padrões de Projeto*. Porto Alegre: Bookman.
3. Assar, S., Boughzala, I., & Boydens, I. (2011). Back to practice, a decade of research in e-government. In S. Assar, I. Boughzala, & I. Boydens (Eds.) *Practical studies in e-government* (pp. 1–12). New York: Springer.
4. Gottschalk, P., & Solli-Saether, H. (2009). *E-government interoperability and information resource integration: Frameworks for aligned development*. New York: IGI Global.
5. UNPAN. (2012). United Nations E-Government Survey 2012—E-Government for the People. United Nations Department of Economic and Social Affairs.
6. Aguair, E. L., Oliveira, J. B. F., Labato, F. H., Carto, K., Silva, T. A. S., & Fernandes, S. G. (2010). Padrões Tecnológicos - o uso na prestação de serviços públicos e no relacionamento com o Governo Federal. In C. S. F. Mesquita & N. L. Bretas (Eds.), *Panorama da Interoperabilidade no Brasil*. Brasília: Orgs. MP/SLTI.
7. UK. (2001). e-Government Interoperability Framework (e-GIF). United Kingdom Government, United Kingdom, Version 6.0.
8. Brasil. (2012). Padrões de Interoperabilidade de Governo Eletrônico—E-PING. Brasília: Comitê Executivo de Governo Eletrônico. Documento de Referência Versão 2012.
9. UNDP. (2007). *E-government interoperability: Guide*. Thailand: UNDP.
10. Erl, T. (2009). *SOA: Princípios de Design do Serviço*. Upper Saddle River: Pearson Prentice Hall.
11. Sommerville, I. (2007). *Engenharia de* Software. São Paulo: Addison Wesley.
12. White, L., Wilde, N., Reichherzer, T., El-Sheikh, E., Goehring, G., Baskin, A., et al. (2012). *Understanding interoperable systems: Challenges for the maintenance of SOA applications*. In 2012 45th Hawaii International Conference on System Science (HICSS), 2012, pp. 2199–2206.

13. Klischewski, R., & Askar, E. (2010). *Success factors of developing G2G services: the case of Egypt*. In Proceedings of the 4th International Conference on Theory and Practice of Electronic Governance, New York, NY, USA, 2010, pp. 152–160.
14. Li, P., Xiangxu, M., Zhiqi, S., & Han, Y. (2009). *A reputation pattern for service oriented computing*. In 7th International Conference on Information, Communications and Signal Processing, 2009, ICICS 2009, pp. 1–5.
15. Fki, E., Tazi, S., & Dupuy, C. S. (2010). *Towards a user intention aware service composition*. In 10th Annual International Conference on New Technologies of Distributed Systems (NOTERE), 2010, pp. 113–120.
16. Tchuta, L., & Chunhua, G. (2011). Atomic new service pattern. *International Journal of Software Engineering and Its Applications, 5*, 1–20.
17. Marks, E. A., & Bell, M. (2006). *Service-Oriented Architecture (SOA): A planning and implementation guide for business and technology* (1st ed.). New York: Wiley.
18. Arsanjani, A., Ghosh, S., Allam, A., Abdollah, T., Ganapathy, S., & Holley, K. (2008). SOMA: A method for developing service-oriented solutions. *IBM Systems Journal*, *47*, 377–396.
19. Gu, Q., & Lago, P. (2007). *A stakeholder-driven service life cycle model for SOA*. In 2nd International Workshop on SERVICE ORIENTED SOFTWARE ENGINEERING: In Conjunction with the 6th ESEC/FSE Joint Meeting, Dubrovnik, Croatia, 2007, pp. 1–7.

Part IV
Semantics for Enterprise Interoperability

Semantic Approach to Automatically Defined Model Transformation

Tiexin Wang, Sebastien Truptil and Frederick Benaben

Abstract Model transformation, regarded as a pillar of model-driven engineering, plays a key role in improving enterprises' interoperability. However, to define the process of a model transformation (within a specific context) needs large amount of human's effort. To reduce human's effort and make model transformation more efficient, automatic model transformation would be a suitable solution. We present an automatic model transformation approach in this paper, which is based on model transformation methodology, using syntactic and semantic check among model elements. In this approach, a generic meta–meta-model and semantic checking rules are proposed. With a simple use case, we illustrate how this approach works.

Keywords Model-driven engineering · Automatic model transformation · Semantic check

1 Introduction

Nowadays, more and more collaborative situations (domains-crossing) are frequently appearing and disappearing. In such a context, interoperability [1, 2] becomes a key competition factor for enterprises.

T. Wang · S. Truptil (✉) · F. Benaben
Centre Genie Industriel, University de Toulouse, Mines Albi,
Campus Jarlard, 81000 Albi, France
e-mail: sebastien.truptil@mines-albi.fr

T. Wang
e-mail: wangtiexin1987@gmail.com

F. Benaben
e-mail: frederick.benaben@mines-albi.fr

K. Mertins et al. (eds.), *Enterprise Interoperability VI*,
Proceedings of the I-ESA Conferences 7, DOI: 10.1007/978-3-319-04948-9_11,

In practice, the interoperability is a big challenge for majority enterprises. In order to solve this problem with a set of recommendations, "EIF" (European Interoperability Framework) [3] was proposed. There are three approaches defined within "EIF", dealing with the interoperability problems. They are "integrated", "Unified" and "Federated". For "Integrated" approach, there exists a common format for all the models (come from the enterprises); for the "Unified" approach, there exists a common format but only at a meta-level; for the "Federated" approach, there is no common format. For the last two approaches (mostly used in practice), numerous users' effort is needed to generate solutions to the interoperability problems, especially on data sharing part; indeed, each enterprise has its own particular date format. To help data sharing, model transformation is used. In order to improve enterprise interoperability, this paper presents a generic solution: automatic model transformation based on the semantic check rules.

This paper is divided into four parts. In the second section, definitions of model, meta-model and model transformation principles are given. Then an overview of our solution is proposed in the third section. The fourth section makes a focus on the semantic mapping approach. Before the conclusion, a case study is presented in the fifth section to illustrate our solution.

2 Model Transformation Overview

With the wide use of model-driven engineering theory in many specific domains, more and more researchers and organizations are becoming interested in finding solutions to effective model transformations.

2.1 Model and Meta-model

Model transformation is based on two basic and crucial concepts: model and meta-model [4].

A model could be seen as a picture of a system, depending on a point of view. This picture is a simplification of this system, which highlights its characteristics. A meta-model defines the characteristics of a valid model.

2.2 Model Transformation Principles

Figure 1 [5] illustrates the model transformation principles.

The two source and target models are built according to their meta-model (MM). The key point is that the source MM shares part of its concepts with the target MM (the two spaces, source and target, have to be partially overlapping in

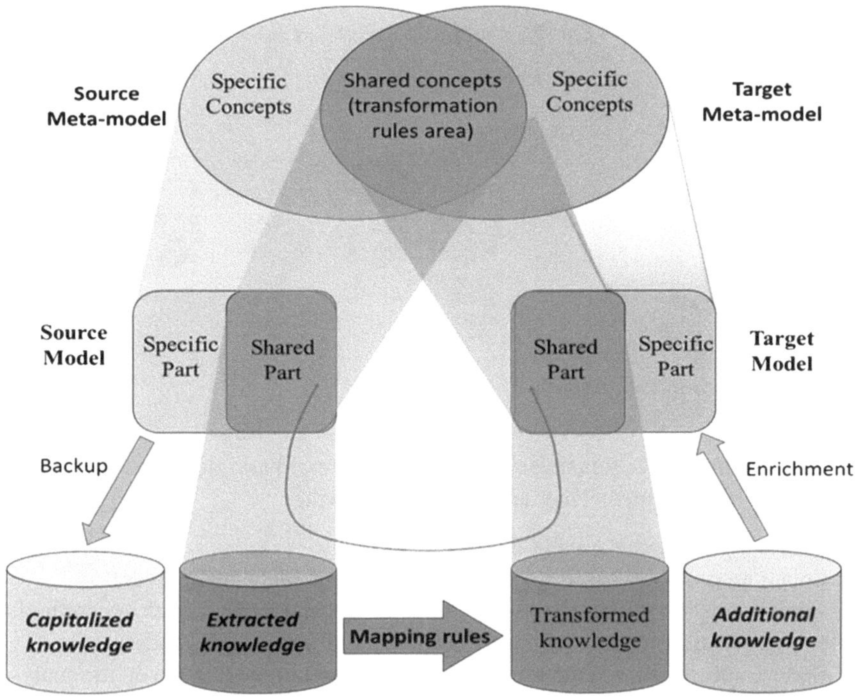

Fig. 1 Model transformation framework

order to allow model transformation). As a consequence, the source model embeds a shared part and a specific part. The shared part provides the extracted knowledge, which may be used for the model transformation, while the specific part should be saved as capitalized knowledge in order not to be lost. Then, mapping rules (built based on the overlapping conceptual area of MMs) can be applied onto the extracted knowledge. The transformed knowledge and an additional knowledge (to fill the lack of knowledge concerning the non-shared part of concepts into the target MM) may be finally used to create the shared part and the specific part of the target model.

3 General Overview of the Solution

3.1 Main Objective

The main objective of this work is to define a process of automatic model transformation.

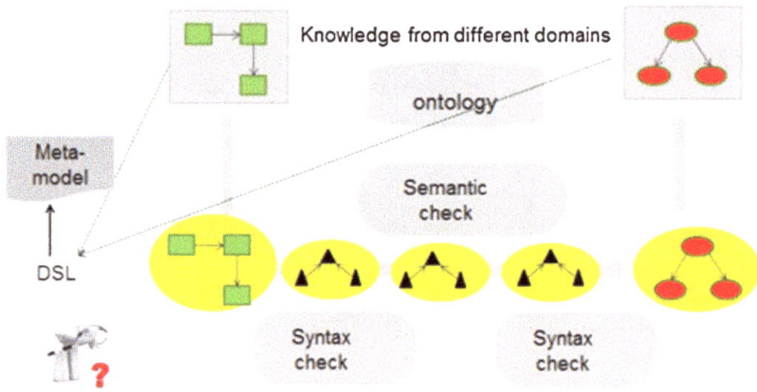

Fig. 2 Architecture of the theoretical solution

In order to achieve this objective, there are several basic function requirements should be implemented. They are listed as following:

- Define a generic meta–meta-model.
- Create an ontology based on the structure of this meta–meta-model.
- Analysis the input from the users (source model and target model; source model and target meta-model; source meta-model and target meta-model).
- Deduce specific source meta-model and target meta-model based on the analysis results and the generic meta–meta-model.
- Apply syntactic and semantic check rules on the definition of model transformation process.

3.2 The Architecture of the Solution

Figure 2 illustrates the architecture of the solution.

The source and target models could be built in different modelling languages ("UML [6]", "BPMN [7]", etc.). In order to ignore the modelling language and use the semantic and syntactic check rules on the definition of transforming process, we suppose to develop several intermediary models (building with a specific modeling language). We deduce the meta-models, which conform to the meta–meta-model, for both the source and target models. Then we can use the semantic and syntactic check rules on the meta-model level to build transformation mappings. During the transformation process, the providers of the source models could check the intermediary models.

To be efficient, all the semantic and syntactic check rules should be used on the same kind of models (intermediary models). So, we divide the transformation process into three steps: from the source model to the intermediary model, among the intermediary models and from the intermediary model to the target model.

The first and third steps just transform the format of the model (the content and concepts do not change). The second transformation step contains three phases:

first, using syntactic check to change the syntactic part of the source model; next, with the help of the "ontology" (which contains domains-cross knowledge), apply the semantic check rules on the intermediary model to transform the content and concepts; finally, using the syntactic check again to transform the intermediary model to its final version. The providers of the source models could check the intermediary models and valid the process. The syntactic and semantic check rules are applied during the transformation process to make the transform mappings.

4 Key Issues

This section illustrates two of the key aspects within our solution. They are: the definition of the meta–meta-model and the semantic check rules used on the transformation process.

4.1 The Meta–meta-model

The explanation of the meta–meta-model will be given with the help of Fig. 3. This meta–meta-model works on the top abstract level of all the other models.

As shown in Fig. 3, there are ten core elements in this meta–meta-model.

- "Environment", describes the context of a system such as crisis situation, supply chain, etc. If two "Environments" describe the same context of a specific situation, the relationship between them is "sameAs".
- "Model" is the core concept in this meta–meta-model. In the context of our solution, every source, target models and their meta-models (deduced or imported) could be regarded as "Model". Every model contains the component: "Element".
- "Element" could contain another "Element" (e.g. in BPMN (White, 2004) modelling context, a pool contains a lane). The "Element" has two inheritance classes: "Concept" and "Link".
- "Concept" stands for an object; it is used to describe a subject that exists in the world.
- "Link" is the relationship between Elements. Every "Link" has two ends (there are two relationships between "Link" and "Element": "from", "to"). "Element" contains "Property".
- "Property" is used to identify and explain the object that contains it. Each "Property" has a "Data Type".
- "Data Type" should be a "Primitive Type" or an "Enumeration".

Element "SemanticRelation" is important. It helps to express the semantic relations between elements. "Environment", "Model", "Element" and "Property" inherit from this abstract class. This means that any items from these class may

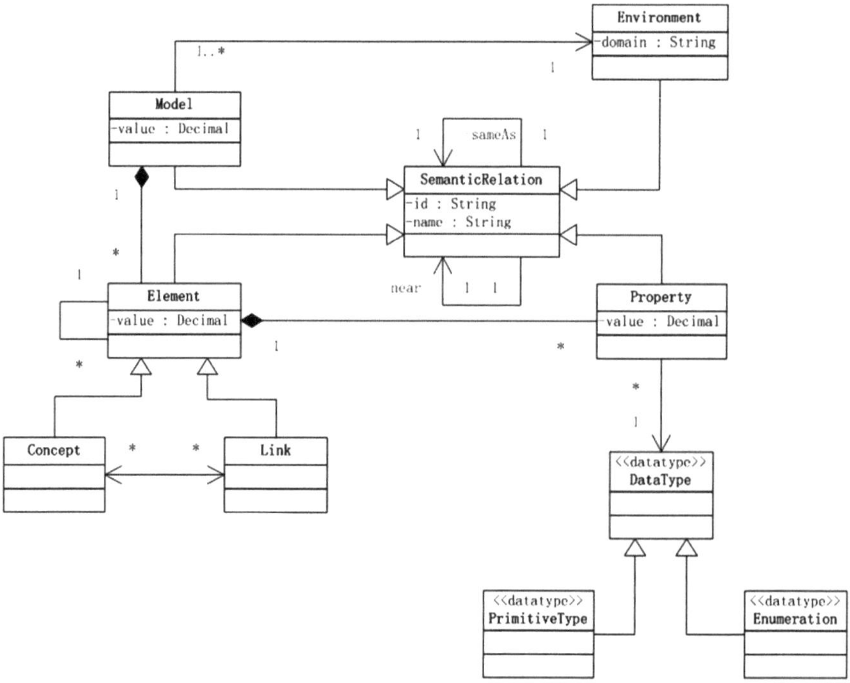

Fig. 3 The meta–meta-model overview

have "sameAs (semantic value: 1)" or "near (semantic value: 0–0.999)" semantic relation with the other relevant items. The value of "near" relationship could be calculated using the syntactic and semantic check rules.

4.2 Semantic Mapping Approach

In practice, semantic check principles have been widely used. For model transformation domain, semantic check methods can also help.

We rely on the existing semantic check rules defined in [8]. The basic idea is: in order to do semantic matchmaking between models from different domains; a common semantic profile [8] should be defined first. According to the ontology we created, we define this semantic profile. Based on the semantic profile, we can compute the semantic distance measurement between the elements from the source model the elements from the target model. After getting the computed results, we can do the matchmaking between the two models.

Here, we define the algorithms to compute the "sameAs" or "near" relationship for the objects of "semanticRelation" class. In practice, we compute the average semantic relation value between source meta-model and target meta-model within five groups: “Environment”, “Model”, “Concept”, “Property” and “Link”.

The factors used in these algorithms are assumed for the first test; the sum of those factors (within one formula) is always “1”. The values of these factors (weights) need to be modified based on more experiments.

(1) For the “Environment”

This classification depends on the users who provide the source and target model. The ontology (create based on the structure of the meta–meta model) records all the categories of the imported “Environment”. The semantic relation value between two “Environments” is calculated using the formula.

$$\text{E\_SR\_V} = \begin{cases} 1 & \text{if “samaAs”} \\ 0.5 & \text{if “near”} \\ 0 & \text{from different domains} \end{cases} \tag{1}$$

If the source model and the target model come from the same “Environment”, then this value is “1”; if similar, this value is “0.5”; if different, this value is “0”.

(2) For the “Model”

The semantic relation value between two models (deduced source meta-model and deduced target meta-model) could be calculated using the formula.

$$\text{M\_SR\_V} = 0.5^{*}\text{E\_SR\_V} + 0.4 * \text{SR\_Name} + 0.1^{*}\text{Num\_Concept} \tag{2}$$

The “E_SR_V” is the value calculated from the first formula. The “SR_Name” is the relation value between the names of the two models; it can be calculated using the semantic and syntactic algorithm method in [8]. The “Num_Concept” is the number of “Concept” involved in a model.

(3) For the “Property”

According to the generic meta–meta-model, “property” is a component of “Concept” and “Link”. Each “property” (in deduced source meta-model) has a semantic relation value with every “property” in the target meta-model, respectively. The formula for this is:

$$\text{P\_SR\_V} = 0.5^{*}\text{SR\_Name} + 0.4 * \text{type} + 0.1^{*}\text{value} \tag{3}$$

Here, the value of the “SR_Name” is calculated in the same way as explained above. If the “type” of the two “property” is the same, its value is “1”; otherwise, its value is “0”. The same calculation rule is used on “value”.

(4) For the “Concept”

“Concept” is the core element in the meta–meta model, the formula for calculating the semantic relation between two “Concepts” is:

$$\text{C\_SR\_V} = 0.3^{*}\text{SR\_Name} + 0.6 * \text{SR\_Pro} + 0.1^{*}\text{M\_SR\_V} \tag{4}$$

In this formula, the “SR_Pro” parameter is calculated using the following formula:

$$\mathrm{SR\_Pro} = \frac{2 * \sum \mathrm{Max(P\_SR\_V1)}}{\mathrm{Num\_SP} + \mathrm{Num\_TP}} \tag{5}$$

In this algorithm, the number of properties of both source model concept “Num_SP” and target model concept “Num_TP” should be calculated first. Then, select the max value of each “P_SR_V” (between the properties of source concept and the target concept), and add them together. The max value of each pair will be selected and added together.

(5) For the “Link”

The semantic relation value computed for the “Link”. The formula for this is:

$$\mathrm{L\_SR\_V} = 0.1^{*}\mathrm{SR\_Name} + 0.35^{*}\mathrm{SR\_FC} + 0.35^{*}\mathrm{SR\_TC} + 0.2^{*}\mathrm{P\_SR\_V} \tag{6}$$

In this formula, the “SR_FC” stands for the semantic relation value between the two “from concept” (every link has two concepts as two ends). The “SR_TC” means semantic relation value between the two “to concept”.

All those pairs selected by the max value will be stored in the ontology within different groups (Model, Concept, Link, and Property); before to do the semantic check, we can search the ontology first, if we cannot find the exact pair, the semantic check algorithms would be used (the new selected pair adds to the ontology). With the help of these six formulas, the definition of model transformation process (mapping rules) could be generated automatically.

5 Case Study

At this moment, we have defined the meta–meta-model, and illustrated the algorithm used to calculate the semantic relation value (which can provide help to automatically define model transformation process).

In this section, a use case, aiming at transforming the “UML [6]” class model to the “OWL [9]” model, will be shown. This case concerns a part of the whole transformation process: using the input models to deduce the meta-models that conform to the generic meta–meta-model and calculating the semantic relation value between the two models. Two very simple models are created for this case (using “UML” and “OWL”, respectively).

The “UML” model shows in Fig. 4.

In this model, there are three classes: “Student”, “Teacher” and “Course”. The relationship between “Student” and “Course” is “to_choose”; the relationship between “Teacher” and Course” is “to_give”. There are a total of eight properties for the three classes.

The “OWL” model is shown in Fig. 5.

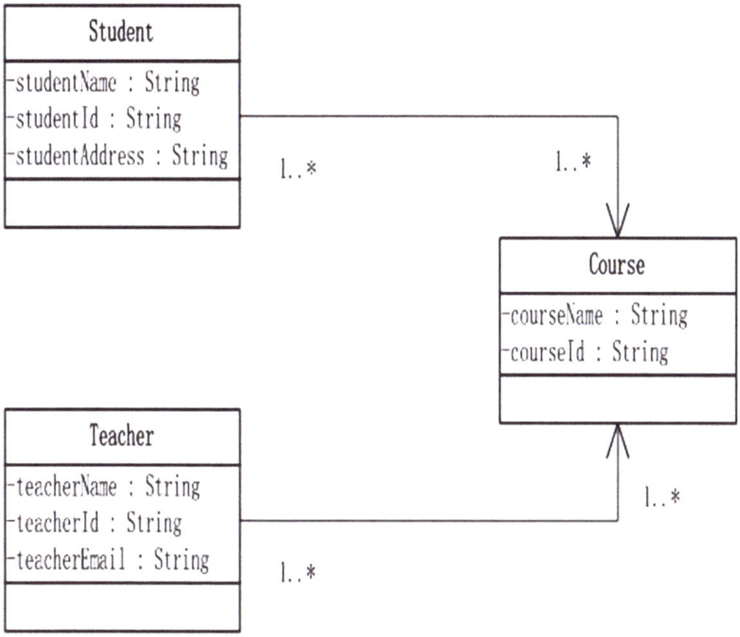

Fig. 4 UML class model

Fig. 5 OWL ontology model

We deduce the specific meta-models (based on the MMM) for both the "UML" model and the "OWL" model. Both of them are "Model"; all the classes stand as "Concept", the properties of these classes could be regarded as "Property" and the "Link" in the meta–meta model replaces the relationship between the classes.

Table 1 E_SR_V value of this case

Environment	UML
OWL	1

Table 2 M _SR_V value of this case

Model	UML
OWL	0.92

Table 3 P _SR_V value of this case

Property	lid	lname	pid	pname	sid	sname
courseId	0.56	0.42	0.56	0.42	0.58	0.42
courseName	0.42	0.52	0.42	0.52	0.44	0.52
studentName	0.42	0.52	0.42	0.52	0.44	0.56
studentId	0.56	0.42	0.56	0.42	0.58	0.42
studentAdd	0.42	0.42	0.42	0.42	0.42	0.42
teacherId	0.56	0.42	0.56	0.42	0.56	0.42
teacherName	0.42	0.52	0.42	0.52	0.42	0.52
teacherEmail	0.42	0.42	0.42	0.42	0.42	0.42

Table 4 C_SR_V value of this case

Concept	Lecture	Student	Professor
Student	0.232	0.548	0.232
Teacher	0.232	0.232	0.472
Course	0.532	0.232	0.232

The "Environments" of the two models are similar. Table 1 shows the E_SR_V.

Table 1 shows the E_SR_V value between "UML" and "OWL" environment is "1".

After calculating the E_SR_V, the next step is to calculate the M_SR_V, Table 2 shows this (the algorithm used is illustrated above).

The most complex step is to calculate the P_SR_V value; for the reason, the number of the "Property" would be very large. We use the formula explained above (as the third formula) to calculate the P _SR_V values. Table 3 shows the result for this case.

In this case study, all the properties' type is "String", and they have no value (just on "class" level, no objects exist. Our purpose is just to find the maximum value pair).

Based on the P _SR_V value known in Table 3, the C_SR_V value could be calculated. Table 4 shows the result of this process.

According to the records of this table, the mapping rules for the model transformation (on class level) could be made. After getting all the C_SR_V values, the

Table 5 L_SR_V value of this case

Link	Select	Teach
to_choose	0.428	0.267
to_give	0.267	0.36

final step is to calculate the L_SR_V value. In this case, there are two links (just has a name, they do not have properties) (Table 5).

Based on all the values recorded in these five tables above, the model transformation process could be automatically defined (search the table above and select the maximum average semantic value for each relevant pair within different categories). Then, build the mappings between such kinds of pairs. A specific ontology will provide the matching pairs (if they have been stored in; if not, add in the new selected pair which got from using the semantic check rules).

6 Conclusion

This paper exposes an approach about automatic model transformation. Comparing with the existing methodologies and principles of this field, the main contribution of our work is to add the semantic check rules on the transformation process. In order to apply semantic check rules on models, we create a generic meta–meta-model. Furthermore, based on this meta–meta-model, we build ontology to store and provide the data basis for making transform mappings automatically.

Automatically defining model transformation process is a big challenge, it can reduce human efforts (avoid the repetitive work). Furthermore, it can also solve the interoperability problems (exchanging information among heterogeneous partners quickly and effectively).

The further work of our proposal focuses on two aspects: fulfilling the ontology and improving the efficiency of the algorithms (doing the semantic check).

References

1. Konstantas, D., Bourrières, J. P., Léonard, M., Boudjlida, N. (2005). *Interoperability of enterprise software and applications*. IESA'05. Switzerland: Springer-Verlag.
2. Pingaud, H. (2009). *Prospective de recherches en interopérabilité: vers un art de la médiation*. CIGI'09, France.
3. Chen, D., Doumeingtsb, G., Vernadatc, F. (2009). Architectures for enterprise integration and interoperability: Past, present and future. *Computers in Industry*, *59*(7):647–659
4. Bézivin, J. (2006). Model driven engineering: An emerging technical space. In R. Lämmel, J. Saraiva, J. Visser (Eds.), *Generative and transformational techniques in software engineering* (Vol. 4143, pp. 36–46). Lecture Notes in Computer Science. Heidelberg: Springer.

5. Bénaben, F., Mu, W., Truptil, S., Pingaud, H. (2010). *Information systems design for emerging ecosystems*. Paper presented at the 4th IEEE International Conference on Digital Ecosystems and Technologies (DEST).
6. Fowler, M. (2004). *UML distilled: A brief guide to the standard object modelling language* (3rd ed.). MA: Addison-Wesley.
7. White, S. (2004). Introduction to BPMN. IBM Cooperation. Retrieved from http://yoann.nogues.free.fr/IMG/pdf/07-04_WP_Intro_to_BPMN_-_White-2.pdf
8. Boissel-Dallier, N. (2012). *Réconciliation sémantique des données et des services mis en oeuvre au sein d'une situation collaborative*. PhD thesis. Les thèses en ligne de l'INP.
9. McGuinness, D., Van Harmelen, F. (2004). *OWL Web Ontology Language Overview*. W3C recommendation.'

Collaborative Mediation Information System Design Based on Model-Driven Business Process Management Approach

Wenxin Mu, Nicolas Boissel-Dallier, Frédérick Bénaben, Hervé Pingaud and Jean-Pierre Lorré

Abstract Driving a BPM (Business Process Management) approach could be dedicated to support the design of IS (Information System). In a collaborative situation, involving several partners, such a BPM approach may be useful to support the design of a Mediation Information System (MIS), in charge of ensuring interoperability between partners' IS (presumed to be service-oriented). For such an objective, there are two main barriers, which are: (i) building the collaborative business process cartography by characterizing the collaborative situation and creating collaborative ontology, and (ii) reducing semantic gap between business activities (from the business process models) and technical web-services (from the physical SOA architecture of ISs). This articles aims at presenting the engineering steps of the whole BPM approach to break the two scientific problems by using a simple example.

Keywords Business process management · Model-driven engineering · Information systems · Interoperability · Mediation · Semantics · Service-oriented architecture

W. Mu (✉)
Beijing Jiaotong University, 3 Shangyuancun Xizhimenwai, Beijing 100044, China
e-mail: wxmu@bjtu.edu.cn

N. Boissel-Dallier
InteropSys, Toulouse, France
e-mail: boisseld@mines-albi.fr

F. Bénaben
Université de Toulouse, Mines Albi, Campus Jarlard 81000 Albi, France
e-mail: frederick.benaben@mines-albi.fr

H. Pingaud
Compus d'Albi, Université Champollion, Place de Verdun 81012 Albi, France
e-mail: herve.pingaud@univ-jfc.fr

J.-P. Lorré
Linagora, 75 Route de Revel 31400 Toulouse, France
e-mail: jean-pierre.lorre@linagora.com

K. Mertins et al. (eds.), *Enterprise Interoperability VI*,
Proceedings of the I-ESA Conferences 7, DOI: 10.1007/978-3-319-04948-9_12,

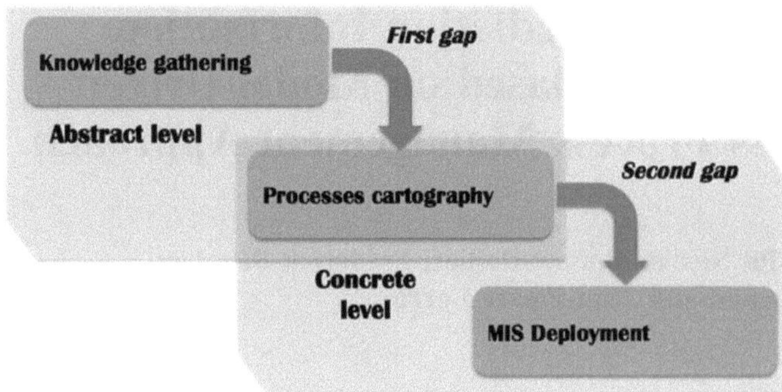

Fig. 1 Overview of the overall BPM approach

1 Introduction

The concept of "interoperability" has been defined in [1] by the InterOp Network of Excellence (NoE) as "the ability of a system or a product to work with other systems or products without special effort from the customer or user". It is also defined in [2] as "the ability of systems, natively independent, to interact in order to build harmonious and intentional collaborative behaviours without modifying deeply their individual structure or behaviour". Consequently, interoperability of organizations appears as a major issue to succeed in building on the fly emerging enterprise networks. Therefore, organizations have to assume the needed interoperability functions: **exchange of information**, **coordination of functions** and **orchestration of processes**. Furthermore, inside these organizations, Information Systems (IS) and computerized systems are assuming both the roles of interface and functional engine. Therefore, IS must be supporting the previously listed interoperability functions. The issue is to ensure that partners' IS will be able to work altogether in order to constitute a coherent and homogeneous set of IS.

The overall objective of the research works presented in this article is to define a mediation information system (MIS) able to connect the whole set of partners' IS in a way that is (i) coherent with the business objectives of the network (effective) and (ii) easy and fast to deploy (efficient). The general purpose of this article is to present concrete results and tools that are in charge of supporting a model-driven approach for MIS design. This approach can be considered according to the following layers (Fig. 1):

- Knowledge gathering (*situation* layer): collect information concerning the collaborative situation.
- Processes cartography design (*solution* layer): design the processes according to the knowledge gathered.
- MIS deployment *(implementation layer)*: implement an IT structure able to run the processes cartography.

The transitions between these layers are the hard-points. Indeed, driving such a model-driven approach requires designing relevant processes cartography (first gap). Furthermore, once that cartography designed, the IS design requires to bridge the semantic gap (second gap) between the *business* processes and the *technical workflows*. The Sect. 2 presents related works and results. The Sect. 3 focuses on the higher level of the approach (business level) while the Sect. 4 presents the lower part of the approach (technical level). Finally, the Sect. 5 concludes and presents some perspectives.

2 State of the Art and Related Research Works

This section presents existing results and research works concerning the two previously presented levels of MISE 2.0. The first sub-section focuses on existing methods of business process management and process modeling tools. The second sub-section aims mainly at defining a way to support the necessary semantic reconciliation between elements of a process model and available web services.

2.1 Abstract Level

The approach of abstract level in MISE 2.0 is greatly motivated by some advancement in Business Process Management (BPM). A business process comprises a "series or network of value-added activities, performed by their relevant roles or collaborators, to purposefully achieve the common business goal" [3]. Van Der Aalst considers that business process management covers the whole life cycle of business processes, including process design, simulation, enactment, monitoring and control, diagnosis, etc. [4, 5]. Application of formal methods in business process management systems is critical to ensure correctness properties of business process definition and furthermore enables the potential analysis [6].

In the world of BPM, many different process modeling notations and tools have been proposed (e.g. IDEF Suite, BPMN, ARIS, UML, Petri Nets, Object Oriented Modeling, CIMOSA). Their functionalities and characteristics vary and can lead to misunderstanding and failure. Furthermore, executable languages used to implement the models (e.g. BPEL or classical programming languages) are also diverse. These identified issues are similar to those identified in the Model-Driven Software Development (MDSD) concept [7]. Patig et al. [8] well summarized the software and tools used to describe business process in sample companies. Patig, Casanova-Brito and Vögeli have conducted a worldwide survey of major public companies to elicit the requirements, which are grounded in the nature of processes and the usage of software. The analysis of 127 responses indicates that human-oriented process modeling languages and BPM tools as well as BPM tools with software integration capabilities are most urgently required.

2.2 Concrete Level

Service composition aims at combining several technical services in order to meet a wider need (such as a business requirement) or, at least, brings in process execution to deal with exchanged messages. WSMX [9] and SUPER [10] are based on WSMO representation. While [9] focus on "1-to-1" logic-based service matching thanks to, respectively, WSMO Choreography and a UPML knowledge model (Unified Problem-solving Method description Language) for process description, [10] uses dedicated ontology, called sBPMN, to express process logic and provides service composition ("1-to-n") in order to deal with granularity differences between business activities and technical services. SOA4All [11] defines a lightweight Semantic Web Service (SWS) representation, called WSMO-Lite and based on WSMO, and its executive environment based on a light WSMX, improving algorithm to provide high-performance composition. Finally, [12] provides service composition based on input/outputs (I/O) matching using SAWSDL formalism in order to ensure runtime message exchanges, whereas [13] takes an interest in service composition regardless of SWS representation, based on both operation and I/O.

In order to reach executable workflows, some of these frameworks also focus on message management. While [9] focuses on a specific semantic data description, [10] deals with message transformation using SAWSDL I/O semantic concepts. Some works, such as S-Match [14] take interest in message matchmaking regardless of process integration and propose effective solutions. Unfortunately, those projects do not handle generation of message transformation.

Business to technical transformation implies "1-to-n" matchmaking in order to handle granularity differences between business activities and technical services. Apart from [10], service composition mechanisms use specific languages to express process logic. This forces organizations to redesign existing processes to align them with information systems. Additionally, they usually focus on only one SWS representation, whereas we try to embrace several standards.

3 Abstract Level Management

For MISE 2.0 abstract level design, the main objective is to build the collaborative process cartography. But what is the collaborative process cartography? And why? The collaborative process of MISE 1.0 [15] is a "mixed" process, which covers the information of strategy, operation and support knowledge [16]. The collaborative process runs among different levels of users. The users have different functional distribution in the enterprise. They concern only part of the collaborative process. It is better to build several small collaborative processes, which present different part of the "mixed" collaborative process. The goal of

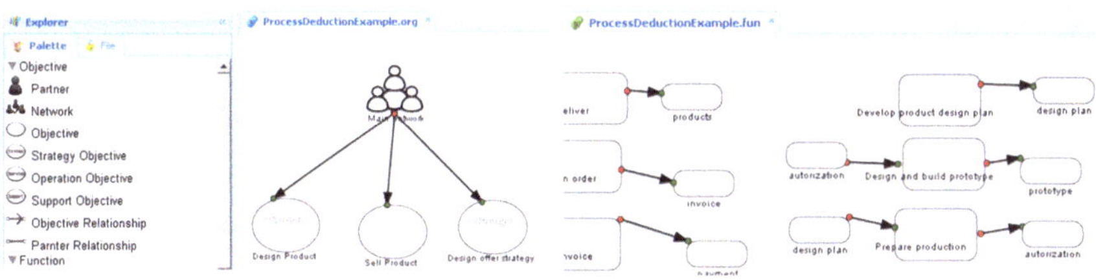

Fig. 2 Step 1: define collaborative network model and function model

collaborative process cartography is to break the "mixed" process into small processes and classify these small processes as strategy, operation or support [16].

To build the collaborative process cartography, the principles are to:

- Gather the essential and minimum collaborative knowledge (e.g. partners, collaborative objective and shared functions) in the mode of model;
- Deduce the missing knowledge with the help of ontology/metamodel;
- Complete the collaborative process cartography with the deduced knowledge and necessary algorithms;
- Extract collaborative process cartography and collaborative processes from the collaborative ontology by using extraction rules.

3.1 Step 1: Knowledge Gathering

The knowledge in this phase covers the target collaborative situation. In MISE 2.0, the collaborative network model and function model represent and define the initial collaborative situation. The collaborative network model (Fig. 2) does not only collect the collaborative network and partners but also sub collaborative network and collaborative objectives. The function model (Fig. 2) represents the information concerning shared partner functions and input/output messages.

3.2 Step 2: Knowledge Transferring

In this phase, the collaborative ontology and transformation rules are defined to transfer the collaboration concepts to the mediation concepts in the collaborative ontology. The knowledge in this phase covers the mediation concepts and instances in the collaborative ontology. There are five groups of transformation rules: create Mediator, create Mediator Relationship, create Generated Mediator Function, link Generated Mediator Function to Mediator, and Create Inter Mediator Function. Table 1 provides one equation as examples of transformation rules.

Table 1 Step 2: transformation rules

Group 2: Create Mediator Relationship Strategy and Operation Objective → Main Function → Business Message → Order
If ∀Strategy Objective (X1) (∀generates (Strategy Objective (X1), Main Function (Y1))) ∧ ∀Operation Objective (X2) (∀generates (Operation Objective (X2), Main Function (Y2))) If ∀Main Function (Y1) (∀out (Main Function (Y1), Business Message (m))) ∋ ∀Main Function (Y2) (∀in (Main Function (Y2), Business Message (m))) →∃ Order (m)(hasMediatorRelationship (Mediator (X1), Order (m))) ∋ ∃ Order (m)(hasMediatorRelationship (Mediator (X2), Order (m)))

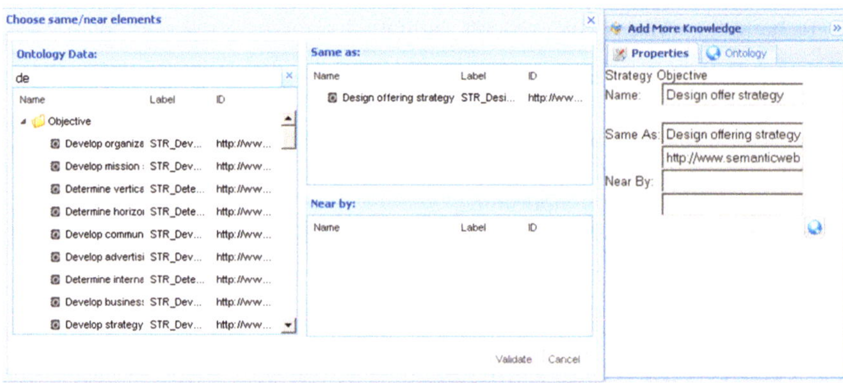

Fig. 3 Step 3: select "same as" and "near by" instances

3.3 Step 3: Knowledge Completing

The knowledge of this phase presents the matching between objective and functions. In this phase, one methodology is developed: business service selection to choose functions to achieve objectives by linking the functions and objectives to the instances of the collaborative ontology by using "same as" and "near by" relations. Figure 3 shows the definitions of the "same as" and "near by" relations.

3.4 Step 4: Knowledge Extracting

The knowledge covers the collaborative process extraction and sequence/gateway deduction. In this phase, the deduction rules are defined to extract the collaborative process cartography and collaborative processes (Fig. 4).

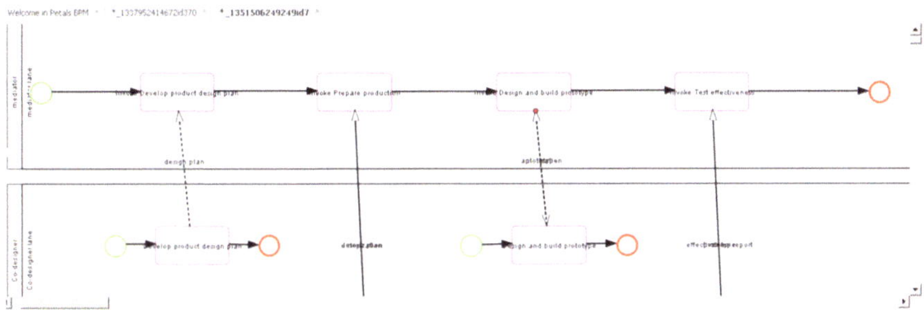

Fig. 4 Step 4: extract collaborative processes

4 Concrete Level Management

This section aims at generating dedicated technical workflows, which fit with business requirements while dealing with interoperability between partners' IS [17]. Our business to technical transformation involve three main steps in order to fit business needs and handle technical specificities:

- The first step aims at completing business information deduced by abstract level transformation. This step rests upon a dedicated BPMN 2.0 extension, called SABPMN.
- The second step focus on "n-to-m" matching between business activities and web services. In this view, we exploit sets of semantic profiles filled with semantic annotation from all models. This step provides for each business activity a list of rated web services.
- Once the real web services chosen by user from previous list, we create expected data transformation. This generation is handled by message matchmaking engine to generate expected XSLT transformation files.

4.1 Step 1: Semantic Annotation of Business Process

Whereas a lot of annotation mechanisms exist for web services, the recent BPMN 2.0 is still devoid of a semantic standard. However, in addition to a higher design range, this second major version brings an XML representation and its extension mechanism. Therefore, the annotation mechanism called called SA-BPMN 2.0 is proposed. This extension adds two XML tags: (i) *SemanticDetails* allows user to describe any activity requirement; (ii) *SemanticElements*, aims at describing messages and sequencing flows, attaching a list of expected messages or elements.

To simplify semantic annotation, the modelling platform embeds annotation tools to allow users to add or edit semantic concept references directly from the business process view (Fig. 5).

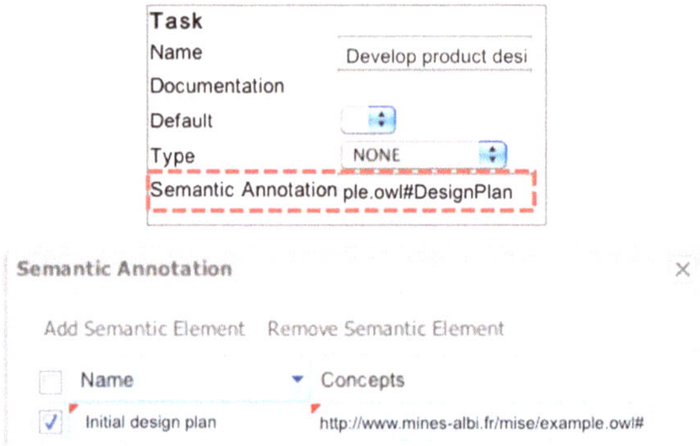

Fig. 5 Semantic annotation of business process (SA-BPMN 2.0)

4.2 Step 2: Service Composition

The proposed approach is based on a "1-to-1" hybrid matchmaking mechanism and focuses on semantic comparison. Semantic distance between profiles is performed thanks to a logic-based reasoning coupled with a syntactic similarity measurement. These measurements use information from operation (service capability or activity requirement) and I/O. In order to perform this service composition, and despite granularity difference of models to match, we use a semantic profile. This profile (represented in Fig. 6), allows us to describe the functional aspects of models. It is filled with semantic annotation from business activities (using our SA-BPMN 2.0 mechanism) or technical services (using SAWSDL or WSMO-Lite for now). This profile also embeds an internal behavior description, composed of a sequence of unit activities. Each of these unit activities is represented by a list of semantic concepts, such as functional description.

Using syntactic and semantic information from business and technical profiles, our matchmaking mechanism then compute semantic and syntactic distance between models. In this view, we first perform a "1-to-1" service matching, comparing semantic concepts and names from both activities and web services profiles (as explained in [18]). If no service fits business requirements of the target activity, we then try to respond to the request using a set of services. At this time, we perform a new service matching using this profile then compute the distance between the proposed sets of services and the initial business activity in order to propose "1-to-n" matching results to users (Fig. 7).

4.3 Step 3: Message Matchmaking

Once the user has selected technical services, we can focus on real data mapping. Semantic business information is not sufficient for message matchmaking. One business concept such as a date can be expressed in many formats. This choice

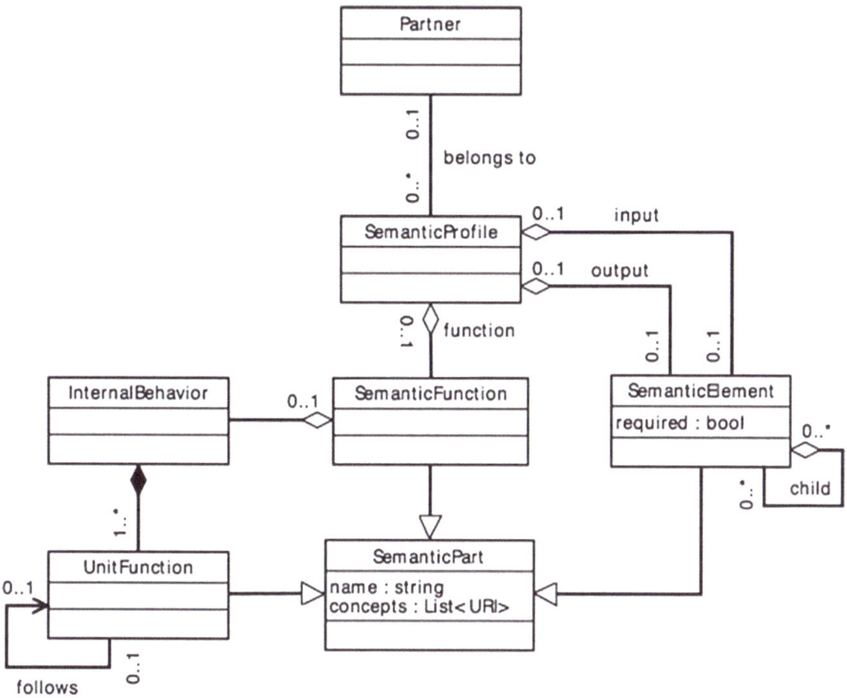

Fig. 6 UML model of our semantic profile

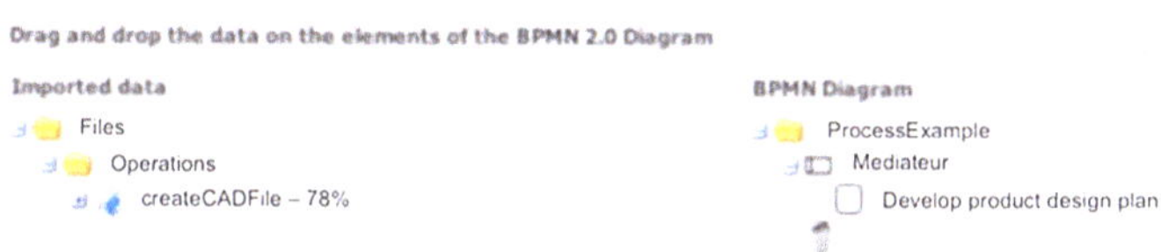

Fig. 7 Service matchmaking validation by user

belongs to the service developer who can also use classic XML Datetime, declared as such in the service description, or choose to use an exotic one, declared as a simple string. In order to solve this problem, we propose a technical ontology focused on format concepts and linked to technical databases filled with syntax representation and conversion formulae.

Thanks to semantic and technical data description of involved messages, we generate data transformations using three main steps for each chosen service:

- First we search for available outputs using process logic.
- Then, using this available data, we try to compute the whole message transformation using semantic links.

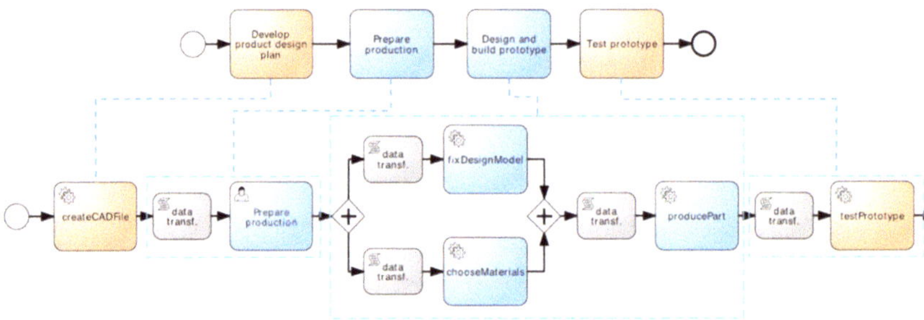

Fig. 8 Comparison between process and generated executable workflow

- If the whole message is not covered by the computed transformation, we first try to find an available transformation.

Once transformation completed, the executable workflow is generated (Fig. 8).

5 Perspectives and Conclusion

The research works presented in this article aims at defining a mediation information system dedicated to support interoperability features among the set of partners' IS. The design method of this MIS is based on a BPM approach and should be considered according to two levels: abstract level and concrete level. The abstract level issues concern mainly (i) the way to gather and to organize effectively the knowledge concerning the target collaborative situation and (ii) the transformation mechanisms able to create collaborative processes cartography from organizational, functional and informational about the target collaborative network. The concrete level issues concern mainly (i) the way to reconcile, through semantic and syntactic mechanisms, activities from collaborative processes cartography with web-services from partners, and (ii) the way to deal, on the fly, with I/O reconciliation through management of contents and formats.

The whole approach presented above, is strongly automated except for the initial collaborative situation characterization, which requires a fully human filling of the different required models. However, the support of collaborative situation requires agility. Considering this agility issue, there is one major point to underline: The tooling of the whole approach is based on web services: each step of the design-time is implemented as a web service and included into an ESB, which run the whole BPM approach through its workflow engine and which benefits from all the agility functions of that environment. Thus, if there is any evolution (change of objectives, resignation of a partner or even a run-time dysfunction of a workflow), the whole system is technically able to change the design workflow and to loop at the appropriate level. The design-time is then included in the run-time thanks to the SOA infrastructure.

Currently, there are several research dedicated to enrich the whole approach. Two points in particular can be highlighted: First, knowledge capitalization is a promising way to improve not only the knowledge base on which the deduction rules run the deducing principles but the content of deduction rules as well. Second, non-functional aspects should also be taken into account, such as quality of services (QoS), monitoring and governance.

References

1. Konstantas, D., Bourrières, J.-P., Léonard, M., & Boudjlida, N. (2005). *Interoperability of enterprise software and applications*. Switzerland: Springer-Verlag.
2. Pingaud, H. (2009). Prospective de recherches en interopérabilité: vers un art de la médiation. In *Proceedings of the Plenary Lecture, 8th International Congress on Industrial Engineering (CIGI), Tarbes*.
3. Ko, R. K. L. (2009). A computer scientist's introductory guide to business process management (BPM). *Crossroads, 15*(4), 4:11–4:18.
4. Van Der Aalst, W. M. P., & Ter Hofstede, A. H. M. (2000). Verification of workflow task structures: A petri-net-baset approach. *Information Systems, 25*(1), 43–69.
5. Ter Hofstede, A., Van Der Aalst, W., & Weske, M. (2003). Business process management: A survey. In M. Weske (Ed.), *Business process management* (vol. 2678, pp. 1019–1019). Heidelberg: Springer Berlin.
6. Van der Aalst, W., & Van Hee, K. M. (2004). *Workflow management: Models, methods, and systems*. Cambridge: The MIT press.
7. (Tom) Stahl, T., & Völter, M. (2006) *Model-driven software development*. New York: John Wiley & Sons.
8. Patig, S., Casanova-Brito, V., & Vögeli, B. (2010). IT requirements of business process management in practice—An empirical study. In R. Hull, J. Mendling, & S. Tai (Eds.), *Business process management* (Vol. 6336, pp. 13–28). Berlin/Heidelberg: Springer.
9. Facca, F., Komazec, S., & Toma, I. (2009). WSMX 1.0: A further step toward a complete semantic execution environment. In *Proceedings of the Semantic Web*: *Research* and *Applications* (pp. 826–830).
10. Hepp, M., Leymann, F., Domingue, J., Wahler, A., & Fensel, D. (2005). Semantic business process management: A vision towards using semantic web services for business process management. In *proceedings of the IEEE International Conference on e-Business Engineering, 2005 (ICEBE 2005)* (pp. 535–540).
11. Lécué, F., Gorronogoitia, Y., Gonzalez, R., Radzimski, M., & Villa, M. (2010). SOA4All: An innovative integrated approach to services composition. In *Proceedings of the Web Services (ICWS), 2010 IEEE International Conference on 2010* (pp. 58–67).
12. Ishikawa, F., Katafuchi, S., Wagner, F., Fukazawa, Y., & Honiden, S. (2011). Bridging the gap between semantic web service composition and common implementation architectures. In *Proceedings of the Services Computing (SCC), 2011 IEEE International Conference on 2011* (pp. 152–159).
13. Guermouche, N., Perrin, O., & Ringeissen, C. (2008). A Mediator Based Approach For Services Composition. In *Proceedings of the Sixth International Conference on Software Engineering Research, Management and Applications, 2008 (SERA 2008)* (pp. 273–280).
14. Giunchiglia, F., Yatskevich, M., & Shvaiko, P. (2007). Semantic matching: Algorithms and implementation. *Journal on Data Semantics*, 9, 1–38.
15. Rajsiri, V., Lorré, J.-P., Bénaben, F., & Pingaud, H. (2010). Knowledge-based system for collaborative process specification. *Computers in Industry, 61*(2), 161–175.

16. ISO 9000. (2012). *ISO 9000 quality management* (September-2005. [Online]). Retreived January 19, 2012, from http://www.iso.org/iso/home/store/publications_and_e-products/publication_item.htm?pid=PUB100224
17. Benaben, F., Boissel-Dailler, N., & Pingaud, H. (2012). Semantic issues in model-driven management of information system interoperability. *International Journal of Computer Integrated Manufacturing*.
18. Boissel-Dallier, N., Bénaben, F., & Pingaud, H. (2012). Model-driven engineering of mediation information system: Application to the ISTA3 use-case. In *Proceedings of the Enterprise Interoperability: I-ESA'12*, 21 Jan 2013. DOI: 10.1002/9781118561942.ch28. Retreived from http://onlinelibrary.wiley.com/doi/10.1002/9781118561942.ch28/summary Valencia, Spain.

An Ontology-Driven Approach for the Management of Home Healthcare Process

Elyes Lamine, Abdel Rahman H. Tawil, Rémi Bastide and Hervé Pingaud

Abstract In the homecare domain, workflows are in the mainstream for supporting the coordination and monitoring of care processes which involve managing a sequence of care workflow (careflow) activities, transmitting the information required for providing care and supporting the invocation of appropriate human and/or IT resources. However, the design of these careflows for later enactment by a Workflow Management System remains a complex task, heavily dependent on patients' profiles and accordingly requiring to be distinctly personalised. This paper proposes an ontology-driven design approach for careflows, to facilitate the construction of personalized careflows. Following an approach grounded in Model-Driven Engineering (MDE), our methodology is based on the matching of ontologies between conceptual models of homecare and a semantic representation of Business Process Modeling Notation (BPMN) which is associated with both Actor and Case Profile ontological models.

Keywords Collaborative process management · Personalised processes · Model driven engineering · Home care ontology · Patient profile · BPMN ontology

---

E. Lamine (✉) · H. Pingaud
Mines-Albi, CGI, Université de Toulouse, Campus Jarlard 81013 Albi Cedex 09, France
e-mail: Elyes.lamine@mines-albi.fr

H. Pingaud
e-mail: herve.pingaud@univ-jfc.fr

A. R. H. Tawil
School of Architecture Computing and Engineering, Docklands Campus,
University of East London, University Way, London E16 2RD, UK
e-mail: A.R.Tawil@uel.ac.uk

R. Bastide
IRIT, CUFR J.F. Champollion, ISIS, Université de Toulouse, Campus Universitaire
81104 Castres, France
e-mail: remi.bastide@univ-jfc.fr

K. Mertins et al. (eds.), *Enterprise Interoperability VI*,
Proceedings of the I-ESA Conferences 7, DOI: 10.1007/978-3-319-04948-9_13,

1 Introduction

Home care services play a major and complementary role in providing the care offer as part of the health systems. They ensure the transfer of an important part of patient care from hospitals to patients' homes by coordinating the activities of all the persons involved in the patient's care process.

Currently, these services structures, as outlined in [1] and several other research work [2–4], are facing many challenges in terms of coordination and continuity of care. From our studies in the homecare area, these challenges are still valid, including the

- lack of inter-organizational communication or coordination of care: there are several associations, organizations and stakeholders that manage the employees working at the patient's homes. However, the lack of reliable and timely inter-communication between these stakeholders seriously hamper their efficiency [2, 4]. Indeed, the only communication means used by most homecare providers is a so-called "liaison logbook" which is a physical notebook, usually unstructured and kept at the patient's home [5, 6].
- deficiency in care continuity: According to [7], the continuity of care is "the fact that medical services are received as a sequence of coordinated and uninterrupted event compatible with the needs of medical patients". In homecare, the need for continuity stands at two levels: (1) interventions level where the stakeholders must ensure continuity and consistency of provided care. (2) Time level, during different steps of patient treatment, particularly when moving from one care institution to another.

To ensure the coordination and continuity of care, and to improve its quality, the caregivers involved in homecare are mandated to: (1) ensure the accuracy of information, by automating the transmission of information and by avoiding several retranscriptions for the same information; (2) Guarantee the timely transmission of information to the appropriate stakeholders, be it information about the patient's condition or other logistical information; (3) Ensure the transmission of information about the actual medical or paramedical interventions performed by the caregivers; (4) Ensure the storage of information about the evolution of patient's health conditions.

Given these needs, workflow management systems appear to be an appropriate tool for supporting the communication and coordination between these different stakeholders through managing a sequence of care workflow (careflow) activities. According to the definition proposed by WfMC (Workflow Management Coalition) [8], the role of a workflow management system is to coordinate the work involved by transmitting the necessary information and tasks, among different stakeholders, depending on the condition of patient's careflow plan and the role of interveners.

Nevertheless, the specific characteristics of processes involved in homecare, make the design of their workflow a real challenge. In order to help in the conception of these workflows, we propose in this paper an ontology-driven approach based on ontology matching between homecare domain models and semantic representation of Business Process Modeling Notations (BPMN).

The core of this paper is structured in three main parts. In the first one, we discuss the need for workflow in automating homecare processes and we point out the challenges for this. In the second part we focus on the presentation of the proposed approach to design a workflow for homecare. In the third part, we discuss related work, the system implementation and our conclusion.

2 Workflow Opportunities and Challenges in Homecare Domain

2.1 *Homecare Process Specificities*

The definition proposed by WfMC (Workflow Management Coalition) [8] for a workflow is "The automation of a business process, in whole or part, during which documents, information or tasks are passed from one participant to another for action, according to a set of procedural rules".

In general, workflow systems are process oriented, where a process represents a set of activities within a course-of-action that needs to take place in a prescribed sequence, to achieve an outcome.

Several work has addressed the area of supporting healthcare process using workflow [9–15], especially at the hospital. Care workflows (careflows) tasks can be of diagnosis, therapeutic, administrative, or decision-making nature. In the study presented in [10], the authors discuss the critical role of workflows in the clinical area, and the challenges encountered in their design. In Song and al [13] the authors note that improving healthcare workflows is very important for improving healthcare quality and efficiency.

For these reasons and to cope with the drawbacks discussed above, we suggest the automation of some homecare processes. These processes, take as an input the person to be taken in charge, and provide as an output the discharged person who is no longer supported by the homecare process. The homecare process contains several activities of different natures. Some are related to the health care, the logistical and administrative aspects, while others are rather related to the socio-economical aspects. Figure 1 shows the main processes involved in homecare, following ISO 9000:2000 which classifies processes into three types: management, operational (management of patient) and support (human resources, purchasing and logistics,…).

In this paper, we focus on the delivery, coordination and monitoring of care. We distinguish two sets of processes: The first set includes the "admission request", "preadmission", "billing stays" and "discharge" processes, which are prone to be automated by a classical workflow management system. The second set of

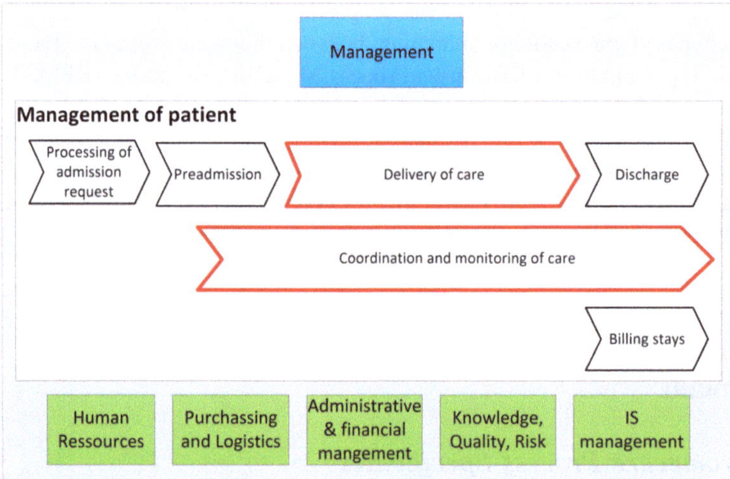

Fig. 1 The main processes involved in homecare, following ISO 9000:2000

processes includes "delivery of care" and "coordination and monitoring of care" which are critical for a successful homecare service and are the main topic of the present paper. The characteristics of these processes make the design of workflows for their automation a difficult task. Typically these set of processes has the following characteristics:

- High degree of customization [11–13]: each patient is a specific case due to his/her particular health conditions, social networking, geographic location, etc.
- Strong human component: The homecare processes is characterized by a strong influence of the actors' competence in achieving the objectives. It is usually long-running and subject to dynamic change [16].
- Collaborative nature: [14, 15]. This collaborative process is built by the assembly of distributed business processes in partner organizations. Its behavior is related to the effective contributions of partners, implemented jointly to achieve a common goal.
- Dynamicity [10]: the process of homecare is continually changing in a very dynamic way and may even get redefined. The main reason for this change is due to the uncertain nature of the environment in which this process is conceived, according to the changing health of the patient and his legal environment. In general, the process it required to quickly and easily adapt to a changing environment.
- Temporal aspects[10]: the tasks in homecare have frequencies and durations which are inconvenient to represent in conventional workflow models.

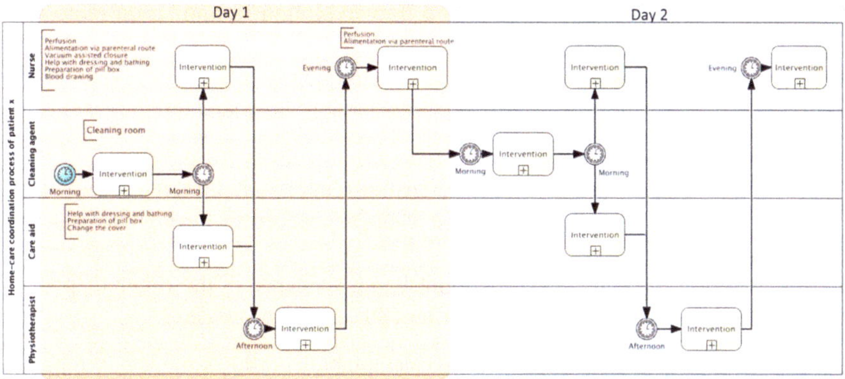

Fig. 2 Home-care workflow

2.2 Workflows Challenges

The specific characteristics of processes involved in homecare make the design of their workflow a challenge. The workflow presented in the Fig. 2 illustrates some of the challenges related to the implementation of a workflow to deliver, monitor and coordinate homecare processes. The workflows consists of monitoring sequences of interventions for a specific patient.

The figure shows a BPMN model that represents the sequence of interventions performed within a two-day timeframe for a specific patient, with the distribution of interventions amongst stakeholders (nurse, physiotherapist, etc.). The intervention plan defines the list of provisional care to be performed, including actions and treatment, prescribed or not, as well as the frequency and schedules, and also specifies the appropriate stakeholders (nurses, cleaning agent,…etc.) to perform each action.

The figure shows one pool representing the patient's home, containing four lines representing four different categories of actors involved in the homecare of a specific patient, including a nurse, cleaning agent, care aid and Physiotherapist.

Such workflow depends on the intervention plan defined by professionals that manages homecare institutions, and also on the profile of the patient such as diseases, the presence of family caregiver, etc. Whereas, conventional workflow models are considered rigid in the sense that they are predefined and not configured to handle specific patients' cases. Accordingly, health and homecare actors can find themselves in situations which are not already planned or may have not been anticipated or pre-considered.

In addition, care workflows (careflows) requires a high degree of customization as each episode of care needs to be considered as a special case with specific patient health conditions, social surroundings, and care provider(s). Therefore, careflow management systems need to support such paradigms whereby flow definition centered patients' profiles and flow evolution become a de facto operation for each episode of care.

Indeed, a set of fundamental questions are frequently tackled in several studies related to the management of care processes having the characteristics mentioned above, including:

- How to reconcile the necessary determinist approach that fixes the logical unfolding of the workflow, with the changes occurring throughout that care processes?
- How to set essential processes without being scattered in the range of possible options? How can this be done with less structured processes?

In order to facilitate customizable and adaptable careflows models, we believe that semantic models enriched with rule-based systems can be applied to conceive and downstream components of workflows.

Taking this into consideration, we have developed an ontology driven approach for homecare workflows conception in order to respond to the above-mentioned questions. This approach is presented in the following section.

3 Proposed Approach to Design Homecare Workflows

Our approach uses ontologies to help in designing workflow models for the delivery, monitoring and coordination of homecare, and guide the definitions of an intervention plan that map to the definition of a workflow model. We define the distribution of tasks (care intervention) to actors (homecare stakeholders) in order to achieve the specific goals to improve the quality of homecare. An intervention plan defines the frequency of intervention and its duration over a certain treatment period.

Figure 3 shows the position of our proposal within the reference architecture of workflow management systems proposed by the WfMC (WfMC, 1999).

This architecture consists of five interfaces with the following components: (1) process definition tools, which are tools for graphical modelling of the process to automate and deploy. (2) Workflow client applications, including any applications to communicate to-do lists, messages, etc. to actors. (3) Invoked applications, any application such as DBMS, web service, etc. which can be called or invoked automatically during the deployment process without intervention or interaction from the user. (4) Other workflow enactment services, which include any external workflow management systems capable of communicating with the workflow management system. This ensure the interoperability between different workflows management systems on the market. (5) Administrative and monitoring tools that provides access to admin tasks, such as suspension of a task, stopping a process, etc.

Based on this reference architecture for workflow management systems, we propose an architecture system where the input is a "patient profile" and the result

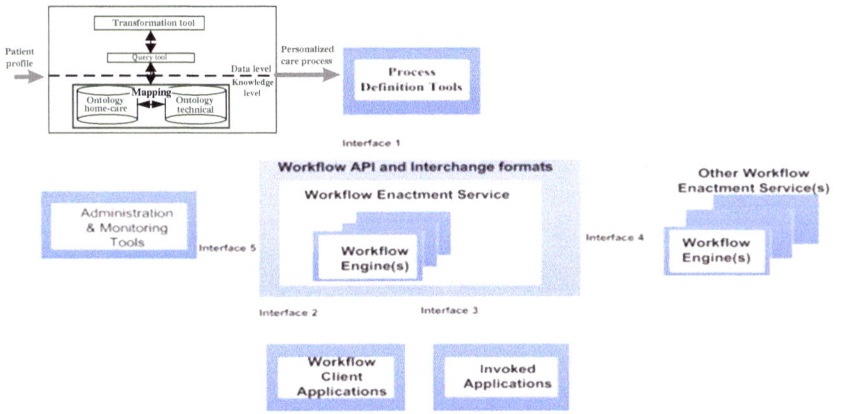

Fig. 3 Position of our proposition in the reference architecture of workflows systems

is a "personalized care process". In fact, the result of our approach must be interpreted by the process definition tool in reference to the workflow architecture. Our approach aims to provide the necessary semantic layer required for customization workflow design and present a query process and model transformation that relies on a knowledge base to generate a BPMN process, based on the patient profile. Specifically, the tool supporting our approach takes as input parameter a patient profile and generates the care process as a BPMN model.

3.1 Transformation Pipeline

Figure 4 shows the steps to design customized BPMN process models for each patient. The resulting process models will be executed on a workflow system.

We use the following components: (1) a knowledge-base of fused homecare ontology and the BPMN ontology described in [16, 17]. This knowledge base contains patients' profiles and rules about the different actors involved in the care process, their expertise and actions involved in treatment of specific diseases. (2) A Java application based on OWLAPI a library for creating, manipulating, serializing and reasoning about OWL Ontologies. The system queries the knowledge-base about the intervention plan for specific patients. The output of this application is an XML file. The XML file defines an intervention plan of specific patient as a BPMN process model. (3) Tools used for model-transformation ready for execution by a workflow engine. The XML file resulting in the last step must be transformed in a format suitable to be interpreted by a workflow engine. Tools such as ATL or XSLT are well suited, and allow transforming the XML file into a standard BPMN 2.0 file.

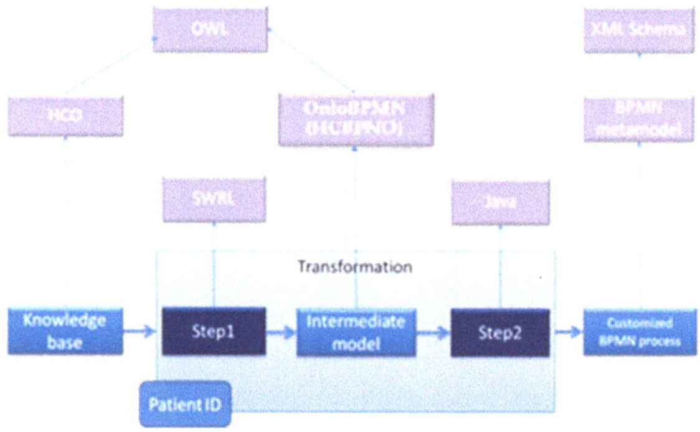

Fig. 4 Methodology for customized workflow design

3.2 HCO, A Homecare Ontology

Few ontologies have been built for the homecare area. The works we mostly refer to is that of [16]. The authors propose ontologies for medical homecare. The design of these ontologies is based on a model of home healthcare defined by a European consortium of homecare professionals, within the European project K4Care.

Based on these ontologies, and our own surveys of current practices in homecare, we propose a high level ontology called HCO (Homecare Ontology) which focuses on the concepts of the coordination in homecare. To develop this ontology of the homecare domain, we followed a synthesis of different methodologies in literature such as the ones proposed in [18, 19, 20]. Ontologies can be represented by a graph whose nodes represent concepts and whose links represent relations between concepts. An excerpt of the proposed HCO is shown in Fig. 5. We use different colors, as specified in the figure legend, to clarify the origin of the concept: CPO, APO or OntoPAD.

HCO includes many rules which describe the conditions and terms relating to concepts defined in advance. They help to design a custom homecare Workflow. The mapping between HCO and OntoBPMN involves two steps: the first step is to merge the two ontologies. The second step is to make the mapping between the two ontologies. The merging of two ontologies is the construction of a new ontology HCBPMNO. This new ontology includes a discrete model of all the concepts for HCO and OntoBPMN ontologies.

4 Implementation of Our Approach

As shown in Fig. 6, our software architecture consists of two distinct parts:

- The knowledge base: the implementation of this part has been developed in previous sections: the construction of HCO and the mapping between HCO and OntoBPMN. The main tool used for developing this section is Protégé 4.1 and

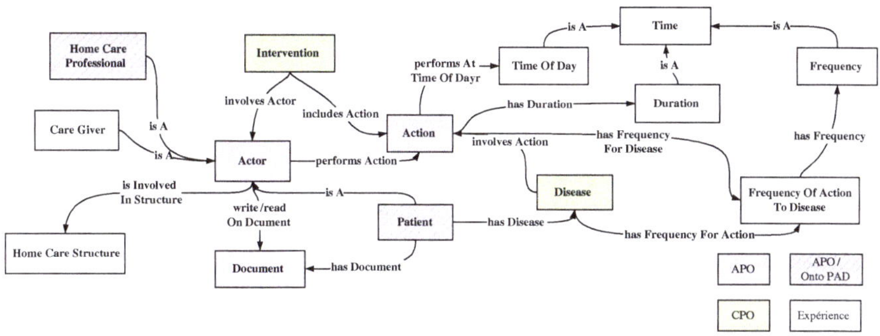

Fig. 5 Main classes and relations of HCO

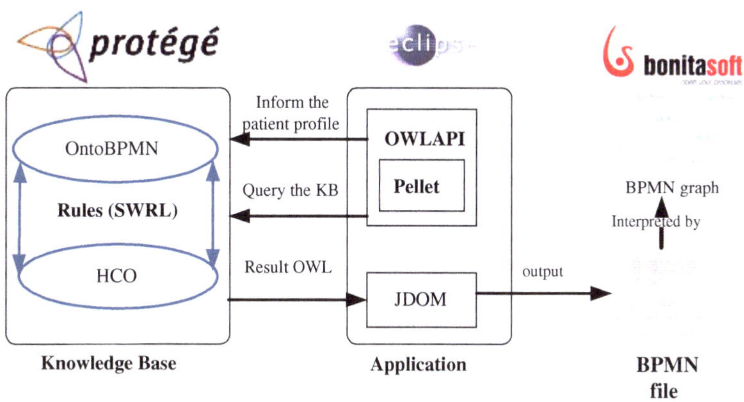

Fig. 6 Software architecture of our approach

its plugins for writing rules and the deductions made from these rules. The protégé plugins used are the SWRL rule language and Pellet Reasoner Plug-in.

- The application: to implement this section, we use the object-oriented language Java. This choice is motivated on the one hand by the advantages of portability and independence from the execution platform brings Java and on the other hand, by the fact that most libraries and APIs for manipulating ontologies are Java based. In our application, we used the following two APIs: OWLAPI[1] that allows the creation and manipulation of OWL ontologies and JDOM[2]: an open source Java API for manipulating XML documents. Figure 6 shows the relationship between selected software components and their interactions.

---

[1] http://owlapi.sourceforge.net/.

[2] http://www.jdom.org/.

5 Conclusion

In order to improve communication and continuity of the homecare process, we proposed a system architecture based on workflow execution to distribute tasks among the numerous homecare stakeholders. The characteristics of homecare processes make the design and enactment of care workflow a challenge. The goal of the presented architecture is to respect the reference architecture for workflow management systems while relying on ontologies to overcome the main challenge of workflow flexibility as the major identified obstacles. We propose an approach using two ontologies (homecare and BPMN ontologies) to help designing homecare workflows. Our on-going work consists of completing the transformation pipeline in order to generate process models suited to be interpreted by a standard workflow engine. Our approach differs from classical workflows-based systems in that we incorporate considerable domain knowledge in addition to organizational knowledge. The fundamental proposition of our approach is its ability to assess a patient case, evaluate and implement a personalized executable care plan.

References

1. FNEHAD. (2009). Fédération nationale des établissements d'hospitalisation à domicile "Livre blan des systèmes d'information en hospitalisation à domicile" 1 édition—Juin (2009).
2. Kun, L. G. (2001) Telehealth and the global health network in the 21(st) century. From homecare to public health informatics. *Computer Methods and Programs in Biomedicine*, 64, 155–167.
3. Woodward, C., Abelson, J., & Hutchison, B. (2001). My home is not my home anymore: Improving continuity of care in homecare. In *The Canadian Health Services Research Foundation*. www.chrsf.ca
4. Helleso, R., Sorensen, L., & Lorensena, M. (2005). Nurses' information management across complex health care organizations. *International Journal of Medical Informatics,76,* 960–972.
5. Bastide, R., Zefouni, S., & Lamine, E. (2010). The homecare digital ecosystem: An information system support architecture. In *IFIP Conference on Personal Wireless Communications*, Knowledge Village, Dubai, UAE, April 12–15, 2010. IEEE Computer Society.
6. Lamine, E., Zefouni, S., & Bastide R. (2010). A system architecture supporting the agile coordination of homecare services. In *IFIP Working Conference on Virtual Enterprises (PRO-VE 2010)*, Saint-Etienne, France, October 11–13, 2010 (Vol. 336). Springer.
7. Shortell, S. (1976). Continuity of medical care: Conceptionalization and measurement. *Medical Care, 14*(5), 377–391.
8. Hollingsworth, D. (1995). *The workflow reference model*, Technical Report wfmctc-1003. Workflow Management Coalition.
9. Lunn, K., Sixsmith, A., Lindsay, A., & Vaarama M. (2003). Traceability in requirements through process modelling, applied to social care applications. *Information and Software Technology*, *45*, 1045–1052.

10. Dadam, P., Reichert, M., & Kuhn, K. (2000). Clinical workflows—the killer application for process-oriented information systems? In *Proceedings of International Conference on Business Information Systems* (pp. 36–59). Pozan, Poland: Springer.
11. Dazzil, L., Fassino, C., Saracco, R., Quaglini, S., & Stefanelli, M. A. (1997). Patient workflow management system built on guidelines. In *Proceedings of AMIA 97* (pp. 146–150). Nashville, TN.
12. Quaglinia, S., Stefanellia, M., Lanzolaa, G., Caporussob, V., & Panzarasab, S. (2001). Flexible guideline-based patient careflow systems. *Artificial Intelligence in Medicine, 22*, 65–80.
13. Song, X., Hwong, B., Matos, G., Rudorfer, A., Nelson, C., Han, M., et al. (2006). Understanding requirements for computer-aided healthcare workflows: Experiences and challenges. In *ICSE'06*, May 20–28, 2006, Shanghai, China. ACM 1-59593-085-X/06/0005.
14. Dwivedi, A. N., Bali, R. K., Wickramasinghe, N., & Gorgui-Naguib, R. N. (2007). How workflow management systems enable the achievement of value driven healthcare delivery. *IJEH, 3*(3), 382–393.
15. Liu, C., Li, Q., & Zhao, X. (2008). *Challenges and opportunities in collaborative business process management: Overview of recent advances and introduction to the special issue*. New York: Springer Science and Business Media, LLC.
16. Valls, A., Gibert, K., Sánchez, D., & Batet, M. (2010). Using ontologies for structuring organizational knowledge in home care assistance. In *International Journal of Medical Informatics, 79,* 370.
17. Ghidini, C., Francescomarino, D., Rospocher, M., Tonella, P., & Serafini, L. (2011). Semantics based aspect oriented management of exceptional flows in business processes. *IEEE Transactions on Systems, Man, and Cybernetics, Part C: Applications and Reviews*.
18. Gruber, T. R. (1993). Towards principles for the design of ontologies used for knowledge sharing. In N. Guarino & R. Poli (Eds.), *Formal Ontology in Conceptual Analysis and Knowledge Representation*. Deventer, The Netherlands: Kluwer Academic Publishers.
19. Uschold, M., & King, M. (1995). Towards a methodology for building ontologies. In D. Skuce (Ed.), *IJCAI'95 Workshop on Basic Ontological Issues in Knowledge Sharing* (pp. 6.1–6.10). Montreal, Canada.
20. Staab, S., Schnurr, H. P., Studer, R., & Sure, Y. (2001). Knowledge processes and ontologies. *IEEE Intelligent Systems, 16*(1), 26–34.

A Semantics-Based Approach to Generation of Emergency Management Scenario Models

Antonio De Nicola, Michele Melchiori and Maria Luisa Villani

Abstract Interoperable companies making business together form large networks. Communication and exchange of goods and documents is permitted by critical infrastructures like the energy network, the railway, and the telecommunication network. These are threatened by several hazards spanning from natural disasters, as earthquakes and tsunami, to anthropic events, as terrorist attacks. An example of such catastrophic events is the Fukushima nuclear disaster causing deaths, destroying buildings and infrastructures and impacting on the supply chains of several companies. Simulation is one of the most promising means to prepare to such events. However, manual definition of emergency management scenarios is a complex task, due to their inherent unpredictability. In this paper an automatic approach to support generation of emergency management scenarios is proposed. This is based on the CEML scenarios modelling language, on the design patterns-based modelling methodology, on the notion of mini-story, and on emergency management ontologies.

Keywords Emergency management · Conceptual modelling · Ontology

A. De Nicola · M. L. Villani
Computing and Technological Infrastructure Lab, ENEA, Rome, Italy
e-mail: antonio.denicola@enea.it

M. L. Villani
e-mail: marialuisa.villani@enea.it

M. Melchiori (✉)
Università degli Studi di Brescia—Dip. di Ing. Dell'Informazione, via Branze 38, 25123 Brescia, Italy
e-mail: melchior@ing.unibs.it

K. Mertins et al. (eds.), *Enterprise Interoperability VI*, Proceedings of the I-ESA Conferences 7, DOI: 10.1007/978-3-319-04948-9_14,

1 Introduction

Interoperable companies have a competitive advantage in facing the challenges of a globalized market with respect to the others. They form a network where goods are bought and sold, documents and data are shared through cloud technologies and company services are provided through the web. Interoperability and collaboration are allowed by enabling infrastructures as the internet, the telecommunication network, the energy network, and the transportation system.

Such infrastructures are constantly threatened by highly unpredictable events as natural events (e.g., earthquakes, tsunami, floods) and anthropic events (e.g., terrorist attacks, environmental disasters). Propagation of effects in case of crisis has disruptive consequences in the society and the productive system of a country. An example is the Fukushima nuclear disaster causing victims and damaging also supply and trade chains from the automotive to the chemical sectors [1–3].

Promptness and reactiveness of enterprises, service providers and institutional operators to face and manage emergency situations are becoming an important issue. A promising approach to reach this goal is computer-based simulation of such crisis events and the related management phase. Existing technologies to such purpose are agents-based simulation [4] and workflow systems [5].

Currently, operators spend a lot of time in defining one single scenario to be simulated. They try to imagine the services and people impacted by an event, what is its intensity and how long it lasts. The result is that they come out with partial solutions and, consequently, they feel they are “tilting at windmills”. We can say that human definition of emergency management scenarios is a loosing strategy since the issue here is to predict as many emergency management (EM) situations as possible. An automatic support may be very helpful in this, but the possible combinatorial explosion of the different evolutions of EM situations has to be faced. Thus, semantics-based techniques may be used for automatic provisioning of meaningful solution models.

In this paper, we propose to use semantics-based languages and tools to enable a bottom-up construction of emergency management scenario models to be simulated by simulation tools. In particular, our proposal is to automatically generate semantically coherent fragments of emergency management scenario models, called mini-stories, to be supplied as input for scenarios creation by composition. For the sake of space, in this paper we focus on the mini-stories generation whereas the composition problem will be treated in future works. In particular, we have chosen CEML [6, 7] as a scenario modelling language, and follow the design patterns-based modelling methodology described in [6]. Furthermore, we have adopted the notion of mini-stories from [8] and we exploit the existence of reference ontologies in the EM domain to provide rules for mini-stories creation and composition.

The paper is structured as follows. Section 2 discusses the related work in the area and Sect. 3 presents an EM scenarios context used as running example to describe the problem addressed in this paper and our objectives. Section 4 presents the EM modelling approach. Then, Sect. 5 describes the EM scenarios generation process. Section 6 focuses on how semantic techniques can support selection of mini-stories and, finally, Sect. 7 provides conclusions and future research directions.

2 Related Work

Ontologies in EM concern mainly three aspects: (i) supporting interoperability among systems [9, 10]; (ii) providing semantics to data and models [11, 12]; (iii) semantic enrichment of simulation models [13, 14]. Supporting interoperability is for example discussed in [9] where an intelligent emergency management system (EMS) for airport EM is presented. The EMS is based on interpreting complex events defined as meaningful correlations of simple events collected and communicated by networks of data acquisition systems operating in the airport. The domain specific ontology is used to integrate and provide meaning to the messages generated by the acquisition systems. The process of building an ontology for EM from glossaries and vocabularies, in order to support communication among software systems for critical infrastructures, is described in [10].

A semantics-based emergency response system is proposed [11] with the aim of supporting emergency decision makers by retrieving and possibly adapting solutions provided for previously events. The system uses real time information provided as short message service (SMS) by on-the-field users. As previously said, ontologies provide semantics to data and models. In particular, they are used both to interpret and classify SMS messages, and for solution retrieval from the solution database. Ontologies have also been recognized to be potentially useful for simulation [13]. In particular, ontologies are used for semantically enriched descriptions of model components in discrete-event simulation modelling [14]. These descriptions are used for performing components discovery and for determining compatibility of components. Compatibility is established based on the data exchanged between them and the fact that their behaviors are aligned.

Our work has some common points with [15] that uses a crisis metamodel and a corresponding ontology in order to support interoperability among the information systems of partners involved in a crisis scenario and that have to collaborate. Specifically, by means of ontology reasoning, the approach in [15] allows the deduction of basic elements (e.g. actors, services) in order to establish a model for the collaborative process.

Finally, our approach is related to all these works but we consider the problem of generation of mini-stories constrained and guided by domain ontologies. Specifically, we combine structural knowledge and ontology knowledge. Structural knowledge defines the admitted relationships among classes of concepts that are relevant in a mini-story. A domain ontology provides instances for the classes of concepts and constraints their co-occurrence in a mini-story.

3 Case Study

The business ecosystem, considered in this case study, consists of a *network of around 100 companies* in Asia making business together in the *automotive sector*. The farms produce *products*, bought and sold in the network. To do this, they share data through the internet and use web services to issue *purchase orders*, *invoices*, *bills of materiales* and other *business documents*. The companies are powered through *renewable energy plants*, i.e., *wind power plants*, *solar panels*, *hydro-power plants*, *geothermal plants*, and other. *Goods* are transported by trains, trucks, airplanes, and ships.

There are several unpredictable events that could affect the infrastructures making interoperability and collaboration of companies possible. For instance, *earthquakes*, *tsunami*, *floods*, *tornado*, *avalanches* can destroy such infrastructures and cause severe damages to cooperative work and, consequently, to the business. In case of disasters, *rescue* and *recovery teams* intervene to restore the previous state of the affairs. Among such teams we cite *fire brigades*, *civil protection*, and *technical operators*.

4 Modelling Emergency Management Scenarios

In this Section we present the conceptual model, illustrated in Fig. 1, that is the basis of the proposed scenarios modelling approach, and an overview of the reference architecture.

Starting from a textual description of the scenario such as the one sketched in Sect. 3, one might be interested in the identification of one or more *EM Scenarios* in order to define an *EM Plan*, which collects all the foreseen scenarios. An EM Scenario is a still narrative situation of an emergency, caused by some unpredictable event, occurring in a certain place and impacting one or more specified real worlds objects (such as people, infrastructures, institutions, companies, and so on), and of the actions taken to solve the emergency. One EM Scenario may be an earthquake of a certain magnitude happening in a certain region, interrupting for some time the electricity and transport infrastructures serving that region where important supplying companies are located, and the related recovery actions.

In order to evaluate the efficacy of various emergency resolution measures, such as the identification of alternative transportation means and/or roads, or the reconfiguration of the electricity network, each EM scenario may be represented through *models*, expressed in machine readable form in order to enable computer supported analysis such as simulation.

In this paper, we face the problem of providing automatic support to the construction of as many as possible EM scenario models to the aim of defining an EM plan for the given business ecosystem. This objective is realized by empowering the work of a general scenarios modelling tool, through which the structure (or

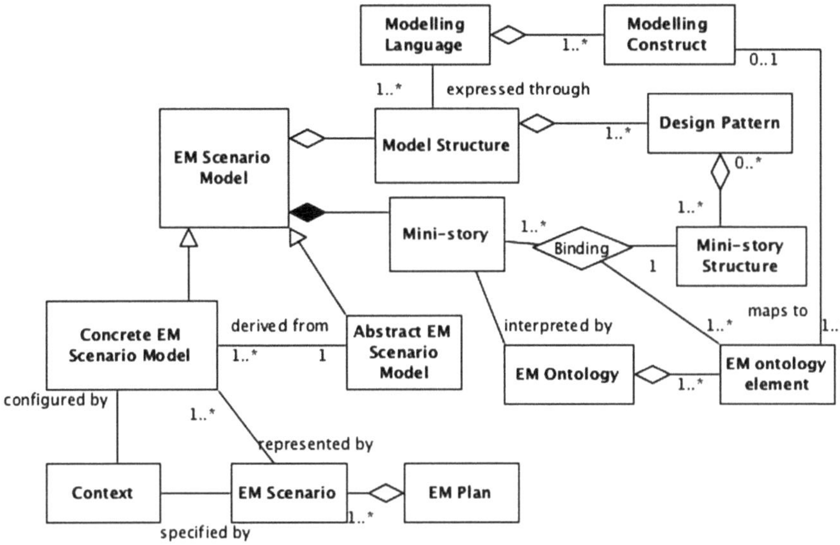

Fig. 1 Conceptual model of the scenarios modelling approach presented as UML diagram

topology) of a scenario is designed, with a support for automatic construction. In this way, one or more EM scenario models can be defined by incrementally instantiating the component types and their connections through semantics-based techniques. This is achieved by using a Knowledge Base containing:

1. an *EM ontology*, covering knowledge about emergencies caused by natural or anthropic events. Such knowledge concerns several domains spanning from descriptions of hazards and events, to critical infrastructures, to services provided to companies and citizens, to recovery and rescue services, and to users. Preliminary examples of emergency ontologies already exist in the literature (for instance [16–18]);
2. specification of the mappings between the EM ontology concepts and the constructs of the modelling language in use. This enables semantics-based creation of scenario models;
3. a repository of *mini-stories,* that are semantically coherent fragments of scenario models, constituting the building blocks for scenario models creation. The mini-stories repository is incrementally populated during execution and the modeller arbitrarily decides the granularity of the mini-stories.

To better illustrate the approach, in this paper we have chosen CEML and the pattern-based methodology presented in [6]. CEML is a domain-specific modelling language derived from SysML [19], an UML's profile, allowing experts to build formally grounded models in a user-friendly way.

5 The EM Scenarios Generation Process

Here we describe a method aiming at automatic generation of EM scenarios models, starting from pre-defined design patterns and by means of mini-stories semantic binding and composition. This method consists of the following steps.

Model structure definition. Given an EM plan definition objective, the first step is to select one or more design patterns to be used as reference structures for the EM scenario models to be constructed. A design pattern is expressed in terms of component types belonging to the modelling language in use, and it is intended as a reusable solution to a set of modelling problems [20]. One such pattern in the CEML language for our case study is the *core EM scenario design pattern* presented in the upper part of Fig. 2. Here the basic elements being part of an emergency scenario are included: an external event affects a service providing some benefits to a user and another service try to recovery from the damage. Starting from this pattern, the modeller identifies one or more structures as modelling building blocks, i.e., mini-stories in the lower part of Fig. 2, to be used to build scenario models.

Abstract Mini-stories creation and composition. Each component type of a mini-story structure is linked to an element of the EM ontology, by automatically choosing from a pre-defined list of variants specified in the knowledge base, in order to generate a mini-story meaningful from a semantic perspective.

The final aim of this step is to allow for a bottom-up creation of *abstract EM scenario models* through patterns instantiations, obtained by semantic composition of abstract mini-stories. An abstract EM scenario model is a representation at conceptual level of one or more design patterns, whose components are abstractions of real objects (objects types), identifiable through concepts and relationships of the EM ontology. In the example above, this abstract EM scenario model would have components such as: *earthquake*, *transportation service* and *electricity infrastructure*, and *business ecosystem*. Figure 2 shows two possible abstract EM scenario models originated from the same design pattern.

Concrete EM Scenario models generation. An *abstract scenario model* can originate one or more *Concrete EM scenario models*, which can differ for: the identification data of the real objects (e.g., name and location) and their other context data; for the magnitude of the emergency; and/or for the response measures (e.g., response time). In our example, from the same abstract model, it is required to simulate the impact of earthquakes of various magnitudes or by varying the epicentre over the given localized component infrastructures and business ecosystem.

6 Semantic Binding of Mini-Stories

In this section we introduce the semantic approach for mini-stories binding through its application on our case study.

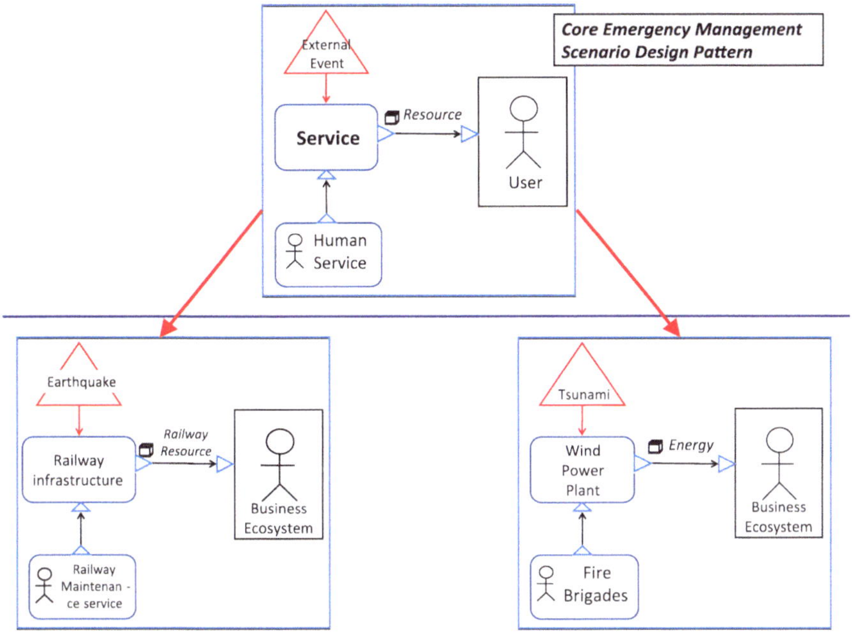

Fig. 2 An example of generation of two abstract scenario models from the core EM scenario design pattern

An EM ontology (see Fig. 3) is used for semantically enriching a CEML model with reference to the EM context of the considered interoperable companies. Generally speaking, an ontology describes static and shared knowledge about a scenario in terms of concepts, properties and relationships.

The ontology is represented using OWL that allows managing and reasoning on it with widely available tools. The OWL representation we adopt follows a standard approach where, in particular, a concept is represented as OWL:class; a property is represented as OWL:DatatypeProperty; an is-a relationship is represented as RDF:SubClass; a domain specific relationship is represented as OWL:ObjectProperty.

In the following, we first introduce a mapping between ontology concepts/relationships and CEML constructs, in order to create the grounding for semantic enrichments of the CEML-based scenario models. Then we explain how this mapping is used for generating semantic bindings to support construction of abstract mini-stories as introduced in Sect. 5.

Given an ontology, we introduce a partial mapping ***Onto2CEML*** between elements of the ontology and the CEML language. In particular:

- a concept a is mapped on a CEML modelling construct α;
- a domain specific relationship r between two concepts a and b is be mapped on a CEML relationship α_r;
- an is-a relationship is not mapped.

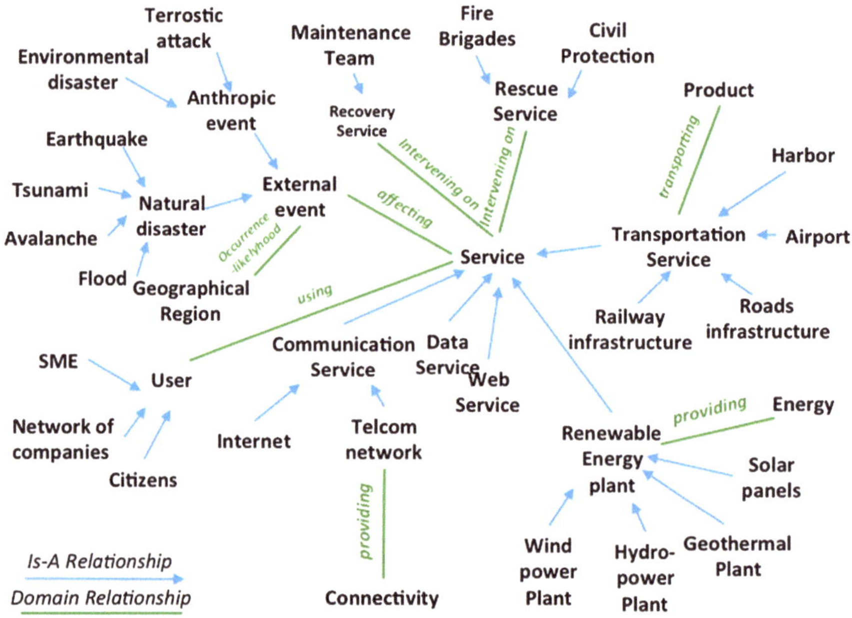

Fig. 3 A graphical representation of an excerpt from the emergency management ontology. Concepts are semantically connected through IS-A relationships (represented as oriented *arrows*) and labeled domain relationships

The scenario designer decides the actual Onto2CEML mapping for the ontology in the considered domain.

Then we introduce a ***CEML2Onto*** mapping where a CEML construct, resp. a relationship, belonging to the range of Onto2CEML is mapped to a set of ontology concepts, resp. relationships, as follows.

CEML2Onto(α) is defined:

- if α is a CEML modelling construct, the set of ontological concepts a_i, such that Onto2CEML(a_i) = α.
- if α is a CEML relationship, the set of triples (a, r_j, b), where r_j is a domain specific ontological relationship such that Onto2CEML(r_j) = α and *a'* has r_j relationship with range *b'* in the ontology. Moreover, *a'* subsumes *a* and *b* subsumes *b'*.

The notion of *subsumption* between ontological concepts can be implemented based on available and effective inference procedures. Simply stated, *a'* subsumes *a* if either they are the same concept or *a'* is more general in the ontology than *a*.

Example. The modelling construct CEML:ExternalEvent represents an active entity (e.g., failure) affecting the effectiveness of a service either human or not. With reference to our domain ontology the designer can classify the concept `Earthquake` through the assignment Onto2CEML(`Earthquake`) = CEML:

ExternalEvent. The concepts `Communication Service` and `Telcom network` can be mapped on the construct CEML:Communi-cationService representing an active entity allowing communication and information exchange between CEML:Services and CEML:Users. Finally, the object property `Affecting` of the concept `External event` having range `Service` can be mapped on CEML:Impact.

Moreover, based on these examples, we have CEML2Onto(CEML: CommunicationService) = {`Communication Service`, `Telecom net-work`} and CEML2Onto(CEML:Impact) = {(`External event`, `Affecting`, `Service`)}.

The CEML2Onto mapping therefore permits to associate each CEML construct occurring in a mini-story structure with a set of ontology concepts and relationships.

Semantic binding. A semantic binding between a mini-story structure S and the ontology is based on CEML2Onto associates:

1. each CEML construct α in S with a concept a of the ontology such that a belongs to the set CEML2Onto(α).
2. each CEML relationship α_r in S, connecting two CEML constructs α_i and α_j in the mini-story, with a relationship r of the ontology such that (a, r, b) belongs to the set CEML2Onto(α) and a, resp. b, is associated with α_i, resp. α_j, in the binding according to the previous point 1.

A semantic binding on a mini-story structure defines therefore an abstract mini-story where every CEML element is associated with an ontology element coherently with the mini-story structure and the ontology. It is possible to generate exhaustively abstract mini-stories from the same structure by building all the possible semantic bindings that are based on a given CEML2Onto mapping. For instance, a semantic binding of the mini-story pattern associates: (i) CEML:ExternalEvent with `Earthquake`; (ii) CEML:CommunicationService with `Telcom network`; (iii) CEML:Impact with (`External event`, `Affecting`, `Service`).

Introducing a *subsumption* relationship between mini-stories may further accelerate this automatic generation process by pruning those mini-stories that are subsumed by more general ones based on the requirements set by the designer. Subsumption of mini-stories is based on the notion of subsumption between ontological concepts.

Subsumption between ministories. Given two mini-stories A and B generated by two semantic bindings on the same mini-story structure S, we say that B subsumes A when:

1. for each CEML construct α in S, the concept a_B associated with α in B subsumes the corresponding concepts a_A in A.
2. for each CEML relationship α_r in S, the binding of α_r in A is the same relationship associated with α_r in the binding of α_r in B.

For example, the mini-story with the meaning `Natural disaster Affecting Communication Service` subsumes the mini-story `Earthquake Affecting Telecom network`.

External rules based on modelling and simulation requirements and/or contextual knowledge (e.g., natural disasters do not directly affect data and web services) could further improve the semantic quality of the automatically generated mini-stories and of their composition.

7 Conclusions

In this paper we presented an approach for generation of emergency management scenarios models. Our aim is to increase preparedness and promptness in managing such situations by providing automatic support to the definition of computable models for simulation tools. Our approach is based on semantics and, in particular, on the design patterns modelling methodology, on the notion of mini-stories, and on the existence of EM ontologies.

As future work, we intend to use semantics-based techniques, i.e., semantic rules, in order to select mini-stories, automatically aggregate them and to generate semantically meaningful EM scenario models.

References

1. Norio, O., Ye, T., Kajitani, Y., Shi, P., & Tatano, H. (2011). The 2011 eastern Japan great earthquake disaster: Overview and comments. *International Journal of Disaster Risk Science, 2*(1), 34–42.
2. Ando, M., & Kimura, F. (2012). How did the Japanese exports respond to two crises in the international production networks? the global financial crisis and the great East Japan earthquake. *Asian Economic Journal, 26*(3), 261–287.
3. Collins, T. (2011). *Great east Japan earthquake: Economic and trade impact*. Retrieved October 1, 2013, from http://www.dfat.gov.au/publications/stats-pubs/great-east-japan-earthquake-economic-and-trade-impact.pdf
4. Collier, N. (2001). Repast: An extensible framework for agent simulation. *Natural Resources and Environmental Issues, 8*, 17–21.
5. Hofmann, M., Sackmann, S., & Betke, H. (2013). A novel architecture for disaster response workflow management systems. *Proceedings of Information Systems for Crisis Response and Management, 2013*, 338–343.
6. De Nicola, A., Tofani, A., Vicoli, G., Villani, M. L. (2012). An MDA-based approach to crisis and emergency management modelling. *International Journal on Advances in Intelligent Systems 5* (1 & 2), 89–100.
7. D'Agostino, G., De Nicola, A., Di Pietro, A., Vicoli, G., Villani, M. L., and Rosato, V. (2012). A domain specific language for the description and the simulation of systems of interacting systems. *Advances in Complex Systems, 15*(1).

8. Thalheim, B. and Tropmann-Frick, M. (2013). *Mini story composition for generic workflows in support of disaster management, Proceedings of 24th IEEE International Workshop on DEXA.*
9. Mijović, V., Tomašević, N., Janev, V., Stanojević, M. and Vraneš, S. (2011). *Ontology enabled decision support system for emergency management at airports. In Proceedings of the 7th International Conference on* Semantic *System (I-Semantics '11)* (pp. 163–166). New York, USA: ACM.
10. Grolinger, K., Brown, K. P., & Capretz, M. A. M. (2011). From glossaries to Ontologies: Disaster management domain(S). *Software Engineering and Knowledge Engineering, 6644*, 402–407.
11. Amailef, K., & Lu, J. (2013). Ontology-supported case-based reasoning approach for intelligent m-Government emergency response services. *Decision Support Systems, 55*(1), 79–97.
12. Li, X., Liu, G., Ling, A., Zhan, J., An, N., Li, L., Sha, Y. (2008). Building a practical Ontology for emergency response systems. *Computer Science and Software Engineering*, *4*, 222–225.
13. Araujo, R. B., Rocha, R. V., Campos, M. R., Boukerche, A. (2008). *Creating emergency management training simulations through Ontologies integration, Proceedings of 11th IEEE International Conference on Computational Science and Engineering Workshops CSEWORKSHOPS '08* (pp. 373–378)16–18 July 2008. USA:Springer
14. Teo, Y., Szabo, C. (2008). CODES: An integrated approach to composable modeling and simulation, *Proceedings of the 41st Annual Simulation Symposium* (pp. 103–110), April 2008. USA: Data Device Corporation.
15. Benaben, F., Hanachi, C., Lauras, M., Couget, P., & Chapurlat, V. (2008). A metamodel and its Ontology to guide crisis characterization and its collaborative management. *Proceedings of Information Systems for Crisis Response and Management, 2008*, 189–196.
16. Wang, W., Zhang, X., Dong, C., Gao, S., Du, L., and Lai, X. (2009). *Emergency response organization Ontology model and its application, Proceedings of ISISE 2009 Symposium* (pp. 50–54).
17. Dong, C. X., Wang, W. J., and Yang, P. (2009). *DL-based the knowledge description of emergency plan systems and its application, Proceedings of MUE'09* (pp. 364–368).
18. Kruchten, P. C. (2007). *A human-centered conceptual model of disasters affecting critical infrastructures, Proceedings of ISCRAM 2007 Conference* (pp. 327–344).
19. OMG-SysML. (2010). OMG Systems Modeling Language Version 1.2. Access: http://www.omgsysml.org
20. Gangemi, A., and Presutti, V. (2009). *Ontology design patterns, In Handbook on Ontologies (2nd ed.,). International Handbooks on Information Systems.* Heidelberg: Springer.

A Methodology to Prepare Real-World and Large Databases to Ontology Learning

Bouchra El Idrissi, Salah Baïna and Karim Baïna

Abstract Several approaches have been proposed for the generation of application ontologies from relational databases. Most of these approaches, propose a fully automatic process based on unrealistic assumption, where the input database is well designed, up-to the third normal form (3NF). Real-World databases may contain irrelevant, missing or erroneous information to the ontology learning process. Preparing databases before ontology learning is quite rare. We propose in this paper a methodology for Database Preparation (DBP), composed of three sub-processes: the extraction of a Business Database (BDB), the cleaning of the BDB, and the enrichment of the cleaned BDB. A proof-theoretical case study shows that the proposed methodology is feasible and useful.

Keywords Ontology · Application ontology · Relational database · Data model · Ontology learning · Database preparation · Semantic enrichment · Database auditing

1 Introduction

Integration of information systems has been always a challenge for the enterprise, despite the advance in Information Technology (IT). Ontology brings a solution to the most challenging level of interoperability, which is the semantic level. It is defined as formal, explicit specifications of shared conceptualization [1]. Explicit

B. El Idrissi (✉) · S. Baïna · K. Baïna
ENSIAS, University Mohammed V Souissi, BP713 Rabat, Morocco
e-mail: bouchra.idrissi@um5s.net.ma

S. Baïna
e-mail: sbaina@ensias.ma

K. Baïna
e-mail: baina@ensias.ma

K. Mertins et al. (eds.), *Enterprise Interoperability VI*,
Proceedings of the I-ESA Conferences 7, DOI: 10.1007/978-3-319-04948-9_15,

means that the specification of concepts is described formally in an unambiguous manner. Specification of shared conceptualization means a common consensus about identified ontology constructs (concepts, properties of concepts, relationships between concepts and axioms) of a domain of knowledge. Thus, ontology provides a rich and formal representation of knowledge to be machine-processable.

The four properties defining the ontology, present also the main barriers for its proliferation: **shared,** because it is difficult to rich a common consensus about a domain; e**xplicit specification,** because it requires an unambiguous description of the ontology constructs and this faces principally the problem of capturing *tacit* knowledge; **conceptualization,** because it involves an abstraction of the studied domain to model concepts and relationships between them; **formal,** because despite of the fact that different open ontology editors have been developed to support the user in ontology formalization (e.g., Protégé [2]), no conceptualization can be entirely formalized due to design ambiguities or to the representation language that cannot entirely represent them [3].

Ontology learning field aims to overcome the ontology acquisition bottleneck, by automatically or semi-automatically generating ontology from some input information sources of types structured (e.g., RDB), semi-structured (e.g., XML) or unstructured data (e.g., text brut). The present paper focuses on the ontology learning from Relational Databases (RDBs). It is an attractive area for developing application ontologies, on one hand due to the pervasiveness of the relational model in industry, commercial and open-source applications and on another hand because a RDB already embeds some approved knowledge of a domain.

From our practical study [4] on ontology learning approaches from relational data models, we distinguished some shortcomings and open issues that are still unanswered, despite of the advance in the learning techniques. One of our remarks was that the majority of the existing approaches are drawn on the assumption that the input database is well-designed and up-to the third Normal Form (3NF). Unfortunately, this unrealistic assumption limits the applicability of these approaches and do not match the state of real-life databases. Further, there are only two approaches from thirty-eight ones (listed in [4]) that propose a limited data preparation process before the ontology learning. These approaches are OWL-miner [5] and RDBToOnto [6]. The approach of OWLminer for data preparation (DP) phase consists of incorporating concept hierarchy, provided by domain expert, as the background knowledge to identify and to select the relevant dataset on the base of a distance-based feature selection algorithm. RDBToOnto incorporates a database normalization step to eliminate data duplication in the source tables through the interpretation of inclusion dependencies defined by the user.

Real-World databases often contain irrelevant and erroneous information to the ontology learning process. According to The Cross-Industry Standard Process for Data Mining (CRISP-DM), DP is one of the most important phases in knowledge discovery which refers to preparing data before extracting knowledge [7]. In order to close the gap of a DP phase in the ontology learning field, we propose through this paper a semi-automatic methodology, which produces a ***Clean and Semantic Business Model*** from an input database.

The reminder of this paper is organized as follows: we provide in Sect. 2 an overview of this study context. The Sect. 3 discusses some relevant issues in real-world databases. The Sect. 4 presents the proposed methodology for DBP. This section is divided in three subsections where, each one describes a sub-process. The Sect. 5 concludes the paper and highlights some future work.

2 The Study Context

Our research project, described in [4, 8] is interested in the interoperability of Enterprise Resource Planning (ERP) systems by Service-Oriented Architecture (SOA). Moreover, our approach depends on the use of ontology to resolve the semantics conflicts in messages of services. The approach relies on the hybrid ontology approach (see Watche et al. [9] for a description of the three ontology-based approaches for information systems integration). The ERPs application ontologies are to be generated (semi-) automatically and the mapping between the local ontologies will be ensured via an Enterprise Ontology (EO) [10].

With no available tool that generates semi-automatically an application ontology, intended for use in semantic interoperability [4], it is clear that we have to implement our proper system. Unlike the other ontology learning approaches that focus on learning techniques without studying in depth the peculiarities of real-life and large databases, we present in this paper a methodology for a pre-processing phase before the learning process. Before going on details on the proposed methodology, we discuss, in the following section, real-world database issues that complicate the learning process.

3 Real-World Database Issues

In addition to the largeness of real-world databases (high volume of data, high number of objects) that influences the performance of ontology learning algorithms and may require parallelization, we detailed, in the following, some additional and serious issues. They are classified on database schema issues and database instances issues. This separation is justified by the fact that both the database schema and instances present the main sources for inferring semantics, as explained in subsequent paragraphs.

3.1 Database Schema Related Issues

The mapping rules involved in the learning techniques, are based on the study of each database table, by considering the relation between its primary-keys and foreign-keys. A simple example is: if a relation contains only descriptive and primary-keys without foreign-keys, this relation is transformed to a concept (OWL [11] class). However, the questions that arise are: what to do if a table has no keys (foreign and primary)?; what to do if a table is technical or temporary without a relevancy to the business domain to be studied?; how to infer the different concepts that may be aggregated in one table (e.g., table *c_order* in *OpenBravo* [12] that combines orders of purchase and sales)?; and the most important is how to drive the accurate meaning of relationships between tables?

All the previous questions are still unanswered by existing approaches and require further investigation. They all, except the last question, depend on the logical model of the database. The last one, however, concerns both the logical model and the conceptual model. For instance, in *OpenERP* [13] there are multiple relationships between the table *product_template* and the table *product_uom* (unity of measure). A product may have a default *uom,* a *uom* for sales and a *uom* for purchase. The existing mapping rules drive three object properties between these tables, but it is not possible to infer automatically the sense of these relationships.

Additionally, some mapping rules (like the one proposed by [14] to drive inheritance relationships) are based on the syntactical comparison of attributes names and data types, to drive equivalence between them. In large, distributed databases and collaborative development environment, it is clear that syntactical comparison is not sufficient to judge such equivalence. Syntactically equivalent attributes could be semantically different and vice versa.

In summary, issues in real-world databases schema that present a challenge for ontology learning, are related to the less expressivity of the relational model, to the complexity of relationships between database tables (e.g., multiple relationships discussed above) and to missing constraints.

3.2 Database Instances Related Issues

The investigation of the database instances, in the learning field, is invariably required to extract some hidden and accurate semantics that cannot be discovered by studying only the database schema. Some examples of these semantics are categorization patterns [15], vertical partitioning of tables [16], cardinalities [14], inclusion and functional dependencies [17]. However, large databases are suspected to contain *dirty* data, which refers to errors and inconsistencies in database content-level, and require data analyzes to be detected.

The problem of data quality has been widely investigated in data mining, data integration and Knowledge Discovery (KDD) domains. In literature, many dimensions were proposed for the assessment of data quality [18, 19]. Batini et al. in [19] present these dimensions definitions and their metrics. For instance, correctness dimension is the degree to which data do not contain errors [18]. These errors could be syntactical or semantic [20]. Some examples of syntactical errors are: domain format errors, if an attribute value does not conform to its format, irregularities or non-uniformity of values for multiple representations of an attribute value. Semantic problems are illustrated in duplicate representation of an entity, in contradictory values and in invalid instances (see [20]). All these problems may prevent from an accurate extraction of semantics. For instance, the proposed algorithm in [14] to extract cardinalities omitted the problem of duplicated or near (inexact) duplicated records (records that represent the same real-world entity, but with different values for all or some of their attributes [20]).

An auditing data phase through parsing and statistical methods, helps detect the kinds of data anomalies in a database [20]. Unfortunately, cleansing data is a time-consuming and expensive task, based on available information and knowledge to perform the correction of detected anomalies [20]. In large databases, it is unrealistic to expect user involvement for a manual correction of database tuples (instances), except if the number of noise is fewer, otherwise the process should be at least semi-automatic.

In short, the process of data cleansing normally never finishes [20], because some anomalies like invalid tuples and semantic ones are very hard to be detected and because data anomalies are dependent. For instance, duplicated records resolution requires that the attributes values are consistent and presented uniformly (see [20] for a hierarchy of data quality criteria).

4 Proposed Database Preparation Process

The rationale behind our approach is the requirement for a ***semantic middleware layer*** between the learning process and the input database. This layer allows conserving only business relations and attributes to the ontology learning process. Further, these relations and attributes must have business and understandable terms, shared between end-users (one of ontology principles).

The process must be preceded by an auditing phase (see Fig. 1a) in order to elicit the types of problems in the database schema and instances. The adoption of the methodology consists to apply appropriate algorithms to the type of data quality problems, found in the database and to enrich it with necessary information.

The whole proposed methodology is depicted in Fig. 1b, illustrating the sequence between the three sub-processes. It requires access to the database and it generates a *Clean and Semantic Business Model,* which will be used as input to the ontology learning process. So, unlike the other approaches that interest more in the automation of ontology learning, without worrying about the quality of the input model or on how to incorporate *tacit* knowledge and additional resources, to upgrade the semantics of the input model and then the quality of the generated ontology, the proposed methodology attempts to overcome these issues.

Below, we describe each sub-process. We precise the problem statement that addresses each sub-process, the objective of the process and how the user is involved in this process. Some implementation guidelines are also provided.

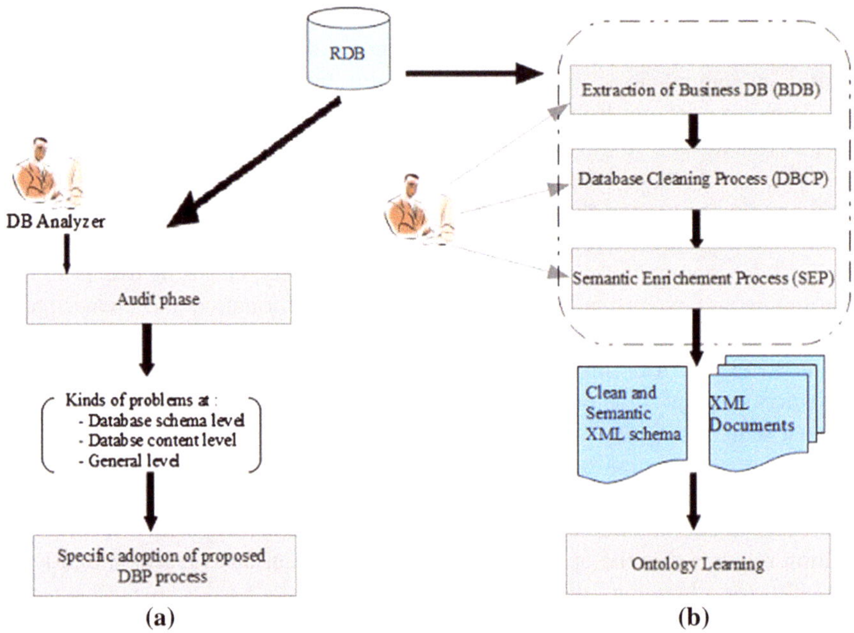

Fig. 1 **a** A pre-database preparation phase for database auditing; **b** The three sub-processes of the proposed methodology

We also indicate some related work that could be investigated to develop the sub-process. A theoretical application of the sub-process is given for a proof of sub-process usefulness.

4.1 Business Database Extraction Process

We described here the first sub-process, which concerns the extraction of a Business Database from the input database.

Problem statement: Many tables and columns in the input database could be without relevancy to the learning process (e.g., temporary, technical).

Sub-process objective: This process has as goal to identify and to eliminate tables and attributes without relevancy to the ontology scope to be studied. This step is prior to any database examination as it helps to economize time and effort. It is important since it is the input for the subsequent sub-processes.

Sub-process user involvement: This sub-process should be carried out with a user who knows well the underlying database. Otherwise, some resources may help in doing this task. HTML forms of front-end application, already designed business services, referential description documents and the data dictionary are examples of these resources. The first step to do is to identify and to remove irrelevant tables. Then, each business table must be studied at alone to eliminate technical, temporary or any irrelevant attribute to the learning process. During this process, it is necessary to maintain all the constraints predefined in the database schema.

Sub-process implementation guidelines: The environment in which to do this process should provide these functionalities: to navigate easily the underlying database objects, to import the selected tables and constraints to a viewer. The tool must enable an import of the database instances from input database to selected tables. The last process must be done automatically at once since there is no transformation applied to the database schema. Moreover, it will be helpful to execute a script with all the necessary deletions, and to apply generic delete requests based on some filters on database tables' names. Several available and open source SQL editors [21] and ETL tools [22] provide similar functionalities. They can present a starting point for further improvements to fulfill the requirement of this sub-process.

Sub-process output: The resulting model is called Business Database (BDB).

Sub-process theoretical application: ERP systems are huge applications with very big number of tables (e.g., there are about 600 tables in OpenERP 7). The naming conventions of database tables differ from one ERP to another. They varied from clear, less explicit, codified and abbreviated ones. For OpenERP as example, the tables are organized by models (account, product, sale, etc.). After an analysis of OpenERP database, we conclude with a large number of tables to be deleted. Examples of these tables are those starting with *base*, *wkfl*, *ir* or including the term *config*. OpenBravo follows another strategy that may help in distinguishing business tables from the other ones. It separates between tables according to the access level (see table *AD_Table* column *accesslevel*): organization for data specific to an organization, system data metadata and so on.

4.2 Database Cleaning Process

Database cleaning is generally the process of detecting, correcting or deleting errors in the database. The well-known term in the literature is data cleaning, but because we are interested in anomalies related to database schema and instances, we use the term database cleaning instead of data cleaning.

Problem statement: The main input on which relies the learning process is the input database. Errors and omissions in this database (at database schema and instances levels) impact heavily the quality of the produced ontology. More the database is clean, more the quality of the ontology augment.

Sub-process objective: To clean as possible the input model. Due to the issues of large databases cleaning (introduced in Sect. 3), we propose another strategy drawing on: the identification of several dataset from the database according to some filters defined by a database analyzer (with appropriate depth to perform semantics discovery). Then, the audit of the different datasets to identify existing anomalies (e.g., missing values, invalid character values, duplicate records). Finally, the DB analyzer has to identify the dataset with less anomalies and which he/she judges more consistent.

Sub-process user involvement: As explained before, the DB analyzer is heavily implicated in this process. In addition to previously outlined tasks, he/she has to judge the possibility for applying some corrections on the dataset either manually, through some personalized scripts or proposed algorithms in the system. Moreover, he/she has to validate discovered duplicated records before their corrections.

Implementation guidelines: Our objective at time is not to develop specific algorithms for data auditing and cleaning but to exploit already developed and available ones. For duplicated records detection, the survey [23] gives an extensive set of duplicate detection algorithms and coverage of existing tools. Many other free tools from [24, 25] could be investigated for data auditing and cleaning.

Sub-process output: The resulting model is called Cleaned Business Database (CBDB).

Sub-process theoretical application: We have already stated the negative impact of duplicated records in calculating the exact cardinalities. Another example is the non-uniformity in attributes values that may prevent the derivation of inclusion and functional dependencies.

4.3 Database Enrichment Process

We describe here the semantic enrichment sub-process. The input of this sub-process is the dataset or the database resulting from the cleaning sub-process.

Problem statement: The resulting model is always a relational model recognized by its limited mechanisms to express the semantics of an application [26]. This sub-process addresses five main issues: (i) Database tables and attributes names could be meaningless; (ii) The exact meaning of relationships between tables could be hidden (see Sect. 3.1); (iii) They could be different and implicit representations of a conceptualization (multi-valued attribute (as simple attribute or in independent table), inheritance (single table, a table per entity, only tables for sub-entities)); (iv) Some schema constraints could be missing in the database schema; (v) Some columns values of type enumeration and categorization could be codified in the database (e.g., *C_order.docStatus* column and *m_product_category* table in openBravo).

Sub-process objective: Semantic Enrichment (SE) is a process that upgrades the semantics of databases [26]. What is needed is a formal representation that enhances the database schema by explicit formalism of hidden and implicit semantics in both the database structure and content. From literature, examples of such representations are the BLOOM model [27] and the approach of Maatuk et al. in [28] for SE of RDB through a Canonical Data Model. However, their intended uses are different from our need.

Sub-process user involvement: To resolve the issues outlined in problem statement, user involvement is required. First, for renaming tables and attributes with clear business terms. For the second and the third issues, it is clear that they cannot be done automatically because they involve *tacit* knowledge. The fourth point requires, at least, user validation of discovered missing constraints (can be through the auditing phase). The last point involves manual discovery of categorization tables and the investigation of other resources, to seek the meaning of codified information, unless the user already has such knowledge.

Sub-process implementation guidelines: For the middle model, we propose the use of XML [29] for different reasons, among them we cite: (i) The features of XML in expressing the semantics of data is greater than the RDB; (ii) its extensibility for designing additional types and extending properties of elements; (iii) several tools exist for migrating RDB to XML (e.g., XMLSpy [30], Oxygen [31], SpringRoo [32]); (iv) Ontology learning from XML is a widely active domain in the scientific community and several tools are already developed (e.g., [33–35]).

Sub-process result: An enriched semantic model in XML format.

Sub-process theoretical application: Several examples that require to be enriched have been already cited in this paper. On XML side, some types to be defined are for example: composite-of for designing an attribute as a composition of other ones, part-of for designing a composition relationship, enumeration type for a categorization table and a column with explicit domain values. For multiple relationships (see Sect. 3.1), the user can define a property that specifies the exact meaning of the relationship.

5 Summary

Ontology cannot be driven automatically and directly from real-world and large RDBs. This is due, on one hand, to the largeness of these RDBs that may include non-business tables, to the complexity of their relationships and to the presence of dirty data. On another hand, the limits of the relational model in expressing the meaning of relationships between tables and in supporting some conceptualizations prevent from the generalization of mapping rules for ontology generation.

The proposed methodology is composed of three sub-processes. The first one removes irrelevant tables and attributes to the domain scope to be studied. The second process aims to clean the extracted business database from noise that may cause the extraction of false semantics and the last sub-process enriches the cleaned database, to explicit hidden semantics. We are now working on the development of an integrated tool to implement all these functionalities.

References

1. Gruber, T. R., et al. (1993). A translation approach to portable ontology specifications. *Knowledge Acquisition, 5*(2), 199–220.
2. Protégé. (2013). Retrieved October 18, 2013 from http://protege.stanford.edu/
3. Cruz, C., & Nicolle, C. (2005). Ontology-based heterogeneous xml data integration. *Journal of Digital Information Management, 3*(2), 133.
4. El Idrissi, B., Baïna, S., Baïna, K. (2013) Automatic generation of ontology from data models: a practical evaluation of existing approaches. In *Proceedings of the 7th IEEE International Conference RCIS'2013* (pp. 241–252). Paris, France: IEEE.
5. Santoso, H. A., Haw, S. C., & Abdul-Mehdi, Z. T. (2011). Ontology extraction from relational database: Concept hierarchy as background knowledge. *Knowledge-Based Systems, 24*(3), 457–464.
6. Cerbah, F. (2008). *Learning highly structured semantic repositories from relational databases. The semantic web: research and applications* (pp. 777–781). Heidelberg: Springer.
7. Liu, H., & Motoda, H. (2001). *Instance selection and construction for data mining*. MA: Kluwer International Series in Engineering and Computer Science.
8. El Idrissi, B., Baïna, S., Baïna, K. (2012). An ontology based approach for semantic interoperability of erp systems within an organization. In *Proceedings of the 2nd International Symposium ISKO-Maghreb 2012*.
9. Wache, H., Voegele, T., Visser, U., Stuckenschmidt, H., Schuster, G., Neumann, H., et al. (2001) Ontology-based integration of information-a survey of existing approaches. In *Proceedings of the IJCAI-01 Workshop: Ontologies and Information Sharing* (vol. 2001, pp. 108–117) Citeseer.
10. Fox M., S., Gruninger M. (1998) Enterprise modelling. *AI Magazine*. 19(3), 109–121.
11. Ontology Web Language. (2013). Retrieved October 18, 2013, from http://www.w3.org/OWL
12. Openbravo. (2013). Retrieved October 18, 2013, from http://www.openbravo.com/
13. OpenERP. (2013).Retrieved October 18, 2013, from http://www.openerp.com/
14. Albarrak, K., & Sibley, E. (2010). An extensible framework for generating ontology models from data models. *International Transactions On System Science and Applications (ITSSA), 6*(2/3), 97–112.
15. Cerbah, F. (2008). Mining the content of relational databases to learn ontologies with deeper taxonomies. In *Proceedings of the IEEE/WIC/ACM International Conference on Web Intelligence and Intelligent Agent Technology, 2008 WI-IAT'08* (vol. 1, pp. 553–557). IEEE.
16. Alalwan, N., Zedan, H., Siewe, F. (2009). Generating OWL ontology for database integration. In *Proceedings of the Third International Conference on Advances in Semantic Processing, SEMAPRO'09* (pp. 22–31). IEEE.
17. Benslimane, S. M., Malki, M., Rahmouni, M. K., & Benslimane, D. (2007). Extracting personalised ontology from data-intensive web application: An HTML forms-based reverse engineering approach. *Informatica, 18*(4), 511–534.
18. Pipino, L. L., Lee, Y. W., & Wang, R. Y. (2002). Data quality assessment. *Communications of the ACM, 45*(4), 211–218.
19. Batini, C., Cappiello, C., Francalanci, C., & Maurino, A. (2009). Methodologies for data quality assessment and improvement. *ACM Computing Surveys (CSUR), 41*(3), 16.
20. Müller, H., Freytag, J., C. (2005). *Problems, methods, and challenges in comprehensive data cleansing*. Paper presented at Conference on Advanced Information Systems Engineering (CAiSE'05), Porto, Portugal.
21. Sql Editor. (2013). Retrieved October 18, 2013, from http://sql-editor.winsite.com
22. ETL Tools (2013). Retrieved October 18, 2013, from http://www.etltool.com
23. Elmagarmid, A. K., Ipeirotis, P. G., & Verykios, V. S. (2007). Duplicate record detection: A survey. *IEEE Transactions on Knowledge and Data Engineering, 19*(1), 1–16.

24. 22 free tools for data visualization and analysis. Computer World. (2011). Retrieved October 18, 2013, from http://www.computerworld.com/s/article/9215504/22_free_tools_for_data_visualization_and_analysis
25. Rahm, E., & Do, H. H. (2000). Data cleaning: Problems and current approaches. *IEEE Data Engineering Bulletin, 23*(4), 3–13.
26. Castellanos, M. (1993). A methodology for semantically enriching interoperable databases. In *Advances in Databases* (pp. 58–75). New York: Springer.
27. Castellanos, M., Saltor, F., Garcia-Solaco, M. A. (1992). Canonical model for the interoperability among object-oriented and relational databases. In *Proceedings of the IWDOM 1992* (pp. 309–314).
28. Maatuk, M. A., Ali, A., Rossiter, N. (2010). Semantic enrichment: The first phase of relational database migration. In *Proceedings of the Innovations and Advances in Computer Sciences and Engineering* (pp. 373–378). Berlin: Springer.
29. Extensible Markup Language (XML) W3C. (2013). Retrieved October 18, 2013, from http://www.w3.org/XML/
30. XMLSpy XML Editor Altova. (2013). Retrieved October 18, 2013, from http://www.altova.com/xmlspy.html
31. Oxygen XML Editor Oxygen. (2013). Retrieved October 18, 2013, from http://www.oxygenxml.com/
32. Incremental Database Reverse Engineering (DBRE) Add-On. Spring source; http://docs.spring.io/spring-roo/reference/html/base-dbre.html
33. Bohring, H., & Auer, S. (2005). Mapping XML to OWL Ontologies. *Leipziger Informatik-Tage, 72*, 147–156.
34. Ghawi, R., Cullot, N. (2007). Database-to-ontology mapping generation for semantic interoperability. In *Proceedings of the VDBLŠ07 conference, VLDB Endowment ACM* (pp. 1–8).
35. Thuy, P. T. T., Lee, Y. K., Lee, S. (2009). DTD2OWL: Automatic transforming XML documents into OWL ontology. In *Proceedings of the 2nd International Conference on Interaction Sciences: Information Technology, Culture and Human* (pp. 125–131). New York: ACM.

Part V
Architectures and Frameworks for Interoperability

Enhancing Collaborations by Assessing the Expected Financial Benefits of Improvement Projects

Romain Miclo, Matthieu Lauras, Franck Fontanili, Jacques Lamothe, Philippe Bornert and Guillaume Revenu

Abstract In the highly competitive world of today, firms try to maximize value for their customers and other stakeholders through the effective management of their networks. That is why companies must adopt changes, generally at a network scale, that often involve high costs and potentially allow high benefits. This paper deals with an original method able to quantify the financial impacts of a planned improvement project. This proposition is instantiated in a three-step methodology: (i) modeling the AS-IS collaborative flows through a Value Stream Mapping (VSM) method and making decisions regarding the potential improvements; (ii) assessing the expected business consequences of the improvement plan through a Discrete Event Simulation (DES) approach; (iii) converting the business evaluation in a financial dimension through a Value Stream Costing (VSC) step. The originality of this research work consists in coupling VSM, DES and VSC techniques in a unique and integrated method able to support decision-making by forecasting financial consequences of any improvement plan, at a network scale.

R. Miclo (✉) · M. Lauras · F. Fontanili · J. Lamothe
Industrial Engineering Department, Toulouse University—Mines Albi, Campus Jarlard
Route de Teillet 81000 Albi, France
e-mail: rmiclo@mines-albi.fr

M. Lauras
e-mail: lauras@mines-albi.fr

F. Fontanili
e-mail: fontanili@mines-albi.fr

J. Lamothe
e-mail: lamothe@mines-albi.fr

R. Miclo · P. Bornert · G. Revenu
AGILEA, Toulouse, France
e-mail: philippe.bornert@agilea.fr

G. Revenu
e-mail: guillaume.revenu@agilea.fr

K. Mertins et al. (eds.), *Enterprise Interoperability VI*,
Proceedings of the I-ESA Conferences 7, DOI: 10.1007/978-3-319-04948-9_16,

A real application case from the construction industry sector is developed at the end of the paper to illustrate the benefits of our proposition.

Keywords Collaborative network · Supply chain · Value stream mapping · Discrete-event simulation · Value stream costing

1 Introduction

Changing technical, economic and business environments force organizations to constantly search for new ways of managing their collaborative networks in order to improve their competitiveness and profitability. Supply chains are probably the most representative systems regarding this issue. They are facing growing pressures due to globalisation, harsh competition, fluctuating energy prices and volatile financial markets. Their strategic goals are set to reduce costs, improve customer services, increase reliability and efficiency of operations, and fast delivery of products to markets.

Because they generally involve high costs and important organizational changes, the potential benefits of such improvements have to be a priori assessed in order to support decision-making process. Through usual methods (Business Process Modeling, Discrete Event Simulation, Value Stream Mapping, etc.), practitioners could evaluate—with more or less facility—the expected business consequences of their improvement plans in terms of lead-time, setting-up, work-in-process, cycle time or quality [1]. However the only criterion that is really important and understandable for all stakeholders of a network is the financial one. For decision-makers, achieving to reduce cycle time by x% would probably be less motivating than reducing working capital by y%. However, very few research works seem to have tried developing a method able to estimate the expected financial result of a given improvement plan.

In accordance with the above, the objective of this paper is to present a complete methodology in order to assess the financial benefits of a planned improvement project. It is organized as follows. First, a review of the literature enables us not only to develop our research framework but also to formulate our research statement. Then we present the integrated methodology and its associated tools: Value Stream Mapping, Discrete Event Simulation and Value Stream Costing. Subsequently, we present the application of our proposal on a construction industry case study and discuss the obtained results. Finally, we draw conclusions, state some limitations and present perspectives for further research.

2 Literature Review and Problem Statement

2.1 Reminder About the Value Stream Mapping Fundamentals

Value Stream Mapping (VSM) is one of the many tools, working methods and concepts regarding improvement projects. VSM was developed by *Toyota* for identifying improvement opportunities to the overall flow from supplier to customer. This is a significant and very popular technique for supporting improvement step. Its principle consists in breaking down a process value stream along the different operations (at a company scale) or along the different installations (at a network scale) in order to analyze each activity that contributes to the overall performance. The "Value" term in VSM does not mean value in economic sense. VSM aims at visually mapping informational and material flows in order to identify waste and suggest changes [2]. The financial flows are not studied through a VSM approach. Since then, we can see that VSM is widespread in the industrial field. VSM is a powerful tool that is applied to practical cases in different sectors of activity within an organization or throughout an entire Supply Chain.

2.2 "Business" Assessment of the Value Chain

Computer simulation is today being used for various applications. The key point of computer simulation is the capability to generate a better understanding of a system by testing its behaviour during the time and regarding several variables. This is a technique that can be coupled with other tools or methods in order to better analyze or design a system [3]. The simulation variant used in this research is Discrete Event Simulation (DES) which deals with flow of parts in a system.

Because the VSM approach is quite static, authors have proposed some coupling approach between VSM and Discrete Event Simulation (DES). The principle of this extent works is the following: when changes are suggested from the current situation, it enables to test and analyze all of them, see in advance the future situation and anticipate future operational benefits. Numerous case studies developed this kind of approach at a local (manufacturing plant) or global (supply chain) level such as [1, 4]. Recent research works have underlined an important limitation of this approach regarding the need to develop the financial dimension in order to discuss cost-benefit analysis of proposed changes (see [5–7] for instance).

To summarize, we could notice that previous research works have already developed solutions to simulate VSM approach through DES technique in order to assess the business benefits of improvement plans, including in a collaborative environment. Nevertheless the question of the cost-analysis dimension through VSM/DES approach has not been treated yet whereas such an approach should be very useful.

2.3 "Financial" Assessment of the Value Chain

A literature review on the financial assessment of the value chain allows determining that some static VSM (i.e. without simulation) approaches have been linked with a benefit cost-analysis. Consequently, it becomes possible to analyse financial interests of improvement projects. Murat Tabanali and Ertay [8] for instance have implemented such an approach regarding a RFID investment project. Other authors [9, 10] went even further as they have proposed mathematical models to evaluate the impact on a value stream. These impacts are gathered following non-monetary or monetary benefits, which make up the "value" of a product [10]. However, [10] have demonstrated that such approaches allow evaluating the cost-benefit for only one scenario of improvement.

That is why authors [9, 10] have suggested developing, in future research-works, complementary simulation approaches to be able to estimate the cost-benefit values of a set of improvement scenarios. Other interests can be considering regarding the association between simulation and cost-benefit analysis. Bosch-Mauchand et al. [9] explain that the simulation allows facilitating the communication between the different stakeholders of the network because this kind of modeling is easily understandable by all the actors. Such an approach can also allow showing and measuring precisely the incidence of this or that improvement project all along the network and to detect potential local dysfunction.

In our research work, we have decided to develop an approach able to associate VSM, DES and a benefit-cost analysis. The question is now to determine the most appropriate methodology regarding our benefit-cost analysis needs.

Maskel and Baggaley [11] and Li et al. [12] addressed the issue. They studied the extent cost-benefits methodologies and evaluate them regarding the needs of continuous improvement project, particularly in Lean Manufacturing environment. They emphasized the inappropriate way to evaluate costs with the traditional standard costs method. They conclude that two main Management Accounting Systems (MAS) can be used for a cost-benefit analysis in such a context: Activity-Based Costing (ABC) and Value Stream Costing (VSC).

On one hand, ABC allocates direct and indirect costs to activities and their necessary resources. ABC needs historical data to identify cause and effects relationships with the aim of assigning costs. On the other hand, VSC targets the costs of each point on the value stream, without distinguishing direct or indirect costs. Besides, it is more adapted to the continuous improvement state of mind because it is more manageable and quicker to adjust. Li et al. [12] have notably demonstrated that VSC appears to be the best alternative to provide a bridge between operational views and financial views during improvement projects.

We propose now a short example considering the operations inside a single company (similar at a network scale) in order to illustrate VSC accounting method: a process including three operations, each one lasting one hour and costing 100€ per hour. Raw material is bought 500€. Between each operation, the product is stocked 14 days with a holding cost of 0.2 % each day (Fig. 1).

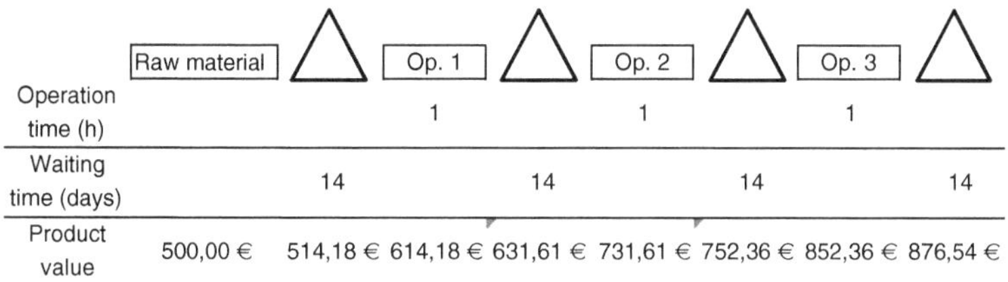

Fig. 1 Value stream costing example

With this example, we can see a 300€ added value and a 76.54€ non-added value (a little more than 25 % of the added value by the company). The formula used to evaluate waiting times is the following:

$$NAV(i) = WT(i) \times PV(i-1) \times HC \tag{1}$$

With: i: column number; NAV: Non Added-Value cost; WT: Waiting Time (in days); PV: Product Value (in €); HC: Holding cost (% per day) (Fig. 1).

2.4 Background Summary and Expected Contribution

Regarding the previous developments, it can be concluded that VSM, DES and cost-benefit analysis should be widespread in a common and integrated approach. In the literature "only" two themes on three that we have retained for our study, seem to have been coupled at the most. That is why elaborating a method gathering VSM coupled with DES and VSC should be an original contribution for both academics and practitioners. The following section develops such a methodology.

3 Proposition

Our proposal consists in integrating in a single step the complementary capabilities of VSM, DES and VSC in order to estimate the business and financial expected results of a set of potential improvements. This approach is developed at a network scale regarding interaction between several companies that are working on a common improvement project. Our proposition is articulated following the step described in the following figure.

Our method, as VSM suggested, focuses on a single flow (a product or a product family) that involves all along the network. If necessary, the method could be repeated for all other relevant flows in order to have an overall view of the expected benefits (physical, informational and financial).

The first step of our method consists in gathering operational data all along the network such as cycle times, available capacities, reliability, Overall Equipment

Effectiveness (OEE) and/or Takt time. Based on this information, it is possible to draw an "As-Is" VSM. According to this map, a first DES model can also be established following the method described in [13] or [3]. We must precise here that the main difference between the VSM model and the DES model relates to the "waiting times". Within a VSM approach, the waiting times are calculated through the takt-time whereas within a DES approach, the waiting times are evaluated through the flow management capabilities evolution (FIFO, critical ratio, etc.). In this research work, the DES approach has been retained. Consequently, contrary to the VSM approach, the waiting time will be an output data of our method.

Once the DES model and VSM map are complete, "validation of the simulation model" must be realized. According to the accuracy expected, simulation model will or not be approved. If not, the simulation model will first be changed. This step can be made several times. If it is not satisfactory enough, VSM can be called into question.

In the following, the financial issue of our approach will be added. First, a cost evaluation must be done to gather the input financial data. These costs for value-added time (during operations) must include energy consumption, maintenance costs, amortisation and charged salaries of operators. Between operations, there are waiting times that are non-value added times. In order to evaluate these costs, an accountant expert must evaluate the inventory holding costs (per year for instance).

Based on this, the next step consists in "Adding cost to the As-Is model following VSC method". To link costs aspect with this suggestion, financial input data must be inserted in the file where operational data is. The model will use this data in order to have the product value line in the financial data output. Within this step the cost-analysis of the As-Is model should be obtained. Analysis can now be done and origins of the main wastes in terms of cost can be detected. Value and non added value all along the network are given and a diagram can be realised to have a graphic view of this evolution.

According to the VSM methodology, experts will directly have suggestions to improve the model and eliminate waste. Improvement suggestions are not discussed in this paper, so we consider that experts have appropriate suggestions according to the current diagnosis on both business and financial dimensions.

As soon as there are enough improvements tried, the best solution(s) that is (are) fully satisfactory will be chosen. Unlike VSM coupled with DES, you will present to all network' stakeholders operational benefits but also financial benefits. Unlike before, precise amounts can be given and the simulation model with the new state (To-Be) can be presented. With this type of presentation, anybody, any decision-maker will understand improvement(s) presented. Most of all, anybody will have the same objectives and they will know why they will put in place such an improvement. Moreover with this method, it would also help to manage improvements. Indeed, goals to attempt are known and quantified. It would be easier to control project progress and compare at the end if these purposes are achieved, and if not analyse why (Fig. 2).

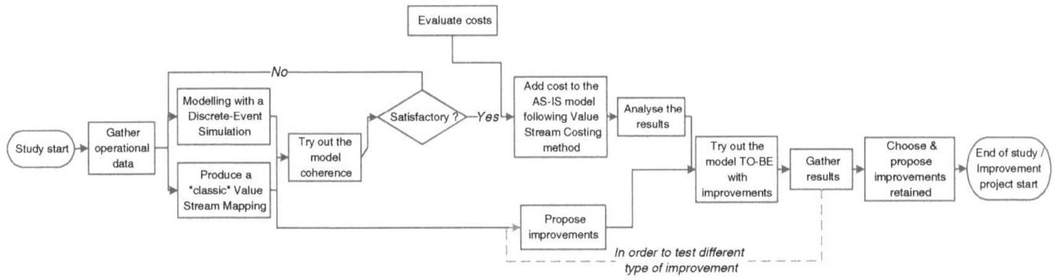

Fig. 2 Method step

4 Application Case

We illustrate our method through a real study, yet simplified in order to facilitate understanding. This case study deals with construction industry Supply Chain (SC). This SC is working on industrial improvements in order to maintain its competitiveness regarding other SC competitors. Several potential improvements are currently considered. The main goal of this SC is to shortly increase its cash flow. In order to attempt this purpose, the SC decision-makers would like to prioritise improvement actions to involve.

4.1 Input data

The SC is composed of 7 actors: "Raw Material Supplier—Rank 3 Supplier—Rank 2 Supplier—Rank 1 Supplier—Manufacturer—Assembler—Fitter". The product family considered here is produced with a 30 lot size (all along the SC). The raw material supplier is supplied every 4 weeks. One lot is delivered each week to the final consumer.

The assembler can use a subcontractor if some complementary capacities are needed. Consequently, two other data inputs are the transfer times: a lot needs 12.5 days to go to the subcontractor and 17.5 days to come back. All the operational data gathered to elaborate the VSM and the DES model are given in the Table 1.

4.2 DES Model

All actors are working with a complete lot (30 units). The subcontractor needs each time the same duration for transporting and operating. There is no lateness in transport, whether it is the supplier or the customer. So this is an analytical model and with this example we can work with 1 article in simulation representing 30 units. The DES model presented in the following figure represents the entire SC

Table 1 Operational input data

Operation	Raw material supplier	Rank 3 supplier	Rank 2 supplier	Rank 1 supplier	Manufacturer	Assembler	Fitter
Cycle time	0.5	2	60	6.6	6	90	7.8
Setting up time	10	15	120	30	0	0	0
Reliability	100 %	100 %	98 %	100 %	100 %	100 %	100 %
Opening time	p1 × 8	p1 × 8	p3 × 8 + SS	p2 × 8	p1 × 8	p3 × 8 + SS	p1 × 8
Operators number	1	2	2	2	5	0	2
OEE	100 %	52 %	80 %	80 %	100 %	100 %	100 %
Cost per hour	41.40€	50.00€	50.00€	51.00€	51.00€	7.3€ (per product)	41.00€

Table 2 Main results

Scenario (number)	Improvement type	In details	Lead time (days)	Product WIP	NVA costs	VA costs	Total (€)	Yearly statement (normal scenario) (€)	Profit (%)
0	As-Is	/	60.9	240.3	12	124	136	/	/
1	(1)	Lot-size = 15	59.7	235.35	12	129	141	−7063	–3.61
2	(2)	"Raw material supplier" setting-up 10 - > 5	60.9	240.3	12	124	136	153	0.08
3	(1) and (2)	Sc. 1 and Sc. 2 improvements	59.7	235.35	12	129	141	−6896	–3.52
4	(3)	Transport time divided by 2	47.0	185.4	8	124	132	5088	2.60
5	(4)	Raw material delivery each 2 weeks	53.3	210.3	11	124	135	1067	0.55
6	(3) and (4)	Sc.4 and Sc. 5 improvements	39.4	155.4	7	124	132	6125	3.13
7	(5)	"Rank supplier 2" cycle time 60 - > 45 min	60.9	240.3	11	112	122	19603	10.02
8	(2)	"Rank supplier 2" setting up 120 - > 60 min	60.9	240.3	11	123	134	2586	1.32
9	(4) and (5)	Sc. 5 and Sc. 7 improvements	53.3	210.3	10	112	121	20863	10.66
10	(1)	Lot-size = 60	75.1	296.4	14	122	136	−65	–0.03

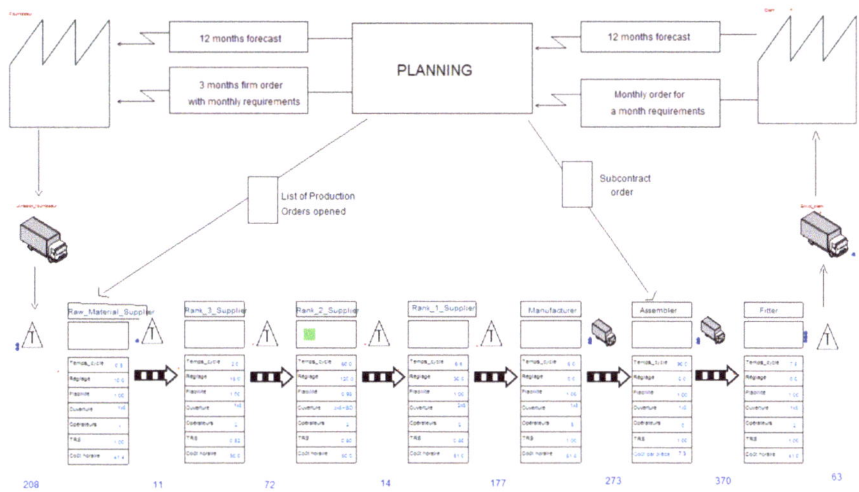

Fig. 3 DES model according VSM standards

accordingly to VSM standards of modeling. Witness® was used to implement the simulations. To be realistic, we have considered for each SC stakeholder only the available capacity dedicated to the studied flow. To do this, other family products were added to the simulation even if they have not been analysis in terms of improvement. In order to have representative results, we made the simulation on a 2 years period. The main output data is related to average waiting periods for each inventory, average lead-times and average work-in-process. Now, we have our VSM and model simulation, the gaps were satisfactory so we can implement the financial data (Fig. 3).

4.3 Cost Implementation and Results Analysis

As explained previously, an accountant expert has to determine the different costs to take into account. All the costs needed are related to cost per unit of time (operators, amortisation, maintenance and energy costs must be included), raw material cost and holding cost.

When costs are implemented in the model and the output spreadsheet is ready, As-Is (current state) simulation model can be calibrated. An output spreadsheet example is shown in Fig. 4. This includes several attributes:

- Added-value (AV) and cumulative AV all along the process for each actor, which is represented by the continuous line (scale on the left);
- Non Added-Value (NAV), firstly in minutes (value at the bottom), and converted into costs. The cumulative NAV cost line is represented by the dotted line (scale on the right) all along the SC.

Operation :	*Raw Material*	Inventory	Raw Material Supplier	Inv.	Rank 3 Supplier	Inv.	Rank 2 Supplier	Inv.	Rank 1 Supplier	Inv.	Manu-facturer	Inv.	Assembler	Inv.	Fitter	Inv.
Operation AV Cost	43,000		0,575		2,083		54,422		6,460		5,100		7,300		5,330	
Cumulative AV Cost	43,00	43,00	43,58	43,58	45,66	45,66	100,08	100,08	106,54	106,54	111,64	111,64	118,94	118,94	124,27	124,27
NAV (min) / takt time		0,0	14,5	0,0	58,0	0,0	1740,0	0,0	191,4	0,0	174,0	0,0	90,0	0,0	226,2	0,0
NAV cost		0,000	0,000	0,000	0,000	0,000	0,014	0,000	0,003	0,000	0,003	0,000	0,002	0,000	0,005	0,000
Cumulative NAV cost	0,00	0,00	0,00	0,00	0,00	0,00	0,01	0,01	0,02	0,02	0,02	0,02	0,02	0,02	0,03	0,03
Operation total value	43,00	0,00	0,00	0,00	2,08	0,00	54,44	0,00	6,46	0,00	5,10	0,00	7,30	0,00	5,33	0,00
Product value	**43,00**	**43,00**	**43,58**	**43,58**	**45,66**	**45,66**	**100,10**	**100,10**	**106,56**	**106,56**	**111,66**	**111,66**	**118,96**	**118,96**	**124,30**	**124,30**

Fig. 4 Results data output

- Then operation value (AV and NAV sum for each operation or inventory period) and cumulative operation value, which is actually the real product value, are the last two lines on the spreadsheet.

4.4 Experiences Plan and Discussion

As soon as the VSM is done, a business expert should have improvements to test. It is useful to try out these improvements at this stage of the method. After testing them, with results and with cost-benefit analysis, we can try out other improvements and especially combine them. Each time a scenario is tried, the main results must be gathered and a list of new improvement ideas may be created. It was done in this case study and the main results are in Table 2. In order to understand it, here is the "improvement type" legend: (1) Lot size change; (2) Single Minute Exchange of Die; (3) Transport time reduction; (4) More raw material delivery; (5) Cycle time reduction. As shown in the following table, we made the comparative-analysis regarding the following criteria: lead-time (in days), Product WIP, Non Added-Value and Added-Value Costs to have the total value (in euros). In the As-Is situation, we have a 60 days lead-time with 240 products (8 lots) in process and a total cost of 136€ (6€ loss from the selling price) with 12€ NAV costs (almost 9 % total value). Even though we experimented 10 different scenarios, we have chosen to focus here only on the most representative ones. The scenarios (1) and (3) consisted in reducing the lot size reduction (15 against 30 before) in order to reduce the NAV costs. But the simulation shows that this approach especially increases the AV costs (so the total value) by 5€, contrary to what we can thought.

This could be explained because the main AV operation is made by the "Rank 2 supplier" (more than 50€). And there, setting-up time is really expensive, so a smaller lot size policy costs more (7000€ each year). As regards lead-time, we only earn one day.

The scenarios (4) and (6) consisted in studying the impact of transportation time reduction (division by 2) regarding the subcontractor as improvements' possibilities were detected on this part of the SC. The simulation shows that such an improvement plan allows earning 12 days lead-time and 55 WIP less. As expected, we only reduce NAV costs by a third. This solution could be expensive and would earn 5000€ yearly.

At this stage, we mostly tried out improvements in order to reduce NAV costs. However we saw thanks to the method that majority costs results in "Rank 2 Supplier". That is why we tried out cycle time and setting up time reduction. In scenario 7, we reduced by 15 min cycle time. As expected, we earn 10 % from the As-Is scenario with a 122€ total value and about 19600€ profit each year. We can remark that lead-time and WIP are equivalent to the As-Is situation.

Our last scenario consisted in increasing the lot-size (30–60 articles for a lot) because with the first scenario, we saw the setting-up time impact. However, with this solution, lead-time and WIP increased a lot and are inefficient as regards costs.

We dealt with sample scenarios, but if we relate improvement benefits with improvement costs, we could try out other scenarios and find the best solution (such as having still more delivery from the supplier or to the customer or evaluate if it will not be interesting to realise the subcontractor operation).

5 Conclusion and Perspectives

The purpose of our research work consisted in proposing an integrated method able to assess the benefits of a potential improvement plans regarding both business and financial dimensions. This tradeoff is reached by using in a common approach three complementary techniques that are Value Stream Mapping, Discrete Event Simulation and Value Stream Costing. The added value of our proposal is positioned regarding extent literature and practitioners' needs. An application case associated to construction industry Supply Chain is shortly develop to illustrate the forces of our methodology. Nevertheless, several perspectives can be addressed regarding this research work. Notably, it would be interesting to manage several flows (and not only one) at a same time and to better support the decision-making by considering quantitatively the uncertainty on data and on simulation model.

References

1. Lian, Y.-H., & Van Landeghem, H. (2002). An application of simulation and value stream mapping in lean manufacturing. In *Proceedings of the14th European Simulation Symposium*.
2. Rother, M., & Shook, J. (1999). *Learning to see: Value stream mapping to add value and eliminate Muda*. Ed. Lean Enterprise Institute.
3. Fontanili, F., Lauras, M., & Lamothe, J. (2013). Pour une ingénierie d'entreprise plus performante par couplage entre modélisation de processus et simulation. In *10ème Congrès International de Génie Industriel CIGI 2013, 12–14 juin*.
4. Lian, Y-H., & Van Landeghem, H. (2007). Analysing the effects of Lean manufacturing using a value stream mapping-based simulation generator. *International Journal of Production Research, 45*(13), 3037–3058.
5. Abdulmalek, F. A., & Rajgopal, J. (2006). Analysing the benefits of lean manufacturing and value stream mapping via simulation: A process sector case study. *International Journal of Production Economics, 107*, 223–236.
6. Singh, B., Garg, S. K., & Sharma, S. K. (2010). Value Stream mapping: literature review and implications for Indian industry. *International Journal of Advanced Manufacturing Technology, 53*, 799–809.
7. Gjeldum, N., Veža, I., & Bilić, B. (2011). Simulation of production process reorganized with value stream mapping. *Technical Gazette, 18*(3), 341–347.
8. Murat Tabanli, R., & Ertay, T. (2012). Value stream mapping and benefit-cost analysis application for value visibility of a pilot project on RFID investment integrated to a manual production control system—a case study. *International Journal of Advanced Manufacturing Technology, 66*, 987–1002.
9. Bosch-Mauchand, M., Siadat, A., Perry, N., & Bernard, A. (2010). VCS: Value chains simulator, a tool for value analysis of manufacturing enterprise processes (a value-based decision support tool). *Journal of Intelligent Manufacturing, 23*, 1389–1402.
10. Sihn, W., & Pfeffer, M. (2013). A method for a comprehensive value stream evaluation. *Manufacturing Technology, 62*, 427–430.
11. Maskel, B., & Baggaley, B. (2003). Practical lean accounting: A proven system for measuring and managing the lean enterprise. *Library of Congrees Cataloging-in-Publication Data*. ISBN 1-56327-243-1.
12. Li, X., Sawhney, R., John, E., & Ramasamy, K. (2012). A comparative analysis of management accounting systems' impact on lean implementation. *International Journal of Technology Management, 57*(1/2/3), 33–48.
13. Solding, P., & Gullander, P (2009). Concepts for simulation based value stream mapping. In *Proceedings of the 2009 Winter Simulation Conference (wsc 2009), 1–4*, 2172–2178.

Interoperability Improvement in Inter-Enterprises Collaboration: A Software Engineering Approach

Mamadou Samba Camara, Rémy Dupas, Yves Ducq and Bacary Mané

Abstract The present research work aims at developing an approach to reach inter-enterprise interoperability and to test its achievement using practices from the software engineering process. Four fundamental activities are identified in the software process: software specification, software development, software validation and software evolution [1]. In this work, interoperability requirements are specified by representing interoperability problems directly on business process models. For the validation activity, an interoperability testing sub-process is defined based on this new form of interoperability requirements specification. It is also demonstrated that the improvement proposed in software specification activity will have positive impact on the software development activity.

Keywords Enterprise interoperability · Software specification · Software validation · Service oriented architectures · Business process modeling

M. S. Camara (✉) · B. Mané
LIRT/ESP, Cheikh Anta Diop University, Génie Informatique ESP-UCAD,
BP: 5085 dakar-fann, Dakar, Senegal
e-mail: mamadou.camara@ucad.edu.sn

B. Mané
e-mail: manebakys@gmail.com

R. Dupas · Y. Ducq
University of Bordeaux, IMS, UMR 5218, 33400 Talence, France
e-mail: remy.dupas@ims-bordeaux.fr

Y. Ducq
e-mail: yves.ducq@ims-bordeaux.fr

K. Mertins et al. (eds.), *Enterprise Interoperability VI*,
Proceedings of the I-ESA Conferences 7, DOI: 10.1007/978-3-319-04948-9_17,

1 Introduction

Interoperability is defined by the ALCTS[1] [2] as the ability of two or more systems or components to exchange information and use the exchanged information without special effort by either system. Improving interoperability depends largely on the implementation of the software engineering process. Four fundamental activities are identified in software process: software specification, software development, software validation and software evolution [1]. A review of the literature was conducted to analyze how these activities were carried out in the interoperability domain. Research works conducted for the development of Collaborative Information System (CIS), which aims at supporting Information Systems (IS) interoperability, have led to two groups of approaches: Model-Driven Interoperability (MDI) approaches [3, 4] and Business Process Lifecycle (BPL) based approaches [5]. However, requirements specification proposed in these approaches does not provide sufficient information to describe interoperability problems and then facilitate the software development activity. Testing techniques have been applied on service-based systems [6] without addressing the issue of interoperability directly. The next section presents the literature of interoperability related to software process. The two following sections are dedicated to the research design and the application respectively.

2 Literature Review Interoperability

2.1 Interoperability Requirements

This section introduces different approaches used to represent interoperability requirements, in the literature. The first approach to define interoperability needs is the maturity models. The maturity models [7–9] represent the interoperability levels using needs expressed in natural language. As explained by Mallek et al. [10], maturity models introduce repetition, ambiguity, imprecision and incoherence in the interoperability needs definition. The second approach consists of formally representing interoperability requirement so as to make it possible the formal verification of the latter (Table 1). The objective of verification is to demonstrate that a set of selected interoperability requirements is satisfied. In an inter-enterprise collaboration context, with several partners related by many collaboration processes, the implementation of this approach may experiment scalability problems. Indeed, the target element for the verification process is the collaboration process in Mallek et al. [10], and is a single entity (people, organization units and material resources) or a couple of entities in Cornu et al. [11].

[1] The Association for Library Collections and Technical Services.

Table 1 Formal representation of interoperability

Technique	Models	Tranformations
Conceptual graphs technique from Mallek et al. [10]	Process models (BPMN), support graphs, fact graphs (conceptual) and constraints graphs	Three ATL (Atlas transformation language) transformations
Model checker based temporal verification from Mallek, Daclin et al. [10]	Process models (BPMN), behavioral model (networks of timed automata), properties (timed computation) tree logic	One ATL transformation and a reformulation the temporal requirements
Mathematical evaluation of interoperability from Cornu et al. [11]	Fourteen typical questionnaires and four types of interoperability scores (equations)	Select questionaire, collect answers, interprete answers and compute score of questionaire

Let us consider the complexity of the verification process's application on the target element (Table 1.): the number of models manipulated, the number of actions (i.e. transform, reformulate, collect or interpret) to process. The main cause of the scalability problems is the fact that each of these actions is to be processed on every target element.

The third approach represents interoperability requirements as problems. Two categories of requirements definition related to interoperability problems can be identified in the literature. In the first one, requirements are specified using models of collaboration processes. These models are used as inputs in MDI or BPL based approaches to define the interoperability solutions [3–5, 12, 13]—Sect. 2.2. The main drawback is that they focus more on functional requirements described partly by the "to-be" processes models which show what the system should do. Therefore, the requirements are disconnected from existing systems and also to the interoperability problems to solve. In the second category of requirements representation, the concept of problem space is used. The problem space is composed of the barrier and the concern dimensions of the INTEROP framework [14, 15]. Chen et al. [14] defined a interoperability matrix combining the three interoperability dimensions (barrier, concern and approach) to allow categorizing knowledge and solutions relating to enterprise interoperability according to the ways of removing interoperability barriers. The main drawback of the interoperability matrix is the fact that the way interoperability problems and solutions are represented is disconnected from enterprise models. In other words, the matrix does not enable identifying a target element in the enterprise (process, activity or organizational unit) where the interoperability problems take place or where the solutions must be deployed.

2.2 Collaborative Information System Architecture

A CIS aims at supporting information system interoperability, that is to say, to satisfy "IS-interoperability" requirements (data conversion, application sharing and process management) [16]. Several research works were conducted in order to find logical and technical solutions for the CIS. Approaches proposed in these works can be categorized in two groups: MDI approaches and BPL based approaches. Recently, MDI for Enterprise Interoperability (EI) has received significant attention in academia and has resulted in various interoperability frameworks such as the IDEAS Interoperability Framework [14] and the ATHENA Interoperability Framework (AIF) [17]. The MDI proposes MDA—Model-Driven Architecture [18]- based model transformations to achieve EI [3, 4, 12, 16, 19]. The second category of approaches aimed at transferring the concept of BPL management to cross-organizational environments [5, 13]. It can be noticed that, in all approaches for CIS development, the proposed platforms are based on Service Oriented Architectures (SOA) [20]. Indeed, interoperability equates to a 'Loosely coupled' degree of coupling [14], and one promising benefit of SOA is to facilitate the construction of flexible and loosely coupled business applications [20].

2.3 SOA Testing

Testing approaches and techniques are developed for a variety of system types such as, traditional monolithic systems, distributed systems, component-based systems (CBS), web applications and service-based systems [6]. Given our problem, we will focus primarily on testing approaches and techniques for service-based systems. For the testing of SOA applications, Wieczorek and Stefanescu [6] identified four distinct testing layers: Unit testing, Service testing, Integration testing and System testing. Lee [21], designed a business-centric SOA test framework based on the execution of business process test scenarios. The business process level is chosen for the system testing because testing of the business layer with which users keep in contact includes indirect testing of the other SOA system layers (service layer and computing resource layer) [21]. Similarly, Vieira et al. [22], advocated for the realization of system testing at process level. The proposed model-based approach is demonstrated and evaluated based on use cases developed for testing a graphical user interface [9].

In our opinion, although interoperability implementation is generally based on SOA, the testing approaches proposed for service-based systems are not suitable for testing inter-enterprise interoperability achievement. Indeed, these approaches [6, 21, 22] do not reference interoperability problems, which makes it impossible to verify the elimination of the latter.

3 Research Design and Hypothesis

The present research work aims at developing an approach to reach inter-enterprise interoperability and to test its achievement using practices from the software engineering process. Four fundamental activities are identified in software process: software specification, software development, software validation and software evolution [1]. For each of the first three activities, the approach proposes to define a sub-goal and determine how to achieve it:

1. **Software specification**. Improve the clarity of interoperability requirement specification.
2. **Software development**. Ease the logical and technical architecture definition by using these interoperability requirements.
3. **Software validation**. Define a method that enables testing the achievement of interoperability improvement.

3.1 Improving the Interoperability Requirement Specification

Software system requirements are often classified as functional and nonfunctional requirements [1]. The interoperability requirement representation proposed in this work consists in representing directly interoperability problems in business process models, mainly in "As-is" ones. This interoperability problems representation will be part of non-functional requirements. Indeed, according to Sommerville [1], non-functional requirements arise through user needs, because of budget constraints, organizational policies or the need for interoperability with other software or hardware systems. The adopted representation is based on a principle that consists of distinguishing between business activities and Non-Value-Added (NVA) activities, mainly inspired by the work done in Camara et al. [23]. Business activities create value in a business process. The NVA activities are defined as the components of business processes that represent efforts between partners to achieve interoperability in information exchange. Interoperability problems are depicted in *Business Process Model and Notation* (BPMN) [24] business process models. The stereotypes [25], generally used for the UML language, will be used to differentiate NVA to business activities in the BPMN process models. BPMN is more suitable for modeling collaborative processes because it helps situate the boundaries of the collaborating companies using pools. The proposed representation technique enables to overcome limitation presented in the Sect. 2.1, since the requirements will relate interoperability problems to target elements (people, organization units and material resources) clearly identifiable in the collaborating enterprises.

3.2 Ease Architecture Definition

Software development is the activity where the software is designed and programmed. The requirements specification results are inputs of the design and implementation processes [1]. In order to develop solution for interoperability problems, the interoperability matrix utilizes the concepts of solution space [14, 15]. The solution space is composed of the three dimensions of the INTEROP framework. The cross of an interoperability barrier, an interoperability concern and an interoperability approach includes the set of solutions to breakdown a same interoperability barrier for a same concern and using a same approach. In order to determine correctly the solution (third dimension), there must be sufficient information to describe interoperability problems (two first dimensions). The requirements representation in process models contains the following information: tasks where interoperability problems arise and the resources (human and non-human) involved in the interoperability problems. This set of information will facilitate the design process and then improve the software development activity.

3.3 Adapt Validation Activity

The test process can be divided into a set of test sub-processes defined to perform a specific test level (e.g. system testing, acceptance testing) or test type (e.g. usability testing, performance testing) within the context of an overall test process for a test project [26]. A test type is a group of testing activities that are focused on specific quality characteristics [26]. In the validation activity of software process, the improvements will consist on considering interoperability as a quality characteristic and developing the interoperability test type. On the basis of the recommendations given in the literature of service-based systems testing, the interoperability testing sub-process will be executed at system testing level using business process models (Sect. 2.3). The main input of the interoperability testing sub-process (Fig. 1) is the "As-is" business process models which represent the interoperability requirements specification (Sect. 3.1). The interoperability testing sub-process will consist in executing each process in order to verify if all the NVA activities it contains in its "As-is" version are effectively eliminated by the implementation of the CIS.

The execution of a business process may reveal the presence of NVA activities. In this situation, the "As-is" business process model gives sufficient information about the interoperability problems related to the concerned NVA activities. The information will be used to fix the interoperability problems.

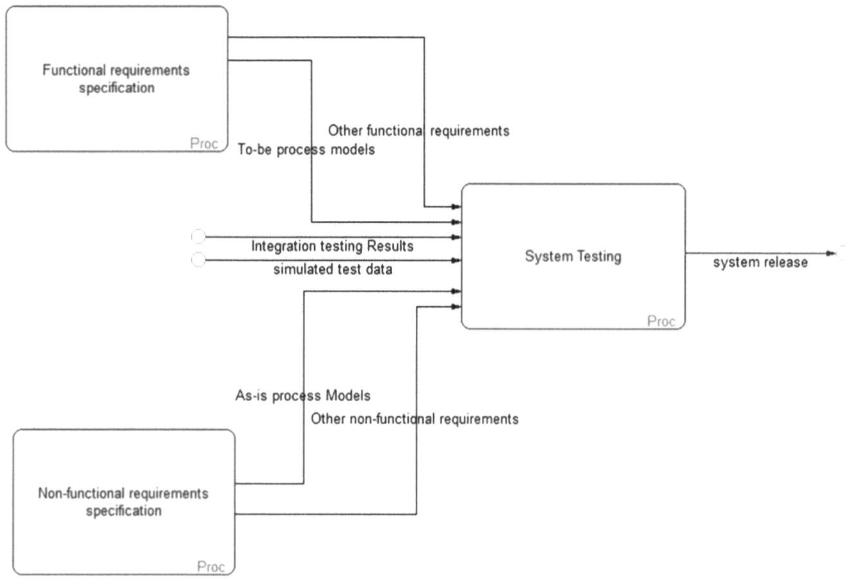

Fig. 1 The interoperability testing sub-process

4 Application Example

An inter-organizational collaboration is used as an illustrative example to demonstrate the applicability of the approach. The collaboration involves two institutions of higher education which offer joint programs: ESP (French acronym for Polytechnic Higher School of Dakar) and ESMT (French acronym for Higher *Multinational School* Of Telecommunications of Dakar). They have undertaken an interoperability project to improve the quality of their collaboration. As recommended by Vieira et al. [22], UML Use Cases and Activity diagrams were used to respectively describe which functionalities should be tested and how to test them.

4.1 Interoperability Requirement Specification

The functionalities of the CIS of the illustrative example are grouped in two modules: deliberation and scheduling. The process "Validate WAMs-Weighted Average Mark" from the deliberation module is chosen to illustrate the interoperability testing sub-process in details. In the "As-is" situation (Fig. 3), the goal of the "Validate WAMs" process was to register the WAMs in a web application called Gepi. The presence of NVA activities is due to the fact that the Gepi Web Application was only deployed for ESMT and only the secretary has access to it. The "To-be" model (Fig. 2) represents the "Validate WAMs" process as it is expected to be executed in the CIS to be implemented.

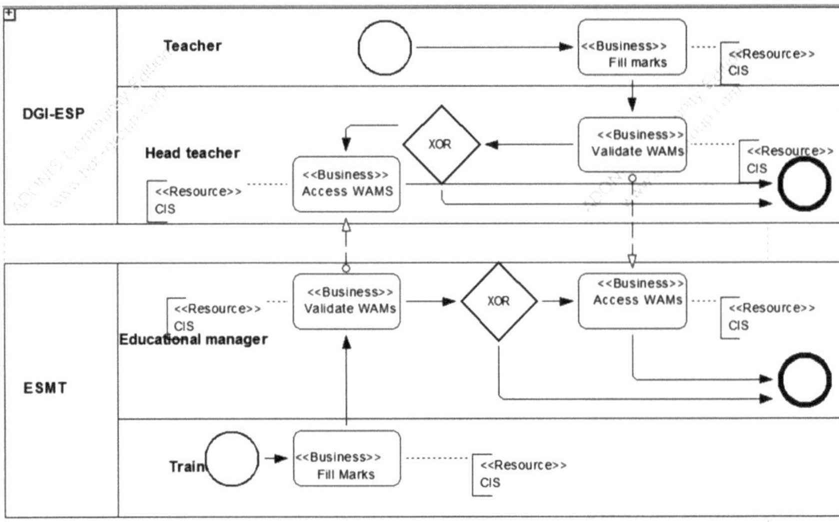

Fig. 2 "To-be" model of the "Validate WAMs" process

4.2 Software Architecture Definition and Implementation

Web services are an industry effort to provide platform-independent SOA using interoperable interface descriptions, protocols, and data communication [27]. In the software development activity, the decision was made to use web services architecture to implement the CIS. The main reason is that web services architecture can support the business activities while removing NVA activities identified in all the "As-is" business process models obtained in the requirements specification. Furthermore, web services will enable the ESP and ESMT to collaborate in an autonomous way, which is important because these institutions are independent. Indeed, interoperability is related to the preservation of autonomy during collaboration, which means that partners can work together (e.g., exchange services) while continuing to follow their own logic of operation [10]. The software development activity was realized using .NET Web Services implemented using Microsoft.NET platform [28]. The Microsoft.NET platform uses technologies involved in the Web service architecture- XML, SOAP, WSDL [29]. Initially, all features of the deliberation module were developed as part of an iterative development process (Fig. 3).

4.3 Interoperability Testing Sub-process Application

The system testing execution covers the four business processes of the deliberation module: "register students", "validate WAMs", "validate absences" and

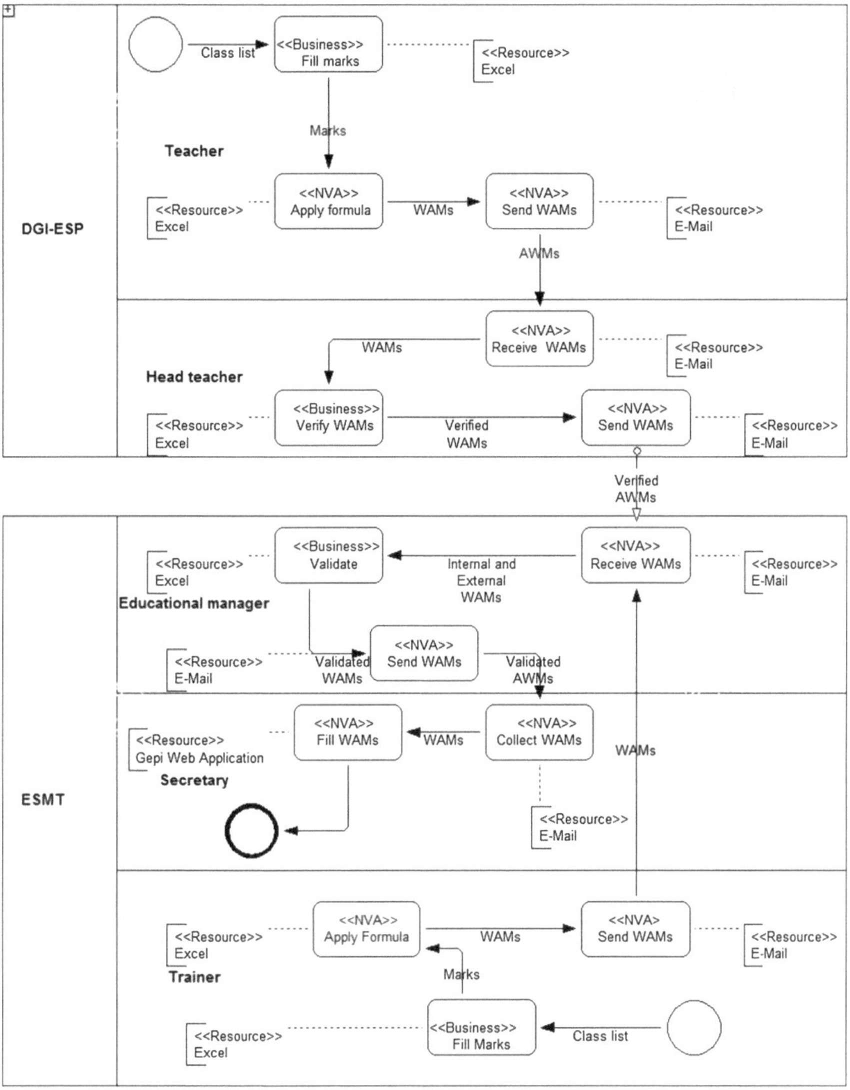

Fig. 3 "As-is" model of the "Validate WAMs" process

"deliberate". The interoperability testing sub-process is executed in this system testing phase after the testing of the functional requirements. The functional requirements testing allowed verifying that the implemented CIS supports the business process as described in the "To-be" models. The interoperability testing conducted reveals that all the NVA activities identified in the "As-is" process models were removed in the implementation of the CIS. This means that the interoperability problems were fully eliminated in this application case.

5 Conclusion

This work was aimed, firstly, to propose a requirements specification that better takes into account the representation of interoperability problems. It has been then demonstrated that this representation of interoperability can positively impact the software development and the software validation activities. Interoperability problems are depicted directly in business process models using the concept of NVA activity. A perspective of this work will consist in improving the interoperability testing sub-process using interoperability measures. The objective will be to prove that the implementation of the CIS has improved the interoperability up to the desired level. The development of this perspective can be based on the interoperability measures from Camara et al. [23] defined as three process Performance Indicators (PIs): the average elapsed time, average cost and percentage of failure.

References

1. Sommerville, I. (2011). *Software engineering* (9th ed.). Addison-Wesley, Boston, USA.
2. ALCTS. (2010). *Description and access task force on metadata* [cited 2011 19 September]. Retrieved from http://www.libraries.psu.edu/tas/jca/ccda/tf-meta3.html
3. Chungoora, N., et al. (2013). A model-driven ontology approach for manufacturing system interoperability and knowledge sharing. *Computers in Industry, 64*, 392–401.
4. Khadka, R., et al. (2013). Model-driven approach to enterprise interoperability at the technical service level. *Computers in Industry*, *64*(8), 951–965.
5. Franco, R. D., et al. (2012). ColNet platform: Resilient collaborative networks through interoperability. In R. Poler, et al. (Eds.), *Enterprise interoperability V: Shaping enterprise*. London: Springer.
6. Wieczorek, S., & Stefanescu, A. (2010). Improving Testing of Enterprise Systems by Model-based Testing on Graphical User Interfaces. In *17th IEEE International Conference and Workshops on Engineering of Computer-Based Systems, IEEE.*
7. Kingston, G., Fewell, S., & Richer, W. (2005). An organizational interoperability agility model. In *The 10th Command and Control Research and Technology Symposium*. Virginia: Command and Control Research Program.
8. Morris, E., et al. (2004). *System of systems interoperability (SOSI): Final report.* The Software Engineering Institute, Carnegie Mellon University.
9. Tolk, A. ,& Muguira, J. (2003). The levels of conceptual interoperability model. In *The 2003 Fall Simulation Interoperability Workshop*. Orlando: IEEE CS Press.
10. Mallek, S., Daclin, N., & Chapurlat, V. (2012). The application of interoperability requirement specification and verification to collaborative processes in industry. *Computers in Industry, 63*(7), 643–658.
11. Cornu, C., et al. (2012). Customizable interoperability assessment methodology to support technical processes deployment in large companies. *Annual Reviews in Control, 36*, 300–308.
12. Lemrabet, Y., et al. (2012). Proposition of business process modelling in model driven interoperability approach at CIM and PIM levels. In R. Poler, et al. (Eds.), *Enterprise interoperability V: Shaping enterprise*. London: Springer.

13. Walter, P., Werth, D., & Loos, P. (2007). Managing the lifecycle of cross-organizational collaborative business processes. In R. J. Gonçalves, et al. (Eds.), *Enterprise interoperability. New challenges and approaches*. London: Springer.
14. Chen, D., Doumeingts, G., & Vernadat, F. (2008). Architectures for enterprise integration and interoperability: Past, present and future. *Computers in Industry, 59*(7), 647–659.
15. Ducq, Y., Chen, D., & Doumeingts, G. (2012). A contribution of system theory to sustainable enterprise interoperability science base. *Computers in Industry, 63*, 844–857.
16. Touzi, J., et al. (2007). Interoperability through model-based generation: The case of the collaborative information system (CIS). In R. J. Gonçalves, et al. (Eds.), *Enterprise interoperability. New challenges and approaches*. London: Springer.
17. Berre, A.-J., et al. (2007). The ATHENA interoperability framework. In R.J. Gonçalves, et al. (Eds.), *Enterprise interoperability. New challenges and approaches* (pp. 569–580). London: Springer.
18. OMG. (2003). *MDA guide V1.0.1*. Object Management Group.
19. Jardim-Goncalves, R., Grilo, A., & Steiger-Garcao, A. (2006). *Challenging the interoperability between computers in industry with MDA and SOA. Computers in Industry, 57*(8–9), 679–689.
20. Pessoa, R. M., et al. (2008). Enterprise interoperability with SOA: A survey of service composition approaches. In *Enterprise Distributed Object Computing Conference Workshops*. IEEE.
21. Lee, Y. (2009). Event-driven SOA test framework based on BPA-simulation. In *First International Conference on 'Networked Digital Technologie*. IEEE.
22. Vieira, M., et al. (2006). Automation of GUI testing using a model-driven approach. In *The 2006 International Workshop on Automation of Software Test*. ACM.
23. Camara, M., Ducq, Y., & Dupas, R. (2013). A methodology for the evaluation of interoperability improvements in inter-enterprises collaboration based on causal performance measurement models. *International Journal of Computer Integrated Manufacturing*.
24. OMG. (2011). *Business process model and notation (BPMN). Version 2.0*. Needham, MA, United States.
25. Frankel, D. S. (2003). *model driven architecture: applying MDA to enterprise. Computing*. Wiley.
26. ISO/IEC/IEEE. (2013). *29119-2-Software and systems engineering. Software testing*. IEEE.
27. Chappell, D. (2009). *Enterprise service bus*. O'Reilly.
28. Ferrara, A., & MacDonald, M (2002). *Programming.NET web services*. O'Reilly Media, Inc.
29. W3C (2004). *Web services architecture*. W3C.

Software Cost and Duration Estimation Based on Distributed Project Data: A General Framework

Safae Laqrichi, François Marmier and Didier Gourc

Abstract Effort estimation is one of the most challenging tasks in the process of software project management. Enhancing the accuracy of effort estimation remains a serious problem for software professionals. Accurate estimation is difficult to achieve. The main difficulty is to collect distributed knowledge as data and information are often dispersed over different services, departments or organisations. Other main difficulty is to propose a model representative enough of this multi-partner behaviour. The objective of this study is to propose a general framework of the estimation starting from the analysis of the available projects database, the choice and establishment of estimation model, up to the use of this model to make estimation for new projects. In this paper, a comparative study between regression models and neural network models is performed. The proposed study is applied on a dataset of an automotive company.

Keywords Neural network · Regression · Duration estimation · Cost estimation · Comparison

1 Introduction

Effort estimation is an important activity in software project management. Estimation in software projects consists in predicting likely amount of effort, time and cost that are required to build a software system. It is used in the whole life cycle

S. Laqrichi (✉) · F. Marmier · D. Gourc
Mines Albi, University of Toulouse, Route de Teillet,
Campus Jarlard 81013 Albi Cedex 09, France
e-mail: safae.laqrichi@mines-albi.fr

F. Marmier
e-mail: francois.marmier@mines-albi.fr

D. Gourc
e-mail: didier.gourc@mines-albi.fr

K. Mertins et al. (eds.), *Enterprise Interoperability VI*,
Proceedings of the I-ESA Conferences 7, DOI: 10.1007/978-3-319-04948-9_18,

of the software project from the bidding until the maintenance of the software. Under estimating the effort and the cost required to develop a project results in budget overruns, while over estimating can lead to miss of biddings. Accurate estimation is then very important for companies' benefit and success.

Effort estimation is a complex activity that requires a high level of interoperability in both steps of model elaboration and estimation for new projects. Indeed, the modeling step needs various data about previous projects that are dispersed over different services and entities (financial data, technical project data…). The estimation step requires various distributed data about new projects to estimate as well. Thus, entities have to communicate and collaborate to provide required information.

Various effort and duration estimation approaches were been developed. Traditional and well-known ones include expert judgment [1], Delphi, COCOMO [2] and Putnam's SLIM [3]. However, estimation methods did not produce sufficiently accurate results, this is one reason why approximately 44 % of software projects according to the Standish Group International fail on meeting the commitment on quality, time and cost.

A set of factors can influence the estimation accuracy and lead to estimations far from the reality. It includes, among others: lack of information about completed previous projects; use of new technologies; lack of experience with similar projects; choice of estimation approach and more [4].

The challenge of improving estimates accuracy has led to the development of several new methods and techniques for effort, duration and cost estimation. These methods are based on artificial intelligence as NN (Neural Network) models [5].

Our work first focuses in formalizing the general estimation framework. This framework enables to compare different models. In this paper, we study the case of regression and NN estimation models applied on a big and diversified case study. This case study does not contain size project that is usually considered to be an important cost driver in estimation model establishment.

The present paper is organized into three sections: the first section presents literature review on effort estimation process, regression and NN models and their comparisons work carried by researchers. In the second section, a general framework for estimation is proposed. Finally, in the third section, the proposed framework is applied to a case study from the automotive industry.

2 Literature Review

2.1 Estimation in Software Projects

The estimation process is based on two principal activities that are: (i) project size measure and (ii) effort, cost and duration estimation.

Project size (i) expresses the size of the software that is derived from the quantification of functional requirements specified by users [ISO/IEC14143].

Project size can be calculated by several methods and techniques of functional measurement such as FPA (Function Point Analysis). It can be expressed in different units such as function points (FP) and source lines of code (SLOC).

The development effort (ii) is a function of the project size; it is expressed in man-hours, man-days or man-months. Duration estimation is either a function of project size or can be derived from the development effort. Effort and duration estimation, once estimated enable to calculate the project cost and staffing.

Various effort estimation methods can be used in the estimation process. They can be grouped in three main categories: (1) experience based methods, which is based on the expert intuition and experience drawn from previous executed project, such as expert judgment and analogy, (2) algorithmic model based methods, which are mainly based on equations expressing the effort as a function of discriminant parameters influencing the effort called effort drivers. Parametric models are established using historical data from complete projects, some of commonly used models are regression based models and Bayesian analysis based models [6]. (3) Non-algorithmic model based methods, which model the relationship between the estimated variable and cost drivers using artificial intelligence techniques like NN and fuzzy logic. The relationship is not assumed to be well known or modelizable to specific shapes or equations [7].

Regression is a widely used modeling technique and NN is a recent and evolutionary modeling technique. In this study, our attention was drawn to these two modeling techniques for the estimation activity of the global process (ii) because they seem to provide good estimates.

2.2 Regression Models Versus Neural Network Models

Regression models are still the most popular models in literature; they include COCOMO [8] and Putnam [9]. Regression aims to model the relationship between inputs and outputs. In software estimation, the inputs are the discriminant parameters influencing the estimated variable called effort or cost drivers, the main cost driver is the software size that is usually expressed in the Source Lines of Code. The output is the estimated variable that can be effort or duration or cost.

There are various types of regression that have been used in effort estimation models namely linear or multi linear regression [10], non-linear regression [8], and ordinal regression [11].

NN is a massively parallel adaptive network of simple nonlinear computing elements called Neurons, which are intended to abstract and model some of the functionality of the human nervous system in an attempt to partially capture some of its computational strengths [12].

NN is used in effort estimation due to its ability to learn from previous data. It is also able to model complex relationships between the dependent (effort or duration or cost) and independent variables (cost drivers). In addition, it has the ability to learn from the training data set thus enabling it to produce acceptable results for unseen data [4]. But NN has one short coming that its estimation reason or relation between inputs and outputs cannot be justified. Different NN based model are proposed to predict and estimate effort and duration as COCOMO based NN model [13, 14] and radial basis function NN model [15].

Several research works have compared NN models with regression models. Finnie [16] compared regression model with two AI based estimation models that are case based reasoning and NN for software development effort. Authors found that AI based models perform better for complex software projects and outliers in training dataset than regression model. However, they gave no justification or explanation for the obtained results.

Heiat [17] experimented FFNN (Feed-Forward Neural Network) with function point and RBNN (Radial Basis Neural Network) with SLOC for a dataset of 67 projects. Author concluded that NN approach is competitive with regression for same cases and significantly more accurate for others. This study presents some limitations: the size of data sets is small, the data sets used varied only in size and complexity varied in terms of language platform.

3 Framework of Estimation of Effort, Duration and Cost

The study presented in this paper relies on the framework shown below (Fig. 1).

This methodology consists of four major steps:

(a) Preparation of data: based on raw database containing information collected from previous achieved software projects
(b) Establishment of estimation models: consists in the establishment of estimation models to compare. For this study, the estimation models processed and compared within this framework are (b1) regression model and (b2) NN model.
(c) Evaluation and comparison of models using evaluation criteria and performance indicators.
(d) Estimates of effort, duration or cost using the selected estimation model.

These four steps are detailed below.

3.1 Data Preparation (a)

Organization's projects database is built over years by projects teams in order to capitalize the experience and information related to completed projects. It contains information about previous achieved projects such as project duration, project cost, project type, and platform development.

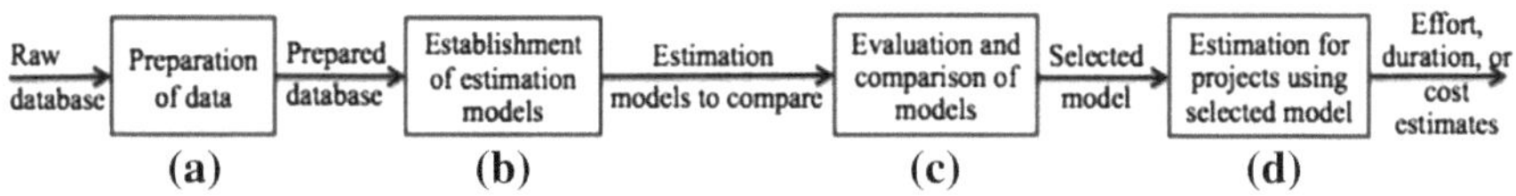

Fig. 1 Framework of estimation for a software project

Steps for database preparation can be summarized as follows:

- Cleaning database: datasets related to irrelevant project parameters, such as parameters concerning information capitalization are discarded. Also, duplicate projects, that are projects with the same parameters but different estimates, are reduced.
- Performing statistical tests: The projects database is explored to determine cost drivers. Cost drivers are parameters that have significant influence on variables to estimate. The statistical test of Pearson correlation and one-way ANOVA can be used to determine cost drivers [18]. These tests enable to examine the significance between the projects parameters and the variables to estimate. The Pearson's correlation test is used for parameters with the ratio scale [19] whereas One-Way Analysis of Variance (ANOVA) is used for parameters with the nominal scale. After determining cost drivers, the other parameters are discarded from the database. Projects with missing values in cost driver fields are discarded. Then, the variables to estimate are adjusted to be normal by discarding projects identified as outliers. A step of data normalization is required for the NN model establishment.
- Dividing database: the projects data should be divided into two segments, one is used to establish and train the effort estimation model and the other is used to test and validate it. The holdout method or the k fold cross-validation [20] approach can be used for this purpose.

In this stage, the projects database is prepared in order to establish effort estimation models.

3.2 Establishment of Estimation Models (b)

(b1) *Regression model establishment*

The regression model establishment consists on modeling the relationship between dependant variables Y and independent variables X_i in the form of a linear equation as:

$$Y = a_1X_1 + a_2X_2 + \ldots + a_nX_n \tag{1}$$

For our study, dependent variables are variables to estimate and independent variables are cost drivers. In order to establish the multi linear regression between these variables based on database, many statistical tools can be used such as XLSTAT.

Fig. 2 Algorithm for NN model establishment

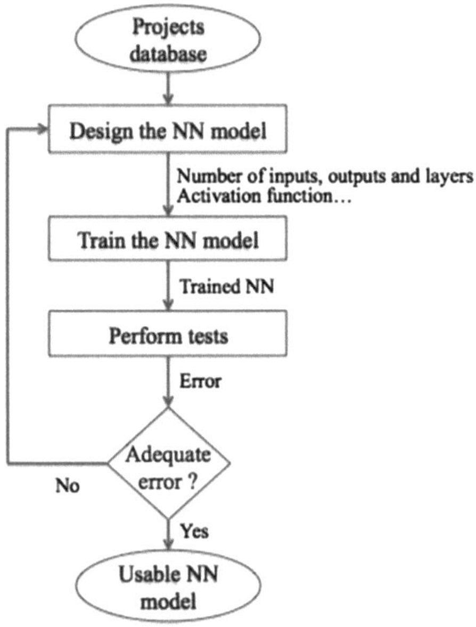

(b2) *NN model establishment*

Steps for NN model establishment can be summarized in the figure below (Fig. 2).

1. NN design consists in defining the architecture of the NN (the number of inputs, the number of output, the number of hidden layers and nodes, activation function). The number of inputs is the number of the projects characteristics, the number of outputs is the number of variable to estimate, the numbers of hidden layer and hidden layer nodes are less than or equal to twice the number of inputs [21]. Activation function is used to transform and squash the amplitude of the output signal of a neuron to some finite value. Some of the most commonly used activation functions are sigmoid, Tanh, and Gaussian [22]. There are a multitude of NN architecture and structure; the most used one is called Multilayer Perceptron (MLP) that is a feed forward artificial NN, i.e. the network is structured in a hierarchical way. It consists of different layers where the information flows only from one layer to the next layer. In this NN, Each node in one layer connects with a certain weight w_{ij} to every node in the following layer. Input nodes distribute the signals to the nodes of the first hidden layer without processing it while nodes of hidden layers are neurons (or processing elements) with a nonlinear activation function [23] (view Fig. 3).
2. Training: The two main techniques employed by neural networks are known as supervised learning and unsupervised learning. In unsupervised learning, the NN requires no initial information regarding the correct classification of the data it is presented with. Supervised training works in much the same way as a human learns new skills, by showing the network a series of examples. The most used supervised training algorithm is back propagation algorithm [17].

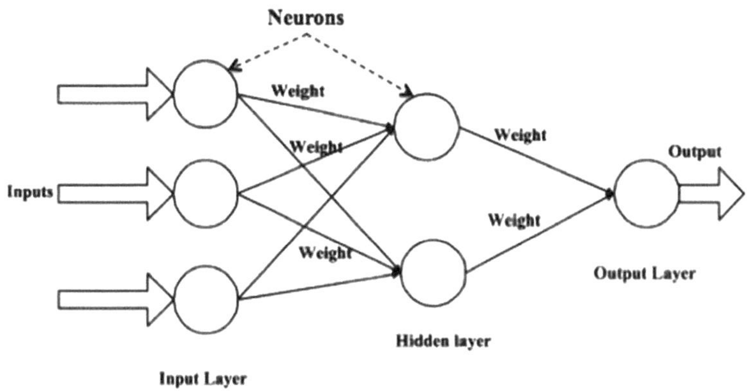

Fig. 3 Architecture of a multilayer perceptron (MLP)

The objective of training is to determine the weights of the NN to bring the outputs of the whole network closer to the desired outputs.

3. Test: After training NN, the test is performed on the database reserved for testing. It makes it possible to measure the potential of success of the trained NN using evaluation criteria. As the figure show, if the results of test are not satisfying, the architecture of the NN model is modified until reaching the most adequate error that is the smallest one.

3.3 Evaluation and Comparison of Models (c)

Model evaluation aims to measure how much the model fits the context of the study. This context is defined by the variables to estimate, the database and the cost drivers used in estimation process.

The evaluating of estimation models makes it possible to compare them in order to choose the most adequate one. For this purpose, different accuracy indicators can be used for this study such as the Mean Magnitude of Relative Error (MMRE), the Pred (0.25) [23, 24].

3.4 Estimation of Effort, Duration or Cost (d)

This last step consists on the implementation of the established models in order to estimate effort, duration or cost for new projects. For estimation using regression model, parameters of the model must be determined for the project to estimate. Then the variable to estimate is simply calculated. For estimation using NN model, parameters must be determined then normalized in the same manner as projects database was normalized during database preparation (a). After that, the variable (s) to estimate is (are) then calculated then converted back to their real scale.

4 Case Study

The experiment described in this paper is carried out on the data provided by an industrial company operating in the automotive sector. The main mission of the company is the design, production and sale of vehicles and mechanical components. The company is also involved in financing vehicle sales and dealership inventories.

The database used in this case study consists on 6078 projects that are either carried or under way. These projects concern several domains of software industry (Cars, finance, commerce…) and their informations are organized into fourteen categories that involve 364 attributes. Due to lack of information about development effort, our focus will be put on the estimation of duration and cost.

4.1 Implementation of the Methodology

The analysis of the database and the statistical tests enables to obtain a database of 214 projects and four attributes that are: project type, project BU (Business Unit), project difficulty, and domain. All cost drivers are qualitative, hence they should be transformed into dummy or binary variables [25].

For regression model establishment (b1), software of data and statistic analysis is used in this work; it makes it possible to automatically establish the regression model using the database. As cost drivers are qualitative in this case study, a special case of linear regression called Analysis of variance (ANOVA) [26] is used to both transform cost drivers into binary variables and establish the estimation model.

For NN model establishment (b2), after transforming cost drivers into binary variables, outputs, that are duration and cost, are normalized to values between zero and one. The resulting database for NN model establishment consists of 27 binary cost drivers and two outputs.

For this case study, MLP architecture is used with 27 inputs and two outputs. In order to determine the hidden layers numbers, training is repeated many times with the variation of the hidden layers number. The best number of hidden layers is that which provide the best performance in test phase. For training, as many experiences have shown that most NN are trained enough in less than 1,000 epochs [27], the number of epochs in this study is set to 1,000 epochs.

Table 1 Comparison of results

Evaluation criteria (%)	Duration estimation models		Cost estimation models	
	Using regression	Using neural network	Using regression	Using neural network
MMRE	69	23	658	14
Pred (0.25)	20	72	7	76

4.2 Experimental Results

The established regression and NN model for estimating duration and cost are applied on the case study, and then the evaluation criteria are calculated. Table 1 presents accuracy indicators calculated for duration and cost estimation using regression and NN.

We use these models to estimate duration and cost of a project for which real duration and cost has been measured at the end of the project by the project management service. The proposed approach gives the results in Table 2.

4.3 Discussion

In this study case, effort estimation model cannot be established because information about effort is not provided in the database.

Table 1 shows that, compared to regression model, NN model provides more accurate estimation for both duration and cost. This can be explained by the capacity of NN to model complex relationship between cost drivers and variables to estimate. A second explanation has to do with the complexity of the case study. This complexity manifests through the lack of relevant parameters due to the lack of information in database and the use of only qualitative cost drivers.

Our approach was applied on an example of a project with known achievement characteristics to concretely observe the estimation results (Table 2), for this example NN model shows better results than regression as well. The differences between the estimated variables and the real ones may be due to the uncertainty in the model parameters and components. For neural network there is a significantly small uncertainty because NN has the ability to deal with the lack of data and cost drivers, in fact, it adjusts the model's weights so that it covers this lack. But for regression models, a complete database and a complete list of cost drivers is necessary in order to achieve good results. Otherwise, there will be a bigger uncertainty in regression model's coefficients.

Table 2 Estimation for a project

Variables to estimate	Real achievement	Estimation using neural network	Estimation using regression
Duration (day)	73	62, 8	36
Cost (K€)	14, 02	17, 18	4556, 5

5 Conclusion

The more accurate estimation is, the better the software project complies with the contractual commitments in terms of budget and duration. The model used for estimation is a crucial factor that affects estimation accuracy.

We presented a framework for estimation starting from the analysis of the available database up to the selection of the estimation model and its use on new projects. This framework is sufficiently flexible to provide estimation of different variables such as effort, duration and cost, depending on the available database about previous completed projects.

The attention was drawn to two models that are regression and NN models. The proposed framework was then applied on an industrial study case that consists of multisite IT projects. This study has shown that NN model is more accurate than regression model even with an important lack of information about previous projects. This lack of information can explain the uncertainty in estimations. Thus, it will be important to be able to measure this uncertainty in order to take it into account in the estimation process.

Future research studies can focus on the need of more realistic estimations by providing not a single value but an interval of estimation and a degree of trust associated to this interval. This can be performed by integrating risk in effort estimation based on research works of [28, 29].

Acknowledgments This work has been funded by the Fund Unique Interministerial (FUI) through the project Projestimate. We wish to acknowledge our gratitude and appreciation to all project partners for their contribution.

References

1. Jørgensen, M. (2004). A review of studies on expert estimation of software development effort. *Journal of Systems and Software, 70*(1), 37–60.
2. Boehm, B. (2000). *Software cost estimation with cocomo II*. New Jersey: Prentice-Hall.
3. Pillai K., & Sukumaran Nair, V. S. (1997). A model for software development effort and cost estimation. *IEEE Transactions on Software Engineering, 23*(8), 485–497.
4. Heemstra, F. J. (1992). Software cost estimation. *Information and Software Technology, 34*(10), 627–639.
5. Dave, V. S., & Dutta, K. (2012). Neural network based models for software effort estimation: A review. *Artificial Intelligence Review*, 1–13.

6. Laqrichi, S., Marmier, F., & Gourc, D. (2013). Toward an effort estimation model for information system project integrating risk. In *International Conference on Production Research (ICPR22)*
7. Idri, A. (2003). Un modèle intelligent d'estimation des coûts de développement de logiciels, Université du Québec à Montréal.
8. Boehm, B. W. (1981). *Software engineering economics*. Upper Saddle River: Prentice Hall.
9. Basha, S., & Ponnurangam, D. (2010). Analysis of empirical software effort estimation models, *International Journal of Computer Science and Information Security IJCSIS*, *7*(3), 68–77.
10. Kok, P. A. M., Kitchenham, D. B. A., & Kirawkowski, D. J. (1990). The MERMAID Approach to software cost estimation, In *ESPRIT'90* (pp. 296–314). Netherlands: Springer.
11. Sentas, P., Angelis, L., Stamelos, I. & Bleris, G. (2005). Software productivity and effort prediction with ordinal regression. *Information and Software Technology*, *47*(1), 17–29.
12. Haykin, S. S. (1999). *Neural networks: A comprehensive foundation*. Upper Saddle River: Prentice Hall International.
13. de Barcelos Tronto, I. F., Da Silva, J. D. S., & Sant'Anna, N. (2007). Comparison of artificial neural network and regression models in software effort estimation. *International Joint Conference on Neural Networks IJCNN 2007*, (pp. 771–776).
14. Idri, A., Khoshgoftaar, T. M., & Abran, A. Can neural networks be easily interpreted in software cost estimation?. *Proceedings of the 2002 IEEE International Conference on Fuzzy Systems FUZZ-IEEE'02*, *2*, 1162–1167.
15. Idri, A., Zakrani, A. & Zahi, A. (2010). Design of radial basis function neural networks for software effort estimation. *International Journal of Computer Science Issues*, *7*(4), 11.
16. Finnie, G. R., Wittig, G. E. & Desharnais, J.-M. (1997). A comparison of software effort estimation techniques: Using function points with neural networks, case-based reasoning and regression models. *Journal of Systems and Software*, *39*(3), 281–289.
17. Heiat, A. (2002). Comparison of artificial neural network and regression models for estimating software development effort. *Information and Software Technology*, *44*(15), 911–922.
18. Huang, S.-J., Chiu, N.-H., & Liu, Y.-J. (2008). A comparative evaluation on the accuracies of software effort estimates from clustered data. *Information* and *Software Technology*, *50*(9–10), 879–888.
19. Neto, A. M., Rittner, L., Leite, N., Zampieri, D. E., Lotufo, R. & Mendeleck, A. (2007). Pearson's correlation coefficient for discarding redundant information in real time autonomous navigation system. In *IEEE International Conference on Control Applications. CCA 2007*, (pp. 426–431).
20. Refaeilzadeh, P., Tang, L., & Liu, H. (2009) Cross-validation. In L. Liu & T. Özsu (Eds.), *Encyclopedia of database systems* (pp. 532–538). New York, USA: Springer.
21. Boetticher, G. (2001). An assessment of metric contribution in the construction of a neural network-based effort estimator. In *Second International Workshop on Soft Computing Applied to Software Engineering*, Enschede, NL, 2001.
22. Karlik, B. & Olgac, A. V. (2011). Performance analysis of various activation functions in generalized MLP architectures of neural networks. *International Journal of Artificial Intelligence and Expert Systems*, *1*(4), 111–122.
23. Moores, T. T. (2001). Developing a software size model for rule-based systems: A case study. *Expert Systems with Applications*, *21*(4), 229–237.
24. Kemerer, C. F. (1987). An empirical validation of software cost estimation models. *Communications of the ACM*, *30*(5), 416–429.
25. Garavaglia, S., & Sharma, A. (1998). A smart guide to dummy variables: Four applications and a macro. *Proceedings of the Northeast SAS Users Group Conference*.
26. Gelman, A. (2005). Analysis of variance: Why it is more important than ever. *Annals of Statistics*, *33*, 1–53.

27. Boetticher, G., & Eichmann, D. (1993). *A neural network paradigm for characterizing reusable software. Proceedings of the first Australian Conference on Software Metrics*, (pp. 18–19).
28. Marmier, F., Gourc, D. & Laarz, F. (2013). A risk oriented model to assess strategic decisions in new product development projects. *Decision Support Systems*, *56*, 74–82
29. Nguyen, T. H., Marmier, F. & Gourc, D. (2013). A decision-making tool to maximize chances of meeting project commitments. *International Journal of Production Economics*. *142*(2), 214–224

A Mediation Information System for Road Crisis Management Unit

Guillaume Macé-Ramète, Matthieu Lauras, Frédérick Bénaben and Jacques Lamothe

Abstract This paper deals with a result of a French project SIM-PeTra aiming at defining an agile and collaborative information decision support system for road crisis management. This paper exoses the global architecture of the information system designed through a model driven engineering approach. The SIMPeTra systems also offers a tool to detect road crisis evolution and the adaptation face to these changes. Method and technologies proposed to tackle this issue are presented in this paper.

Keywords Crisis management · Model driven engineering · Agility

1 Introduction

In March 2013 and December 2010 North West of France was unusually hit by hard winter conditions with heavy snow falls. This snow episode disorganized the everyday life and provoked a major road crisis. Consequences were (i) traffic disruption (hundreds kilometres of traffic jam) (ii) trucks accidents, (iii) drivers blocked in their cars all night long. Furthermore, economics consequences were

G. Macé-Ramète (✉) · M. Lauras · F. Bénaben · J. Lamothe
Centre de Génie Industriel, Ecole des Mines d'Albi, Carmaux,
Campus Jarlard Route de Teillet 81000 Albi, France
e-mail: guillaume.maceramete@mines-albi.fr

M. Lauras
e-mail: lauras@mines-albi.fr

F. Bénaben
e-mail: benaben@mines-albi.fr

J. Lamothe
e-mail: lamothe@mines-albi.fr

K. Mertins et al. (eds.), *Enterprise Interoperability VI*,
Proceedings of the I-ESA Conferences 7, DOI: 10.1007/978-3-319-04948-9_19,

disastrous because a lot of Christmas deliveries has been delayed. Aiming at solving this crisis situation, a crisis cell was set up and led by the West CRICR (Centre Régional d'Information et de Coordination Routière, regional traffic coordination and information centres). It had three main functions in crisis management: (i) inform drivers about situation, (ii) ensure coordination between all heterogeneous actors and (iii) take decision to manage road traffic (allowance to circulate, activate truck storage area, define diverting itinerary). A feedback [1] considering these events emphazised some crisis management problems. Firstly a lack of coordination between actors was underlined. For instance, two conflicting decisions to forbid circulate on a road were taken on the one hand by the CRICR (open the road) and on the other hand by the department prefect (close the same road). This organizational problem highlights a communication gap between different crisis cells. Secondly, some information about crisis situation on "the field" were missing to the Crisis Management Unit. For instance the crisis cell was not aware about new events on the road like accidents or road blocks. As a consequence, the CRICR have difficulties to take clear and good decisions facing this misunderstanding situation.

A solution to overcome this issue is to set up an information decision support system (IDSS) for crisis management. Such a system should be able to [2] (i) gather the information (collect, update and share), (ii) analyze and be helpful to understand the current situation (represent, monitor, detect and alert) and (iii) support decision-making (decision elaboration, resources allocation, planning and analyze the impact of the decision). In order to be efficient this system should detect crisis changes and adapt his behavior in an acceptable lead time. This ability to be flexible and adaptable could be defined as the agility of the IDSS. On the other hand, in order to face the lack of actor coordination, the IDSS should be able to ensure interoperability between all actor services which is according to [3], "the ability of a system or a product to work with other without special effort from the customer or user".

Thus, the main research issue is how can we define an Information Decision Support System helping heterogeneous actors to collaborate and being agile for road crisis management. In the following, the first part deals with a literature review on agile tools for crisis management. Then we propose a design method for an agile decision support system for road crisis management.

2 Literature Review

A possible solution to support crisis response is to set up a MIS coordinating actors. The ISyCri project (Interoperability of System in Crisis situation) deals with such an information system. The ISyCri Home Page [4] Explains how to design and implement a MIS supporting a crisis response based on a Service Oriented Architecture (SOA). The MIS implementation starts by deducing collaborative process cartography from a crisis characterization and a description of each actor service. The crisis and actors descriptions are done respecting a generic crisis metamodel defined in [5]. This metamodel describes the crisis as a complex system divided into three complementary dimensions (i) crisis characterization, (ii) studied system (crisis environment and crisis risks) and (iii) treatment system (actor and services used to respond to the crisis situation). Once the metamodel is instantiated, the collaborative process deduction is done respecting rules described by [6]. *A major lack of this deduction is that it only takes into account crisis characteristics and actor's services but does not care of means dimensions and possible processes options.* In the second phase of MIS implementation, the collaborative process is defined as a workflow following a transformation described by [7]. Then a new transformation is done to implement technically the workflow into an Enterprise Service Bus (ESB). This phase is done automatically increasing system agility in terms of flexibility (it can support crisis or actor evolutions) and reactivity (short design lead time).

We defined previously the agility in crisis context as "the ability of a system to detect an evolution in crisis situation and to adapt its behavior in a short lead-time". Consequently, the main goal of an efficient detection consists in monitoring the ongoing process in order to know if the system behavior is relevant still regarding the current situation. In a French funded project SocEDA [8] (Social Event Driven Architecture) report, a three steps methodology based on a comparison of models was proposed to detect changes in a collaborative situation. The first step consists of calculating the different models that will be compared. The two following models are defined:

- The field model: it is a representation of the ongoing field situation. This model is updated by information gathered by sensors, observations and reports. A major problem is that this information comes into different format. A big effort should be done in order to match data from sensors to metamodel concepts. The resulting model is a representation of the "real situation".
- The expected model: it is a "should be" representation of the system. It depicts the awaited situation applying the response process on the first situation model. It is updated by activity status of the ongoing process.

The second step of detection method consists on calculating the divergence between the two models. As explained previously, the two models are extracted from an ontology i.e. they are expressed in OWL language [9] that is an XML based language. Each model (field and expected) matches with an ontology i.e. two XML files are used. Thereby, the differentiation between the two models is based on an XML file comparison. An adaptation of the XML Unit framework [10] provides a tool that can compare and identify all differences between two XML files. Finally, once the comparison is made, it is necessary to analyze the possible differences between expected and current model. This analysis could lead the crisis-cell to adapt the global crisis response process. In our context of road crisis management (due to bad weather conditions), we find a weakness of this detection mechanism as it is done in the present with the current information. It doesn't take into account the future situation.

3 Scientific Proposition

In the following we propose an information system design using a model driven architecture. We focus on each step of the model driven approach, from the Computer Independent Model level to the Platform Specific Model. The last part of this section deals with a framework to improve agility.

3.1 Computer Independent Model Level

The main objective of this part is to gather enough information about crisis situation in order to deduce a collaborative process response of crisis situation. The information is structured following a metamodel defined in [11]. This metamodel is a UML class model organized in three layers: a collaborative core, a domain layer and an application field over layer. The collaborative core contains collaborative concepts that are useful for describing a collaborative situation. Concepts are organized in five families:

- *Collaboration environment* (components and characteristics): the subject of the collaboration, implying on what the collaboration is going to be.
- *Collaboration object* (opportunity, threat and facts that impacts environment component and collaboration objective) defines why people are going to collaborate.
- *Collaboration treatment system* (actors, actor capabilities, resource) defines who is involved and their roles in the collaboration.
- *Collaboration dynamics* (process, activity, event and message) aims at defining in what way people are going to collaborate fulfilling the collaboration object.
- *Collaboration performance assessment* (performance objective, Key Performance Indicator) assesses the collaboration dynamics and deals with how people collaborate.

All these concepts take part of the collaborative core and are firstly extended by a domain specific layer. For instance the ISyCri crisis metamodel propose a crisis domain layer by specifying crisis concepts such as risk, consequences, goods, civilian society, treatment system… A major objective of layers is to provide a specific set of concepts to information system final users. Secondly, the SIMPeTra road crisis metamodel extends the ISyCri Metamodel with concepts more specific for a road crisis situation such as highway, sections, storage area, accident, traffic jam risk to be compliant with concepts and semantics used by the CRICR.

The metamodel aims at defining concepts and vocabulary to be instantiated into models. Those models have two mains utilisations. Firstly they provide to the crisis cell a snapshot of the current crisis situation and are useful to take good decisions. Models can be instantiated "manually" using a human modeller informed by crisis response actors or "automatically" been interfaced with other information system (for instance: meteorological information, driving conditions information…). A first version of a crisis modeller has been developed for SIM-PeTra project. It takes into account Meteo France previsions (vigilance level represents by colour department: green, yellow, orange and red levels) and driving conditions provided by road manager IS (represented by road colour: green, yellow red and black levels) and proposes to the crisis cell a possibility to add an accident on the map.

Secondly, having a model of the current crisis situation is useful in order to deduce a crisis response process. This is the second utilisation of crisis situation model. The defined model is integrated into an ontology [12]. This domain specific ontology is structured according to the metamodel definition. We choose to use OWL format as language ontology. Links are made between crisis problems (studied system) and actors capabilities (treatment system). Ontologies are put into an inference engine to deduce a collaborative process. The process deduction is made of three steps. Firstly, a set of rules is applied on the ontology in order to enrich it by making new links between concepts. For instance if a concept A is close to a concept B and a concept C is in relation with concept A then we could deduce that concept C could be in relation with B. In crisis terms, if we have a risk A and we know that a risk B is similar to A and that a service C prevents A we can deduce that C could prevent also risk B. After this first set of rules we have a list of services that can reduce our problems. The second set of rules is about the service feasibility. It concerns if a service is available and underline the resource allocation issue. Once this second set of rules is executed, a list of feasible services is made. The next step consists in organize all services to deduce a process depending on conditions to execute service (for instance a service A needs an other service B to be executed).

3.2 Platform Independent Model Level

Once the collaborative process is defined, it might be transformed into a workflow to be executed by an orchestrator. Then we need to make a semantic association between business activities and technical IT services provided by actor. In other words a correspondence should be done between an activity as defined in a BPMN process [13] and a Web Service that can be directly invoked. In some case and especially in crisis response context, a business activity doesn't have any correspondence with an IT service. For instance the activity "evacuate the area" can not be done by an Information System. However, in order to follow the process execution we need to know the status activity (started, ended, work in progress, failed…). Thus, we propose to generate automatically a Web Service interface for concerned actor to inform the crisis cell about the status activity. The PIM level concerns the transformation of the collaborative process designed in BPMN language into an executive file designed in BPEL language.

On the other hand in order to assess the collaborative process in real time a Performance Measurement System is proposed. Firstly a set of Key Performance Indicators (KPI) is defined to evaluate the current process. It permits to give to the crisis cell information about how activities are processed according to several dimensions: effectiveness (does the process fulfils entirely the main objective), efficiency (are the resources engaged well used) and relevance (are the resources engaged well dimensioned regarding the main objective). A formula and an objective value are defined for each KPI. Then during the process execution, KPI are calculated and monitored on a dashboard for crisis management.

3.3 Platform Specific Model Level

The PSM level consists in the workflow implementation into an Entreprise Service Bus. The workflow is running into the bus through a workflow engine. This one ensures the link and the coordination between all services and actors. It acts like a conductor who plays a score (BPEL file) to an orchestra composed by actors and services associated.

The workflow model is projected on the ESB Meta Model in order to ensure the service connexions and to provide all information needed to make the system functional. Now the ESB chosen is not yet defined but it should take into account the SOAP protocol and provide a workflow engine compliant with BPEL format.

3.4 Agility

According to Lagadec definition [14], a crisis can be seen as a disordered state of a reference universe. So, crisis can be considered as an evolutive phenomenon. Thus crisis cell may face these changes and adapt his behaviour. Three changes reasons can be identified [15]: (i) crisis nature changes (for instance in Fukushima tsunami crisis to nuclear one), (ii) modification of response composition (an actor can leave or join the crisis cell), (iii) activity dysfunction during crisis response (the activity result expected is not good). The information system should be able to support the crisis cell by detecting quickly variations and provides solutions to adapt crisis response behaviour. Thus we define this capability to detect and adapt quickly as the system agility. The information system needs to implement the two functions: detection and adaptation in an acceptable leading time.

In order to detect rapidly changes in the crisis situation, we propose to adopt an Event Driven Architecture for the Information System. Devices and services emit events during the crisis response. Gathering this information into models permits to detect evolution, on the first hand, of the crisis situation and on the other hand, of the crisis response. Events are a good way to show in real time crisis situation. But be able to gather information in real time using sensors, devices and services is a first part of the changes detection process. In order to have a good detection, our purpose is to compare three different models that represent three different views of the situation. First model is a representation of the field state, what really occurs in the crisis situation. It is based on captors, sensors and devices information. We call it field model. The second model represent the crisis situation supposed state after an activity execution. For instance a process activity is chosen because it prevents a risks. At the end of the activity we expect that the risks has been prevented and diseapear on the crisis model. This second model is called expected model. The detection is made by comparing this two first models and underlining divergences. Next step is to qualify the divergence to readapt (i) at the CIM level with a new process deduction if new risks or new consequences appears (ii) at the PIM level with a new workflow definition if an actor changes his services (iii) at the PSM level if a service/activity execution failed.

This detection/adapt mode is the easiest way to detect divergences. But we can think that divergences could have been detected earlier by anticipating future states of the system. This is the purpose of the third model called projected model. This model is a snapshot of the future situation taking into account on the first hand, situational events forecasted (meteo weather, traffic conditions) and on the second hand, services events expected (activity may end so risk disappears). KPI are also projected into the feature and measure with events forecasted. The evaluation of the future situation through KPI projection is a good way to assess the collaborative process deduced. Indeed if the projection is too bad regarding crisis situation, it means that the current process should be changed.

4 Use Case

The following section shows the implementation of the Information System in the context of the SIM-PeTra project. This use cased is based on events that occurred during the road crisis in March 2013. This major crisis was managed by the CRICR. It had to take decision about traffic management dealing with the real traffic situation. In order to be prepared to cope with a transport crisis situation, some plans are defined during the preparation phase. This plan is called PIZO (Plan Intempéries Zone de l'Ouest). This is divided in four levels depending of the crisis gravity. Each level is a step in a crisis response. Thus, level 1 means that the crisis cell is awake and vigilant to all events that could happen. It lasts from the 15th of November to the 15th of March. Level 2 corresponds to a pre-alert phase according to a future event. In level 3 some traffic management measures are taken in order to prevent a risk. In PIZO level 4 curative traffic management measures are taken.

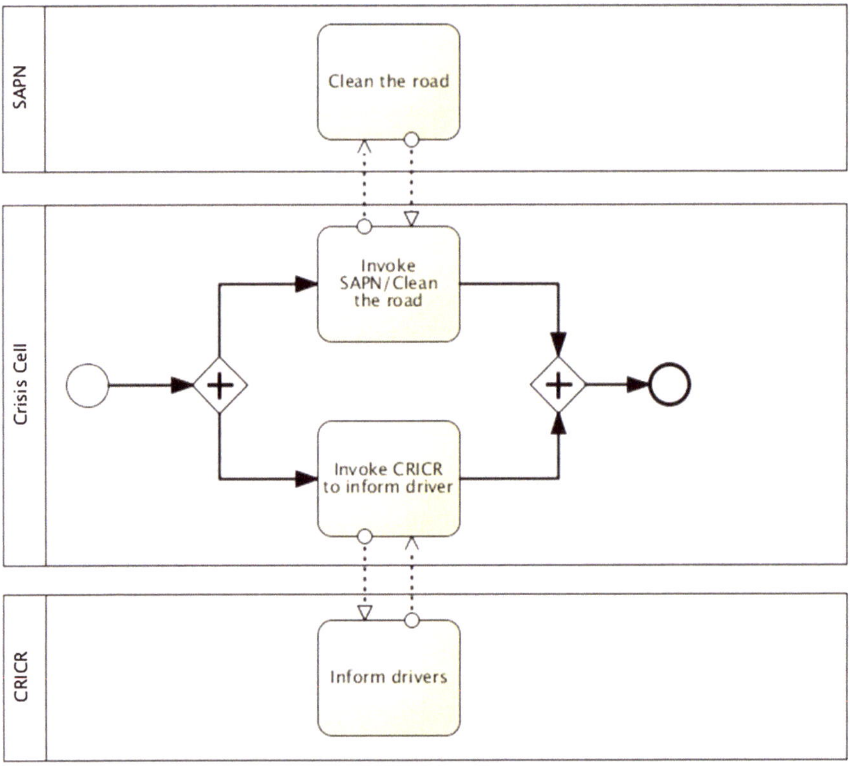

Fig. 1 PIZO L2 process

The 10th March of 2013 at 8:00 P.M, Meteo France emits a new weather report that announced a meteorological alert. It forecasted snowfalls on cold asphalt for Manche and Calvados departments at a level of 3 on 4 (orange vigilance level). The Mediation Information System is connected with the Meteo France information system and an update of the ontology is made. Thus, the system detects that a process has to be launched thanks to the following rule:

$$\text{Forecast of snowfalls L3} \rightarrow \text{PIZO L2}$$

The PIZO L2 (Fig. 1) is only a pre alert for all actors to get ready to face the crisis. For example, snowploughs are deployed alongside the road by public works and a new collaborative process is engaged (process to pass to PIZO L2).

The process contains on the first hand, "manual" business activities that only can be made by humans (validate the department) and on the second hand, "automatic" business activities that can be made by information systems (inform all actors). Each partner involved in the crisis response has his own dashboard helping him to know what he has to do.

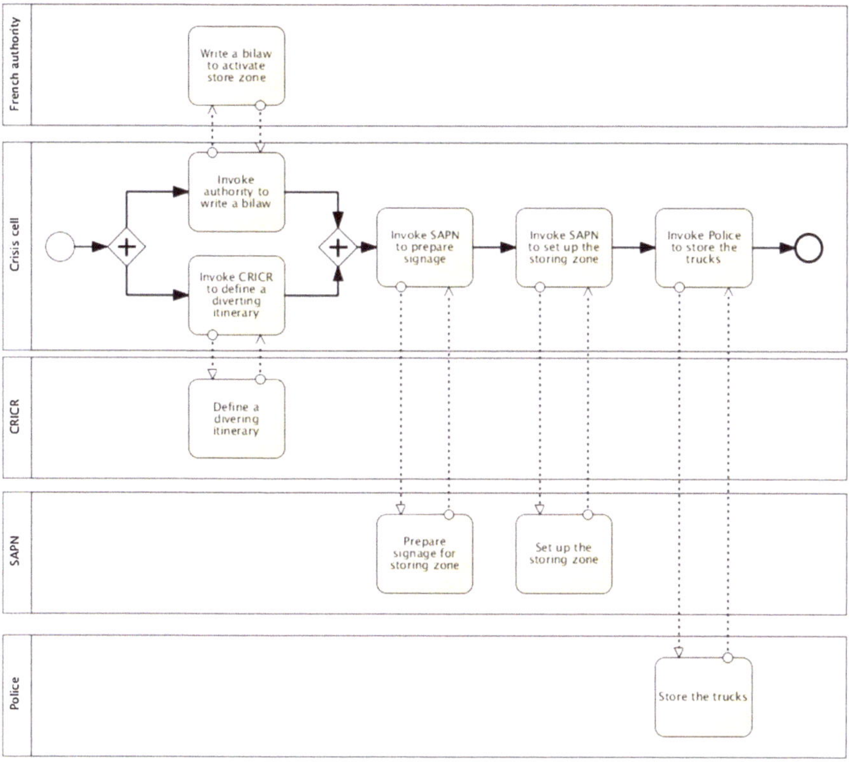

Fig. 2 New process proposition

The 11th of March, at 10 AM crisis cell gets two additional information. Firstly snow is falling and staying on the asphalt in Manche and Calvados. Secondly traffic conditions are difficult on the A84 motorway (motorway with high slopes). Those information are updated on the current model which represents the "on field" situation. After this a first set of rules on the model are ran by an inference engine in order to deduce new information about the situation.

Snow falls && slopes on motorway → new risk of overturned truck.

Then thanks to the last rule a new risk has appeared on the current model. The next step is to compare it with the expected model (blank model for the moment). The result of this comparison is the risk of overturned truck on the A84 motorway. Thus a the deducing rules are applied on the current model in order to find solutions for this problem. The following rule is then applied.

Risk of overturned truck → close motorway to trucks

Then the system popups an alert to crisis managers notifying them of the new situation and purpose to launch the process of close motorway to trucks. The decision-maker can accept this proposition and the process is launch or modify and customize it to launch a new process (Fig. 2).

Once all bylaws are set up, the crisis management cell can follow the situation evolution thanks to the update of the different models. Using this architecture increase the agility of the response to the crisis situation. The real situation is well-represented and analysed thanks to the different model updates (current and expected models). The crisis cell gathers field data in quasi real time. On the other hand thanks to business rules and process deduction the lead-time to take decision and coordinate actors is reduced.

5 Conclusion

In this paper we exposed a method to define an agile and collaborative Mediation Information System for road crisis management. The crisis management unit is able to coordinate all actors involved in crisis response thanks to the orchestrator and might be able to be aware of new crisis evolutions gathering informations from other information system. Those evolutions are considered and analysed and proposed adaptation of the crisis response using a rule-based system. Thanks to the Model Driven Approach, the Mediation Information System can be redesigned in order to be relevant to the crisis situation.

Future works includes a development of a new use case with more actors involved in the crisis situation and a part of the system is going to be implemented in the CRICR.

References

1. Panhaleux, J. David, D. Labia, P. Picquand, J-L. Rivière, D. Terrié, F., & Leyrit, C. (2011). Retour d'expérience à la suite de l'épisode neigeuc survenu le 8 décembre 2010 en Ile-de-France, Conseil général de l'Environnement et du développement durable.
2. Aligne, F. (2009). Which Information and Decision Support System for the Crisis Management. *Information Systems and Technology Panel (IST) Symposium*, Bucharest, 11–12 May 2009.
3. Konstantas, D. Bourriere, J-P. Leonard, M. & Boudjlida, N. (2004). Preface. In *Proceedings of the First Conference on Interoperability of Entreprise Software and Applications*, Springer.
4. The ISyCri Home Page. http://www.irit.fr/IsyCRI
5. Truptil, S. (2011). Etude de l'approche de l'interopérabilité par médiation dans le cadre d'une dynamique de collaboration appliquée à la gestion de crise, INP Toulouse.
6. Benaben, F., Touzi, J., Rajsiri, V., Truptil, S., Lorré, J.P., & Pingaud, H. (2008). Mediation information system design in a collaborative soa context through a mdd approach. *Model Driven Interoperability For Sustainable Information Systems (MDISIS'08).*
7. Truptil, S., Bénaben, F., Couget, P., Lauras, M., Chapurlat, V., & Pingaud H. (2008). Interoperability of information systems in crisis management: Crisis modeling and metamodeling. *I-ESA'08*, Berlin.
8. The SocEDA project (SOCial Event Driven Architecture): https://research.linagora.com/display/soceda/SocEDA+Overview

9. Dean, M., Schreiber, G., Bechhofer, S., van Harmelen, F., Hendler, J., Horrocks I., McGuinness, D. L., Patel-Schneider, P. F., & Andrea Stein, L. (2004).OWL Web Ontology Language Reference, W3C Recommendation 10 February 2004.
10. Bolaños Alonso, D., Sierra Alonso, A., Alarcon Rodriguez, I., & Pruebas, de. (2008). Software y Junit un análisis en profundidad y ejemplos prácticos, Pearson Educación, Madrid.
11. Macé Ramète, G., Lamothe, J., Lauras, M., & Bénaben, F. (2012). A road crisis management metamodel for an information decision support system, In 6th *IEEE International Conference on Digital Ecosystems Technologies (DEST)*, Campione d'Italia, Italy, June 2012.
12. Gruber, T. (2001). What is an ontology? http://www-ksl.stanford.edu/kst/what-is-anontology.html
13. Owen, M., & Raj, J. (2004). "BPMN and business process management: An introduction to the new business process modelling standard. *Business Process Trends.* March 2004.
14. Lagadec, P., (1992). La Gestion Des Crises—Outils de réflexion à l'usage des décideurs, Paris, Ediscience International, 1992.
15. Pingaud, H. (2009). Rationnalité du développement de l'intéropérabilité dans les organisations, Management des technologies organisationnelles. *Journées d'études, 2009*, 19–30.

Part VI
Future Internet and Enterprise Systems

Capturing and Structuring Interoperability Requirements: A Framework for Interoperability Requirements

Nicolas Daclin and Sihem Mallek

Abstract The main objective of this communication is to discuss and present a framework for interoperability requirements. More precisely, the here presented research focuses on the dimensions to consider in order to capture and structure interoperability requirements in a precise way for parnters that want, further, to verify the truthfulness of these requirements. First, the concept of interoperability is presented according its main characteristics and the need to express it as a requirement. Then, the dimensions of the framework are highlighted and related in order to be suitable and usable. Finally, some interoperability requirements and their positionning in the framework for interoperability requirements are introduced to show the interest of such approach.

Keywords Interoperability · Requirements engineering · Interoperability requirements · Framework for interoperability requirements

1 Introduction

The "*ability of enterprises and entities within those enterprises to communicate and to interact effectively*" [1]—in other words interoperability—has become, over the past years, a major issue and a key factor of success for collaborating enterprises. As a consequence, numerous research works was proposed to characterize

N. Daclin (✉)
Laboratoire de Génie Informatique et d'Ingénierie de Production, ENS Mines Alès,
Parc scientifique G. Besse, 30035 Nîmes cedex 1, France
e-mail: nicolas.daclin@mines-ales.fr

S. Mallek
Laboratoire de Génie de l'Environnement Industriel et des Risques Industriels Naturels,
ENS Mines Alès, 7 rue J. Renard, 30319 Alès cedex, France
e-mail: sihem.mallek@mines-ales.fr

K. Mertins et al. (eds.), *Enterprise Interoperability VI*,
Proceedings of the I-ESA Conferences 7, DOI: 10.1007/978-3-319-04948-9_20,

[2], implement [3, 4] and improve [5, 6] this aspect of a partnership. Although, these works are fully adapted to develop enterprise interoperability, few of those take an interest in the definition and the structuration of interoperability requirements that partners can verify during partnership. Nevertheless, requirements engineering is an important aspect, offering the possibility for partners to dispose of a clear repository of interoperability requirements that guide them in their characterization and their use for verification in order to know the possible defects of a partnership in term of interoperability. In this context, it is interesting to propose a framework allowing to guide the definition, the capture and the structuring of interoperability requirements in a good way *i.e.* respecting approaches of requirements engineering proposed in other fields (e.g. System Engineering [7]). Indeed, disposing, in the end, of a set of interoperability requirements clearly defined and structured has to enable that selected solutions are perfectly eligible to satisfy requirements and further, to assess their satisfaction throughout a partnership lifecycle. The here presented research, focuses on the development of a framework in order to guide partners in their interoperability requirements engineering according to basic dimensions related to the concept of interoperability and analysis levels.

This paper is structured as follow. After this brief introduction, the problematic and expected result of this research work is presented in Sect. 2. A survey related to this research is given and discussed in Sect. 3. Section 4 presents the proposed framework for interoperability requirements regarding its dimensions, and their relationship as well as the way to exploit the framework. The final section presents the conclusions and the prospects of this research.

2 Problematic and Expected Results

There are two main ways to make out when talking about requirements engineering practice [8]. On the one hand, the requirements engineering includes the elicitation, the writing, the refinement… of requirements to verify. On the other hand, the requirements management includes the support for user access, the versioning, the change, the traceability… of requirements. Although different in nature, these two aspects are complementary and are a crucial phase for stakeholders that want to obtain a system (e.g. technical, product, service, organization…) that meet their expectations initially expressed. Indeed, these requirements will have to be verified (e.g. analysis, inspection, test, simulation…), all along their life cycle from the engineering phase to the integration phase via the production phase, in order to show that a given system satisfies what is required and to avoid problems (e.g. drift from the original cost and quality, delay, cancellation in worst cases…). In the same manner, some requirements will have to be verified, as well, during the operational phase of the system in order to ensure and to uphold the satisfaction of these requirements until the retirement (or dismantling, recycling) of the system.

Although this step is a key step for the success of a project and, conversely, a factor of project failure [9], this aspect remains too often and unfortunately neglected by actors (stakeholders, acquirer…) due to several points such as [10]:

- Actors do not know what to do (e.g. they do not know how to write requirements, they do not understand the process);
- Actors do not understand why requirements are necessary (e.g. they do not understand the impact, they think requirements are just obviousnesses, they think it is just a document);
- Actors prefere rather do something else (e.g. they do not have time, they think the review process will highlight errors/omissions).

Beyond the classical functional requirements and in the limited frame of interoperability, this one is clearly identified as a non functional requirement (NFR) that have to be considered all along system's life cycle (from the engineering for its definition and also, after the initial use for its satisfaction) [11, 12]. Once more, even if interoperability is identified as a NFR which can have a strong impact on the functioning of a system and furthermore, its quality of service, it falls often in the category of the so called "forgotten'-illities" [13].

Thus, according to the statements mentioned hereinbefore, capturing, defining and structuring requirements—in our case interoperability requirements—is a lock for partners that want to implement interoperability in their partnership and further, to make sure that interoperability is properly reached while partnership is existing. Two major issues have to be considered for those that want to dispose of requirements. On the one hand, they have to face up with the expression of requirements. Authors of requirements dispose of languages, each one with their strengths and weaknesses from natural language (readability, richness, extensibility versus low consistency and precision) to formal language (strong precision and consistency versus low readability and constructability). They dispose also of practical rules guiding and allowing to write, review and share requirements in a proper way (we can mention here the well know acronyms M.U.S.T[1] and S.M.A.R.T[2] to ensure that a requirement is well expressed). This is not the purpose of the paper and reader may wish to refer to numerous works related to this field for more details [14]. On the other hand, requirements baseline is not sufficient to manage them and use it. They have to be grouped within a repository namely a "*set of requirements*". The main purpose of this set is to perfectly structure and precise the baseline of requirements in order to dispose of requirements that are easier (1) traceable in order to link a requirement to its origin and trace it throughout its lifecycle (e.g. defined, verified, allocated, satisfied…), (2) modifiable/removable/addable, (3) usable to implement relevant solutions that meet

[1] M.U.S.T is a mnemonic device that stands for Measurable, Useful, Simple and Traceable.

[2] S.M.A.R.T is a mnemonic device that Specific, Measurable, Achievable, Relevant and Traceable.

expectations and, (4) identifiable. Thus, this repository has to allow to manage properly the identified requirements.

Furthermore, the use of a clear set of requirements can guide and help stakeholders to elicit other requirements from a well-defined basis rather than with a jumbled list of requirements. In this way, a set of requirements has to possess some characteristics such as [15]:

- *Complete*. The set of requirements includes complete definition of expectations. This means that the set includes all stakeholders' requirements (no more, no less).
- *Consistent*. The set of requirements is a consistent expression of expectations. This means there are no conflicts between requirements leading to an empty solution space.
- *Feasible*. The set of requirements is a feasible expression of expectations. This means that all requirements can be satisfied simultaneously.
- *Bounded*. The set of requirements takes place within a defined scope. The set clearly communicates expectations. Furthermore, the set is used by actors in charge of the implementation of solutions. Thus, requirements that are irrelevant with the scope have to be removed.
- *Structured*. The set of requirements is entirely understandable and assimilable, without undue cognitive loading for a given reader. Furthermore, each requirement is placed in a context in such a way that it is easily identifiable.

Thus, disposing of a framework of interoperability requirements helps actor to better capture and to structure their interoperability requirements in order to facilitate their implementation and their evaluation during partnership lifecycle. In this case, this framework has to consider several dimensions related to interoperability concept. First, the framework has to consider the concept of interoperability. This means that it is required to define precisely what interoperability is and, by extension, what interoperability is not. This aspect focuses mainly on the definition of the characteristics of interoperability during a partnership. Indeed, depending of the phase of the partnership (*i.e.* when interoperability is susceptible to occur and to be implemented within a partnership) interoperability can have more or less properties that define it. Then, framework for interoperability requirements has to consider the main problems of interoperability. Indeed, a requirement finds its origin in an expectation, a lack or dissatisfaction. Hence, the framework has to include this aspect in order to show which problem interoperability requirement has to satisfy. Finally, the framework for interoperability requirements has also to consider what "*perspectives*" can be impacted by an interoperability requirement. The objective is to identify precisely what criteria are concerned by interoperability in order to show the importance of the satisfaction of a given (set of) requirement(s).

All these dimensions have to be consistently related in order to offer a framework for interoperability requirements that respect the characteristics—of a set of requirements–as presented previously. This framework has to ensure completeness of interoperability requirements. With such framework actors are able to improve completeness by having a global view of interoperability requirements.

Then, it has to ensure the consistence of requirements as well as the feasibility. The fact to dispose of a framework allows (1) to identify easier possible conflict between requirements and (2) to select easier the requirements to satisfy and to avoid requirements that are difficult (or even impossible) to perform. Finally, it has to be bounded and structured. The here proposed framework focuses on interoperability requirements. In this way, all requirements that are not relevant to this field have to be removed, and have to be easily identifiable in agreements with all dimensions of the framework.

3 Interoperabilityand Requirements: A Quick Scan

Numerous works dealing with interoperability, in different application domains (e.g. medicine, military, computer science), have been developed in recent years. As far as enterprise interoperability is concerned, these works endeavored to (1) clearly define and structure interoperability [2], (2) develop methods/tools to evaluate interoperability [5] and (3), propose methodologies [3] for its implementation trough existing and identified solutions. Some of these works are directly, or can be related to interoperability requirements aspects.

Interoperability frameworks focus on the structuration of the concept of interoperability by the consideration and the arrangement of its different aspects (problems, approach…). Thus, the major part of existing frameworks [16, 17] highlights that interoperability problems are related to conceptual, organizational and technological issues. Other frameworks go further and integrate more dimensions in order to capitalize interoperability solutions [18]. In that sense, frameworks for interoperability take, more often, an interest in the "*how*" (*i.e.* solution) rather than in the "*what*" (*i.e.* what we have to do?). Although, interoperability frameworks seem separated to the interoperability requirements consideration, they allow to structure perfectly interoperability and its aspects have to be fully integrated into a framework which aims to facilitate the capture and the structuring of interoperability requirements.

Methods to evaluate interoperability, whether in term of maturity [19, 20] or operational interoperability [6, 21], underlying the notion of requirement. Indeed, regarding maturity evaluation, each layer represents key points that enterprise has to satisfy in order to evolve throughout maturity levels. In that sense, each level describes, in natural language, what is requested to reach a given level of interoperability. These descriptions are no more no less expected conditions in order to guarantee the satisfaction of a given level. However, as mentioned just above, the notion of requirement is implicit and not formalized, but these models represent a strong contribution and can be used to express and elicit requirements. In the same manner operational measurement tools define and give equations in order to evaluate interoperability. The effective results of this measurement are often compared to expected results. Once more, expected results represent, in some ways, requirements in term of operational interoperability.

Fully related to interoperability requirements aspects, [22] proposes a definition and a characterisation of fundamental interoperability requirements. Nevertheless, these requirements take place at high level of abstraction (first level of requirements, neither derived requirements nor refined requirements) and do not consider their structuration. In the same way, the last version of TOGAF [23] highlights the necessity to define clearly interoperability requirements. This necessity is characterized by guidelines that have to help to define and establish interoperability requirements. Besides, let us note [24] which proposes a set of interoperability requirements (related to Gas and Electricity Smart Metering Systems) respecting a defined and common format to express them.

Regarding the consideration of interoperability requirements structuration we can mention [25] which proposes interoperability requirements and proposes a model as a formalized graph including 3 dimensions related to interoperability. Additionally, this work proposes some interoperability requirements that can be verified. This work is relevant but more oriented towards the verification of requirements thanks to formal techniques and does not consider the impact of interoperability on partnership and on a given partner. We can also note the interoperability framework proposed in [18] which includes an additional dimension (interoperability engineering phase) that consider requirements definition. However, this dimension is no more developed and none explanations are given on the way to exploit this aspect. Hence, it can be interesting to extend this dimension in order to fully consider interoperability requirements.

It emerges two points of interest from these considerations. First, the necessity to dispose of a clear definition and expression of interoperability requirements. Even if interoperability requirements can be perceived, they are not well expressed *i.e.* following a strict guideline to write requirements in a good way and that really exploitable. Second, the necessity to clearly structure each of identified interoperability requirements according to (1) the problems of interoperability, (2) the definition of interoperability and (3) the different "*perspectives*" that can be impacted by interoperability. If numerous interoperability requirements can be extract from existing works (beyond the correctness of their expression), the first point to consider is to dispose of a set of requirements that respect defined characteristics in order to provide feasible solutions that satisfy stakeholders' expectations.

4 Framework for Interoperability Requirements

4.1 Fundamental Dimensions

The framework for interoperability requirements organizes and structures requirements to guide the selection of the best solution of interoperability but also to verify their satisfaction continually during a partnership. In other words, it has to make available a set of interoperability requirements that can be easily identifiable, verifiable and traceable. As a consequence, the framework has to consider:

- The problems of interoperability to which partners face up;
- The characteristics impacted by the non-satisfaction of requirements;
- The moment when interoperability occurs during a partnership.

At the end, the framework has to provide adapted requirements for each dimension in order to verify their satisfaction according to stakeholder's expectations. The proposed framework for interoperability requirements defines three basic dimensions such as: abstraction, analysis and interoperability levels. *Abstraction levels* represent the categories of interoperability that can be developed in enterprise and thereby the problems of interoperability. A requirement comes from dissatisfaction, a lack, thus, the definition of requirement belongs to the definition of problem space (that a solution will have to satisfy further). Hence, it is required to consider the identified problems of interoperability in the framework. *Analysis levels* represent the characteristics that are impacted by the implementation of interoperability. The satisfaction of a given requirement can impact criteria either on partnership or on a given partner. Hence, requirements have to be defined according to these identified criteria in order to highlight what is expected in term of interoperability and, vis-à-vis of the partner and partnership. *Interoperability lifecycle levels* represent the requirements of interoperability according to what is expected in term of interoperability and all along a partnership. This means that requirements, to verify, evolve and are not necessarily the same as we go along the evolution of partnership. For instance, requirements at the beginning can be related to the interfacing while requirements at the end can be related to the unplugging in a proper way and with a limited impact on partner.

The dimension of *abstraction levels* takes up the classical categorisation of interoperability problems. For the record, these categories are defined such as [1]:

- **Conceptual**. This category is related to syntactic and semantic problems of interoperability. It can also integrate the expressivity problem.
- **Organisational**. This category is related to the definition of organisation structure, management techniques and policies within partners. This dimension can be refined, taking into consideration the responsibilities, authorities decision-making processes, policies, organisational processes and regulatory.
- **Technological**. This category is related to the problems inherent to the use of Information and Communication Technologies. Beyond the logical aspects this category can also concern physical aspects (e.g. compatibility problem of physical interfaces).

The dimension of *analysis levels* identifies the main characteristics that can be impacted by the implementation of interoperability namely, performance, stability and integrity.

- **Performance**. It refers to the ability of a system to reach its objectives. In this case, requirements positioned on this characteristic are about the expected

performances in term of interoperability but also about the impact of interoperability on partner.
- **Stability**. It refers to the ability of a system to maintain its reliability and to get used to its environment.
- **Integrity**. It refers to the ability of a system to stay coherent and to be able to ensure its functions (or to get back to a functioning mode which is known) in case of modifications (e.g. loss of resource).

The dimension of *interoperability life cycle levels* is related to the decomposition of interoperability in agreements with partnership evolution. Interoperability requirements can evolve during a partnership, depending if it is the beginning (e.g. connection of partner), the operational phase (e.g. exchange, sharing…) or the end (e.g. dismantling). Thus, each phase has to be clearly defined in term of interoperability. To do that, this dimension is concerned by compatibility, interoperation, autonomy and reversibility.

- **Compatibility**. It is related to the beginning of collaboration. Requirements positioned on this group are related to the interfacing aspect of the collaboration (logical as well as physical).
- **Interoperation**. It is related to the effective aspect of collaboration. In this case, requirements are related to the operational aspects of interoperability (e.g. performance).
- **Autonomy**. It is related to the effective aspect of collaboration as well. Requirements are related to the fact that partners are interoperable while retaining their operational thinking.
- **Reversibility**. It is related to the end of collaboration (i.e. dismantling). In this group requirements are related to the achievement, by a partner, of original objectives, functions… after a partnership and despite adaptations or changes.

The following figure presents the framework for interoperability requirements and the space requirements are positioned according its dimensions Fig. 1.

4.2 Additional Dimensions

In order to better structure interoperability requirements, the framework for interoperability can be extended and consider further dimensions. At this stage, one other dimension is identified such as "*partnership/partner*". As briefly addressed in previous sections, interoperability requirements can impact either the partner ship itself or partners themselves. For instance, the requirement "*the duration to connect application is less than x time units*" will impact the partnership and not necessarily a given partner. This requirement is positioned in the framework at the compatibility, performance and technological levels. Conversely, the requirement "*function f, performed by resource r involved in partnership, is even though performed*" will impact directly a partner (see Fig. 2 for its positioning in the framework). It is important to consider this aspect to know what is

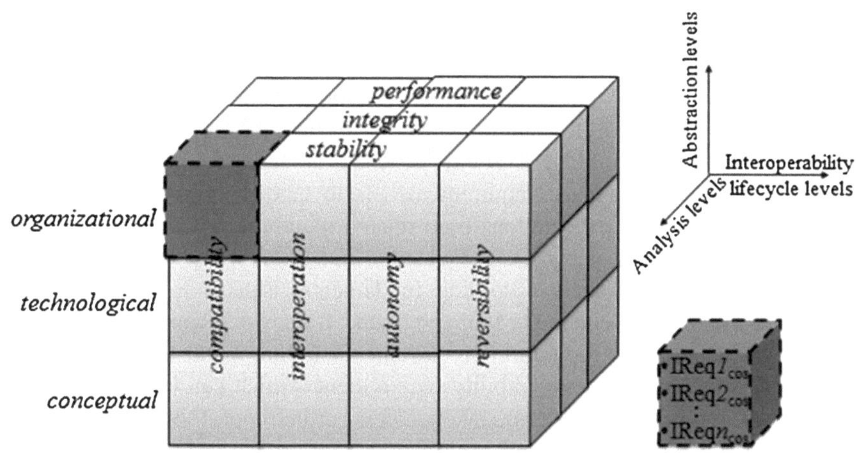

Fig. 1 Basic dimension of the framework for interoperability requirements

	Compatibility	Interoperation	Autonomy	Reversibility
Performance	$Req_{partnership}$: *"resource r required by service has the required aptitude"* Verif: conceptual graphs			
Stability		$Req_{partnership}$: *"resource r performs expected services despite perturbing environment"* Verif: test		
Integrity			$Req_{partner}$: *"function f, performed by resource r involved in partnership, is even though performed"* Verif: expertise	$Req_{partner}$: *"function f is performed despite the loss of its resource r"* Verif: expertise

Fig. 2 Interoperability requirements for the organizational level

impacted by the requirements to find/adapt solutions or untighten requirements. Another dimension could be related to the verification means of requirements. In that sense, verification methods and means must be defined to guide user in the verification phase. Finally, Fig. 2 shows some examples of interoperability requirements within the framework.

Other dimensions can be defined and added by users for a specific purpose or usage of the framework. However, the basic dimensions previously presented will have to be keeped since they represent the fundamental aspects of the interoperability requirements.

5 Conclusion and Prospects

In a collaborative context, having interoperability requirements can allow to verify that a selected interoperability solutions will meet original expectations and further to ensure that interoperability is always reached. Nonetheless, do not possess a clear framework—to manage requirements—is to take the risk of not meeting certain interoperability expectations (non related requirement, conflict, omission, no traceability…). This paper has presented a first framework to better structure interoperability requirements according fundamental aspects to consider when interoperability is implemented. Moreover, this framework can also used as a guide to elicite other interoperability requirements. Future works are related to the positioning of existing interoperability requirements which can be present in the literature or coming directly from stakeholders. Furthermore, this framework will be also technically implemented with works performed in [25] to allow the selection and the verification of requirements.

References

1. ISO/DIS 11345-1, (2009) *Advanced automation technologies and their applications*. Part 1: Framework for enterprise interoperability.
2. Chen, D., Doumeingts, G., & Vernadat, F. (2008). Architectures for enterprise integration and interoperability: Past, present and future. *Computers in Industry, 59*(7), 647–659.
3. Daclin, N., Chen, D., & Vallespir B. (2008). Methodology for enterprise interoperability. In *17th IFAC World Congress*, Seoul, South Korea, July 2008.
4. Bourey, J.P., Grangel, R., Doumeingts, G., & Berre, A.J. (2007). Report on model driven interoperability. INTEROP Network of Excellence IS 508011, deliverable DTG 2.3.
5. C4ISR Architecture Working Group (AWG). (1998). Levels of Information Systems Interoperability (LISI). USA Department of Defense.
6. Ford, C.T. (2008) Interoperability measurement, PhD Thesis, department of the air force Air University. *Air Force Institute of Technology*.
7. INCOSE. (2011). Systems engineering handbook—A guide for system life cycle processes and activities—v.3.2.1, INCOSE-TP-2003-002-03.2.1, January 2011.
8. Favaro, J., De Koning, H. S., Schreiner, R., & Olive, X. (2012). *Next generation requirements engineering, 22th Annual INCOSE International Symposium*. Rome: Italy.
9. The Standish Group. (1995). The chaos report. The Standish Group International, Inc.
10. Bochmann, G.V. (2010). *Writing better requirements*. University of Ottawa, slide only.
11. Ross, A. M. (2008). Defining and using the new "ilities", Systems Engineering Advancement Research Initiative (SEARi). working paper series.
12. ATHENA Integrated Project. (2003). Advanced technologies for interoperability of heterogeneous enterprises networks and their applications. Integrated project description of work.
13. Willis, J. (2011). *Systems Engineering and the forgotten '-Illities, 14th Annual Systems Engineering conference*. San Diego: CA, the United States of America.
14. ISO/IEC/IEEE. (2011). Systems and software engineering—Life cycle processes—Requirements engineering. ISO/IEC/IEEE 29148:2011.
15. INCOSE. (2012). *Guide for writing requirements*. v1.0, Requirements WG-INCOSE, INCOSE-TP-2010-006-01.

16. IDEAS. (2003). Interoperability Development for Enterprise Application and Software, IST-2001-37368, IDEAS Project Deliverables (WP1-WP7), Public reports.
17. ATHENA Integrated Project. (2005). Requirement for interoperability framework, product-based and process-based interoperability infrastructures, interoperability life-cycle services, ATHENA deliverable A4.1.
18. INTEROP. (2006). Enterprise Interoperability-Framework and knowledge corpus-Advanced report, INTEROP Deliverable DI.2, DI (Domain Interoperability), Network Of Excellence IS 508011.
19. ATHENA Integrated project. (2005). Framework for the Establishment and Management Methodology, ATHENA Integrated Project, Deliverable DA1.4.
20. C4ISR. (1998). Levels of Information Systems Interoperability (LISI), USA Department of Defense, Architecture Working Group (AWG).
21. Kasunic, M., & Anderson, W. (2004). Measuring systems interoperability: challenges and opportunities, Software engineering measurement and analysis initiative, Technical note CMU/SEI-2004-TN-003.
22. Daclin, N. (2012). Embedding interoperability in System of System: characterization and definition of fundamental requirements. In *13th IFIP Working Conference on Virtual Enterprises* (PRO-VE'12), Bournemouth: United Kingdom.
23. The Open Group Architecture Framework. (2011). TOGAF version 9, chapter 29 Interoperability requirements.
24. Harrison, S., & Brogden, J. (2007). Interoperability requirements, Supplier Requirement for SMART Metering (SRSM) project report.
25. Mallek, S., Daclin, N., & Chapurlat, V. (2012). The application of interoperability requirement specification and verification to collaborative processes in industry. *Computer in Industry, 63*(7), 643–658.

A Non-functional Framework for Assessing Organizations in Collaborative Networks

Aurélie Montarnal, Matthieu Lauras, Frédérick Bénaben and Jacques Lamothe

Abstract The French project OpenPaaS aims to support collaborative process by first deducing the process from the collaboration objectives and then orchestrating it. In order to design the process, a functional matching is established in order to find which sets of organizations are able to fulfil the objectives. Then a non-functional selection has to be executed in order to find the "best" process, with the most adapted partners. This paper presents a framework that has been settled for evaluating the organizations through non-functional criteria. Based on various cases of partner selection, this framework is intended to be the most exhaustive possible: it should allow the system to evaluate organization as a human would do in the case of a request for proposal. A structure of framework is first proposed, that fits with the OpenPaaS utilization. Then, non-functional criteria are classified according to it.

Keywords Non-functional criteria · Quality · Collaborative process · Partner selection

---

A. Montarnal (✉) · M. Lauras · F. Bénaben · J. Lamothe
Mines Albi, University of Toulouse, Campus Jarlard, Route de Teillet 81000 Albi, France
e-mail: aurelie.montarnal@mines-albi.fr

M. Lauras
e-mail: matthieu.lauras@mines-albi.fr

F. Bénaben
e-mail: frederick.benaben@mines-albi.fr

J. Lamothe
e-mail: jacques.lamothe@mines-albi.fr

K. Mertins et al. (eds.), *Enterprise Interoperability VI*,
Proceedings of the I-ESA Conferences 7, DOI: 10.1007/978-3-319-04948-9_21,

1 Introduction

The OpenPaaS project aims to support collaborative relationships between organizations. When an organization proposes a new collaboration opportunity the system proposes a set of partners and corresponding processes, which means that it can deduce the sequence of activities to execute in order to fulfil the objectives and the actor of each activity. Then the platform orchestrates the process and continues to support the collaboration during the run time.

In order to select partners among the organizations, there are two global steps: (i) which organizations have the required capacities, (ii) which organizations to select in order to obtain the 'best' process. This second question implies the assessment of the processes based on non-functional criteria.

The partner selection in virtual enterprise environments has been widely discussed in the literature and cost, delivery time and quality are the most often used criteria in such a problematic. However this triptych is neither adapted to specific collaborative contexts nor representative of the human reasoning when it comes to choose the ideal partner.

Based on a literature review this article proposes a new referential of non-functional factors that allows a broker to better specify its expectations concerning the collaborative context and the quality of the final service or product. Concretely these factors will appear on each organization profile that will help to characterize its business and will be visible by any other organization.

2 Proposal of a Three Dimensional Framework

2.1 Three Ways to Inform the Non-functional Criteria

The first thing to care about is how the criteria should be informed. It is indeed important for the broker to know measurable criteria values as costs or delivery time, but the fuzzy term of quality is a subjective judgement that is rather informed by the customers of an organization. Three ways to attribute values to non-functional factors are finally highlighted:

- (a in Fig. 1): The organization gives the criteria value on its own profile.
- (b in Fig. 1): An actual or former partner shares its working experience with the organization and therefore gives value to the criteria.
- (c in Fig. 1): The system automatically measures or calculates values and displays them on the organization profile (considering the fact that it supports the collaboration in design time, i.e. when the process is deduced, as well as in run time, i.e. when the process is orchestrated).

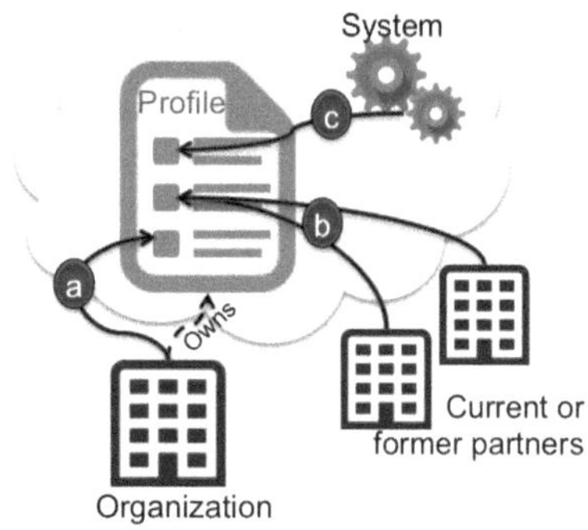

Fig. 1 Three ways to inform non-functional criteria

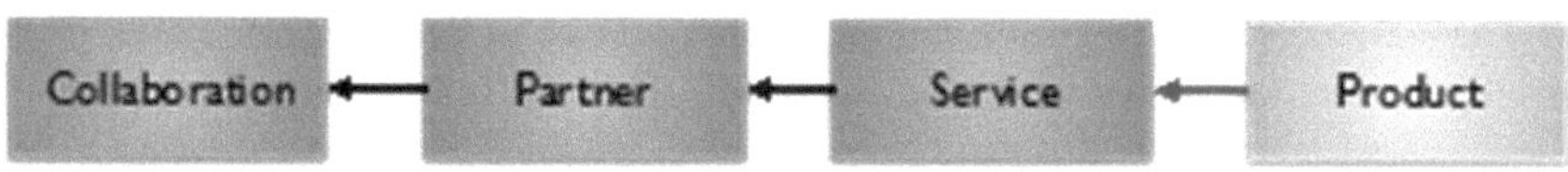

Fig. 2 Inheritance levels

2.2 Inheritance Levels

The criteria can be defined on four levels: the collaboration, a partner, a service or a product. As the OpenPaaS platform aims to support collaborations, it is important that partners share the same collaborative constraints that are frequently provided by an agreement. Thus an inheritance of the non-functional criteria is established. On each level, each individual partner inherits from the previous level which means that if a criteria is required for the collaboration, it is a fortiori for each partner too. Then, partners provide either a service or a product, but a product obviously comes with a linked service (for example the payment or the delivery). Consequently the eventual product level inherits from the service level which inherits from the partner (Fig. 2).

2.3 Non Functional Categories

The literature provides frameworks that aim to assess the quality of any work provided by an organization to a customer i.e. either a service or a product.

The service oriented SERVQUAL referential [1] is based on data gathered from enterprises and defines the service quality through five dimensions established on subjective trust from the customer as well as on technical skills reliability: Tangibles, Reliability, Responsiveness, Assurance, Empathy.

Garvin [2] tries to answer to the question: "what is quality for a product?" As product quality can be seen through a high cost, it can also be seen just through its characteristics and attributes or it can be a correlation between the performance of the product and an acceptable cost. Garvin's objective is to aggregate all the different definitions of quality in order to establish a global framework for better understanding the main element of quality. The author has based the framework on eight dimensions: Performance, Features, Reliability, Conformance, Durability, Serviceability, Aesthetics, Perceived quality.

To these 13 axes of study about quality, Hansen and Bush [3] add one more dimension: cooperativeness.

Table 1 brings a summary of all these dimensions and a definition of them. A correlation between some of them seems to be intuitively done, that is why they appear to be gathered in the table. Note that the term of reliability has been taken from the SERVQUAL framework and the definition of this term given by Garvin has rather been used for a criteria definition (cf. Fig. 4).

The six emerging categories that are expressed as I to VI in the remainder of the article are the categories we chose to classify the non-functional criteria.

3 Classification of Relevant Criteria for the Openpaas Project According to the Proposed Framework

3.1 Six Ways to Select Partners Based on Non-functional Criteria

In order to define non-functional criteria, the literature review has been oriented towards the usual factors for selecting partners in various contexts.

3.1.1 OASIS Standard

OASIS (Organization for the Advancement of Structured Information Standards) is a worldwide consortium whose role is to work on the standardization of formats. The WSQF (Web Services Quality Factors) Standard [4] aims to establish the functional and non-functional factors that define the quality of a web service. This standard is particularly relevant since it is commonly used in the selection of web services when orchestrating a process, for example [5].

The Fig. 3 illustrates the structure of the quality factors. As the web service can be considered as a very technical level of the collaboration, the factors are not oriented towards a "business" level as it is the case in this paper. However two groups seem to be relevant in the OpenPaaS case: the Business Quality Group and the Variant Quality Group. Most of the criteria of these groups have been kept and redefined to fit a "business" level.

Table 1 Definition of Garvin, Parasuraman et al. and Hansen and Bush dimensions

	Criterion	Parasuraman et al. [1]	Garvin [2]	Hansen and Bush [3]
I	Tangibles	"Physical facilities, equipment, and appearance of personnel"		
II	1. Reliability	(SERVQUAL)	1. "Ability to perform the promised service dependably and accurately"	
2.		Assurance	2. "Knowledge and courtesy of employees and their ability to inspire trust and confidence"	
III	1.	Responsiveness	1. "Willingness to help customers and provide prompt service"	3. "Speed, courtesy, and competence of repair"
		2. Empathy	2. "Caring, individualized attention the firm provides its customers"	
3.		Serviceability		
IV	1. Performance		1. "Primary operating characteristics of a product"	
	2.		Characteristics	
2.	"Secondary	characteristics that supplement the procust's basic functioning"		
V	1. Aesthetics		1. "How a product looks, feels, sounds, tastes, or smells"	
	2. Conformance		2. "Degree to which a products' design and operating characteristics match pre-established standards"	
	3. Perceived quality		3. "evaluation of the product depending on its image, advertising, or brand name"	
VI	Cooperativeness			Services or special efforts that a supplier could concede to a customer

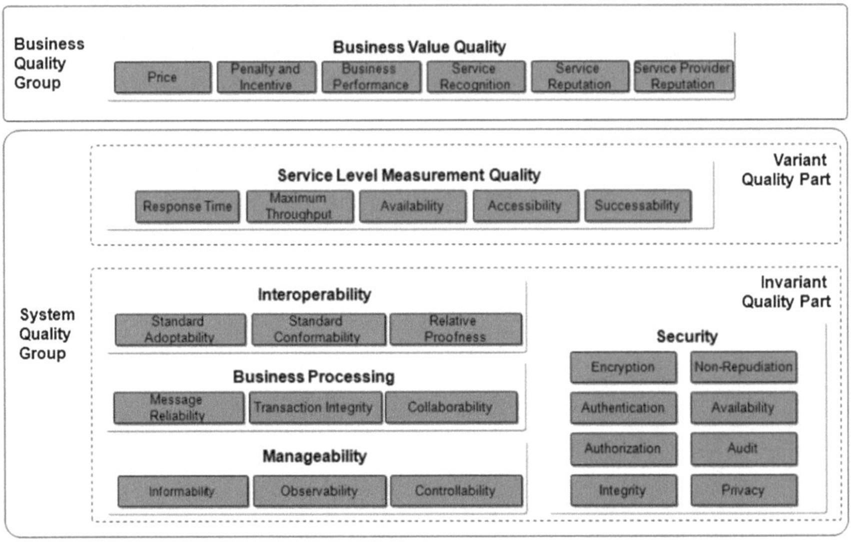

Fig. 3 Structure of web services quality factors, [4]

3.1.2 Web Service Selection Through Non-functional Features

Badr et al. [6] proposes a classification for an ontology that aims to allow web service selection through non-functional properties. It is divided into two parts corresponding to the properties of the context and of the Quality of Service (QoS). Most of the criteria are the same as the OASIS previous standard, but some are new: the *location*, the *payment method* and a last concept of *organization agreements* which allows to detect preferences of partnerships for the organization—based on the current and previous collaborations.

3.1.3 NFR (Non-functional Requirements) in the Software Engineering

In Roman and Boehm et al. research [7, 8] the authors have worked on NFR frameworks applied to the software engineering problematics. Among many others, the following criteria can be found: *performance requirements*, *economic requirements*, *functionality*, *usability* or *efficiency*. However, these NFR are rather oriented towards their specific application. Following the example of the OASIS Standard, the definitions of the NFR can not be used as they are, but must be adapted to the more generic context of OpenPaaS.

	I	II	III	IV	V	VI
Collaboration						
Own profile		*Penalty/ incentive *Confidentiality				
Partner						
Own profile			*Quick on short notice delivery			*Shipping arrangements *Just in time delivery offered *Credit terms offered *Long term price agreements *Payment methods
Partner profile		*On time delivery *Accuracy of system billing	*Reputation *Contact *Rapidly responding & solving the problem *Eagerness to meet the needs			*Ability to understand special orders
Automatically						*Organizations agreements
Service						
Own profile	*Equipment			*Price *Total costs *Delivery lead time		*Authorization *Location
Partner profile			*Knowledge/ expertise assessment		*Reliabability (Garvin)	*Agility
Product						
Own profile		*Large/ small orders capacity		*Product technical characteristics *Cost of ownership *Guaranteed life		*Product availability
Partner profile					*Expected quality *Real cost of ownership *Expected life	

Fig. 4 Classification of the selected non functional criteria

3.1.4 Supplier Selection Criteria

Davidrajuh and Deng [9] propose three criteria to constitute the basis when selecting suppliers:

- *Agility*: ability of the organization to react quickly and efficiently to the unexpected.
- *Quality*: the partner should be at least ISO certified and with the possibility to make audits.
- *Leanness*: set of the effective costs of the supplier.

3.1.5 Multiple Criteria Method to Evaluate Suppliers

Xia and Wu propose [10] a way to select suppliers using both qualitative and quantitative criteria. The authors divide the supplier evaluation into three parts, which eventually contain sub-criteria:

- *Price.*
- *Quality*: the *technical* quality of the products, their *defects* and their *reliability.*
- *Service*: the services the supplier is able to provide concerning its products, e.g. *on-time delivery*, *supply capacity*, *repair turnround time* and *warranty period.*

3.1.6 Use Case of a Supplier Selection in the Industry of Wood

Hansen and Bush [3] base their research on a survey they conducted among organizations of the industrial area. They finally obtain a set of 80 criteria classified according to the SERVQUAL and Garvin's dimensions. As the use-case is oriented towards wood purchasing, the framework is rather product-oriented. However it can easily be adapted to our three dimensional framework previously proposed since it does not only take into account the product but also its acquisition and the customer/supplier relationships. *Most of the criteria that are not specific to the field of the use-case have been kept.*

3.2 Selection of Non functional Criteria

Based on all these six ways to select partners, a set of the most relevant has been established and then classified according to the three dimensional framework proposed in the 1.2 part. The Fig. 4 finally summarize the entire paper by illustrating the set of criteria on the framework.

1. Collaboration
   - *Penalty/Incentive*: Financial penalty or incentives to be contractualized and measured on run time.
   - *Confidentiality*: Each partner signs a confidentiality agreement.
2. Partner
   - *Reputation*: Reputation of the service provider.
   - *Payment methods*: Accepted methods of payment.
   - *Organizations agreements*: Preferences and history (ongoing partnerships).
   - *Accuracy of billing system*: Accuracy if the organizations billing system, from the point of view of the partners: was there mistakes? (can lead to serious business issues).
   - *On time delivery*: Has the product or the service been delivered on time?

- *Quick on short notice delivery*: Is the organization efficient enough to deliver on short notice
- *Eagerness to meet the needs*: Eagerness of the organization to understand and answer correctly to the partner's needs.
- *Rapidly responding & solving the problems*: Is the organization able to respond rapidly to partner's problem?
- *Contact*: General assessment of the relationship between the partner and the organization.
- *Ability to understand special orders:* Efficiency of the enterprise to respond to special/exceptional order.
- *Shipping arrangement*: Ability to offer shipping arrangement to the partner.
- *Credit terms offered*: Does the organization accept credit? What are the terms?
- *Long-term price agreements offered*: Being recognized as regular customer. Long term business relationship arrangement.
- *Just in time (JIT) delivery offered*: Capacity/ability/coordination of the organizations to deliver products JIT.

3. Service

- *Price*: Estimated price of the service.
- *Delivery lead time (business performance):* Estimated time to complete the order.
- *Authorization:* Accessibility to the available capacities. (i) Monitoring: Ask for the advancement, anytime. (ii) Observability: subscribe to advancement notifications.
- *Location:* Execution location.
- *Agility:* Ability to react quickly and effectively to a sudden situation.
- *Total cost:* Every effective cost.
- *Equipment:* Equipment used to execute the activity.
- *Knowledge/expertise assessment*: Professionalism of the organization.
- *Reliability:* Does the service conform to the expected and guaranteed accuracy and capacities?

4. Product

- *Large/small orders*: Capacity of the expected: willingness to respond to small orders, capacity to respond to large.
- *Product technical characteristics.*
- *Guaranteed life*: Claimed life.
- *Cost of ownership:* Claimed costs of use.
- *Product availability:* Current availability.
- *Expected Quality*: Does the product conform to the expectations (outwardly & on use)?
- *Expected life*: Does the product conforms to the guaranteed expected life?
- *Real Costs of Ownership:* What does effectively cost the product on use?

4 Conclusion

To conclude, this paper proposes a new non-functional framework for selecting a partner in the case of a request for proposal. This framework aims to be as exhaustive as possible and should be sufficient to describe any organization through the four levels: in a collaboration, as a partner and as an organization that sells service or products. The next step will consist in deducing the "best" processes, according to the expectations of the broker of the collaboration. Thus processes could be ranked for letting the broker organization make its own final choice among them. Then comes naturally a second question: how to make a smart deduction that would avoid to find all the potential processes before assessing them? These two issues will eventually lead to the establishment of a unique algorithm that will explore solutions, assess them and always go to a better one until reaching the most ideal solution.

Acknowledgments The authors wish to acknowledge the French Project OpenPaaS for its support. We also wish to acknowledge our gratitude and appreciation to all the OpenPaaS Project partners for their contribution during the development of various ideas and concepts presented in this paper.

References

1. Parasuraman, A., Zeithaml, V., & Berry, L. L. (1988). SERVQUAL: A multiple-item scale for measuring customer perceptions of service quality. *Journal of Retailing, 64*(1), 12–37.
2. Garvin, D. A. (1984). What does 'product quality' really mean? *MIT Sloan Management Review, 26*(1), 25–43.
3. Hansen, E., & Bush, R. J. (1999). Understanding customer quality requirements: Model and application. *Industrial Marketing Management, 28*(2), 119–130.
4. Bosak, J., McGrath, T., & Holman, G. K. (2006). Web services quality factors version 1.0. Organization for the Advancement of Structured Information Standards (OASIS), Standard.
5. Zribi, S., Bénaben, F., Lorré, J.-P., & Pingaud, H. (2013). Enhancing services selection by using non-functional properties within BPMN in SOA context. In *Collaborative Systems for Reindustrialization* (Vol. 18, pp. 305–313). Dresden, Germany.
6. Badr, Y., Abraham, A., Biennier, F., & Grosan, C. (2008). Enhancing web service selection by user preferences of non-functional features. In *4th International Conference on Next Generation Web Services Practices, 2008. NWESP'08* (pp. 60–65).
7. Roman, G. (1985). A taxonomy of current issues in requirements engineering. *Computer, 18*(4), 14–23.
8. Boehm, B. W., Brown, J. R., & Lipow, M. (1976). Quantitative evaluation of software quality. In *Proceedings of the 2nd International Conference on Software Engineering* (pp. 592–605).
9. Davidrajuh, R., & Deng, Z. (2000). Identifying potential supplier for formation of virtual manufacturing systems. In *Proceedings of 16th IFIP World Computer Congress.*
10. Xia, W., & Wu, Z. (2007). Supplier selection with multiple criteria in volume discount environments. *Omega, 35*(5), 494–504.

A Flexible Monitoring Infrastructure for a Cloud-Based ESB

Amira Ben Hamida, Julien Lesbegueries, Thomas Morsellino, Sarah Zribi and Jean-Pierre Lorré

Abstract Nowadays, cloud platforms are widely used both in-premise in so called private Cloud or through public Clouds. Multiple models are available: IaaS platform to take benefit of ressources virtualisation, PaaS to provide development and deployment of Cloud dedicated IDE, SaaS to provide end-user ready applications. Main added value comes from the elasticity and multi-tenancy necessary for accessing largely distributed services to build complex application. Meanwhile, this capability raises the need for a monitoring infrastructure that would ensure services are behaving as expected. The control should be operated in a non intrusive and efficient way. At the same time, Event Driven Architectures (EDAs) are considerably growing thanks to the loosely coupling paradigm they provide. The SocEDA project investigates innovative solutions to provide a Cloud-based platform for large scale social aware Event-Driven Architecture (EDA). In this context, we take benefit from both Enterprise Service Bus (ESB) and EDA technologies in order to provide a flexible monitoring infrastructure for a cloud-friendly ESB. In this paper, we present the main aspects of the implemented middleware.

Keywords SOA · ESB · EDA · SLA · ELO · Cloud · Monitoring · Middleware

---

A. Ben Hamida · J. Lesbegueries · T. Morsellino (✉) · S. Zribi · J.-P. Lorré
Linagora GSO, 75 route de Revel 31400 Toulouse, France
e-mail: thomas.morsellino@linagora.com

A. Ben Hamida
e-mail: amira.benhamida@linagora.com

J. Lesbegueries
e-mail: julien.lesbegueries@linagora.com

S. Zribi
e-mail: sarah.zribi@linagora.com

J.-P. Lorré
e-mail: jplorre@linagora.com

K. Mertins et al. (eds.), *Enterprise Interoperability VI*,
Proceedings of the I-ESA Conferences 7, DOI: 10.1007/978-3-319-04948-9_22,

1 Introduction

One of the main challenges for the Future Internet (FI) is the processing of a large amount of data constantly sent throughout the networks and among the users. In such environments, several entities with ultra large scale quantities such as services, things, sensors, and applications come into play. Event-driven Architectures (EDA) are particularly adapted to similar highly distributed platforms. Actually, they rely on a loosely coupled paradigm where events are produced, brokered and consumed asynchronously by the interested stakeholders. For instance, interested services can subscribe and process relevant and different events simultaneously. We are firmly convinced that this feature considerably leverages the capabilities of Service-Oriented Architectures (SOA)-based applications to process relevant events. Meanwhile, handling an ultra large scale number of events faces the applications and their users to challenging issues, such avoiding bottlenecks and security failures and preventing the relevant events from being drown out. In the context of the SocEDA Project,[1] we address these issues thanks to both EDA and Social Networks paradigms. In this scope, we present our contribution of an elastic cloud-based lightweight service bus granting access to services, and an adapted monitoring framework acting as an event consumer. Describing the social aspects is beyond the scope of this paper.

2 Overview

The core proposition of this paper is twofold. It leverages the capabilities of ESB technology with an event-based, architecture that captures the produced events and processes them. Hereafter, we briefly describe both infrastructures.

- Petals EasyESB

We build Petals EasyESB prototype in order to natively support FI-adapted capabilities, besides, the common known ESB integration and orchestration functionalities. Petals EasyESB is a research prototype inspired from the Petals ESB product, which is natively distributed and open source.[2] Atop of the bus we implement specific aspects that confer to the bus the ability to face both the monitoring and cloud requirements. More details are given in Sect. 1.

- A Business Service Monitoring Framework.

Atop of this service runtime middleware, we implement a monitoring framework called EasierBSM. We rely on governance and monitoring layers able to capture,

[1] French National Research Agency funded project http://www.agence-nationale-recherche.fr.

[2] http://petals.ow2.org/.

process and control the traffic of events produced by services interactions. In order to achieve a high-level monitoring, we inspire from the commonly used Service Level Agreement Standard (SLA) and its related standards (WS-Agreement[3] and WS-CAP[4] [1]), and introduce a new analogical paradigm, targeting the events, namely the Event Level Agreement (ELA). ELAs allow to define agreements on the frequency of events supposed to be received from a given producer, topic or content. More details are provided in Sect. 1.

3 An Elastic Lightweight Service Bus

We enhance Petals EasyESB with specific intrinsic facilities that move it forward to next ESB generation. First, we design Petals EasyESB in a manner that eases its deployment on top of a decentralized cloud infrastructure, by relying on the Service Component Architecture (SCA) paradigms. Second, we provide a lightweight and flexible bus able to be deployed on constrained and decentralized cloud nodes. Third, we exploit the SCA philosophy to adding new components embedding a monitoring behaviour. Hereafter, we detail the cited capabilities as well as their implementations.

3.1 Cloud-Friendly Facilities

Considering the increasing proliferation of the cloud platforms, the FI environments are more and more relying on specific hardware infrastructures virtualized thanks to cloud-based facilities. Such platforms relieve the systems administrators from maintaining the platforms and handling the hardware failures. The cloud-based platforms represent a step forward that promises to leverage ESBs. Furthermore, in order to achieve an efficient deployment on top of the cloud, a natively distributed ESB is required. Petals EasyESB adopts a suitable topology allowing the architecture flexibility and ability to be deployed atop of decentralized hardware architecture. This is obtained thanks to the modularity gained thanks to the adoption of the SCA specification; Petals EasyESB enables a dynamic and elastic deployment over a cloud infrastructure. Elasticity in terms of deployment is enabled thanks to the dynamic adding and withdrawal of the bus nodes. We implement the needed configurations in order to automate the deployment of the bus on top of a multi-organization cloud solution. Indeed, thanks to a specific

[3] http://www.ogf.org/documents/GFD.107.pdf.

[4] http://docs.oasis-open.org/emergency/cap/v1.2/CAP-v1.2-os.htmland.

deployment engine, each ESB node is deployed separately by operating independent and iterative tasks. Thanks to the setting of specific deployment scripts we achieve an ultra large-scale deployment of the bus nodes over the cloud. These scripts can be executed in a parallel way to obtain a better coverage of the underlying cloud platform. In order to ease the deployment task, we implement a console client for generating the deployment scripts for the cloud. Furthermore, in order to obtain more elasticity the bus topology can be updated at runtime without stopping the bus nodes. For instance, Petals EasyESB allows adding new bus nodes to the existent topology. This is achieved by a simple call of the function *addNeighbourNode()*. Using this feature enables the dynamic discovery of the services exposed on the bus nodes. This way a service deployed on a given bus node, itself deployed on a cloud machine, is able to access a service that is deployed on another bus node regardless of their physical addresses. The technical registries of each bus node where services are referenced are aware of the added services of the newly joining neighbour nodes.

3.2 A Lightweight and Flexible Bus

Petals EasyESB topology relies on the association of interoperating services, rather than a single centralized and rigid container as commonly encountered. It implements the Service Component Architecture (SCA)[5] specification. As such, Petals EasyESB relies on the basic SCA Assembly Models that structures the middleware into modular composites dynamically and easily pluggable. Consequently, Petals EasyESB relies on specific unit deployments, holding services and abstracting their physical locations. Business functionality is afforded thanks to the exposed services implemented separately. The lightweight aspects reside in the ability of holding uniquely the needed functionalities relieving the bus from hosting all the services, as it is the case in common ESBs. Such modularity brings the ability for the bus to be deployed on memory-constrained machines. For instance, in terms of memory usage,[6] Petals EasyESB can be stored on only 50 MB of memory in comparison with existing solutions such as, Mule ESB with 75 MB, IBM Websphere with 550–600 MB, and 1.1 GB for Oracle Service Bus. Moreover, when started in the minimal version, Petals ESB uses only 300 MB while, Mule ESB consumes 768 MB, IBM Websphere uses 1–2 GB and Oracle recommends 4 GB. Finally, Petals EasyESB is a standalone bus that does not require the installation of further software such as for IBM and Oracle Solutions. Similar features are worthwhile when targeting a cloud infrastructure.

[5] http://oasis-opencsa.org/sca.

[6] http://www.mulesoft.com/mule-esb-small-footprint.

3.3 Abilities for Components Support

Petals EasyESB takes benefit from the modularity promoted by the SCA principles and exploits the services facilities for easily supporting new components. These features are available at both deployment and execution times. Considering the deployment time, Petals EasyESB provides a Standard Development ToolKit for easily creating and generating components and deployment units for the bus. Actually, thanks to the adoption of the Service Provider Interface approach,[7] we enhance the bus capabilities of dealing automatically with the components life-cycle. At runtime, atop of each bus node we provide a high-level administration layer proposing the needed functionalities for components support. At runtime, the functionalities are exposed as Web Service functions towards a Web administration. Useful operations such as binding, deploying, exposing, and proxifying services, are supported. More advanced features related to the management of the bus topology are also available and described later in the paragraph. We strongly believe that adding the abstract high-level layer for dealing with the bus administration eases the interaction between the bus nodes and leverages the interoperability between heterogeneous services and applications. This feature is exploited in order to add monitoring-enabled components to the bus (Sect. 1).

4 Implementation of a Monitoring Solution

Monitoring capabilities are of the utmost importance especially when dealing with a very large and heterogeneous number of services. It is a common fact that services fail or misbehave in performing their required objectives. Such cases need to be reported in order to put in place the needed measures to correct the occurring issues. In particular, this functionality is greatly needed, due to the fact that as the number of services increases, the probability of encountering failures increases as well. In order to achieve the services QoS monitoring, we implement a mechanism based on the Event-Driven Architecture (EDA) paradigms. Moreover, exploiting the facilities of Petals EasyESB of supporting additional components as described in the previous section, we deploy on top of a native Petals EasyESB, components that convey the bus with a monitoring behaviour. These components take in charge the collection of the exchanged data and the management of the services QoS. More precisely, profiling Petals EasyESB with a monitoring behaviour enables the subscription to the services exchanges. In the following, we briefly describe the ELO concept and give further details of the implemented interception and monitoring mechanism.

[7] http://docs.oracle.com/javase/tutorial/sound/SPI-intro.html.

Table 1 Possible event-level objectives (ELO)

Property	Per Subscription	Description
Min/Max number of notifications received over time	Yes	It may be possible for an event consumer to define the maximum number of notifications that can received in a window period (e.g.'at most 10 messages/second')
Maximum number of different events sources	No	What is the limit of the platform in terms of listening to different event sources?
Privacy	Yes	Restricts the ability of a user to subscribe to a stream (directly or for a user-specified event pattern). Should not be alterable later because we only enforce privacy a subscribe-time
Maximum lag between subscription time and start time of receiving notifications	Yes	What is the maximum time that it is needed in order to start receiving notifications after you have been registered to an event source?

4.1 Event Level Agreements

In the same way agreements can be defined for services in Service Level Agreements (SLA), they can be defined for events too. Indeed, for instance, a commonly used Service Level Objective (SLO) for services is the latency. The service must answer in less than a given time. In a similar way, an Event Level Objective (ELO) can be defined on the frequency of the produced events (see Table 1 for examples of ELOs). In an FI environment where a very large number of events can be triggered, the ELA management allows to control the events flow traffic and prevents the relevant one from being drowned. Alterable property of an ELO is a property that can be updated at any time (unalterable otherwise). An ELA is composed of three main elements: (i) a Producer, (ii) a Topic and (iii) a Content.

4.2 Events Capturing and Processing

The monitoring layer is in charge of collecting technical and business information about services and events. Our contribution of a distributed monitoring framework is carried out on the basis of the facilities that are conveyed by the WS-DM [2] and WS-Agreement standards. In order to achieve ELA and SLA assessment, we realize an experimental framework able to monitor service providers or event producers thanks to a QoS manager, and detect agreement violations thanks to an agreement manager. We implement the corresponding open source prototype called EasierBSM (for Business Service Monitoring [3–5]). Figure 1 illustrates the architecture of EasierBSM as well as the reporting mechanism.

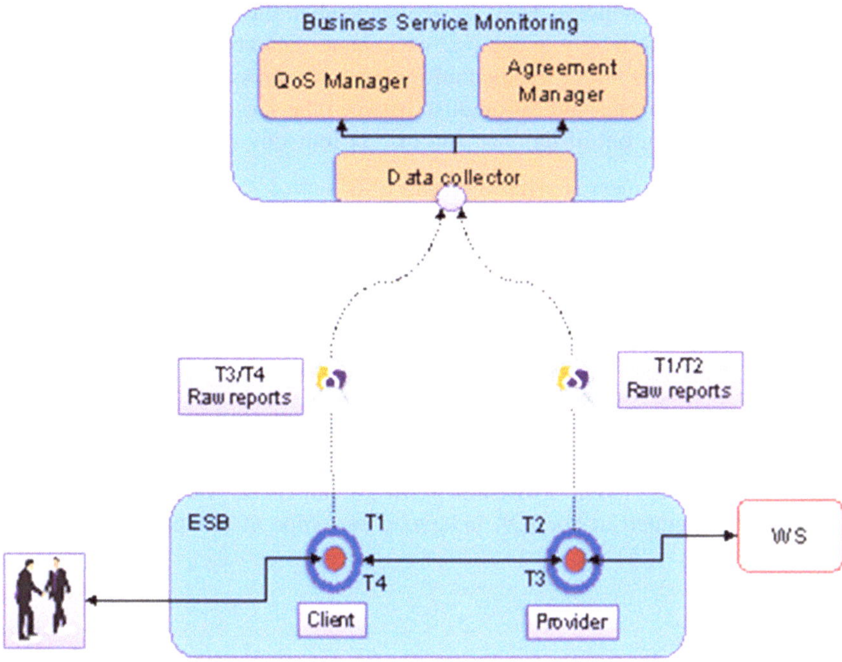

Fig. 1 Monitoring reports: Raw reports to monitor components activity

The monitoring is composed of three components as follows:

- a ***Data Collector*** that plays the role of a broker for incoming events. It is in charge of forwarding these events to the other components,
- A ***QoS manager*** that computes statistics on services and events, builds QoS reports (MetricsCapabilities from the WS-DM standard) and sends them as added value events,
- An ***SLA Manager*** that accepts WS-Agreements, watches involved services or events, and sends violation alerts if they fall short in fulfilling their service level objectives.

Once EasierBSM is started, it is federated to Petals EasyESB, thanks to a connection process involving automated subscriptions. EasierBSM fetches then all the available service endpoints, and creates for each endpoint a corresponding monitoring endpoint. The next step is the subscription of EasierBSM to the creation of new endpoints on the service bus. A new endpoint is created when a new service is exposed to the bus. For instance, when new endpoints are created, EasierBSM is notified and a new monitoring endpoint is created. Next, the monitoring endpoint subscribes to the monitoring information produced by the service bus. Monitoring information is transmitted through the bus and intercepted thanks to the monitoring interceptors. Interceptors are supported by the ESB and they surround each endpoint. They are activated when EasierBSM is connecting to the ESB. Thanks to these interceptors, we capture data about the interacting services as Raw Reports where detailed information is expressed. We encapsulate the Raw Reports into events. We distinguish between the service calls entering to the bus and those going out of the bus, and extract four timestamps,

1. **T1** when sent request is arriving in the ESB
2. **T2** when the request is going out of the ESB
3. **T3** when the answer, sent by the provider, is arriving in the ESB
4. **T4** when the answer is going out of the ESB.

The Data Collector collects the Raw Reports at T2 and T4 and forwards them to the QoS Manager that computes the real duration of the service execution. These reports can also be forwarded to the SLA Manager that compares the real values with those of the SLA or the ELA. When these values do not correspond, the SLA Manager creates a violation alert and sends it to the concerned component, in our case the Complex Event Processor (CEP) component. Describing the CEP mechanism integrated in SocEDA is beyond the scope of this paper.

5 Conclusion

In this paper, we illustrate our solution of providing a monitoring infrastructure on top of a cloud-friendly bus. We expose the main aspects that enable the implementation of such and infrastructure. We rely on Petals EasyESB as an enhanced ESB able to be deployed atop of a cloud infrastructure and to at the same time to be profiled with monitoring behaviours. More precisely, we exploit the bus the facility of adding new nodes we write automated deployment scripts to install the bus nodes on top of a cloud layer. Furthermore, exchanges between services bound to the bus are intercepted, processed and filtered thanks to an EDA solution. The suggested solution is used within the SocEDA project and showcased thanks to a nuclear simulation use case.

Acknowledgments This work was partially supported by ANR funding number ANR-10-SEGI-013 (SocEDA)10. Besides, the authors would like to thank Dr. Nicolas Salatgé for his helpful support especially about continueous integration and several development tasks involved within the SocEDA platform.

References

1. Common Alerting Protocol (CAP) v1.2, http://docs.oasis-open.org/emergency/cap/v1.2/CAP-v1.2-os.html. 2010.
2. Web Services Distributed Management: Management Using Web Services (WSDM-MUWS) v1.1 part 2—http://docs.oasis-open.org/wsdm/wsdm-muws2-1.1-spec-os-01.htm. 2006.

3. Ben Hamida, A. Bertolino, A. Calabrò, A. De Angelis, G. Lago, N & Lesbegueries, J. (2012) *Monitoring service choreographies from multiple sources.* In *4th International Workshop on Software Engineering for Resilient Systems, SERENE* (pp.134–149).
4. Ben Hamida, A. Lesbegueries, J. Salatge, N.& Lorré, J-P. (2012). A toolkit for choreographies of services: Modeling, enactment and monitoring. In *OTM Workshops*, (pp.60–63).
5. Lesbegueries, J. Ben Hamida, A. Salatge, N. Zribi, S. & Lorré, J-P. (2012). Multilevel event-based monitoring framework for the petals enterprise service bus: Industry article. In *DEBS*, (pp.48–57).

Enterprise System Architecture to Sustain Cross-Domain e-Healthcare Applications

Natalia Costetchi, Cristian Danila and Aurelian Mihai Stanescu

Abstract Romania's Healthcare Complex Adaptive System of Systems is in progress to support a very ambitious process of structural, but behavioral also, transformation. Healthcare domain, e-Health systems and e-Healthcare subsystems are facing, worldwide, with objective phenomena: ageing population, increasing prevalence of chronic diseases, degenerative diseases, healthcare costs rising, and so on. One topics to be under debate due to present paper contributions is concerning with care diseases information capturing, but knowledge management by both sharing best practice and explicit recommendations within a holistic approach for cross-domain investigations. Information Technology and Communication Tools (ICT) within the new Future Internet Enterprise Systems (FInES) paradigms is a major key to reach high Quality of Services, efficient implementation of Web interoperability oriented new methodology to develop complex composite Web services and last but not least, new System Architecture of e-Healthcare systems. A research programme so called E4H CAS [Environment, Economics, Education and Entrepreneurship to sustain e-Healthcare Complex Adaptive Systems] is aiming at populating the e-Health framework with focused prototypes and applications like knowledge capture, repository archiving and best practices of rare diseases multidisciplinary diagnoses and therapy.

Keywords Information system · Service architecture · e-Healthcare framework

N. Costetchi (✉)
Automatics and Computer Science Faculty, University Politehnica of Bucharest,
313 Splaiul Independentei, Bucharest, Romania
e-mail: natalia_costetchi@yahoo.com

C. Danila · A. M. Stanescu
Department of Automatic Control and Industrial Informatics,
University Politehnica of Bucharest, 313 Splaiul Independentei,
Bucharest, Romania
e-mail: danila.cristian.84@gmail.com

A. M. Stanescu
e-mail: amstanescu@yahoo.com

K. Mertins et al. (eds.), *Enterprise Interoperability VI*,
Proceedings of the I-ESA Conferences 7, DOI: 10.1007/978-3-319-04948-9_23,

1 Introduction

"Human development is a process of enlarging people's choices; the most critical of these wide-ranging choices are to live a long and healthy life, to be educated and to have access to resources for a decent standard of living" [1]. By focusing the key concept "healthy life", the worldwide accepted definition for health is "A state of complete physical, mental and social well-being and not merely the absence of disease or infirmity" [2].

All societies are deeply concerned within health domain reforming process sustained by new "wave" ICT technology under the FInES paradigm shift. This topic is actual and burning in entire context: local, national, transnational, international and global. Due to globalization process the issue of health became global.

Romania's Healthcare Complex Adaptive System of Systems is in progress to support a very ambitious process of structural, but behavioral also, transformation. Healthcare domain, e-Health systems and e-Healthcare subsystems are facing, worldwide, with objective phenomena: ageing population, increasing prevalence of chronic diseases, degenerative diseases, healthcare costs rising, and so on. One topics to be under debate due to present paper contributions is concerning with care diseases information capturing, but knowledge management by both sharing best practice and explicit recommendations within a holistic approach for cross-domain investigations.

1.1 e-Healthcare Application with Respect to Rare Disease Patient

One of the complex problems to be solved urgently worldwide is the domain of rare diseases [3]. This area is now one of the priorities in the second programme of Community action in health field [4]. According to [5] rare diseases, including those of genetic origin, are life-threatening or chronically debilitating diseases which are of such low prevalence that special combined efforts are needed to address them so as to prevent significant morbidity, perinatal or early mortality, or a considerable reduction in an individual's quality of life or socio-economic potential. EU considers diseases to be rare when they affect not more than 5 per 10,000 persons [6].

In Romania the estimated number of persons affected by rare diseases is about 1,300,000 but approximately 1,250,000 persons have not been diagnosed properly or didn't receive the appropriate therapy [7].

Based on European Council recommendation [8], the Nation Health System (NHS) of Member States (MS) is engaged to: concentrate on supporting and strengthening the adoption of national plans and strategy for responding to rare diseases by the end of 2013; improve recognition and visibility of rare diseases, encourage more research into this area, establish link between centres of expertise and professionals in different countries through European reference networks in order to share knowledge and expertise; identify where patients should go when such expertise cannot be made available to them at local or national level. European action aims to support and help patients and professionals collaborate across MS to share and coordinate expertise and information through linking centres of expertise in different countries and making use of new information and communication technologies such as e-Health.

E-Health plays a clear role and is a key to achieve significant improvements in access to care, quality of care, and the efficiency and productivity of the health sector. There are many examples of successful e-Health developments and deployments [9]. The definition of e-Health is "the use of information and communication technologies (ICT) for health to, for example, treat patients, peruse research, educate students, track diseases and monitor public health" (WHO).

Romania as part of European Union starts to apply the European policies into national strategy programmes. In this respect was established a National Program for Rare Diseases (2010–2014) [10] in order to meet the requirements for collaborative relationships in rare diseases area. However, at national Romanian level there is a lot of work to be performed. In 2012 a qualitative study of quality of life of people diagnosed with rare diseases in Romania [11] was highlighted a set of difficulties for patients that suffer from a rare disease.

A Romanian patient diagnosed with a rare disease based on health legal framework [12, 13] has the chance to be treated outside the country if NHS couldn't assure appropriate treatment. The issue of delays and breakpoints in the health information flow of health system should be addressed and solved.

Our hypothesis is whether the e-Health solutions should be developed and integrated in health care system based on an agile e-Health Enterprise Architecture (EA) Framework. Our aim is to propose a system architecture that could be applied for macro vision of National e-Health strategy and/or micro vision of specific domain strategy, like e-Health for rare diseases. Health Information Exchange (HIE) should be based on an iterative implementation of collection and aggregation of all legal requested health information for Electronic Health File Patient Rare Disease to evaluate and release cross-border appropriate treatment.

To integrate stakeholders and e-Health solutions in a citizen-centred e-Healthcare agile Complex Adaptive System is a very promising challenge within "Horizon 2020" Digital Agenda.

1.2 Preliminary Research

A research methodology based on mixed techniques was selected in order to evaluate the possibility of HIE about rare diseases patients in Romanian NHS.

Our research has been initially based (phase 1—empirical study) on: domain literature and legal framework; up-to-date establishment according to both quantitative and qualitative research methods. Through quantitative research we have been focused on discovering and understanding of experience, perspective of health professionals in informational health system context. As regarding qualitative research we aim at exploring a real perception of already using e-Health solution implemented in the health units or data interexchange in National Integrated Informational System (NIIS).

1.2.1 Quantitative Research

In order to perform quantitative research there was created a survey for understanding personal involvement and experience of using communication and information technologies by healthcare professionals. The statements were grouped by following main areas: information about health information system (HIS) in the organization (components, level of data collection, patient related information, standard of implementation, integration of HIS with other health systems like: Diagnosis Related Groups (DRG) [14] and NIIS [15] in order to interexchange data; and perception of health care professionals about current e-Health tools.

1.2.2 Qualitative Research

For qualitative research there were selected two tools:

- Interview

The first goal was to collect valuable data and opinions from cross disciplinary and sectorial professionals about e-Health public solution implemented and outcome of its use. Participants were from following domain: IT Health Information Systems provider, statisticians from public hospitals, family and specialist doctors, CFO from private hospital, expert from District Health Insurance Houses, expert from National School of Public Health, Management and Professional (NSPHMPDB), former president of health Statistic Institute, experts from Romanian National Alliance for Rare Diseases and National Committee for Rare Diseases.

All of them are linked together, through health informational flow used in the day-to-day business.

The second aim was to analyse the possibility of integration of stakeholders in the Enterprise Architecture e-Health Framework and interoperability issue to be addressed.

- Case study

We have studied a patient, diagnosed with a rare disease (di George complete syndrome) and we observed the patient pathway from the first interaction with Romanian health care system up to the approval by Health Ministry the thymus transplant in paediatric hospital in London [16]. This study emphasizes the breakpoints and delays in health care system that could be critical for a patient with the mentioned disease.

The study was performed based on all discharged medical documents and files, laboratory analysis, the documents released from Health Information System Paediatric Hospital "Gr. Alexandrescu" and actual interoperability of data for health information exchange across actors.

Each stage for data collection contributed to redefine the question asked and problematic of rare disease domain. The mistrust of patients and professionals in Romanian Healthcare System made difficult to achieve a holistic approach in this research.

Some findings obtained by empirical traditional quantitative research pointed that 45 % of participants do not have a valid email address on the official hospital site.

In qualitative research based on interviews, we found a lack of connection point between the decision bodies and the units that hold the primary information that is needed to the decisional process. This connection point is vital for validation of the decisions with some values and real economic results collected during the medical act that is provided by the medical units. The information is spread out and split into varied interest domains, in some cases, hold by different public institutions, and the treated aspects do not make a system, a whole.

We propose a paradigm shift from service oriented to patient centred Health System. The paradigm changes will allow developing future cross-domain and interdisciplinary approach for address appropriate treatment for patient with rare disease. Moreover, after implementing and integration of Electronic Health Record (EHR) in National Health Information System, a next step is Electronic Health File Patient Rare Disease.

The rest of the paper is organized as follows. Section 2 presents basic requirements e-Health Enterprise Systems. In Sect. 3, an Enterprise Architecture based e-Healthcare Framework is proposed. Section 4 represents the conclusions.

2 Basic Requirements for e-Health Enterprise Systems

Numerous standards have been developed in the e-Health domain. There is a need for quality assessment models that not only evaluate the performance of the e-Health Enterprise Systems but also cover important aspects such as: interoperability, adaptability and security. Two aspects should be considered: product quality and process quality. We propose a hierarchical model for quality of applications that takes into consideration all these factors.

The model has three quality dimensions [17] which assess the e-Health Enterprise System from a certain point of view: business, operational or systemic. Each quality dimension is composed of one or more quality aspects. For each aspect we consider one or more quality items, which can be measured based on a

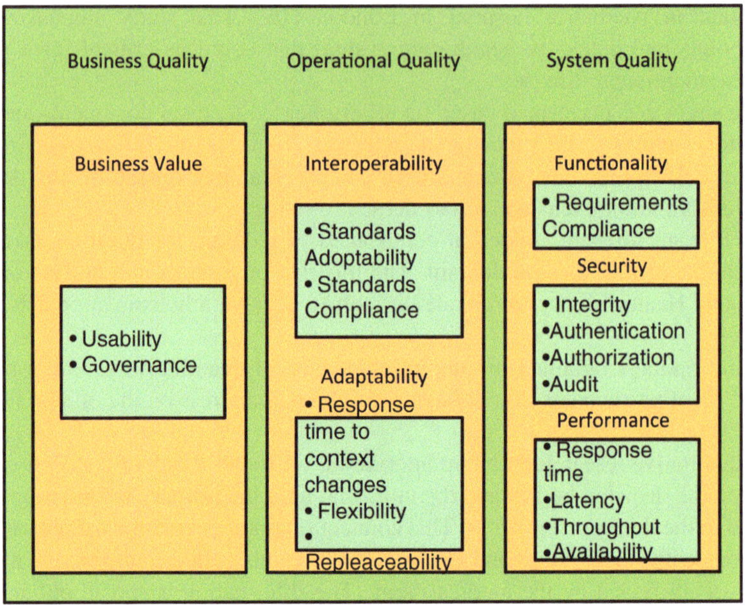

Fig. 1 Quality model for e-Health enterprise systems

quality criteria. The measurement can be either qualitative or quantitative. The three quality dimensions are: Business, Operational and System as represented in Fig. 1. The business quality dimension describes quality aspects that belong to the business value of the system. The operational quality dimension describes quality aspects related to interactions between applications while the system dimension includes quality aspects that apply to the entire e-Health Enterprise System.

Business quality dimension contains the following quality items: usability and governance. These items can be qualitatively evaluated based on a history of the application usage experience and user feedback.

Operational quality model includes application interoperability and adaptability quality aspects. Interoperability quality aspect is concerned with the evaluation of the compatibility level between applications. Although communication standards have been created, applications are still developed on different platforms and according to different specifications. Standards adoptability and standard compliance are quality items that evaluate the message exchangeability and conformity to a certain standard in order to make the applications interoperable. Adaptability quality aspect evaluates the ability of an application to respond to external stimuli. Adaptability implies three abilities: the ability to recognize an environmental change, the ability to determine the change to be applied to the application and the ability to effect the change.

System quality model evaluates the applications compliance with user defined requirements both functional and non-functional. It contains the following quality aspects: application functionality, application security and performance. The quality of the security incorporated into the applications has the following quality items: authentication, authorization, encryption, non-repudiation, audit and integrity. The performance quality aspect of the applications can be composed of quality items like: response time, throughput, availability, accessibility, latency, accuracy.

The list of quality items is not exhaustive, but only describes most common quality features desired in an e-Health Enterprise System. The quality model includes not only product related quality characteristics but also process specific quality attributes. The quality model can be easily extended with more quality items according to the context of the applications.

3 Enterprise Architecture Based e-Healthcare Framework

E-health represents the use of information and communication technologies (ICT) for supporting heath. The e-Health Enterprise System Architecture (EA) provides a framework to support e-Health in delivering coherent and interoperable e-Health solutions that can form together a true integrated e-Health System aiming to deliver shared data and applications to healthcare participants.

In this paper we introduce an e-Healthcare Framework (Fig. 2) for guiding the development and evolution of an integrated National Health Information System, inspired from Enterprise Architecture concepts [18–20].

In order to achieve a comprehensive e-Healthcare Framework several dimensions are addressed:

1. At the contextual level the following are addressed: Strategy, Objective, Goals. This allows a careful consideration and definition of what is appropriate and marks the highest priority for e-Health area.
2. The scope and strategy should be maintained by a legislative framework that supports and creates an appropriate behaviour for implementing and acting according to defined strategy. In order to integrate e-Health tools in health care services, political and policy background is necessary to meet nation's needs and capacities (conceptual level).
3. Business Architecture modelling is seen as a solution to bring technological innovations to successful deployment of the business. In order to define the Business architecture, one has to define: business domains, business processes and value streams, capabilities, models (domains, functions) and events. According to SAMBA (Structured Architecture for Medical Business Activities) project, a method to depict a process model was developed, which elucidates all essential parts of the process [21].

   This component is one of the most critical from EA due to an importance of organizational goals/strategy and Informational Technology alignment.

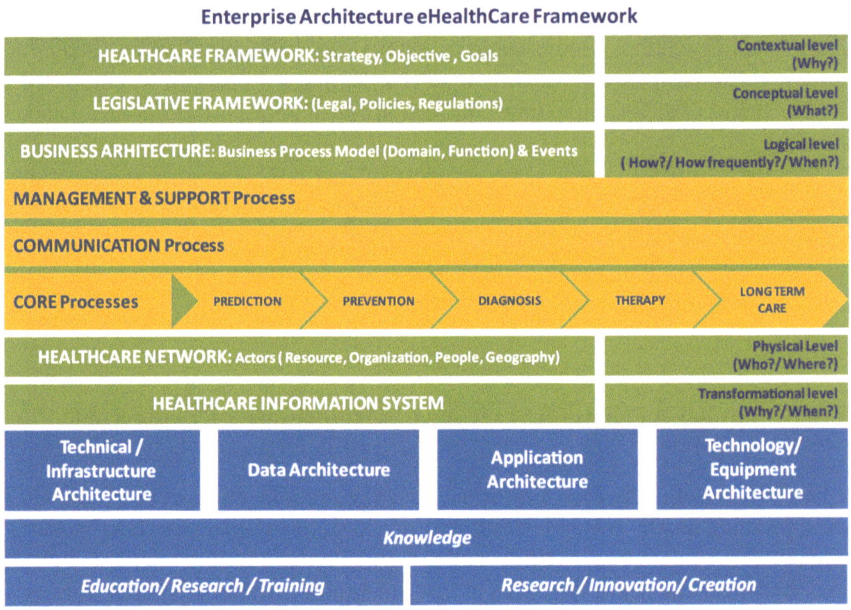

Fig. 2 E-Healthcare framework overview

The research findings [22] show that developing the BA can improve business and IT alignment and provide inputs to other architectures, i.e. information, application and technical components of EA.

4. Health systems are networks of actors: define resources, organizations, people, geography, actors, parties and roles.
   This section identifies the resources used to define EAs and architecture solutions.
5. Healthcare Information System includes: Information management Principles Information Management Policies and Rules.
   Information Architecture describes the components for managing information across the enterprise. High-level enterprise data model and the data flow patterns around a particular architecture scope.
6. Technical/Infrastructure Architecture includes: Hardware platforms, Data centres, Local and wide area network, Operating system interoperability (semantic and syntactic), Security Infrastructure, Client technology platforms, Protocols and technology choices.
   Technology Architecture describes how the infrastructure underlying the business, application, and information architectures is organized.
7. Data Architecture includes: Data model, Meta data dictionary, Classification.
8. Application Architecture includes: Software applications, Interfaces, User.
   Application Architecture describes the architecture for enterprise applications, information systems, and automation processes that support the business architecture.

9. Technology/Equipment Architecture include the convergence of medical equipment based on new technology insight and IT systems. Medical devices no longer operate independently within the healthcare environment. They provide critical patient care information to clinical and physician staff and must be compatible to achieve optimal efficiencies, not only within departments, but facility and system-wide.
   Accommodating advanced technology solutions directly impacts on operational efficiency, quality of care and patient satisfaction.
10. Knowledge—is accepted widely that we moved into knowledge era. Since the paradigm shift was realized, it is mandatory to manage knowledge more effectively. The knowledge "is a fluid mix of framed experiences, values, contextual information and expert insight that provides a framework for evaluating and incorporating new experiences and information. It originates and is applied in the minds of the people that hold the knowledge. In organizations, it often becomes embedded not only in documents or repositories but also in organizational routines, processes, practices and norms [23]. In this context, knowledge is supporting layer for e-health framework in order to enhance the value of stakeholders' personal capital solely by deployment of systems, processes and technologies.
11. Educational/Research/Training and Research/Innovation/Creation are important steps to support and continue the improvement and changes of e-Healthcare system.

Especially for Health, which has a holistic approach in the System, it is demanded to share and coordinate expertise for better and more efficient solution and health services to be meet.

4 Conclusions

In order to meet Healthcare Informational objectives of a society it is important as first step to establish an Enterprise Architecture for e-Healthcare Framework. We proposed a representation which integrates components tailored to implementation of e-Health strategies in a holistic view. This framework can be applied to meet both general and specific objectives such as the Romania Healthcare framework for Rare Diseases.

Future development of Electronic Health File Patient Rare Disease and integration in actual National Integrated Informational System would make possible to streamline the health information exchange between actors. Aggregated digital patient rare disease information based on legal requirements and health standards would essentially solve the problem of delays and breakpoints in releasing files for cross border treatment and facilitate cross network's communication.

Acknowledgments The work has been funded through the European Social Fund, Sectoral Operational Programme Human Resources Development 2007-2013 of the Romanian Ministry of Labour, Family and Social Protection, Financial Agreement POSDRU/107/1.5/S/76813.

References

1. United Nations Development Programme. (1990). Concept and Measurement of human development. Human Development Report. Retrieved September 15, 2013, from http://hdr.undp.org/en/media/hdr_1990_en_chap1.pdf
2. WHO Definition of Health. Retrieved April 6, 2012, from http://www.who.int/about/definition/en/print.html
3. Commission of the European Communities. Communication from the Commission to the European Parliament, the Council, the European Economic and Social Committee of the Region on Rare Diseases: Europe's Challenges. In Brussels (Ed.). Retrieved from http://ec.europa.eu/health/rare_diseases/portal/index_en.htm
4. European Communities. Retrieved October 15, 2014, from http://eur-lex.europa.eu/LexUriServ/LexUriServ.do?uri=OJ:L:2002:271:0001:0011:EN:PDF
5. EUCERD. Retrieved October 15, 2012, from http://www.eucerd.eu/upload/file/Reports/2013ReportStateofArtRDActivitiesIII.pdf
6. The Council of the EU. (2009). Council recommendation on an action in the field of rare diseases. Official Journal of the EU. Brussels.
7. Health Ministry and National Alliance for Rare Diseases Romania. *National plan of rare diseases* (2010–2014). Retrieved October 20, 2013, from http://bolirareromania.ro/node/212
8. European Commission. Retrieved October 11, 2012, from http://eur-lex.europa.eu/LexUriServ/LexUriServ.do?uri=OJ:C:2009:151:0007:0010:EN:PDF
9. Silber, D. (2003). The case for eHealth. European Institute of Public Administration 2003. In *Presented at the European Commission's First High-level Conference on eHealth*, May 22/23 2003.
10. Romanian National Alliance for Rare Diseases. Retrieved October 11, 2012, from http://www.edubolirare.ro/sites/default/files/reviste/plannational/plan_nat_br.pdf
11. Quality of Life Journal. Retrieved August 13, 2013, from http://www.revistacalitateavietii.ro/2012/CV-2-2012/01.pdf
12. National Insurance House. Retrieved August 13, 2013, from http://www.cnas.ro/furmulare-e/formulare-e
13. Health Ministry. Order Nr. 50 from 23 January 2004 Retrieved August 13, 2013, from http://www.dspbihor.ro/wp-content/uploads/2011/10/Legislatie/Ordin%20nr.%2050%20din%202004%20al%20Ministerul%20Sanatatii.pdf
14. NSPHMPDB. Retrieved April 2, 2012, from http://www.drg.ro/
15. National Insurance House, & SIVECO. (2012). *Interface specification SIUI—Application Reports for vendor*. Version 2.2 from 12.01.2012. Retrieved June 30, 2012, from http://193.151.30.188/cnas/servicii_online
16. Pirvu, S. (2013). *Daniel's story and patient pathway*. Rare Disease Di George. Retrieved October 15, 2013, from http://danielpirvu.ro/ro/
17. Stegaru, G. (2013). *Modeling framework for quality driven web service composition*, Ph.D. Thesis.
18. Stănescu, A. M., Moisescu, M. A., Sacală, I. S., Burlacu, G., Nedelcu, R. A., & Cojocaru, L. E. (2012). Concurrent innovation-based eEnterprise. In *International Conference on Engineering, Technology and Innovation*, 18–20 Iunie 2012, Munchen, ISBN: 978-1-4673-2273-7, indexed in IEEE Xplore. doi: 10.1109/ICE.2012.6297697
19. Moisescu, M. A., Sacală, I. S., Stănescu, A. M., & Serbanescu, C. (). Towards integration of knowledge extraction from process interoperability in future internet enterprise systems. In

14th IFAC Symposium on Information Control Problems in Manufacturing, 23–25 May, 2012, Bucharest, Romania, ISSN: 1474-6670, ISBN: 978-3-902661-98-2. Doi: 10.3182/20120523-3-RO-2023.00395

20. Burlacu, G., Stănescu, A. M., Moisescu, M. A., Sacală, I. S., & Cojocaru, L. E. (). Development of a modeling framework for future internet enterprise systems. In *16th International Conference on System Theory, Control and Computing*, ICSTCC 2012, 12–14 October 2012, Sinaia, Romania, Paper ID: 178, ISBN 978-606-8348-48-3.
21. SAMBA. *Structured architecture for medical business activities*. Retrieved August 20, 2011, from http://www.ljungskilerevyn.se/fogare/samba/english.htm
22. Li, C., & Steenkamp, A. L. (2011) *Conference for information systems applied research*. Wilmington North Carolina, USA
23. Davenport, T. H. (1998). *Working knowledge: How organizations manage what they know*. Retrieved October 12, 2012, from http://www.kushima.org/is/wp-content/uploads/2013/09/Davenport_know.pdf

Part VII
Platforms for Enterprise Interoperability

Digital Business Ecosystem Framework for the Agro-Food Industry

Luiza-Elena Cojocaru, Joao Sarraipa, Ricardo Jardim-Golcalves and Aurelian Mihai Stanescu

Abstract The paper introduces concepts concerning Digital Business Ecosystem (DBE), its involved actors and their roles. It presents a framework to establish guidelines in a composition of a DBE. In additional, an ontology about its domain knowledge is proposed, which is accomplished with concepts, properties and rules related to the DBE actors. Finally, a branch of fruit and vegetables, with typical situations that have been identified in Romania was chosen as a case study to exemplify the usefulness of various IT solutions, performing an instantiation of the proposed framework main functions to which the members of the DBEs can benefit.

Keywords Digital business ecosystem · Agro-food · Ontology

L.-E. Cojocaru · A. M. Stanescu
Faculty of Automatic Control and Computer Science, University "Politehnica" of Bucharest, Splaiul Independentei nr. 313, Sector 6, Bucharest, Romania
e-mail: luiza_cojocaru85@yahoo.com

A. M. Stanescu
e-mail: amstanescu@yahoo.com

J. Sarraipa (✉) · R. Jardim-Golcalves
CTS, UNINOVA, Departamento de Engenharia Electrotécnica, Faculdade de Ciências e Tecnologia, FCT, Universidade Nova de Lisboa, CTS—UNINOVA, Campus da FCT 2829-516 Caparica, Portugal
e-mail: jfss@uninova.pt

R. Jardim-Golcalves
e-mail: rg@uninova.pt

K. Mertins et al. (eds.), *Enterprise Interoperability VI*,
Proceedings of the I-ESA Conferences 7, DOI: 10.1007/978-3-319-04948-9_24,

1 Introduction

Currently, new technologies and methodologies can support new business networks and opportunities establishment. As an example, a Digital Ecosystem is a persistent digital environment populated by digital components that evolve and adapt to local conditions with the evolution of its constituents [1]. Consequently, Digital Ecosystem endows its participants, especially SMEs, with accessible information technology, supporting and reinforcing their competitiveness [2].

On the other hand, the concept of "business ecosystem" as defined J. F. Moore is "an economic community supported by a foundation of interacting organizations and individuals—the organisms of the business world" [3]. Such economic community produces goods and services valued by clients, which are members of the ecosystem.

As a result of the structural dependence between digital and business ecosystems, it has been created the DBE (Fig. 1) [1]. In such new ecosystem, digital and business characteristics are interconnected and co-exist forming a dynamic innovation system. In a DBE there is a clear requirement, which intends to perform an explicit connection regarding the domain lexicon between two or more entities that could participate in such system kind. This point is even more relevant if we consider that digital ecosystems include digital representation of economical aspects of business ecosystems, used to search and to recommend potential business partners.

The research work presented in this paper is about the introduced concepts accomplished by guidelines to its implementation. Thus, in the following chapter it is presented a framework to provide the main functions for the proposed DBE establishment. Additionally, in this process the members of this DBE and the relation between them are highlighted. Since ontologies are typically used to represent a common view of the domain knowledge, it was the technologic information element chosen by the authors to handle semantics especially with the purpose to contribute to interoperability maintenance of such systems. At the end, a case study is presented to explain the IT solutions chosen, from which the members of the DBE can benefit.

2 The DBE Framework

To create a DBE, the main functions provided by such system has to be established. Accordingly to this, authors propose a framework, which functions are represented in (Fig. 2).

The Business Ecosystem is represented by the business relations that occur in the Collaborative Networks. A collaborative network is a network consisting of a variety of entities (e.g. organizations and people) that are largely autonomous, geographically distributed, and heterogeneous in terms of their operating

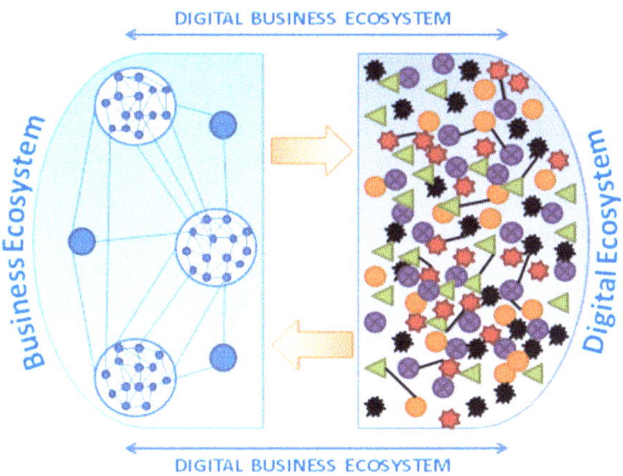

Fig. 1 Structural coupling between digital and business ecosystems (adapted after [1])

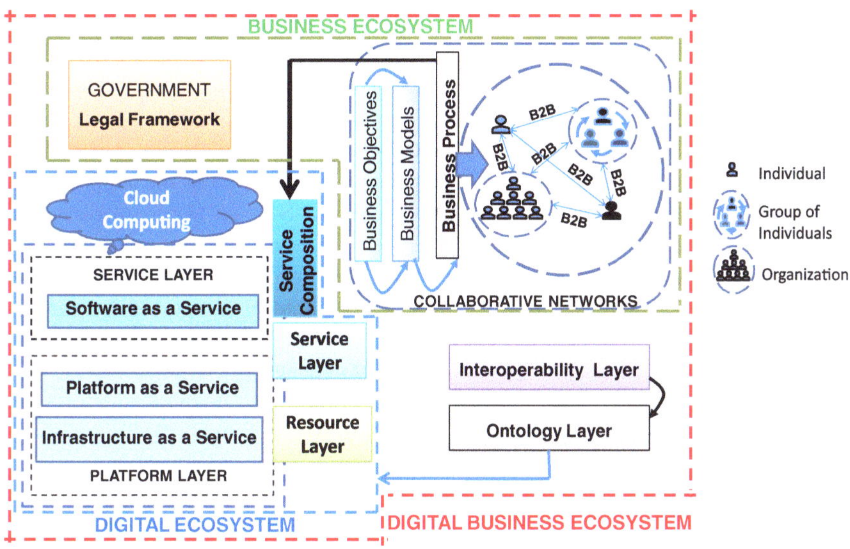

Fig. 2 DBE framework

environment, culture, social capital and goals, but that collaborates to better achieve common or compatible goals, and whose interactions are supported by computer networks [4].

The business relations in a business ecosystem are characterized by the Business Objectives. These lead to the creation of Business Models, which results in building Business Processes. Business relations start with the simple B2B networks between individuals and evolve to collaborative networks between different several entities—organizations, institutions, or individual groups that have expertise in a particular domain (e.g. Agriculture; Economy; Computer Science). These business relations take place in a legal framework provided by the Government, which provides specific policies, rules and subsidies allocation procedures. Also, the Government ensures a safe environment for all the business actors.

The Digital Ecosystem is represented by the ICT solution used to facilitate the interaction between business actors and also to facilitate the access to shared knowledge. Due to its capacity of reducing overall client side hardware requirements and complexity [5], cloud computing was used as the main digital ecosystem component. Cloud Computing use is also justified by its capability to handle big amount of data and because such data can be securely stored yet accessible from anywhere, anytime. Data availability has been triggering numerous new value-added services, which in turn have created more data and services [6].

A Service Layer that provides services via Service Composition also composes the digital ecosystem, enabling through the Cloud the use of the Software as a Service (SaaS) model.

- SaaS provides the capability of using the provider's applications running on a cloud infrastructure. The applications are accessible from various client devices such as a web browser. In this case, the user does not manage or control the cloud infrastructure, including network, servers, operating systems, storage, or even individual application;

The Resource Layer component is the layer between the business environment and a database or external services, and in this case it can be used in the support of the Platform as a Service (PaaS) or Infrastructure as a Service (IaaS) infrastructure models.

- PaaS allow the created or acquired applications implementation onto the cloud infrastructure using programming languages and tools supported by the provider. In this case, the user has only control over the deployed applications and possible application hosting environment configurations;
- IaaS provides capabilities like processing, storage, networks, computing resources. The users will be able to deploy and run operating systems and applications and also have control over these.

An ontology layer and an interoperability layer compose the DBE. Through the Ontology Layer, ontologies can be used for the explication of activities semantics in general, as well as in the support of complex systems that could provide information to a specific people about a specific domain. The interoperability Layer sustains the Ontology Layer. Interoperability is another aspect that must be considered in these ecosystems. In the knowledge network whose digital content is created in different ways by different members, interoperability is about ensuring automation and automated management of the ecosystem. Interoperability is mainly addressed at the level of transfer, exchange of information, mediation and integration of content [7]. Due to the support of ontologies in interoperability resolution establishment, these two layers are inter-connected.

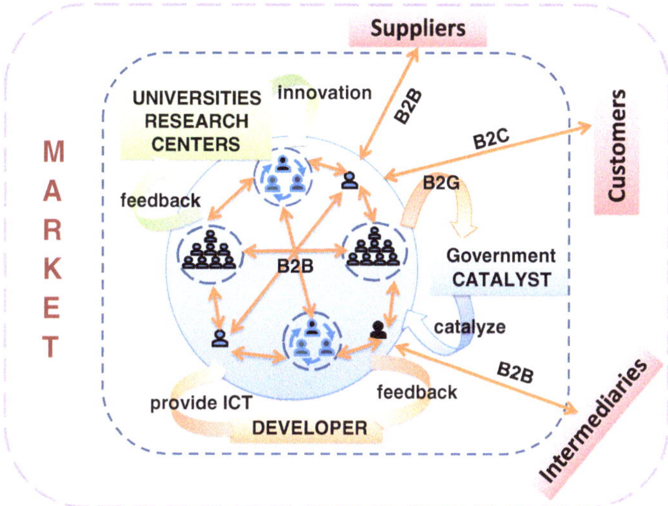

Fig. 3 Business relations between DBE's members

2.1 Key Members in the DBE Proposed

DBE initiative aims to share best practices and cultural exchange not only within the organizations networks, but also among common entrepreneurs, and mainly refers to three aspects: technology, business practices and knowledge. In Fig. 3 it is represented the business relation between the DBE members and the Market, how they interfere and what are their roles.

The Government plays the role of Catalyst, meaning that enables and facilitates the interaction between the members of the local ecosystem, providing the legal framework and the food security policy and the necessary funds for the development of the ecosystem. At the same time, the Catalyst coordinates all regional development projects and establishes criteria for granting funds.

The Developer is the entity in charge with developing the IT services used in the interactions between members of the local ecosystem. This meaning the development of peer-to-peer networks and also to provide the infrastructure required to these communications. The Developer can be an enterprise with a high level of experience in this sector.

Universities and Research Centers have the role to provide to community the latest technological progress, technology adoption and innovations in the field and also to inform what the best practices to follow.

The enterprises benefit from the sharing of information regarding best practice, the possibility of collaboration with other enterprises in order to a better respond to market demands and new challenges, including widening their own affairs.

Also, simple entrepreneurs or producers as community members have advantages. They are able to form small groups with other simple entrepreneurs, and collaborate for a specific goal, in order to have a better chance and opportunity in business. Enterprises, Organizations or simple entrepreneurs can easily access the funds and subsidies or the regional development projects in which they benefits for free of government assistance. Thus, they become stronger in the market and can face easier the competition. Once the DBE is well developed, its members can extend their activities. They are now a kernel, and know to work together, so the next step would be the accessing to other markets.

3 DBE Ontology

The DBE members need to agree on a common lexicon to be used in their collaboration communications in order to understand each other. Thus, an ontology has been developed with the aim of representing the knowledge to represent a common view and to facilitate the understanding of the concepts and relations regarding the DBE domain actors and elements.

The ontology is composed of over 50 concepts, which between them, there are more than 35 relations. This ontology could also give support to a system to provide different type of information/support to enterprises, simple entrepreneurs (e.g. farmers), as for instance, about what they can do to become ecosystem members; or how they can access to regional development funds. In a kind of conclusion, the ontology helps people and machines to use the same terms for expressions and thus to achieve better mutual understanding.

The proposed ontology was built using Protégé-OWL Editor, OWL Full version. Protégé is a public and free tool that uses OWL and other languages to built ontologies [8]. The Web Ontology Language (OWL) is one of the most known languages for creating ontologies. This language can be used along with information written in RDF [9]. The Resource Description Framework (RDF) is used for conceptual description or modelling of information that is implemented in web resources, using a syntax notations and data serialization formats [10]. OWL Full is appropriate for the ones who need maximum expressiveness and the syntactic freedom of RDF with no computational guarantees [11]. Consequently, authors used OWL Full to represent the knowledge in the DBE developments. Figure 4 illustrates an excerpt of the developed ontology using Protégé, namely the roles and interactions between the main entities forming DBE, and which intends to formalize the knowledge represented in Fig. 2.

The ontology was built on the top of a clear understanding of the concepts used to represent this domain knowledge. As an example of such approach, in the following it is presented one of its 50 concepts definitions: "A Virtual Organisation comprises a set of (legally) independent organizations that share resources and skills to achieve its mission/goal, but that is not limited to an alliance of for profit enterprises. A Virtual Enterprise is therefore, a particular case of VO" [12]. The ontology developed is able to answer questions such as (but not only):

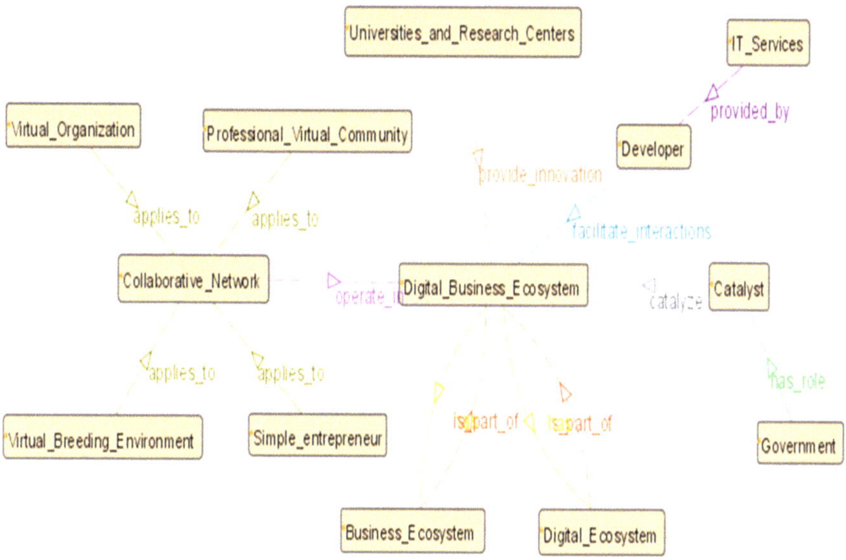

Fig. 4 General representation of DBE

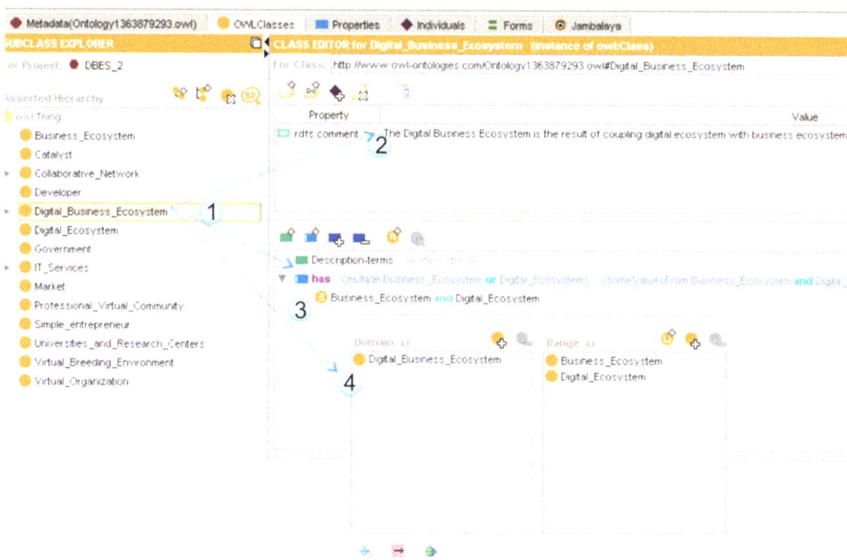

Fig. 5 Description of DBE in Protégé

Q1. What is DBE? Fig. 5 shows that DBE is a class (1), it has a description (2) and is composed of other two different classes Business Ecosystem and Digital Ecosystem (3) and (4).

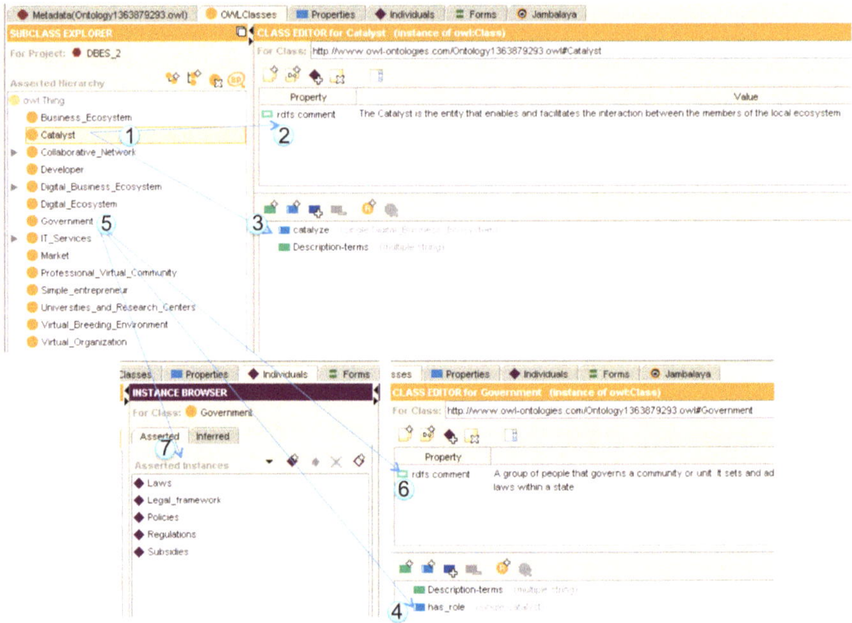

Fig. 6 Description of the catalyst in Protégé

Q2. Who is the Catalyst? The Catalyst is a class (1), with the description (2) and has the property to catalyze the DBE (3). In this ontology, the role of Catalyst is played by the Government (4), which is also a different class (5), has a term description (6) and instances (7) like: legal framework, laws, policies, regulations and subsidies (Fig. 6).

Q3. What is the difference between Ad-Hoc Collaboration and Long Term Strategic Network? In Fig. 7 it can be seen that Ad-Hoc Collaboration is a subclass of Collaborative Network (1), has a term description (2) and is disjoint to Collaborative Network Organization, which is another subclass of Collaborative Network. The Long Term Strategic Network is a subclass of Collaborative Network Organization (4), also has a term description (5) and is disjoint of Goal Oriented Network, which is a subclass of Collaborative Network Organization (6).

The answers to these questions show that there are two directions in which the ontology has been developed—defining the structure and the relations between DBE members—describing the participants and their roles.

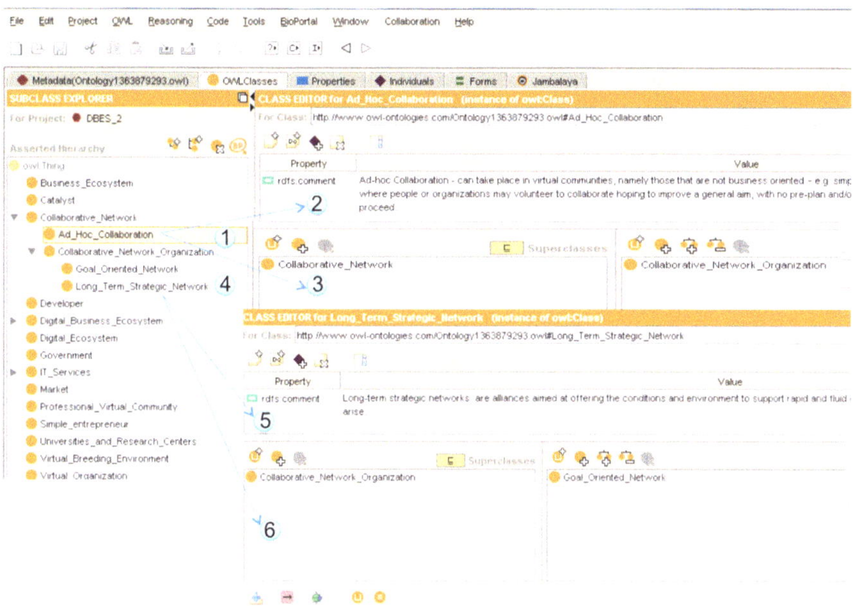

Fig. 7 Description of the Ad-Hoc collaboration by comparison to long term strategic network

4 An Agro-food Case Study

In Romania, the agriculture played an important role, with great economical importance. After few years of decline, the agriculture has started to gain its main role. The usage of DBE can offer new possibilities for development of this domain. Thus, DBE implementation in Romania will take into account several specific aspects like: Romanian legal framework and development degree concerning natural resources, technological resources and financial resources. In order to better understand the interaction between DBE members, a case study for fruits and vegetables branch of Agro-food Industry is presented in the following. In this case study are presented IT solutions for the members of the DBE, regarding typical situations that it have been identified in Romania.

The DBE for the branch of fruit and vegetables relates experts such as enterprises and farmers with agricultural scientists, computer scientists or economists, under the supervision of the government, with the aim to lead to knowledge growth in the field and to increase the regional development. Enterprise X deals with the manufacture of canned fruit and vegetables. Its objective is to buy fruits and vegetables with an acceptable quality at a lower price and in such a way to obtain a larger profit. Enterprise Y has the object of sale/marketing of fruit and vegetables. Wants: to have clients as many as possible, to have information about the competition and to expand its operations in other areas. Enterprise Z deals with the supply of fruits and vegetables at the market. It has only so far supplied goods to small stores in the region, but wants to increase its revenue by signing up for supply contracts with large chain stores (hypermarkets). Farm F deals with growing vegetables and fruits, different varieties, in different quantities and different ways of cultivation (outdoors, in greenhouses, bio, using natural or chemical fertilizers, in different soils), because it wants to better respond to the market demands. Also, it would like to be known by many people, to have more clients and to increase its revenues.

Enterprises X, Y, Z and Farm F play the roles of Suppliers, Intermediaries or even Customers in the DBE presented in Figs. 3 and 4. All these enterprises are independent and derive incomes more or less satisfactory, but are willing to cooperate, to create attractive packages for customers, so that they can better cope with the competition in the market. Sure that these enterprises or farms would not exist on the market, if there were no customers, to whom they sell products. Food is part of our daily life and there are a large variety of preferences (that mainly take into account two aspects: health and money). That is why it is important to understand what the customers want. Customers want good, healthy and safe food, which they want to buy at the lowest price. But, depending on the money they have, they are still willing to give up some of their "expectations". Therefore, the offer of the market must respond to all requests. The ICT solutions provided are:

For the cans marker, can provide access to a database containing all the suppliers of fruits and vegetables, selling prices and product quality. Thus, the marker save time and can quickly negotiate contracts with suppliers, taking into account their preferences in terms of quality and price. Depending on the goods purchased, a large variety of cans can be obtained in order to satisfy both demands of customers who want to eat healthy and, those who want a cheap product.

For the Seller/Trader, it can provide access to a database containing information about other sellers offering similar products in the market and their offers. Also, the Seller can publish its offer through the infrastructure provided by the Developer, can negotiate contracts with suppliers and can sell goods to customers on the Internet. Thus expanding its activities in the virtual environment and can offer other integrated services such as delivery to the customer and provide a history of the products purchased.

In the same way, the Developer provides access to information for the Supplier. And the Supplier also can negotiate and sign its contracts through the secure infrastructure. In the virtual environment he can meet different costumers, with different expectations and also he can bind business partnerships with others suppliers. The farmer is the direct producer of fruit and vegetables. He can attract the customers by providing all information related to production: variety, cultivation method, method of taking care and any other information that could make him special, compared to other producers. Customers can easily find the history of the products they purchase and are also protected by a secure environment, guaranteed by the government. The purpose of the government is to encourage the development of this ecosystem, from which to obtain incomes (in the form of fees and taxes). Then, a part of this revenue will be invested in the ecosystem infrastructure, resulting a continuous development and expansion. For the Developer, such a contract signed with the government, aiming to not only the development of the applications and necessary tools for the ecosystem, but also to provide and stability and maintenance, is definitely a contract that brings considerable income and, a reputation in IT field. The Universities also gain from this cooperation, because they have to implement, to test and to validate their innovation in the Agro-food branch of fruits and vegetables.

5 Conclusions and Future Work

DBE is an innovative concept [2], especially in a sustainable context. The DBE implementation stimulates creativity and competition, offers equal opportunities and stimulates local knowledge development through the adoption of ICT. After a few years of decline, the agriculture has now an great importance in Romania. In this context, DBE implementation offers new possibilities for development in this domain. In this paper, a framework is proposed and it describes the business relation and their digital representation provided by the ICT solutions. Also, the DBE members have been identified and have been highlighted the relation between them.

An ontology representing the knowledge used to understanding the concepts, relations between different members and their roles regarding the DBE domain is presented. It facilitates specific reasoning over the knowledge providing the tool for explicit context awareness, which is something important for a business creation. This research has followed solutions able to support new business networks and opportunities establishment.

As future work, authors will work on the: (1) Definition of standards for DBE architecture; (2) Ontology definition improvement for handling different aspects related to interoperability; (3) Modelling the business process using Business Process Models and Notation; and (4) A new Case Study implementation addressing all these future features. In additional it is intended to integrate all the presented information systems elements in a platform able to give direct support to its members in a structured way as the proposed framework.

Acknowledgments This article results has been developed within the EXPERT POSDRU/107/1.5/S/76903 (http://plone.cempdi.pub.ro/BurseDoctoraleID76903) and FITMAN N° 604674 (http://www.fitman-fi.eu) projects, that has been funded respectively from the European Social Fund and EC 7th Framework Programme.

References

1. Dini, P., & Nachira, F. (2007). The paradigm of structural coupling in digital ecosystems. In *Toward digital business ecosystems*. Boston: Edward Elgar.
2. ONE, SEAMLESS, VISP and CONTRACT projects Consortiums (2008). Digital ecosystem village—ecosystem oriented technologies and platforms for SMEs evolution and

development in the digital age. Retrieved from the web at April 2013: http://www.visp-project.org/docs/DE_Village_leaflet_19_Nov_08.pd
3. Moore, J. F. (1996). *The death of competition: Leadership and strategy in the age of business ecosystems*. New York: Harper Business.
4. Camarinha-Matos, L. M., & Afsarmanesh, H. (2008). *Collaborative networks: Reference modeling*. Springer, ISBN 978-0-387-79425-9.
5. Zissis, D., & Lekkas, D. (2012). Addressing cloud computing security issues. *Future Generation Computer Systems*, *28*, 583–592. (Elsevier).
6. Bojanova, I., Zhang, J., & Voas, J. (2013). Cloud computing. *IT Professional, 15*(2), 12–14.
7. Burlacu, G., Stanescu, A. M., Sacala, I. S., Moisescu, M. A., & Cojocaru, L. E. (2012). Development of a modeling framework for future internet enterprise systems. In *16th International Conference on System Theory, Control and Computing, Sinaia.*
8. Stanford Center for Biomedical Informatics Research (2013). *Protégé Overview*. Retrieved August 2013, from http://protege.stanford.edu/overview/index.html
9. W3C (2012). *OWL 2 web ontology language document overview (2nd ed.)*. Retrieved August 2013, from http://www.w3.org/TR/owl2-overview/
10. World Wide Web Consortium. (2013). *Semantic web*. Retrieved August 2013, from http://www.w3.org/standards/semanticweb/
11. World Wide Web Consortium. (2004). *OWL web ontology language guide*. Retrieved August 2013, from http://www.w3.org/TR/owl-guide/
12. Camarinha-Matos, L. M., & Afsarmanesh, H. (2005). The emerging discipline of collaborative networks. *Journal of Intelligent Manufacturing, 16*(4–5), 439–452.

Integrating Smart Objects as Data Basis in Virtual Enterprise Collaborations

Sebastian Zöller, Ronny Hans, Dieter Schuller, Ulrich Lampe and Ralf Steinmetz

Abstract Small and Medium Enterprises of the manufacturing domain have to cope with a highly competitive market today. To establish flexible and efficient collaborations with partners in such an environment, new collaboration concepts and corresponding IT architectures are required, such as Virtual Manufacturing Enterprises. Therefore, we provide in this paper an overview of a generic IT architecture for realizing collaborations within Virtual Manufacturing Enterprises. However, besides an adequate IT architecture, a sound and up-to-date data basis is an essential necessity for inter-company collaborations to be successful. Smart Objects constitute a promising technology to gather and transmit a huge diversity of different process-relevant data in real time and can thus act as valuable data source in order to achieve such a comprehensive and up-to-date data basis. In consequence, we describe in this paper how Smart Object technology can be employed and integrated in our architecture for Virtual Manufacturing Enterprises in order to enable efficient data provisioning in such collaboration scenarios.

Keywords Virtual manufacturing enterprises · IT architecture · Data provisioning · Smart objects

S. Zöller (✉) · R. Hans · D. Schuller · U. Lampe · R. Steinmetz
Technische Universität Darmstadt—Multimedia Communications Lab (KOM), Rundeturmstr. 10 64283 Darmstadt, Germany
e-mail: sebastian.zoeller@kom.tu-darmstadt.de

R. Hans
e-mail: ronny.hans@kom.tu-darmstadt.de

D. Schuller
e-mail: dieter.schuller@kom.tu-darmstadt.de

U. Lampe
e-mail: ulrich.lampe@kom.tu-darmstadt.de

R. Steinmetz
e-mail: ralf.steinmetz@kom.tu-darmstadt.de

K. Mertins et al. (eds.), *Enterprise Interoperability VI*, Proceedings of the I-ESA Conferences 7, DOI: 10.1007/978-3-319-04948-9_25,

1 Introduction

In our globalized world, manufacturing companies have to cope effectively with various challenges, for example shorter product life cycles, quick response times, and a higher degree of customization [1]. Especially Small and Medium Enterprises (SMEs), which do not possess the financial strength to face all the upcoming challenges individually, need to focus on flexible and efficient collaborations in order to successfully address these challenges and remain competitive in today's markets.

The mentioned challenges can be addressed with a popular concept named Agile Manufacturing (AM) [2]. This concept focuses on setting up whole organizations for production across enterprise boundaries [3]. One core concept to achieve AM is the Virtual Manufacturing Enterprise (VME) [2, 3]. Such VMEs are created to address specific market opportunities [4]. A VME is founded by at least two companies and puts the partners into a position to easily collaborate in order to use common processes to achieve common goals [5, 6]. In such a collaboration form, the partners share costs, skills, and core competencies [4].

A basic building block for VME-based collaboration is Information and Communication Technology (ICT) [7]. In contrast to traditional ICT architectures, a high degree of flexibility and interoperability is required to seamlessly integrate business processes between collaborating partners [6]. An indispensable necessity for all affected processes and interactions in such collaboration is an up-to-date data basis. In this respect, Smart Objects provide a promising technology, which can be used to achieve a corresponding up-to-date data basis. Smart Objects possess sensing, processing, and communication capabilities and can thus be employed as one of the major information sources in VMEs [8, 9]. In the considered context of manufacturing and particularly collaboration in the form of VMEs within the manufacturing domain, Smart Objects can, for example, be machines or means of transport which are, e.g., equipped with technology to measure different environmental parameters, detect critical deviations, and wirelessly transmit the measurement results. In the light of the briefly sketched benefits that Smart Objects could provide to VME-based collaborations, we describe the integration and communication possibilities for Smart Objects in the context of VMEs in this paper.

The paper proceeds as follows: In Sect. 2, we present our approach for an ICT architecture for VMEs. We elaborate on possibilities of how to employ Smart Objects in a VME and integrate them in the proposed VME architecture in Sect. 3. Section 4 concludes the paper with a summary of our findings and an outlook on future work.

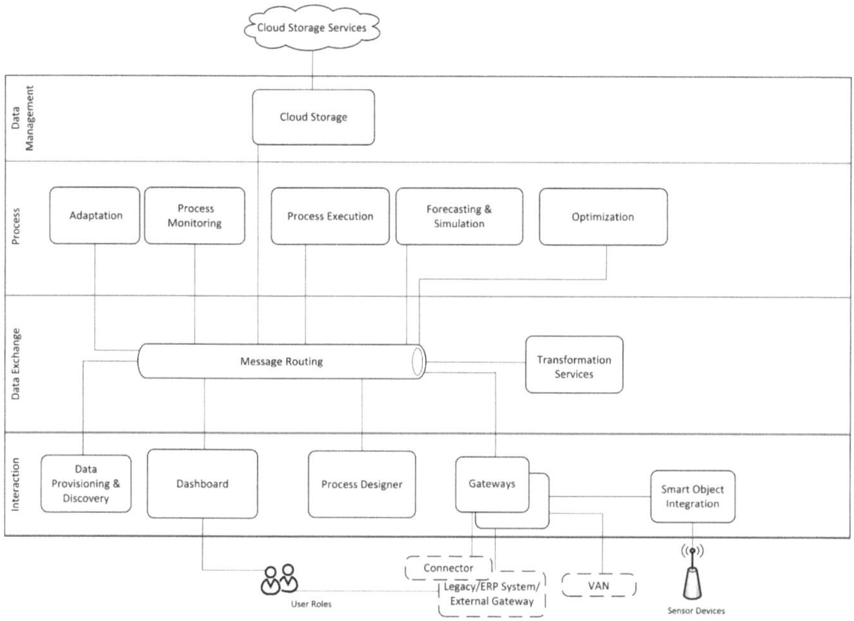

Fig. 1 Virtual manufacturing enterprise ICT architecture [11]

2 ICT Architecture for Virtual Manufacturing Enterprise-Based Collaboration

The following section describes the architecture model for VMEs, which we take as a basis for all our considerations (cf. [10]). The proposed architecture model is depicted in Fig. 1. Within the architecture, we distinguish different components, which are grouped in layers. For the data exchange between these different components, a corresponding message routing infrastructure is foreseen as central element. To ensure the required degree of performance, reliability, and stability for the data exchange via this message routing infrastructure, the correspondingly required Message Routing component is deployed as a Cloud-based service.

Besides the central message routing component, another vital component of our architecture is the Dashboard component. It constitutes the central point for user interaction. It provides the user interfaces to all other components and offers users the functionality to configure, monitor, and access all VME information, whereas we understand users in the context considered here as responsible employees of companies participating in a VME or actively pursuing the participation in a VME.

The initial step to deploy a VME is designing the required processes. For this process design, a corresponding Process Designer component is used. It enables a broker, understood simultaneously as the process designer as well as the person who controls a process during its execution, to design all production steps including all relevant parameters for individual activities. Suitable partners for participation in the VME and execution of single process activities can be found with the Data Provisioning and Discovery component. Furthermore, the Data Provisioning and Discovery component allows users to provide various data about their company, e.g., descriptions of factories, products, delivery time, and costs. A Cloud Storage component serves as the central data storage within the architecture for such and other data. This component allows to create and to use different storage spaces for different data, e.g., binary, structured, semi-structured, and semantic data stores (cf. [10]).

To find individually fitting partners or a combination of partners who are able to fulfill previously designed process steps, a Forecasting and Simulation component is included. This component allows to conduct simulations of the devised process and the corresponding forecasting of process behavior and process results based on the retrieval of information provided from partner enterprises, e.g., via their ERP systems, and process step-specific information like possible lead times. The results of such process forecasting and simulation support the broker to decide which partner should be included in the process and assigned to which tasks within the process. Further, based on the forecasting values, the broker can estimate the robustness of a process. In addition, recommendations of a set of best fitting partners are provided based on mathematical optimization models, which are realized within the Optimization component.

After the process design has been conducted and partners have been assigned to all required activities, the process execution is initiated by the Process Execution component. To monitor the running manufacturing processes within the VME and compare them with pre-defined values—for example, to detect any deviations—a Process Monitoring component is provided. This Process Monitoring component uses and aggregates information from internal and external data sources. In case unexpected events are discovered, it may become necessary to change running processes. Correspondingly, a Process Adaption component is included in the architecture. Examples of such deviations can be changed prices or updated delivery times of partner services.

As already mentioned, a sound and up-to-date data basis is a fundamental requirement for successful collaboration. In consequence, it may be necessary to provide means to integrate external third-party systems, like an ERP system, within the VME. To enable the required communication with such external third-party systems, Gateways are required. These Gateways provide the connection between the VME's ICT infrastructure and external systems serving as data sources, for example, ERP systems or Smart Objects. Due to the possible heterogeneity of the different systems employed by the various participants within a VME, translation of data formats might be required. Such data transformation is covered by the Transformation Services component within our architecture.

Information is the basis for successful collaboration, as already mentioned, and information is processed in all the above described components. For gathering external information, our proposed architecture includes two major parts: Smart Objects, e.g., in the form of (wireless) sensor devices, and the corresponding Smart Object Integration component. These will be described in detail in the following section.

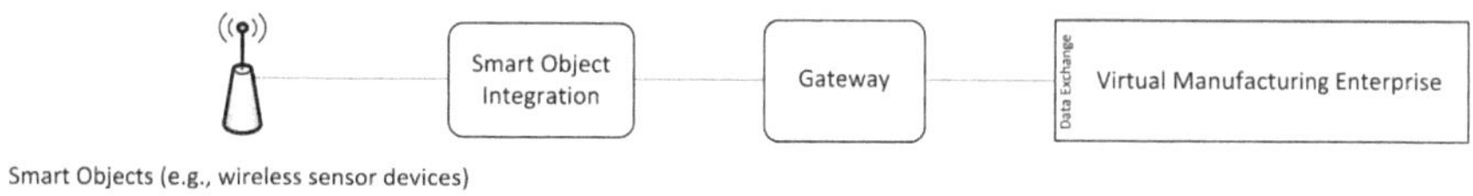

Fig. 2 Integration of smart objects in the proposed overall VME architecture

3 Smart Objects as Integrated Data Sources in VMEs

Smart Objects, being objects with sensing, processing, and communication capabilities [8, 9], can provide diverse context data in real-time in a broad range of relevant areas within VME-based collaborations, e.g., in logistics transport processes [12] as well as in storage processes [13], and the manufacturing process itself [14]. Hence, they can contribute to the provision of in-depth and up-to-date process-relevant data. In this context, Smart Objects can be constituted of, e.g., machines or means of transport which are for example equipped with wireless sensor technology to measure different environmental parameters like temperature, shock, or tilt values during production and transportation. However, to be of any use the data gathered by Smart Objects must be efficiently integrated in the overall VME architecture. Therefore, the provisioning of adequate communication possibilities and means to integrate a corresponding wide range of possibly different Smart Objects is required. Thus, in the following, we provide a detailed description of how the integration of heterogeneous Smart Objects in our proposed overall VME architecture can be realized.

From an overall view of the VME architecture, Smart Objects can be treated like (external) third party systems. Consequently, Gateways (cf. Sect. 2) as mediators between these third party systems and the VME can be employed as a basic means to integrate Smart Objects within VMEs and act as technical bridges for establishing the required communication possibilities, as well. Nevertheless, against the background that Smart Objects can be realized in very different ways, using different hardware platforms, different communication channels and communication protocols and consequently possess quite distinct characteristics, the corresponding heterogeneity of Smart Objects has to be explicitly accounted for. This implies, for example, that different communication protocols as well as different physical communication technologies and data formats have to be supported. Thus, to account for this technological heterogeneity, the usage of dedicated Smart Object Integration components is proposed (cf. Fig. 2).

Depending on the characteristics and capabilities of the employed Smart Objects, Smart Object Integration components can be realized on the Smart Object itself or, in case it is technologically not capable of this, a realization in a distributed manner can be used by implementing selected functionalities on the Smart Object itself and realizing remaining functionalities on a Gateway. In this context, specifically the establishment of a physical communication channel has to be provided as well as adaptation of data to different data formats employed by

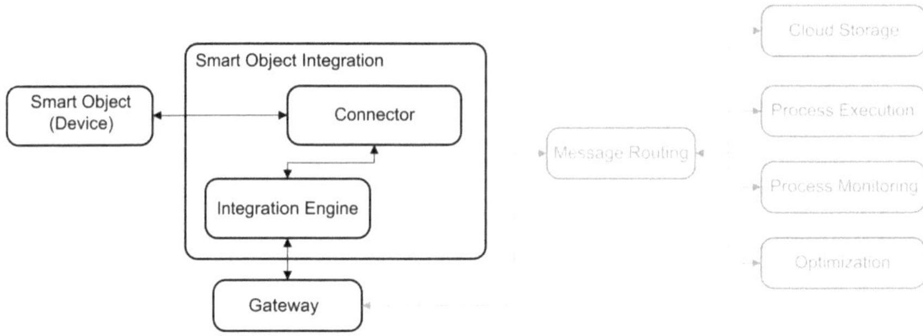

Fig. 3 A detailed view on smart object integration

different Smart Objects and Gateways. Thus, the Smart Object Integration component has to comprise a Connector, which establishes the (physical) communication channel between Smart Object and Gateway, and an Integration Engine, which can provide required data transformation operations (cf. Fig. 3).

For example, in the case of monitoring storage or transport sub-processes in the context of a manufacturing process, e.g., realized by intelligent packaging using wireless sensor network (WSN) technology [12, 13], it has to be considered that WSN technology is usually designed for short range communication, for example making use of IEEE 802.15.4-based or Bluetooth-based communication. As a consequence, in such a context wide-range communication possibilities and corresponding protocol and data transformations have to be provided to enable data exchange with other components in the VME architecture, for example, to store data in the Cloud Storage. One realization possibility for this is the usage of smartphones as technological bridges, which provide protocol and data transformation as well as wireless communication channels to Smart Objects and into the VME [14].

For the specific data exchange between Smart Objects and the other architecture components, the realization of two communication paradigms is proposed (cf. Fig. 4).

Since users within a VME, respectively system components of the VME architecture, might need certain data from Smart Objects at certain points in time, the possibility for a request/response-based pull communication must be provided. This means that in case a user or a software component requires data, a query can be issued for this data and a corresponding response with the requested data is returned. In this context, Web Service technologies can be used so that a Web Service is called and a corresponding query is issued to the appropriate Smart Object, which in return provides the queried data via the Smart Object Integration component and the corresponding Gateway as Web Service response. In case smartphones are used as technical bridges between the VME and Smart Objects—as described in the example above—mobile Web Service technologies could be used in this context, like i-jetty [15] or the Sphinx Mobile Web Server [16].

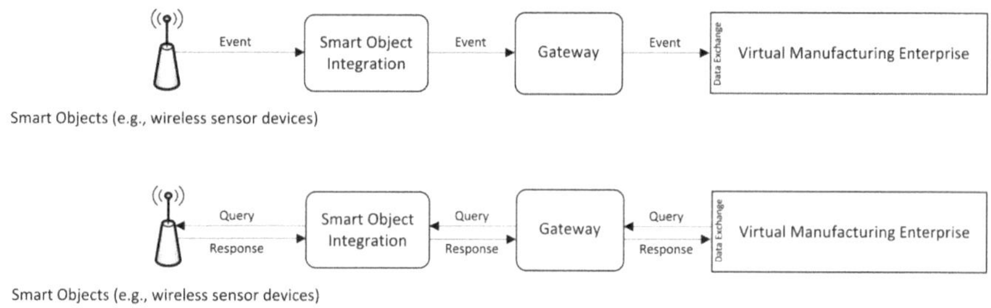

Fig. 4 Access to smart objects data via push and pull communication

In addition, usually not only data as answer to specific requests is required, but in the sense of real-time monitoring and data availability, the transmission of alarm messages should be provided in case critical situations have been detected by Smart Objects autonomously. Consequently, besides the described request/response-based pull communication, event-based push communication should be possible as well. This accounts for situations where critical deviations of normal behavior shall be detected locally by Smart Objects and corresponding users and systems should be informed automatically. In such circumstances, Smart Objects make use of their processing capabilities and perform regular checks of their measured status data and compare it by target-performance comparisons to given thresholds, which can be previously specified by domain experts. In case a deviation (usually called an "event", cf., e.g., [17]) is detected, the Smart Object pushes a corresponding alarm message into the VME system in real time. With such early warnings, corresponding early process adaptations and countermeasures are made possible. Furthermore, the described event-based push communication reduces communication overhead, because no regular polling with high polling rates of data from Smart Objects is required and data is only transmitted if it is necessary. So, appropriate real-time monitoring with required alarm messages and sufficient data fidelity can be ensured employing the described event-based push communication.

The described approach for Smart Object integration allows the exploitation of an always accessible up-to-date data basis of process-relevant data provided by Smart Objects. Thus, it constitutes a central building block for real-time process monitoring and corresponding control within VME-based collaborations.

4 Conclusions and Future Work

Small and Medium Enterprises (SMEs) in the manufacturing domain usually face strong competition within the market. To remain competitive, new and agile forms of collaboration, such as Virtual Manufacturing Enterprises (VMEs), are required. For the realization of such collaborations the underlying Information and

Communication Technology (ICT) plays an outstanding role. Because current approaches in ICT do not fulfill the requirements of VMEs, we presented an architectural model to enable VME-based collaborations across the boundaries of different SMEs. It consists of different components, with each of them offering specific functionality contributing to enabling a more efficient and effective collaboration within VMEs.

A sound and up-to-date data basis constitutes a major foundation for any kind of successful collaboration. Thus, means to gather and provide a wide range of process-relevant data are required as well in VME-based collaborations. In order to provide a correspondingly broad range of real-time process data, we propose to make use of Smart Objects as data sources. This allows for a continuous process monitoring in real-time and furthermore enables active data retrieval by user queries when data is needed as well as active pushing of immediate alarm messages by Smart Objects themselves in case critical deviations are detected. In consequence, we described means of how to employ and integrate Smart Objects as data sources in VMEs.

In the future, we plan on prototypical implementations of the described components in cooperation with industry partners. Especially for the Smart Object integration, we strive to evaluate different technologies and implementations, ranging from different smartphone platforms to specialized routers and a prototypical shop floor deployment.

Acknowledgments This work was partially supported by the Commission of the European Union within the ADVENTURE FP7-ICT project (Grant agreement no. 285220).

References

1. Dowlatshahi, S., & Cao, Q. (2006). The relationships among virtual enterprise, information technology, and business performance in agile manufacturing. *European Journal of Operational Research, 174*, 835–860.
2. Yusuf, Y., Sarhadi, M., & Gunasekaran, A. (1999). Agile manufacturing: The drivers, concepts and attributes. *International Journal of Production Economics, 62*, 33–43.
3. Jain, S. (1995). Virtual factory framework: A key enabler for agile manufacturing. In: *Proceedings of Symposium on Emerging Technologies and Factory Automation* (pp. 247–258).
4. Park, K. H., & Favrel, J. (1999). Virtual enterprise—information system and networking solution. *Computers & Industrial Engineering, 37*, 441–444.
5. Corvello, V., & Migliarese, P. (2007). Virtual forms for the organization of production: A comparative analysis. *International Journal of Production Economics, 110*, 5–15.
6. Martinez, M., Fouletier, P., Park, K., & Favrel, J. (2001). Virtual enterprise—organisation, evolution and control. *International Journal of Production Economics, 74*, 225–238.
7. Faisst, W. (1997). Information technology as an enabler of virtual enterprises. In *Proceedings of the European Conference on Virtual Enterprises and Networked Solutions.*
8. Vasseur, J.-P., & Dunkels, A. (2010). *Interconnecting smart objects with IP: The next internet.* Burlington USA: Morgan Kaufmann.

9. Kortuem, G., Kawsar, F., Fitton, D., & Sundramoorthy, V. (2010). Smart Objects as building blocks for the internet of things. *IEEE Internet Computing, 14*, 44–51.
10. Hans, R., Zöller, S., Abels, S., Miede, A., & Steinmetz, R. (2013). Enabling collaboration in virtual manufacturing enterprises with cloud computing. In *Proceedings of the 19th Americas Conference on Information Systems*.
11. ADVENTURE Project Team. *D3.1 Global architecture definition document.* Retrieved October 04, 2013, from http://www.fp7-adventure.eu/resources/
12. Jedermann, R., Behrens, C., Laur, R., & Lang, W. (2007). Intelligent containers and sensor networks. In M. Hülsmann & K. Windt (Eds.), *Understanding autonomous cooperation and control in logistics* (pp. 365–392). Berlin: Springer.
13. Marin-Perianu, M., Meratnia, N., Havinga, P., de Souza, L., Müller, J., Spiess, P., et al. (2007). Decentralized enterprise systems: A multiplatform wireless sensor network approach. *IEEE Wireless Communications, 14*, 57–66.
14. Zöller, S., Reinhardt, A., Meyer, M., & Steinmetz, R. (2010). Deployment of wireless sensor networks in logistics. In *Proceedings of the 9th GI/ITG KuVS Fachgespräch Drahtlose Sensornetze* (pp. 67–70).
15. Bartel, J. *I-Jetty: Webserver for the android mobile platform.* Retrieved October 04, 2013, from https://code.google.com/p/i-jetty/
16. Sphinx Software. *Sphinx mobile web server.* Retrieved October 04, 2013, from http://sphinx-soft.com/MWS/
17. Bretzke, W.-R., & Klett, M. (2004). Supply Chain Event Management als Entwicklungspotenzial für Logistikdienstleister. In Beckmann, H., (Ed.), *Supply chain management*. Berlin: Springer.

Infusing Verification and Validation in ICT Solutions in Manufacturing: The FITMAN V&V Method

Fenareti Lampathaki, Dimitrios Panopoulos, Panagiotis Kokkinakos, Christina Bompa, Sotirios Koussouris and Dimitris Askounis

Abstract As a plethora of technological ICT solutions invade the manufacturing domain under the umbrella of Future Internet, the need to ensure that such solutions are built in the right way and address the real needs of an enterprise becomes more and more crucial. In this context, this paper introduces an innovative way of performing Verification and Validation activities, which brings together both the technical and the business viewpoints and can be applied throughout the whole lifecycle of the software product. It describes the necessary steps, the techniques to be applied for each step, the roles to be involved, and suggestions for crowd assessment. It is developed taking into account its application in manufacturing settings, yet it is reusable to any project that has software development tasks and follows either the traditional waterfall method or invests on agile software building patterns.

Keywords Software engineering · Verification · Validation · Evaluation · V&V · Future internet · Manufacturing

---

F. Lampathaki (✉) · D. Panopoulos · P. Kokkinakos · C. Bompa · S. Koussouris · D. Askounis
Decision Support Systems Laboratory, National Technical University of Athens, 9 Iroon Polytechniou str. 15780 Zografou, Greece
e-mail: flamp@epu.ntua.gr

D. Panopoulos
e-mail: dpano@epu.ntua.gr

P. Kokkinakos
e-mail: pkokkinakos@epu.ntua.gr

C. Bompa
e-mail: chbompa@epu.ntua.gr

S. Koussouris
e-mail: skous@epu.ntua.gr

D. Askounis
e-mail: askous@epu.ntua.gr

K. Mertins et al. (eds.), *Enterprise Interoperability VI*,
Proceedings of the I-ESA Conferences 7, DOI: 10.1007/978-3-319-04948-9_26,

1 Introduction

Nowadays, Europe is facing major societal and financial challenges and is called upon to support growth in an ever changing environment while, simultaneously, providing lasting solutions. Manufacturing is the fuel of economic and business growth, with "ICT-enabled intelligent manufacturing" emerging as one of the four pillars to support European manufacturing industry in the challenging transition from post-crisis recovery to European STEEP (Social, Technological, Economical, Environmental and Political) sustainability [1]. In order to boost innovation and effectiveness, it becomes imperative to build the right software product, that supports the everyday manufacturing operations, and build this software product right from its conception to final release.

Verification and Validation (V&V) as a concept is actually not something new, as scientists, engineers and developers have been using for years this term to assess that a project/initiative has been tested against predefined requirements. V&V constitutes the subject of a considerable number of books, publications and other efforts over the last 10–15 years [2], yet the need for V&V is not deeply instilled in the community [3].

In general, V&V accomplishes the proper documentation and global assessment of a model, a software project/component or a theory [4] and the international literature is replete with various Validation and Verification definitions. In the context of this paper, the ANSI/IEEE Std 1012–2012 [5] definition of V&V is adopted, where:

- **Verification** is the process of providing objective evidence that the software and its associated products conform to requirements (e.g., for correctness, completeness, consistency, accuracy) for all life cycle activities during each life cycle process (acquisition, supply, development, operation, and maintenance); satisfy standards, practices, and conventions during life cycle processes; and successfully complete each life cycle activity and satisfy all the criteria for initiating succeeding life cycle activities (e.g., building the software correctly).
- **Validation** is the process of providing evidence that the software and its associated products satisfy system requirements allocated to software at the end of each life cycle activity, solve the right problem (e.g., correctly model physical laws, implement business rules, use the proper system assumptions), and satisfy intended use and user needs.

The present paper aims at developing a novel V&V method in order to provide industrial organisations with a structured and clear set of approaches and guidelines for testing, verifying, validating and assessing software products they plan to embed in their manufacturing procedures. The proposed V&V method incorporates and adapts best practices from the international bibliography, while infusing new, innovative, aspects relevant to the agile philosophy and the crowdsourcing trend. The proposed method has been elaborated in the context of the EU FP7 FITMAN project [6] and will be applied in trials from the manufacturing domain.

To this direction, the structure of the paper is as follows: Sect. 2 outlines the methodology followed in order to develop the V&V method proposed in this paper. Section 3 presents the results of an extensive desk research that took place in order to identify and review international standards and existing V&V approaches. Section 4 describes in detail the steps and the stakeholders of the V&V method leading to Sect. 5 that concludes this work.

2 Methodological Approach

In order to conceptualize and elaborate on a generic, yet holistic, V&V Method, the following approach was followed.

Initially, the scope of the V&V method was defined taking into account the wide range of technical and business needs and the ICT-driven mentality in the manufacturing domain. To avoid confusion in terminology, consensus was reached on a common glossary around V&V activities.

As a next step, an extensive state-of-the-art analysis took place in order to understand and map the landscape of V&V and evaluation standards, methods and tools. During this iterative process, multiple sources were consulted including standardization organisations, electronic publications, academic search machines, general searching machines etc.

Upon carefully studying the lessons learnt from the V&V state of play analysis, the conceptualization phase of the V&V method started by brainstorming on the various phases of the method, as well as the relations and interdependencies amongst them (providing all necessary inputs, outputs, etc.). The most relevant underlying approaches upon which the V&V method is built are presented as well as the stakeholders' envisaged role in each step is anticipated.

Iterative methodology updates were performed so as to incorporate feedback from domain experts, before concluding in a generic and inclusive V&V method.

3 State of the Art Analysis

In order to gain a thorough and complete vision on the existing V&V approaches/ methodologies, an extensive desk research had to take place. In the process of reviewing the state-of-the-art in verification and validation methodologies, especially when dealing with software development, the waterfall model and agile software development came to the foreground.

3.1 V&V in Software Engineering

Today, the waterfall model [7] constitutes a well-established software engineering philosophy promoting a step-by-step, sequential and highly structured process for building software. Indicative steps of the waterfall software development model include Requirements Identification/Specification, Design, Implementation/Coding and Integration. The need to verify and validate software developed following the waterfall model has led to one of the most well-known approaches for software verification and validation: the V-model [8].

The V-model practically constitutes an extension of the aforementioned waterfall model; in fact, its left branch practically is the waterfall model, while its right branch includes approaches/techniques for the verification and validation of the software under development. Unit Testing, Integration Testing, System Testing, Operational Testing, Acceptance Testing constitute members of an indicative, yet not exhaustive, list of such verification and validation techniques (Fig. 1).

Today, though, agile software development [10] is gaining more and more momentum. In contrast to the waterfall model, agile software development is based on an iterative and incremental workflow, aiming to rapid, collaborative and flexible end-product development. The main idea behind agile software development is breaking the main project into a larger number of relatively small increments; each one developed in short, efficient and effective iterations.

Unlike the V-model, in agile software development, there is no special procedure (or set of procedures) dedicated to verification and validation activities. Each iteration involves working in all functions: planning, requirements analysis, design, coding, unit testing, acceptance testing etc. At the end of the iteration, a working product (regardless of its size and importance) is demonstrated to stakeholders who provide the necessary feedback.

3.2 Key Take- Aways from State-of-the Art

During the detailed V&V and Evaluation state-of-the-art analysis, nine standards (e.g. [5]), 110 techniques (and their tools) (e.g. [4, 11]) and plenty of approaches (e.g. [12–18]) were studied and a set of insightful conclusions were extracted:

- There is an active on-going research on the fields of V&V throughout the years with V&V processes closely coupled with almost any business activity.
- A plethora of tools and techniques are available but without clear distinction between V&V and Evaluation.
- No "Universal" or Holistic V&V Approach covers any potential need, with most of the identified and analysed methodologies/approaches not being mature and widely established.

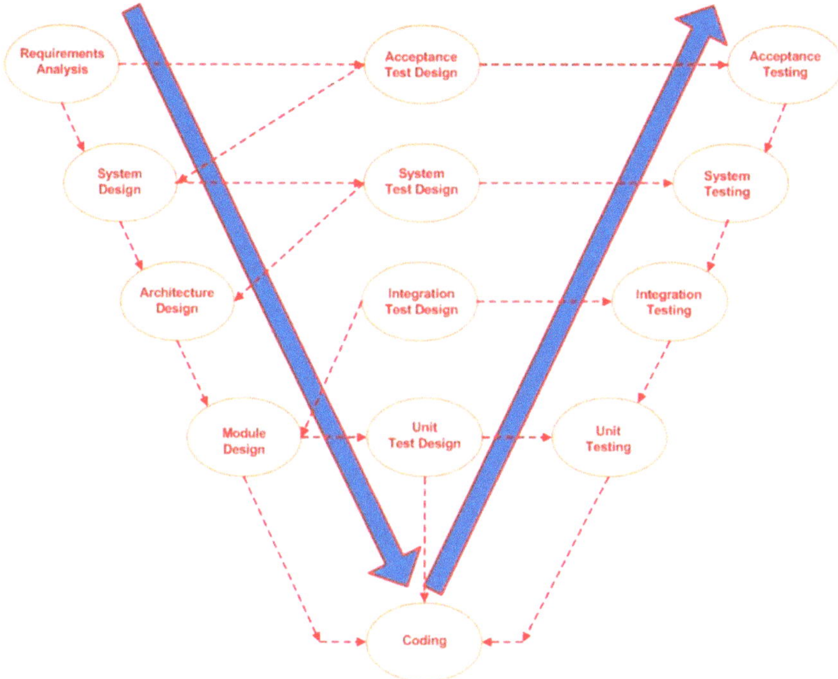

Fig. 1 Example of traditional V-model [9]

- The Waterfall model in software V&V prevails as most of the resources referring to agile V&V activities appear in blogs and online resources rather than rigorous academic work presented in papers.
- Crowdsourcing aspects cannot be easily located in existing/established V&V and/or evaluation approaches with these procedures being "closed" to specific stakeholders that are typically involved in a highly structured way (like questionnaires).

4 V&V Method

The V&V methodology which is presented in this paper introduces an innovative way of performing V&V activities in various ways by:

- Bringing together and getting the best of breed of the agile and waterfall software engineering philosophies which have been essentially dealt as contradictory trends and schools of thought until now.
- Infusing a crowd assessment mentality within the V&V activities, under each methodological step from code verification to business validation.

- Balancing and bridging the business and technical perspectives in an effort to assess the software and the "fit for purpose" requirements and to evaluate the overall software's added value.

4.1 Stakeholders

Due to the span of a software product development, the V&V method requires various actors to be involved, in order to effectively perform the V&V activities. These roles are the following:

- End Users as representatives of the customers, who state the business needs and translate those needs into end users visions.
- End Users Supporting Team who is responsible for ensuring that the integrated solution delivered to the end users is in alliance with their vision.
- The Sprint Master, the Product Owner and the Development Team composing the self-organizing, and self-accountable Sprint Team [19] with:
  - The Sprint Master held accountable for the communication inside-out and vice versa of the development team, ensuring that the team is not distracted and is focused on the tasks at hand to deliver value in a short time-box, for following the rules of agile and ensuring each sprint/iteration and the whole process is performed as planned and in lines with the methodological directions and finally for the protection, mentoring and guidance of the Development Team.
  - The Product Owner being the Sprint Team member, who serves as customers' representative and is responsible for creating the product backlog in a way that customer needs are met through technical robust solutions.
  - The Development Team, consisting of software analysts, programmers and testers and being in charge of delivering shippable extended product versions per sprint and to cover the necessary expertise.

Finally, this method aspires to involve external stakeholders, such as the general public or stakeholders working on similar projects, to bilaterally exchange views on the software product under development. By engaging with a considerable number of experts, the need to move towards a community based V&V methodology becomes even more evident.

4.2 Perspectives and Steps

The proposed V&V method is essentially divided into two perspectives:

- The trial specific perspective (T) which assesses whether the IT and business requirements and domain's needs are met, and

- The product-specific perspective (P) which describes how to verify and validate the product during its development.

Breaking down the V&V method step-by-step, the potential techniques to be employed, the stakeholders to be engaged, and the potential crowd engagement methods to be applied are recommended. In detail, the V&V method steps include:

4.2.1 The Trial Specific Perspective (T)

The trial specific perspective (T) bears the following V&V steps (Fig. 2):

- The **Business Validation (T-2)** step assesses whether the overall end user solution eventually offers sufficient added value to the end users. This step is performed by the end user team and the proposed technique to perform this V&V activity is the Simplified ECOGRAI Methodology for decisional models [20]. A crowdsourcing notion that could be incorporated here is the open collaboration and innovation platforms, online deliberation and feedback tools, physical and online workshops, traditional online survey tools, game-like applications and/or social networks.
- The **Trial Solution Validation (T-1)** guarantees that the overall end users solution satisfies intended use and user needs. The End User Solution Owner and the End User team are held accountable for this activity and the proposed technique to be used is the User Acceptance Testing. Proposed crowd engagement and involvement aspects to be considered are enterprise collaboration environments and other traditional methods (like questionnaires).

4.2.2 The Product-Specific Perspective (P)

The product specific perspective (P) bears the following V&V steps (Fig. 3):

- The **Product Validation (P-5)** step examines whether the product satisfies intended use and user needs. The Product Owner is responsible for the proper conduction of this step and the proposed technique to do so is the Black Box Testing for Validation. Crowd assessment suggestions to be incorporated are online deliberation and feedback tools, open calls for testing scenarios, workshops and specific working groups, traditional online survey tools and social networks.
- The **Release Verification (P-4)** step determines whether the requirements of the final product release are met and the **Backlog Verification (P-3)** one determines whether the requirements of the product after each sprint are met. For both activities responsible are the Development Team, the Sprint Master and the Product Owner and the techniques which is suggested to efficiently perform these V&V activities is Regression Testing. In order to involve the general public deliberation and feedback tools for ex-ante crowdsourcing and dedicated

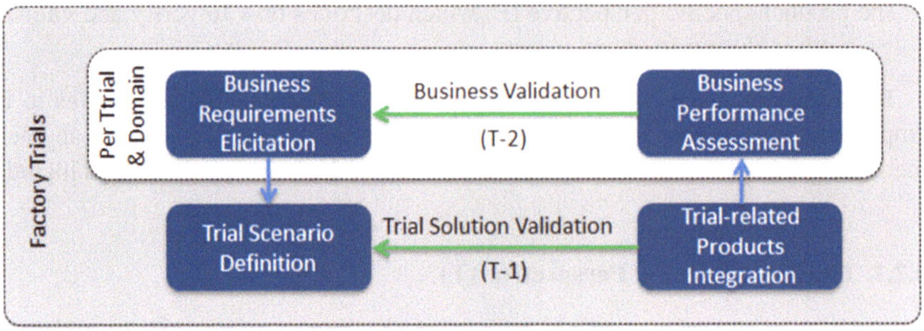

Fig. 2 V&V Steps under the trial specific perspective (T)

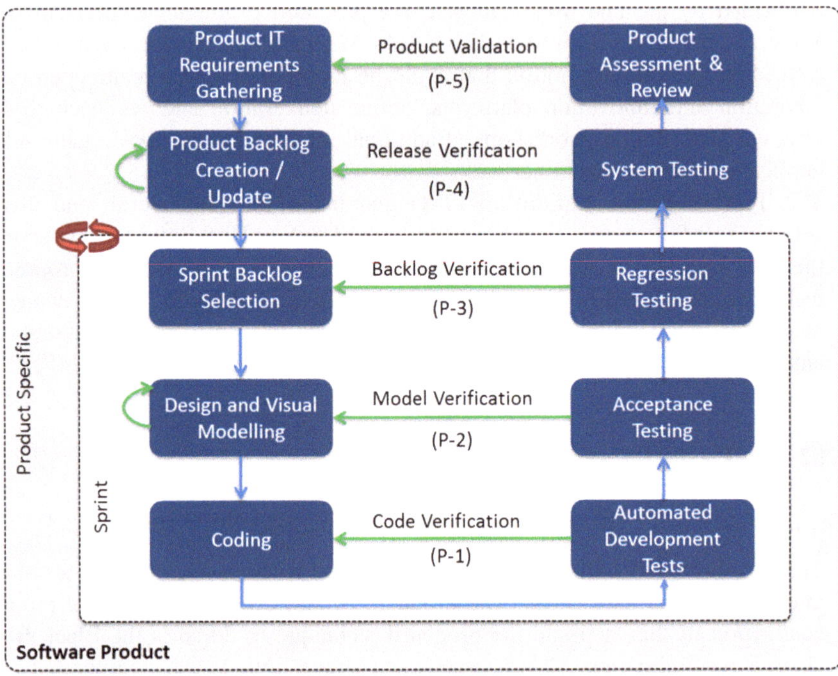

Fig. 3 V&V Steps under the product specific perspective (P)

IT tools for ex-post crowdsourcing are suggested to be incorporated as crowdsourcing methods.

- The **Model Verification (P-2)** step envisages to coordinate the alignment between design and requirements, as well as between design and code. Traceability Analysis is considered the appropriate technique here and the Development Team and the Sprint Master are the stakeholders responsible for this V&V activity. Physical or online workshops, readily-available prototyping applications and social deliberation platforms are some propositions for involving the general public during the conduction of this activity.

- The **Code Verification (P-1)** step ensures functionality, correctness, reliability, and robustness of code. The proposed technique to do so is considered the White Box Testing and the Development Team is the responsible stakeholder to put through this activity.

The proposed methodology capitalizes on lessons learned from the state of the art analysis while, in parallel, going beyond the state-of the art by introducing a set of innovative aspects. In particular, at first sight the methodology's structure, especially concerning the product specific perspective, seems to follow very closely the V-model's concepts. By following an agile way in each product's development and incorporating the sprint concept, agile development and classic V&V procedures are blend together. Additionally, this method introduces the higher level of the trial specific perspective, which addresses the evaluation and business assessment needs of a manufacturing environment. Finally, for each V&V activity and step, not only a proper technique is proposed, but a crowd assessment method as well, which is a fact not commonly met in the literature and of high value to all stakeholders. All the above can be noticed with a thorough examination of the proposed model and all combined are resulting in a newly introduced method in comparison to the usual V&V approaches.

5 Conclusions and Next Steps

The V&V Method presented in this paper aims at verifying, validating and evaluating a software product from its conception to final release and implementation in real-life settings. The added value of the proposed methodology can be summarized as:

- Bridging between the business and technical perspectives, as the proposed method assesses a software product during its whole lifecycle, from the code modules development stage to the after release, fit for purpose assessment and business evaluation stage.
- Combining the two most cited software engineering philosophies, the agile and waterfall one, which in most cases were dealt as controversial. The FITMAN V&V methodology adopted the best and most useful in a manufacturing environment aspects of the aforementioned philosophies in order to be complete and effective.
- Infusing crowd assessment in the V&V activities in order to tap the crowd intelligence power and the opinion of a wider experts' community base.

The FITMAN V&V Method is developed in order to be accompanied with the proper business and technical criteria which go hand in hand with each step's implementation. Those criteria are to be coupled with proper indicators and formulas in order to be measured. In addition, guidelines have been developed in order to facilitate each stakeholder involved apply the method in real life cases. Those criteria with their indicators and the guidelines as well as a developed V&V package are already defined and described in the context of the work conducted in the FITMAN project and were not part of this paper.

With regard to its application in the context of FITMAN project [6], the scope is to verify that the FI-WARE [21] generic and FITMAN specific enablers (as well as Trial Specific Components) satisfy the technological, platform and architectural integration requirements imposed; validate that the FI-WARE generic and FITMAN specific enablers (as well as Trial Specific Components) satisfy the requirements of Smart-Digital-Virtual use case trials; and identify the evaluation and assessment criteria to be used in all Use Case Trials, taking into account sustainability (STEEP) and future business benefits.

Finally, the proposed V&V method is general enough and domain-agnostic, thus it may be applied both to the manufacturing domain, as it happens in the FITMAN trials, and to any other domain. The driving principles of the V&V method are completeness and applicability: it provides all interested stakeholders with the baseline framework and the theoretical background to effectively undertake any V&V activity; it is thus up to each trial and each development team to streamline the method according to their own needs.

Acknowledgments This work has been partly funded by the European Commission through the Project FITMAN: Future Internet Technologies for MANufacturing (Grant Agreement No. 604674). The authors wish to acknowledge their gratitude and appreciation to the FITMAN project partners for their contribution during the development of various ideas and concepts presented in this paper.

References

1. EC. (2010). *Factories of the future (FoF) multi-annual strategic roadmap*. Retrieved September 15, 2013, from http://ec.europa.eu/research/industrial_technologies/pdf/ppp-factories-of-the-future-strategic-multiannual-roadmap-info-day_en.pdf
2. Allemang, R. J. (2013). *Editorial: Verification and validation. Sound and vibration*. Retrieved September 10, 2013, from http://www.sandv.com/downloads/1301alle.pdf
3. Tolk, A. (2012). *Engineering principles of combat modeling and distributed simulation*. New York: Wiley.
4. Rakitin, S. R. (2001). *Software verification and validation for practitioners and managers*. Boston, London: Artech House.
5. IEEE. (2012). *1012–2012—IEEE standard for system and software verification and validation*. Retrieved July 12, 2013 from https://standards.ieee.org/findstds/standard/1012-2012.html
6. FITMAN Project. (2013). Retrieved September 15, 2013, from http://www.fitman-fi.eu/
7. Waterfall Model. (2013). Retrieved June 10, 2013, from http://www.waterfall-model.com/v-model-waterfall-model/
8. Larman, C. (2004). *Agile and iterative development: A manager's guide*. Addison-Wesley Professional

9. V-Model. (2013). Retrieved September 15, 2013, from http://www.waterfall-model.com/v-model-waterfall-model/
10. Boehm, B., & Turner, R. (2004). *Balancing agility and discipline: A guide for the perplexed* (Vols. ISBN 0-321-18612-5 Appendix A, pp. 165–194). Boston, MA: Addison-Wesley.
11. Spanoudakis, G., & Zisman, A. (2005). Software traceability: A roadmap. In Chang, S. K. (Ed.), *Handbook of software engineering and knowledge engineering* (Vol. 3). Singapore: World Scientific Publishing.
12. International Atomic Energy Agency. (1999). *Verification and validation of software related to nuclear power plant instrumentation and control*. Technical Reports Series No. 384, Vienna.
13. Thacker, B., Doebling, S., Hemez, F., Anderson, M., Pepin, J., & Rodriguez, E. (2004, October). *Concepts of model verification and validation*. Retrieved July 10, 2013, from http://www.ltas-vis.ulg.ac.be/cmsms/uploads/File/LosAlamos_VerificationValidation.pdf
14. Pelliccione, P., Inverardi, P., & Muccini, H. (2009, May–June). *Charmy: A framework for designing and verifying architectural specifications*. Retrieved July 11, 2013, from http://ieeexplore.ieee.org/xpl/freeabs_all.jsp?arnumber=4711062&abstractAccess=no&userType=inst
15. Ganesan, D., Lindvall, M., Ackermann, C., McComas, D., & Bartholomew, M. (2009). *Verifying architectural design rules of the flight software product line 2009*. Retrieved July 11, 2013, from http://www.few.vu.nl/~rkrikhaa/adam/architecture_analysis_splc09.pdf
16. Kung, D., & Zhu, H. (2008, January 15). *Software verification and validation*. Retrieved July 11, 2013, from http://cms.brookes.ac.uk/staff/HongZhu/Publications/swvv3.pdf
17. Jackson, M., Crouch, S., & Baxter, R. (2011, November). *Software evaluation: Criteria-based assessment*. Retrieved July 11, 2013, from http://software.ac.uk/sites/default/files/SSI-SoftwareEvaluationCriteria.pdf
18. Sridharan, M. (2013, June 2). *Crowd sourced software testing*. Retrieved July 12, 2013, from http://it.toolbox.com/blogs/mithuns-memoirs/crowd-sourced-software-testing-52763
19. Dubinsky, Y., & Hazzan, O. (2004). Roles in agile software development teams. In J. Eckstein, H. Baumeister (Eds.) *Extreme programming and agile processes in software engineering* (Vol. 3092, pp. 157–165). New York: Springer.
20. Doumeingts, G., Clave, F., & Ducq, Y. (1995). ECOGRAI—a method to design and to implement performance measurement systems for industrial organizations—concepts and application to the maintenance function. In A. Rolstadås (Ed.) *Benchmarking—theory and practice IFIP advances in information and communication technology* (pp 350–368).
21. FI-WARE Open APIs for Open Minds. (2013). Retrieved September 15, 2013, from http://www.fi-ware.eu/

Collaborating Multiple 3PL Enterprises for Ontology-Based Interoperable Transportation Planning

Muhammad Ali Memon, Agnès Letouzey, Mohammed Hedi Karray and Bernard Archimède

Abstract Today enterprises have to distribute their final products to far away consumers. It is difficult and not cost effective for these enterprises to manage their own transport vehicles. Thus, they outsource their transportation tasks to third party logistics (3PL) companies. These 3PL companies take transport orders from several clients and try to group them in the vehicles to utilize their resources at maximum. An issue of interoperability arises, when 3PL companies have to process different transport orders arriving from several clients in different formats and terminologies. Secondly, how 3PLS will collaborate with other 3PL companies following different working standards and also for collaboratively delivering transport orders which single 3PL cannot deliver alone due to its limited operational geographic area. Interoperability to achieve collaborative transportation planning is our concern in the context of this paper. Interoperability is a key issue for collaboration, especially in case of heterogeneous environment, when entities trying to collaborate have different ways of functioning and follow certain standards specific to their organizations. So the objective of this paper is to present a distributed and interoperable architecture for planning transportation activities of multiple logistics enterprises aiming at a better use of transport resources and by grouping transport orders of several manufacturers for each effective displacement.

M. A. Memon (✉) · A. Letouzey · M. H. Karray · B. Archimède
University of Toulouse, INP-ENIT, 47 Avenue d'Azereix, BP1629 65016, Tarbes Cedex, France
e-mail: muhammad-ali.memon@enit.fr

A. Letouzey
e-mail: agnes.letouzey@enit.fr

M. H. Karray
e-mail: mohamed-hedi.karray@enit.fr

B. Archimède
e-mail: bernard.archimede@enit.fr

K. Mertins et al. (eds.), *Enterprise Interoperability VI*,
Proceedings of the I-ESA Conferences 7, DOI: 10.1007/978-3-319-04948-9_27,

Keywords Interoperable and distributed scheduling · Multi-agent systems · Collaborative transportation planning · Third party logistics · Ontology

1 Introduction

More often, companies wishing to reach the far away customers could not possibly purchase their own fleet of vehicles to transport their goods. These companies contact third party transportation companies to ship their products, same as a courier company. This need led emergence of Third Party Logistics enterprises (3PL) [1]. In that case, suppliers can outsource their complete transportation tasks to 3PL enterprises and then these 3PLs take charge of whole transportation process. To fulfil customers' demands and improve the performance of supply chains, 3PL must manage its own resources and collaborate with other 2PL (carriers) and 3PL companies to reach far away customers at lower price. Additionally these 3PLs will group together several transport orders sharing similar origins or destinations in vehicles to deliver them collectively. Eventually minimizing number of transport travels and minimizing environmental pollution. This collaboration involves a good understanding of exchanged information between clients and 3PL and between 3PLs, especially about locations, product constraints, vehicles type, etc.

Clients will generate their transport orders by their own specific ways, which will not be understandable by 3PLs. There is a need of an interoperable mechanism which can transform the information for that 3PL in an understandable form. This transformation should deliver correct information and without any distortion. Similarly for communication in case of multiple 3PLs, they need an intermediate mechanism to understand each other's working methods in order to collaborate. One solution is to let each entity work in its own manner by using their terms, but defining them using their local ontologies and let interoperable service utilities (ISU) handle the transformations on the basis of common semantics [2, 3]. Thus, the schedule of all transport orders has to be achieved by several interoperable scheduling systems.

The work presented in this paper proposes ontology based interoperable framework to support collaborative transport planning for 3PL enterprises in distributed manner. The objective is to describe the I-POVES Interoperable (Path Finder, Order, Vehicle, Environment, and Supervisor) for improving collaboration and interoperability between 3PL enterprises and clients. After a state of the art on the latest research on interoperability for transportation planning, we describe the interoperable architecture of the POVES model and the associated ontologies. Finally, we conclude with future work.

2 State of the Art

Several approaches have been proposed to solve transportation planning problem. Sauer and Appelrath [4] proposed a centralized approach with a global scheduler, which schedules transportation planning activities. They model the problem by a

5-tuple (R, P, O, HC, SC), where R denotes the set of required resources, P the set of products, O the set of actual orders and HC and SC the sets of hard and soft constraints, respectively. They use a rule-based approach and heuristics to produce several scheduling strategies. This approach is centralized and is limited to the planning of transportation activities of a single enterprise. The need for confidentiality limits the scope of centralized approaches.

Baykasoglu and Kaplanoglu [5] proposed a multi-agent approach to address collaborative transportation problem. This approach is based on cooperation between transport order agents and truck agents, which proposes grouping multiple orders together in a vehicle. In this approach, transport order agent is bound to accept the proposition from one truck agent, which provides a nonstop delivery from origin to destination. However, in reality a truck rarely alone transports a transport order. A transport order requires, most often, several trucks.

Takoudjou et al. [6] propose a multi-agent heuristic to address the transport problem with transhipment. Their methodology is decomposed in four steps. In first step, they calculate PDP (pickup and delivery solution) without transhipment/ Cross Docking solution for all random requests. In second step, they try to optimize the PDP solution with VND (Variable Neighbourhood Descent method) using Path Relinking. In the third step, they calculate PDPT (PDP with transhipment) solution, compare it with PDP solution and keep the best one. This whole procedure is repeated to the number of iterations. This work makes an assumption that number of vehicles is not fixed and if no vehicle can satisfy a request because of the noncompliance with the constraints (vehicle capacity, time windows, etc.), a new route is created with a new vehicle to welcome the considered request. It is not realistic to create a new vehicle each time a transport order needs one. Moreover, this method calculates PDP solution without transhipment improves it by optimizing it and then destroys the PDP solution to obtain PDPT solution with transhipment. In further studies a simulation framework is presented by Sprenger and Mönch [7] for assessing the performance of cooperative transportation planning and isolated transportation planning. Mes et al. [8] study the interaction between intelligent agent strategies for realtime transportation planning. A multi-agent theoretical approach on dynamic transportation planning is given in [9].

Above mentioned papers are interested only for transportation planning and do not take into account the interoperability aspect. Following papers discuss the interoperability for exchanging information. Niarki and kim [10] propose an ontology based personalized route planning system which uses multi-criteria decision making from user/decision maker. Whether, a certain route is better than other strongly depends on environmental situations and user preferences. In addition to these criteria, the impedance of road plays a very important role in route planning. Impedance factors involved in determining the travel time are the volume of the traffic, the type of the road, the road width, number of junctions and turns etc. This approach models the decision making criteria using ontology and apply an impedance function in route finding algorithm to find the personalized route for the user(s).

Paul Davidson et al. [11] have developed an adapter based open source freeware to exchange information between business systems. System was tested on two case studies for improving transport activities for small medium enterprises, based in Sweden. Smith and Becker [12, 13] propose an ontology based toolkit for constraint based scheduling system called ozone. The ozone ontology provides a framework for analysing the information requirement of a given target domain, and a structural foundation for constructing an appropriate domain model. We here are interested to use the aspect of interoperability for solving collaboration problem in transport planning domain.

3 P.O.V.E.S Multi-Agent Model

3.1 Description of Model

The POVES multi-agent model (Fig. 1) is developed for collaborative transportation planning activities. It is inherited from SCEP multi-agent model, which is being used with success for manufacturing planning since years. Limitations restricted SCEP for transportation planning [14], due to that POVES emerged after overcoming these limitations. POVES introduces an indirect cooperation between two communities of agents, Order agents called (O) and vehicle agents called (V), leading to a high level of co-operation. Each order agent manages one transport order from first party logistics (1PL). Each vehicle agent manages one vehicle of the organization. A supporting agent "Path Finder" elaborates for a transport order the traveling route between pickup and delivery locations. The cooperation between order agents and vehicle agents is performed synchronically through the background environment agent E. The supervisor agent S controls the model functioning. The detail working procedures and functioning of POVES model is given in [15].

3.2 Limitations of the Model

This model is well suited for transportation planning in case of only single 3PL enterprise with a fleet of its own vehicles. However a single 3PL enterprise operates in a limited region and it is unlikely possible that it can fulfil TO entirely. It is more often that, TOs have to be delivered to faraway clients in a region outside the reach of this 3PL. It must collaborate with other 3PL enterprises that operate in other regions to make the delivery of the products to faraway clients. Moreover, if more than one 3PLs operating in the same region entirely or partly, they will increase the chance of TO delivery on time, as one of either must have a vehicle available to deliver the order on time. Additionally in POVES, TOs arrive

Fig. 1 POVES model

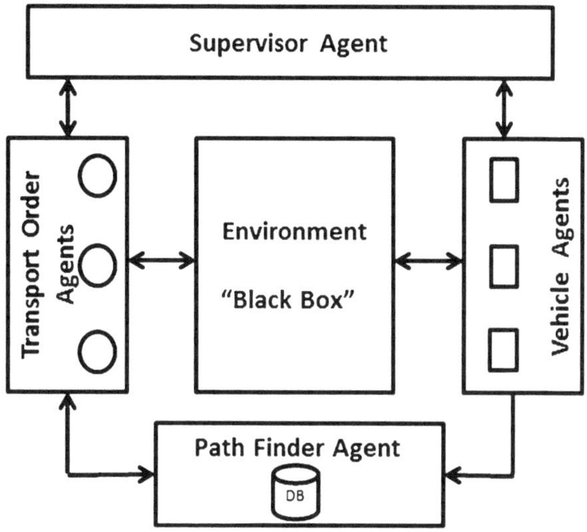

from several clients, but these TOs should be in the same format that should be compliant with the format understandable by Path Finder agent to find the elementary activities. In reality, customers generate their TOs in different formats that are not interoperable by the 3PL enterprises, because each of the customers has its own way of interpreting and representing locations, paths, etc. Similarly, each of 3PL enterprises defines their elementary activities and vehicle parameters according to their own methods and formats. There is need of some intermediary mechanism that could understand all the formats in order to produce better results, which is currently not available in the POVES previously presented. For these reasons POVES model must be evolved.

4 I-POVES

In order to take into account limitations in the preceding section, we propose an evolved version of POVES model shown in Fig. 2. In this model, we add ontologies and interoperable service utilities (ISU) in order to achieve interoperability to treat different formats. Ontology provides a shared vocabulary (terminologies), which can be used to model a knowledge domain. Each customer who generates TOs, defines them using the terminologies from their local ontology. Similarly, each transporter (3PL) has also its local ontology to define its elementary activities and vehicle parameters, which is comprehensible only by this transporter. In addition to that, we add a global ontology, which provides a federation of concepts. Those concepts map with the concepts of all the local ontologies on the basis of common semantics [3]. Global ontology has consistent and coherent information and has standard and shared terminologies.

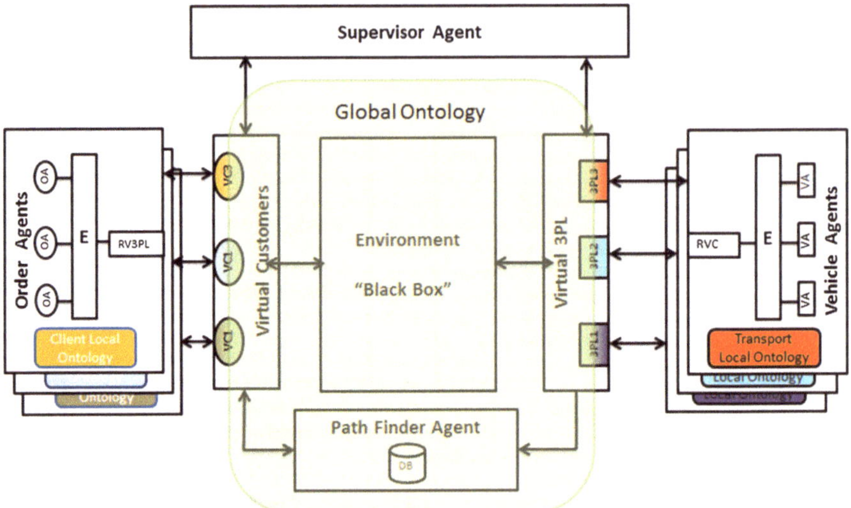

Fig. 2 I-POVES model

Furthermore, in order to work on that global ontology standard terminologies, we require interoperable service utilities, which will be used for matching and translation of terminologies from local ontologies to global ontology and vice versa. In I-POVES, virtual customer presents an ISU that will match and translate the enterprise's "Customers local ontologies" terminologies to global ontologies' terminologies and then communicate with Path Finder to find the route. Path Finder finds a route based on the terminologies of global ontology. Then, virtual customer will send computed route tasks to environment for planning. Similarly, virtual 3PL is also an ISU that will retrieve the tasks from environment. Those tasks are represented in the form of global ontology standard and translated and sent into the format of local ontologies for respective 3PLs by ISU. After 3PL planning, virtual 3PL sends back to the environment propositions received by 3PLs after translating them from transporter's local terminologies to global ontologies. Virtual 3PL will also translate the elementary activities of each 3PL and their vehicle parameters and send them to Path Finder to update its database before commencement of the planning process, each time Path Finder agent is activated.

The use of local and global ontologies will provide liberty to customers and transporters to work on their own standards without bothering of everybody else. Similarly intermediary ISUs will provide interoperability between them to work together in order to provide collaborative transportation planning.

Ontologies presented in this paper are loosely inspired from the Ozone ontology developed by Smith and al [12, 13]. Figure 3 presents an example of a local ontology of a transporter. Blue circles represent domain classes, red rectangles represent properties and green rectangles represent class instances and orange rectangle represents class attributes. Transporter owns fleet of vans represented by class 'Van'. These vans have facilities, which are represented by a class 'Facility'

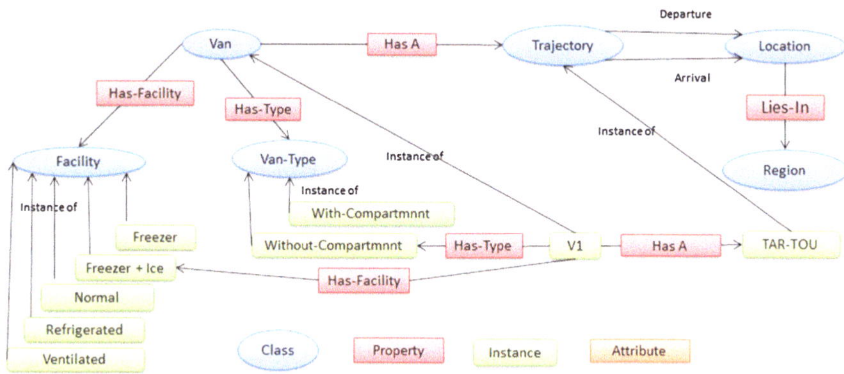

Fig. 3 Example of transporter local ontology

which have five instances to represent facilities type: 'Freezer', 'Freezer + Ice', 'Normal', 'Refrigerated' and 'Validated'. Association between class 'Van' and class 'Facility' is represented by a property called 'Has-Facility'. Vans with compartments are used to transport different type of products with different temperature requirements. There is another class called 'Van-Type', which has two instances called 'With-Compartment' and 'Without-Compartment'. Relation between class 'Van' and 'Van-Type' is represented by property 'Has-Type'. Each van is associated with certain nonstop trajectories assigned by transporter. Each trajectory has location of departure and location of arrival. Each location lies in certain region, where transporter provides its logistics services. Class 'Van' has a relation called 'Has A' with class 'Trajectory'. Class 'Location' is associated with class 'Region' with the property 'Lies-In'.

V1 represents one of the instances of Van which has facility of 'Freezer-Ice' and is of type 'Without-Compartment'. 'TAR-TOU' is an instance of class 'Trajectory' which has Tarbes as location of departure and Toulouse as location of arrival. Instance V1 is associated with this trajectory and is represented by the property 'Has-A'.

Transporters even following the same standards but residing in different countries will express locations differently as each country has its specific way of defining geographical divisions. For example in France there are regions and departments and in Pakistan there are provinces, divisions and districts. Furthermore, transporters operating even in the same region to their clients have heterogeneous way of using locations name. In our example transporter defines TAR and TOU as abbreviations of Tarbes and Toulouse, which may not be understandable by the client placing the order. This antagonism generates the need of a global ontology. It represents the concepts that are semantically similar used in local ontologies to follow a single standard during the planning process. Figure 5 illustrates the global ontology.

Similarly Fig. 4 represents an example of client local ontology. Client generates set of TOs represented by a class 'Transport-Order' which is created to deliver a

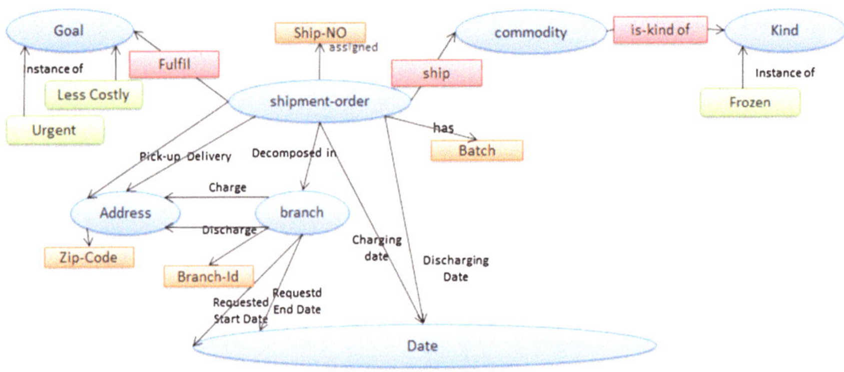

Fig. 4 Example of client local ontology

product P1 of type 'Has-Type' Live animal of quantity 'Quantity'. TOs have city of pickup and delivery in global, which are then decomposed into elementary nonstop travels called 'Tasks' by Path Finder agent. Client attaches objective to each TO in order to define its priority which needs to be fulfilled. Here 'Objective' class have three instances; 'Less Costly','Urgent','Less Distance'. Client proposes requested start date and requested end date for these tasks and in return receives potential dates and effective dates from transporter (Fig. 5).

Global ontology has the concept Vehicle that corresponds to concept 'Van' and 'Truck' for local ontologies. Vehicle performs 'Activity' similar to 'Trajectory' and 'Travels' in local ontologies. Vehicle has 'category' partitioned and 'whole space' similar to 'Without-Compartment' and 'With-Compartment' in ontology for transporter.

There have to be matching criteria between local ontologies and global ontologies embedded in ISU of both the sides; clients and transporters. Table 1 shows the concepts of alignment between local client ontology and global ontology, while Table 2 shows the alignment of transporter local and global ontology. The alignment mechanism used here is constructed manually but can be automatized using approach proposed by Song Fugi in [16]. He developed an ontology alignment technique to contribute federated enterprise data interoperability approach at semantic level. ISU will use this matching mechanism to make the transformation possible from local ontologies concepts to global concepts and vice versa. Local ontologies are subjected to evolve when new vehicles and new travels are added or clients progress from local to global. This evolution will cause the enrichment of these local ontologies, also forcing the enrichment of global ontology at the same time in order to continue keeping the compliance. This evolution is independent of the planning mechanism making this framework of distributed nature. Additionally transporters and clients joining or leaving the model do not affect its interoperable nature. It will require only the enrichment of the global ontology and updating the matching and transformation mechanisms in ISUs of POVES model.

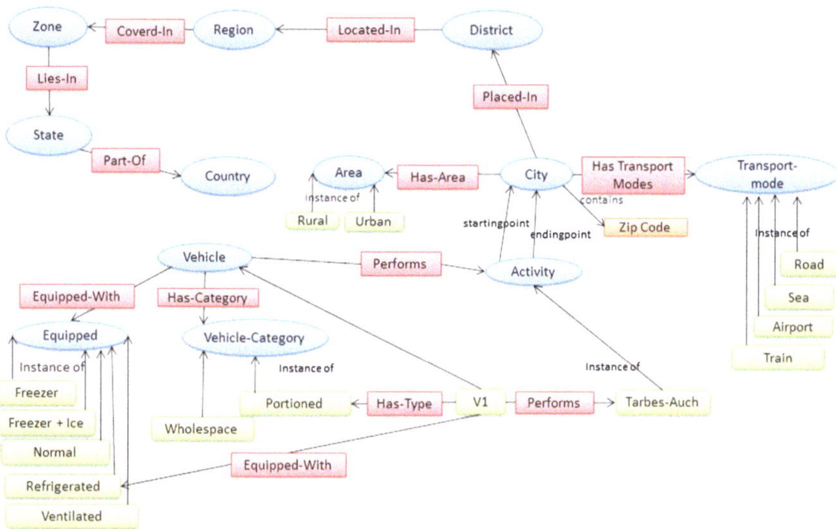

Fig. 5 I-POVES global ontology for transport

Table 1 Concepts alignment between global and client ontology

Global	Client
Transport order	Shipment order
Order-No	Ship-No
Objective	Goal
Product	Commodity
Type	Kind
Origin	Charge
Destination	Discharge
Pickup date	Charging date
Delivery date	Discharging date
Quantity	Batch
Task	Branch
Task-No	Branch-ID
City	Address

Table 2 Concepts alignment between global and transporter ontology

Global	Transporter
Vehicle	Van
Activity	Trajectory
Equipped	Facility
Category	Van-Type
Origin	Departure(location)
Destination	Arrival(location)
Department	Region

5 Conclusion and Future Work

In this paper we presented ontology based a collaborative and interoperable framework called "I-POVES" for transportation planning problem In I-POVES, each order agents has its own local ontology and describes its transport orders using the concepts from local ontology. Similarly vehicle agents also have their own local ontologies to describe their vehicles and activities. We used a federated approach based on global ontology that maps all shared concepts used by local ontologies on both the sides. There are two ISUs, one on the order agents' side and one the vehicle agent side, to perform transformation between local and global ontologies. In I-POVES, firstly Path Finder Agent elaborates, when solicited for each order the traveling routes between pickup and delivery locations. Secondly Order agents offer transport jobs through sequential auctions and vehicle agents compete with each other to serve those jobs. Vehicle agents propose grouping these jobs together to execute them simultaneously. Multiple 3PL enterprises collaborate through this framework to propose the delivery of transport orders together. One of the future directions is, in case when 3PLs have their own planning mechanism and they just want to use I-POVES for collaboration with other 3PLS. How much transport ISU will be capable to handle not only ontology transformations but also the transformation from I-POVES planning mechanism and 3PL's local planning mechanism and vice versa.

References

1. Marasco, A. (2008). Third-party logistics: A literature review. *International Journal of Production Economics, 113*(1), 127–147.
2. Zbib, N., Archimède, B., & Charbonnaud, P. (2012). Interoperability service utility model and its simulation for improving the business process collaboration. In *Enterprise Interoperability V* (pp. 403–413). London: Springer.
3. Karray, M. H., Morello B., & Zerhouni N. (2010). A contextual semantic mediator for a distributed cooperative maintenance platform. *IEEE INDIN Conference* (pp. 181–188).
4. Sauer, J., & Appelrath, H. (2000). Integrating transportation in a multi-site scheduling environment. IEEE System Sciences. *Proceedings of the 33rd Annual Hawaii International Conference*. 9 pp.
5. Baykasoglu, A., & Kaplanoglu, V. (2011). A multi-agent approach to load consolidation in transportation. *Advances in Engineering Software, 42*(7), 477–490.
6. Takoudjou, R. et al. (2012). *A Hybrid multistart heuristic for the pickup and delivery problem with and without transshipment 9th International Conference on Modeling, Optimization and SIMulation, 2012*.
7. Sprenger, R., & Mönch, L. (2008, December). A simulation framework for assessing the performance of cooperative transportation planning algorithms. In *Proceedings of the 40th conference on winter simulation* (pp. 2769–2776). Winter Simulation Conference.
8. Mes, M., van der Heijden, M., & Schuur, P. (2010). Interaction between intelligent agent strategies for realtime transportation planning. *Central European Journal of Operations Research, 203*(1), 1–22.

9. Kozlak, J., Créput, J., Hilaire, V., & Koukam, A. (2004). Multi-agent environment for dynamic transport planning and scheduling. *Computational Science-ICCS, 2004*, 638–645.
10. Niaraki, A. S., & Kim, K. (2009). Ontology based personalized route planning system using a multi-criteria decision making approach. *Expert Systems with Applications, 36*, 2250–2259.
11. Davidsson, P., Ramstedt, L., & Törnquist, J. (2005). Inter-Organization Interoperability in Transport Chains Using Adapters Based on Open Source Freeware. *Interoperability of enterprise software and applications*, Berlin, Germany: Springer Verlag.
12. Becker, M. A., & Smith, S. F. (1998). *An ontology for multi-modal transportation planning and scheduling*. The Robotics Institute: Carnegie Mellon University.
13. Smith, S. F., Cortellessa, G., Hildum, D. W., & Ohler, C. M. (2004b) "Using a Scheduling Domain Ontology to Compute User-Oriented Explanations", In Planning and Scheduling. In L. Castillo, D. Borrajo, M. A. Salido & A. Oddi (Eds.), *Frontiers in artificial intelligence and applications series*, IOS Press, Forthcoming.
14. Memon, M. A., & Archimède, B. (2013). Towards a distributed framework for transportation planning: A food supply chain case study. *IEEE International Conference on Networking, Sensing and Control*, France.
15. Memon, M. A., & Archimede, B. (2013). *A Multi-Agent Distributed Framework for Collaborative Transportation Planning in IFACE Conference on Manufacturing Modelling, Management and Control, MIM 2013 Saint Petersburg*, Russia.
16. Song, F., Zacharewicz, G., & Chen, D. (2013). An ontology-driven framework towards building enterprise semantic information layer. *Advanced Engineering Informatics, 27*(1), 38–50.

Part VIII
Services for Enterprise Interoperability

A Service Selection Approach in Cloud Manufacturing for SMEs

Haijiang Wu, Dan Ye, Shanshan Liu, Yan Yang and Lin Bai

Abstract Small- and Medium-sized Enterprises benefits much from Service-oriented manufacturing, which utilizes the internet and service platform to arrange manufacturing resource and provides service according to the customers' demands. On this platform, service selection is one of the key steps for customers to get the best services. This paper introduces a service selection approach in cloud manufacturing for Small- and Medium-size Enterprises. First, we build a service selection model, including service evaluation and service constraints. And then a service selection algorithm is presented based on the service selection model to identify the best service for a service buyer. Finally, a case study is given to illustrate how this approach works in Cloud Manufacturing Platform for Small- and Medium-sized Enterprises (CMfg-SME).

Keywords Cloud manufacturing · Service selection · Small- and medium-sized enterprises

H. Wu · D. Ye (✉) · S. Liu · Y. Yang · L. Bai
Institute of Software, Chinese Academy of Sciences, 4 S. 4th Street of Zhongguancun, Beijing, China
e-mail: yedan@otcaix.iscas.ac.cn

S. Liu
e-mail: liuss_1988@163.com

Y. Yang
e-mail: yy@otcaix.iscas.ac.cn

L. Bai
e-mail: bailin@otcaix.iscas.ac.cn

H. Wu
University of Chinese Academy of Sciences, 4 S. 4th Street of Zhongguancun, Beijing, China
e-mail: wuhaijiang12@otcaix.iscas.ac.cn

K. Mertins et al. (eds.), *Enterprise Interoperability VI*, Proceedings of the I-ESA Conferences 7, DOI: 10.1007/978-3-319-04948-9_28,

1 Background

Cloud manufacturing (CMfg) [1] is a service-oriented manufacturing model proposed recently, which utilizes some new information science technologies rose in recent years, such as cloud computing, internet of thing, high performance computing, virtualization and service-oriented architecture (SOA), etc. Service model is the base of CMfg platform. All the manufacturing resources in CMfg platform are virturelized as manufacturing services, so that they can be shared and traded on CMfg platform. CMfg is characterized as providing service on the demand of the service user and paying as you go. Thus finding the appropriate service and providing it on the demand of the service user is one of the key issues in CMfg platforms.

As we all known, the small- and medium-sized enterprises (SME) are the majority of the manufacturing enterprises, which can provide most of the manufacturing services. However, there are some problems faced by SMEs [2]. The manufacturing capability resource of SMEs is abundant, but most of them are at the bottom of the value chain. The lack in self-innovation and design capability, the deficiency of credibility of manufacturing resource transaction and commercial reputation evaluation system, and the infancy capability in providing follow-up service, hinder the additional value creation. Although some exiting cloud service platforms [1, 2] have been built for SMEs, but few works considers the service selection in CMfg-SME.

While there are some significant differences between web service selection and manufacturing service selection. First, the object of web service selection is a web service, which is a computer program and has no off-line work. But the cloud manufacturing service is more than a web service, and need human beings to assist it accomplishing product manufacturing. For example, if an enterprise provides an automobile parts manufacturing service, it has to manufacture the automobile off-line. Second, the service evaluation model is different. In web service, the service evaluation considers the attributes of itself only, such as cost, time, and usability, while the off-line activity of cloud manufacturing service cannot be evaluated in this way. Last but not the least, the feedback of the service is different. When a web service is called by a user, the service provider can get the feedback from the return of the return value, but it is not so easy for the CMfg-SMEs service provider because of the off-line operation. For SMEs, there are more off-line activities, because it is more difficult for them to visualize their equipment, and connect it to the cloud manufacturing platform.

In this paper, we introduced an SMEs oriented cloud manufacturing service selection approach. First, a service selection model is built to define the service selection problem. In this model, we defined the manufacturing task, the services in the task, the relations between the services and the constraints on service selection. Second, based on the service selection method, we describe the service selection algorithms. Finally, a case study is described to illustrate the service selection approach.

The rest of the paper is organized as follows. In Sect. 2, we list the related works. The service selection model in CMfg-SME is introduced in Sect. 3. After describing the detail of the service selection algorithm in CMfg-SME in Sect. 4, we give a case study in Sect. 5, and then conclude in Sect. 6.

2 Related Work

Most of existing works about service selection focus on web service selection, which is also named as web service composition. Hiroshi et al. [3] gives a detail define of web service composition,and proposed a web service composition approach based on Genetic Algorithm called E3, it first divide all the individuals into two groups (feasible solution and unfeasible solution), and then find the relationships between different individuals according to their QoS scores. Kuzu and Cicekli [4] proposed an automated web service composition approach based on an AI planner that is designed for working in highly dynamic environments under time constraints. Feng et al. [5] study QoS-aware and multi-granularity service composition, based on the behavioral signature model they defined, they make new web service plans by substituting the service with another service or a set of services of finer or coarser grain, until the service has better quality. Haddad et al. [6] study Multicriteria Evaluation Component (MEC), which takes as input a set of composite Web services and a set of evaluation criteria and generates a set of recommended composite Web services.

Most existing related work about manufacturing service selection focus on either grid manufacturing model and other other manufacturing model, or on big enterprise service cloud manufacturing, so their works concentrate on the web service selection. Huang et al. [7] proposed a comprehensive performance evaluation metric for service-oriented manufacturing, and developed an optimal service selection algorithm considering efficiency and effectiveness. In manufacturing grid system, Tao et al. [8] takes the user's feeling into account in the resource service optimal-selection (RSOS), by using an intuitionstic fuzzy set. When evaluating the QoS, they used a time-decay function. Based on this, they illustrated their resource service optimal-selection approach. Weiming et al. [9] proposed an agent-based service-oriented integration architecture to leverage manufacturing scheduling services on a network of virtual enterprises, they also implemented the inter-enterprise manufacturing resource sharing, service selection and integration of inter-organizational business processes.

3 Service Selection Model

In order to establish the collaborative manufacturing chain for the cloud manufacturing environment, service selection is used to identify a service node or a set of service nodes of this chain. To describe the scenario of service selection, we define the service selection model as follows.

3.1 Term Definition

3.1.1 Concrete Manufacturing Service

Enterprises can provider manufacturing service, it will be visualized as a web service, although some SMEs cannot visualize all the equipments as web services, but its manufacturing capabilities can be described in the web services. Here the manufacturing service provided by an enterprise is defined as Concrete Manufacturing Service (CMS).

Note that, for SMEs, the concrete manufacturing service can describe its capabilities, but the web service execution cannot reflect the real manufacturing process, which is different from the manufacturing service execution of big enterprises.

3.1.2 Abstract Manufacturing Service

Unlike big Enterprises, SMEs are in a highly competitive environment, many SMEs will provide manufacturing services of the same content, so this service can be grouped as an Abstract Manufacturing Service (AMS).

$$\mathrm{AMS} = \{\mathrm{Set}(\mathrm{CMS})\} \tag{1}$$

3.1.3 The Relationship Between Cloud Manufacturing Services

For AMS, there are six kinds of relations between CMfg services(as shown in Fig. 1): parallel, exclusive, sequential, cyclical, parallel aggregate and exclusive aggregate. When a CMfg service has a parallel relationship with another, it means that these two services can be executed in parallel, and the exclusive relationship means that only one of this service can be executed. If a service has a sequential relationship with another one, they should be executed in turn. Any service that has cyclic relation, will be executed more than once (Figs. 2, 3).

For CMS, there are only two service types: parallel relation and exclusive relation.

3.1.4 Manufacturing Task

When an SME enterprise wants to manufacture a product, but it cannot process all the parts of this product, the best way to achieve this product manufacturing is collaborating with other enterprises. Here we abstract this procedure as a manufacturing task as follows:

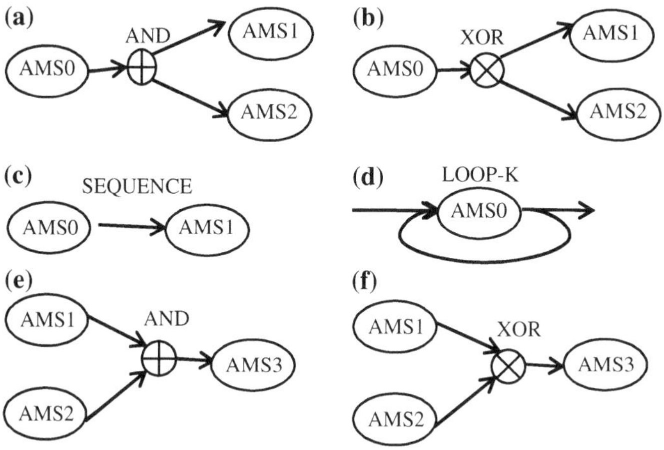

Fig. 1 The relations between AMS: **a** Parallel relation; **b** Exclusive relation; **c** Sequential relation; **d** Cyclical relation; **e** Parallel aggregate relation; **f** Exclusive aggregate relation

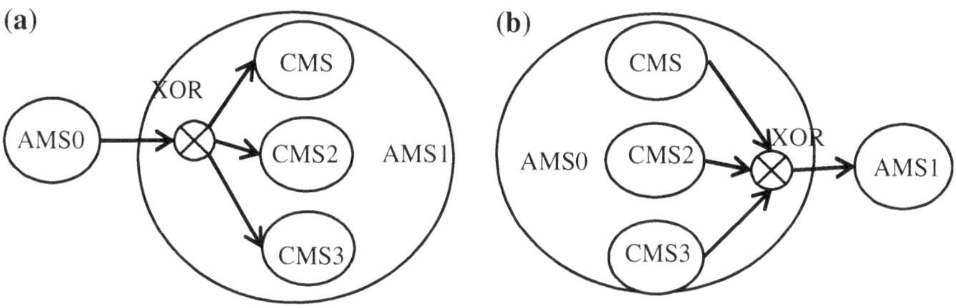

Fig. 2 The relations between CMS: **a** Parallel relation; **b** Exclusive relation

Definition 1 A cloud manufacturing task is a collaborative manufacturing chain, in which different enterprises want to manufacture a product cooperatively. A manufacturing task T is in the form of:

$$T = (\mathrm{Set}(\mathrm{AMS}), \mathrm{Set}(\mathrm{Relation}), \mathrm{START}, \mathrm{END}) \tag{2}$$

START indicates the begin of a manufacturing task, and has no incoming degree. END is the end node of the collaborative manufacturing chain, so it has no out degree. Set (AMS) is set of abstract manufacturing service, each AMS represents a service node of the collaborative manufacturing chain, the CMSs in each AMS are the candidates of the service nodes. Before an AMS can be executed, some other AMSs should be executed ahead or later. Set (Relation) is a set of relationships in the collaborative manufacturing chain.

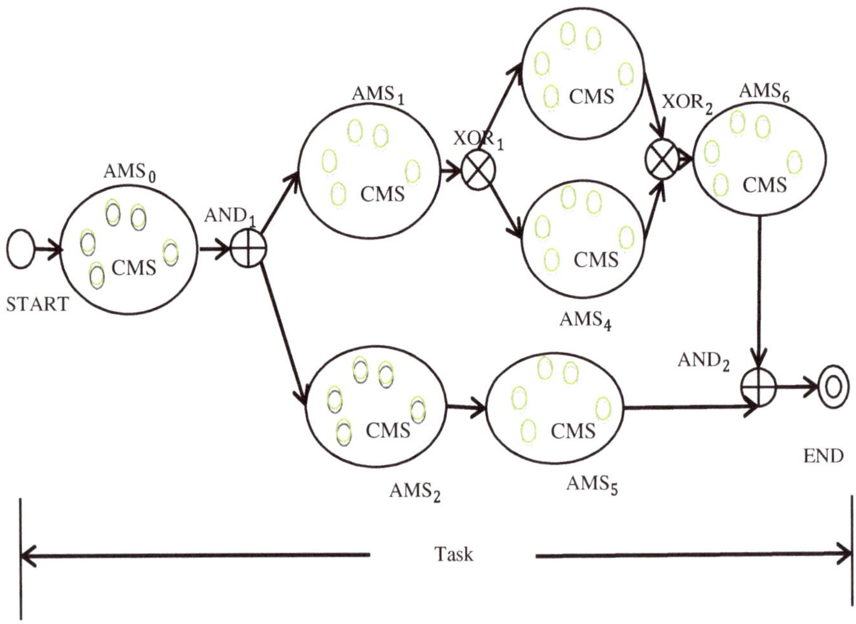

Fig. 3 Example of manufacturing task

3.2 Service Evaluation in CMfg

There are many features can be used to evaluate a CMfg service, but we think the following ones are the most important.

The first one is Service Matching Score (SMS), a high SMS value means the candidate service is more likely to serve as they want. The second is Service During Time (SDT), SDT value indicates how long will the candidate service take to achieve its work. And then are the credit of service (Cs), the service provider (Cp) and the experience of the service provider (Ep), which are computed based on the Dirichlet distribution [10]. We also use the price the candidate service demand (P) and the freightage (F), which can be obtained from the service description.

Above all, some of the features are also used in the CMfg service selection, while freightage is used for SMEs specifically.

3.3 Service Constraints in CMfg

Definitely, the goal of service selection is to find a service with high SMS, low SDT, high Cs, high Cp, high Ep, and low P and F. Usually, the importance of features will not be the same, some are mandatory constraints, and the others are non-mandatory.

Definition 1 A service constraint in CMfg is a quad:

$$\text{Constraint} = (\text{SCOPE},\ \text{NodeFrom}, \text{NodeTo}, \text{ATTRIBUTE}) \tag{3}$$

$$\text{Attribute} \in \{\text{SMS}, \text{SDT}, \text{Cs}, \text{Cp}, \text{P}, \text{F}, \text{Ep}\} \tag{4}$$

$$\text{Scope} = \left(\text{r}_\text{i}, \text{t}_\text{j}\right)\ \text{r}_\text{i} \in \text{Relation}, \text{t}_\text{j} \in \text{Interval} \tag{5}$$

$$\text{Relation} = \{>, \leq, \geq, <, =\} \tag{6}$$

$$\text{Interval} = \text{R}^+ \cup \{\text{Max}, \text{Min}\} \tag{7}$$

SCOPE is a set of AMS and the combination of AMS that the constraint will take effect on. ATTRIBUTE is a set of features we use to evaluate a service. NodeFrom is the start node that the constraint takes effect on, and NodeTo is the end node that the constraint takes effect on.

A cloud manufacturing task may have more than one constraint:

$$\text{ConstraintS} = \{\text{Constraint}_1, \text{Constraint}_2, \ldots, \text{Constraint}_\text{k}\} \tag{8}$$

According to the definition of service constraint in CMfg, an example is given in Eq. 9:

$$\text{Constraint} = (\text{AMS}_1, \text{XOR}_2, \text{SDT}, (<, 80)) \tag{9}$$

This constraint represents that the service user wants to find a service composition whose total SDT is less than 80.

4 Service Selection Algorithm in CMfg-SME

4.1 Single Service Selection (SSS)

In SSS, there is no constraint on the relation between the services. We only need to choose the best CMS for an AMS, so the SMS feature is the most important one. After that, we can order the services according to the feedback of the users. The feedback of the users includes: customer satisfaction (CS), timeliness of delivery (TD) and the accuracy of the description (AD). All these feedbacks are endowed with weights, the sum of these weights is 1.

$$\text{W}_{\text{CS}} + \text{W}_{\text{CD}} + \text{W}_{\text{AD}} = 1 \tag{10}$$

The SSS algorithm is described as follows: First, the query conditions are used to find the most matching services. Then the matched service will be ordered according to its feedback score. Finally, the service with the highest score will be returned to the service user.

4.2 Multi-Constrains Service Selection

Multi-constrains service selection is an NP problem. It is not easy to find the optimal answer to the problem. Genetic algorithm (GA) [11] is a heuristic search that mimics the process of nature selection, it can generate solutions to optimization problems using techniques, such as inheritance, mutation, selection and crossover. There are some key steps in GA design including: genome encoding, fitness function design, genetic operators design. We use a fitness function like the Ref. [12], and use the proportional model in the selection operator. The genome encoding and crossover operator are designed as follows.

4.2.1 Genome Encoding and Representation

The encode of the genome in this paper is an array with length N, N is the number of the AMS. Each gene represents an AMS, and it also contains some CMS, as shown in Fig. 4. For the task that all the relation nodes are AND relations, we need to select a CMS for each AMS.

If both AND relation and XOR relation exist in a task, the generated gene should satisfy the constraints introduced by XOR. In this paper, we use an array FLAG to mark the AMSs which already have CMSs selected for them. The first element of FLAG is used for START node, and the last element of FLAG is used for END node, both of them are assigned 1. Other values of FLAG are initialized to NULL.

For AMSs that have AND relation, the corresponding element of them in Flag are assigned to 1, because only all of them have been executed, could the next AMS start to work. In a task, only one of the AMSs that have XOR relation will be executed, so in a FLAG, one of the AMSs that have XOR relation is assigned to 1, and the left is assigned to 0. A genome encoding example is shown is Fig. 4.

4.2.2 Crossover Operator Design

After genome representation, a service composition can be represented as a genome, and the genome has a FLAG array, which indicate whether an AMS in the task is selected in the service composition.

The result of crossover operator is the generated genes that are legal, so we will not cross the genes whose corresponding values are 0 in FLAG. Obviously, when both of the corresponding values are 1 in FLAG, two gens can be crossed. If one gene whose corresponding value in is 1 FLAG, and that of the other is 0, the relation of these two genes is XOR, but only one of them can be selected in a service composition. It is impossible to cross these two genes.

In a word, we can only cross the genes whose corresponding values are 1 in FLAG.

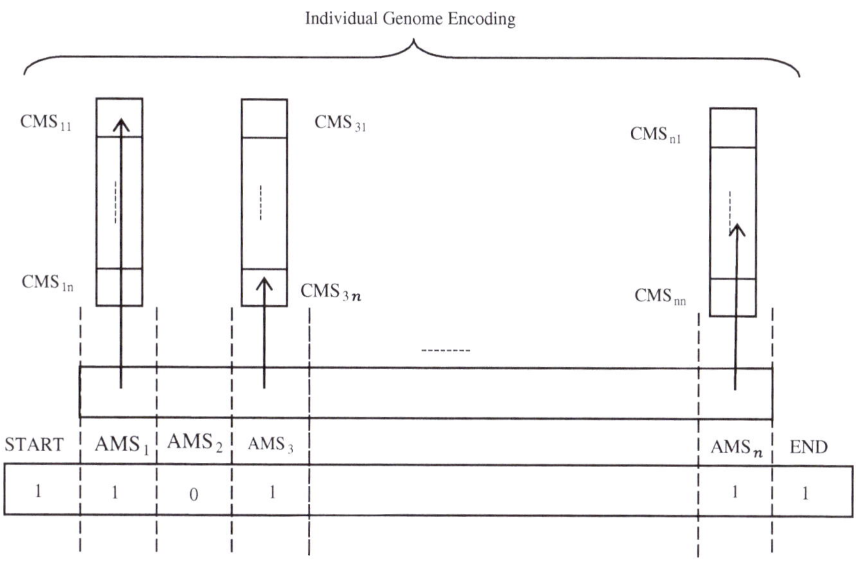

Fig. 4 Encoding a task

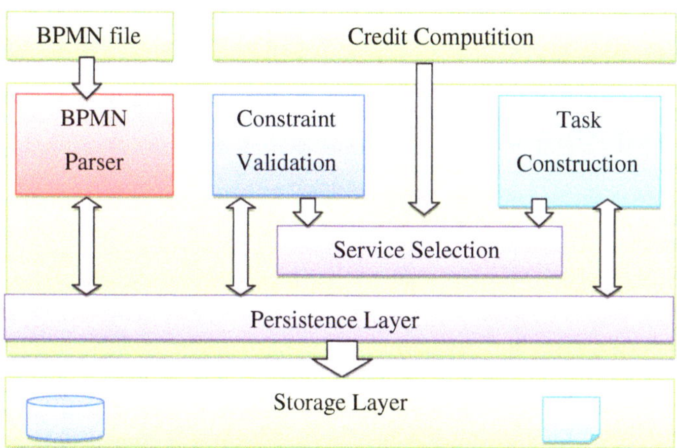

Fig. 5 Workflow of the service selection tool for CMfg-SME

5 Case Study

In order to validate the thought and the application of service selection in CMfg-SME, a service selection tool for CMfg-SME is built as Fig. 5 illustrated. This tool is a prototype, which take BPMN files and the transaction record as input, and output the service selection result.

The workflow of this tool is introduced as follows: the tool first take BPMN file as input, parse it and store the result; then the parsed result is used to new a task, including the AMSs in the it, and find all the CMSs for each AMSs; after that, the credit of the service provider and service buyer are computed, so is the experience of them; finally, based on the result of constraint validation, the service selection is used to find service composition for the users.

The persistence layer is used to unify the actions which persists information from high layers, and interact with the storage layer. The storage layer is used to execute the concrete information saving actions and store the information to database or files.

6 Conclusion and Future Work

We analyze the characteristics of CMfg-SME, then build a service selection model. Considering the relation between services, we define the constraints on the service selection. Finally, a GA based service selection algorithm is introduced.

However, there are still some directions to explore: (1) this approach have not run in big real systems: (2) more features will be used to evaluate the cloud manufacturing services.

Acknowledgments This work has been partly funded by the National Natural Science Foundation of China under grant 61170074, the Chinese National Hi-Tech. R&D Program under grant 2012AA011204, and the National Key Technology R&D Program under grant 2012BAF11B04–4. The authors wish to acknowledge the Commission for their support.

References

1. Tao, F., Cheng, Y., Zhang, L., et al. (2011). Cloud manufacturing. *Advanced Materials Research, 201*, 672–676.
2. Huang, B., et al. (2013). Cloud manufacturing service platform for small-and medium-sized enterprises. *The International Journal of Advanced Manufacturing Technology*, *65*(9–12), 1261–1272.
3. Wada, H., et al. (2012). E^3: A multiobjective optimization framework for SLA-aware service composition. *IEEE Transactions on Services Computing*, 5(3):358–372.
4. Kuzu, M., & Nihan Kesim, C. (2012). Dynamic planning approach to automated web service composition. *Applied Intelligence*, *36*(1), 1–28.
5. Feng, Z., et al. (2013). QoS-aware and multi-granularity service composition. *Information Systems Frontiers*, *15*(4), 553–567.
6. Haddad, S., Mokdad, L., & Youcef, S. (2010). Selection of the best composite web service based on quality of service. *ISSS/BPSC*, *10*, 255–266.
7. Huang, S., et al. (2011). Optimal service selection and composition for service-oriented manufacturing network. *International Journal of Computer Integrated Manufacturing*, *24*(5), 416–430.

8. Tao, F., Zhao, D., & Zhang, L. (2010). Resource service optimal-selection based on intuitionistic fuzzy set and non-functionality QoS in manufacturing grid system. *Knowledge and Information Systems, 25*(1), 185–208.
9. Shen, W., et al. (2007). An agent-based service-oriented integration architecture for collaborative intelligent manufacturing. *Robotics and Computer-Integrated Manufacturing, 23*(3), 315–325.
10. Shanshan, L. (2012). *Research and implementation of cloud manufacture service selection for medium-sized and small enterprise*. Master thesis. University of Chinese Academy of Sciences.
11. Goldberg, D. E., Korb, B., & Deb, K. (1989). Messy genetic algorithms: Motivation, analysis, and first results. *Complex Systems, 5*(3), 493–530.
12. Zheyuan, J., Jianghong, H., & Zhao, W. (2009). An optimization model for dynamic QoS-aware web services selection and composition. *Chinese Journal of Computers, 32*(5), 1014–1025.

Multi-dimension Density-Based Clustering Supporting Cloud Manufacturing Service Decomposition Model

Jorick Lartigau, Xiaofei Xu, Lanshun Nie and Dechen Zhan

Abstract Recent years, the research on Cloud Manufacturing (CMfg) has developed extensively, especially concerning its concept and architecture. Now we propose to consider the core of CMfg within its operating model. CMfg is a service platform for the whole manufacturing lifecycle with its countless resource diversity, where organization and categorization appear to be the main drivers to build a sustainable foundation for resource service transaction. Indeed, manufacturing resources cover a huge panel of capabilities and capacities, which necessarily needs to be regrouped and categorized to enable an efficient processing among the various applications. For a given manufacturing operation e.g. welding, drilling within its functional parameters, the number of potential resources can reach unrealistic number if to consider them singular. In this paper, we propose a modified version of DBSCAN (Density-based algorithm handling noise) to support Cloud service decomposition model. Beforehand, we discuss the context of CMfg and existing Clustering methods. Then, we present our contribution for manufacturing resources clustering in a CMfg.

Keywords Cloud manufacturing · Clustering algorithms · DBSCAN · Cloud service decomposition model

---

J. Lartigau · X. Xu (✉) · L. Nie · D. Zhan
School of Computer Science and Technology, Harbin Institute of Technology, 92 Xi Dazhi Street, Xinjishu building, room 504, Nangang District, Harbin 150001 Heilongjiang, China
e-mail: xiaofei@hit.edu.cn

J. Lartigau
e-mail: jorick@hotmail.fr

L. Nie
e-mail: nls@hit.edu.cn

D. Zhan
e-mail: dechen@hit.edu.cn

K. Mertins et al. (eds.), *Enterprise Interoperability VI*,
Proceedings of the I-ESA Conferences 7, DOI: 10.1007/978-3-319-04948-9_29,

1 Introduction

In our days, business opportunities are mainly linked with the rapid development of IT capabilities. The connectivity and interplay among all kind of devices is rising exponentially and open new technological doors for various applications in numerous domains. For instance, manufacturing industries sense changes in the way they run their business from a lot of different perspectives e.g. control and management, productivity, interoperability. The Internet, incorporating computers and multimedia, has provided tremendous potential for remote integration and collaboration in business and manufacturing applications [1]. However, the technological gap between large and small middle size manufacturers is growing as the IT innovations speed up. To compete with others, manufacturers often invest in better equipment overqualified for their production rate. Therefore, they rightly rely on collaboration to expand their business along with their resources occupancy. During recent years, new manufacturing models emerged to enhance and facilitate the collaboration among manufacturers and also the share of resources e.g. Manufacturing Grid, Network manufacturing; while others aimed to reduce the cycle and cost of product development e.g. Web-based manufacturing. By now it has become clear that increased volatility of market conditions disfavors rigid and hierarchical architectures [2]. Flexibility in the infrastructure is an essential driver for a responsive and adaptive manufacturing environment.

From an IT perspective, Cloud computing has been identified and largely applied to redefine the way business can be appreciated. It can be perceive as a way to increase capacity or add capabilities without investing in new infrastructures, training the personnel, or licensing new software. It gives small businesses access to technologies that previously were out of their reach to compete with larger ones [3]. From a manufacturing perspective, Cloud Manufacturing (CMfg) has been recently proposed [4] as an extension of Cloud Computing, aiming to enhance resource sharing and occupancy among manufacturers along with their capabilities. CMfg is a service-oriented manufacturing model designed to combine manufacturing capabilities for dedicated manufacturing services. Considering the operating model of CMfg (Fig. 1), the setup of the virtual enterprise is a circular relationship, where the service to be manufactured meets a specific design.

From customer's service requirements, the agent-based platform i.e. denoted Cloud Platform, is responsible for the service decomposition into several cloud services that can be executed by a singular manufacturing resource. Through, resource evaluation, selection, composition and scheduling, the cloud platform build virtual enterprise with its unique supply chain.

This paper focuses on manufacturing resource clusterization to speed up the processes along the operating channel. Indeed manufacturing resource organization in service clusters is a vital feature to insure an efficient CMfg structure. Accordingly, we propose to review the existing research on clustering methods and propose a CMfg clustering framework based on a density-based approach.

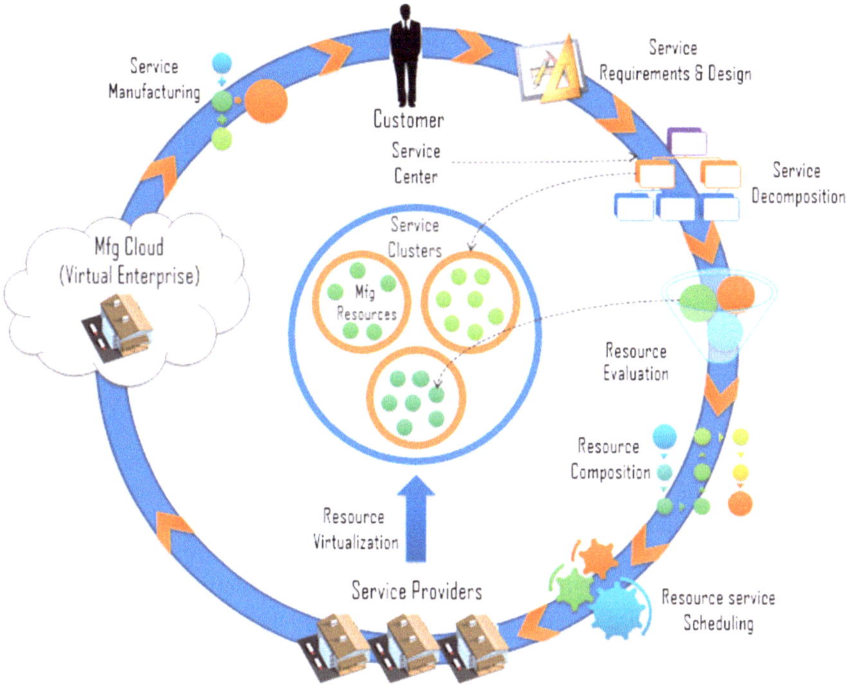

Fig. 1 CMfg operating model

2 CMfg Clustering Framework

2.1 Clustering Algorithms

When it comes to clustering, the notion of "cluster" is not precisely defined [5]. Consequently, many methods have been developed to meet general and more specific problematic. Meanwhile, they can be divided in two main groups i.e. hierarchical and partitioning methods. Hierarchical methods are based on the decomposition of the resources space S into single cluster for one single resource i.e. *agglomerative* and *divisive* hierarchical clustering. Jain et al. [6] identified more categorization according to their similarity approach i.e. Single Link, Complete Link and Average Link Clustering. These methods presents several drawbacks by building creating discordances in the clustering. Partitioning methods require an initial partitioning of the space to define clusters and finally allocate resources. For optimal resource allocation, the space definition of the partition is an exhaustive exercise, which theoretically requires browsing all possible solutions. The handicap comes with the initial setups which can strongly impact the result of the process and also significantly increase the computation time.

Other methods exist like Fuzzy clustering or density-based clustering. Fuzzy clustering is one of the most applied clustering methods. This method derived from Fuzzy set is built on membership degree. But the veracity of membership or non-membership degrees cannot fully be justified. Density-based algorithms create clusters according to the resource density in the space. The main idea of density-based approach is to find regions of high density and low density, with high-density regions being separated from low-density regions [7]. The advantage of density-based methods enables clusters in arbitrary space shapes and a fair balance between cluster space size and population size.

2.2 CMfg Clustering Framework

In this paper, we propose a modified DBSCAN (Density-based clustering handling noise) algorithm to match our space definition process and to create clusters according to criterions vector of priorities. Some recent researches proposed modified DBSCAN using different structure e.g. [8–10], mainly focusing on computation time improvements especially for large population of, and also differ in the handling of noise points. However the matter of our research is motivated by more practical perspectives where the number of resources in the space never reaches millions and where the common DBSCAN computation time is proven reasonable. Also it highlights the distinction of common capabilities and specific ones to offer advantages for our clustering framework. In this case, noise points refer to resources with features and capabilities different from common resources.

The CMfg clustering framework proposed is illustrated by the Fig. 2. To speed up the research process we want to distinguish common resources from singular ones for more specific demands e.g. drilling of special metal like iridium.

Beforehand, it is essential to build the resource space. Unlike Fuzzy clustering, based on membership degree, we insure the veracity of the space definition by linking this process directly to the resource features and metrics.

3 Resource Space Definition

3.1 Cloud Service Decomposition Model

In clustering, the space definition is often related to similarity measure. But we aim at a different angle, focusing directly on the manufacturing resource features. Our goal is to strengthen the link between the space definition for the clustering process and the CMfg service decomposition model from the resource layer (Fig. 3).

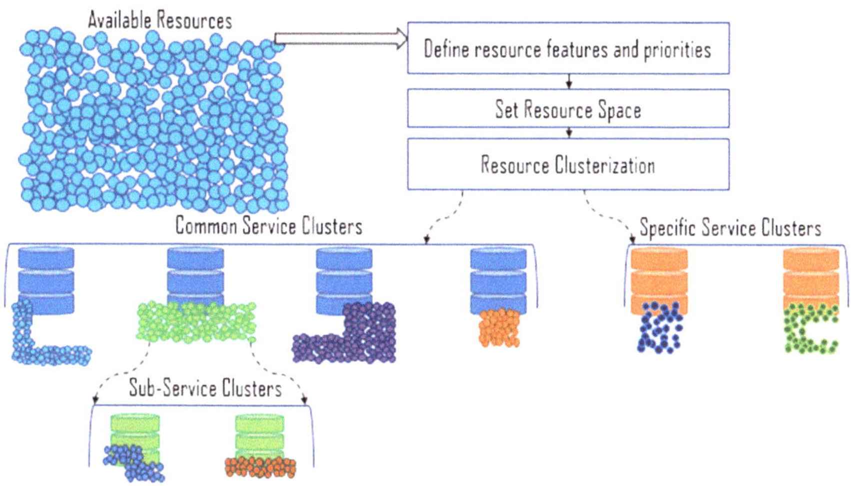

Fig. 2 CMfg clustering framework

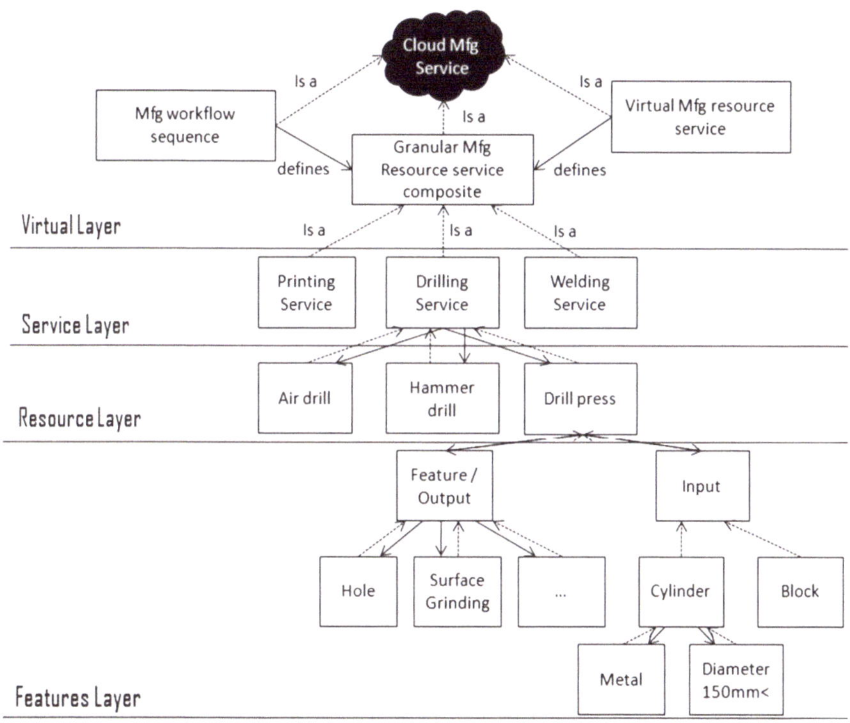

Fig. 3 CMfg service decomposition model

Tao et al. [11] proposed a classification of resources in the Manufacturing Grid eco-system, also applicable in CMfg. But in our case we only focus on manufacturing resources and their related features.

3.2 Resource Space Definition Presets

As mentioned, our approach in clusterization is to link the clustering method directly to the decomposition model of the resources, where the features or capabilities stand for criterions and vectors of the final clusters. Since we are focusing on space definition, we must consider N-dimensions, where each axis is related to a given criterion. Once the space settled, the density-based clustering algorithm can be launched to generate clusters. The process can be further repeated to define sub-clusters using other criterions.

Now let's consider CRT the set of criterions vector of the space definition with $i = 1, 2, 3, \ldots, N$ and $N \in \mathbb{N}$.

$$CRT = \{crt_1, crt_2, crt_3, \ldots, crt_i, \ldots, crt_N\} \quad (1)$$

Each criterion is attached to its given weight to translate its importance and priority in the characterization of the resource. Therefore let's consider the set ω of N weights related to the criterions such as with $\omega_i \in \mathbb{R}_0^+$ and $\sum_{i=1}^{N} \omega_i = 1$.

$$\omega = \{\omega_1, \omega_2, \omega_3, \ldots, \omega_i, \ldots, \omega_N\} \quad (2)$$

In our approach, we only consider functional parameters as criterions although our method can also encompass non-functional parameters e.g. reliability, price. Also, for the sake of readability we represent the space definition as a tri-dimensional Euclidian space where the ordered triple is associated to $(crt_i, crt_{i+1}, crt_{i+2})$. A given criterion crt_i can whether be semantic or numeric. Indeed, the associated metrics can be considered from a semantic or numeric perspective.

3.2.1 Case 1: Semantic Criterion

If crt_i is a semantic criterion, we consider the whole set of resources to be positioned in order to gather the all set of semantic values s_i, with $j = 1, 2, 3, \ldots, M$ and $M \in \mathbb{N}$.

$$s_i = \left\{s_1^i, s_2^i, \ldots, s_j^i, \ldots, s_M^i\right\} \quad (3)$$

Each $s_{i,j}$ will be graduated on the axis p_i in order to position the resources. We also want to note that a given resource can take several $s_{i,j}$. Consequently this resource will be duplicated in the space definition and can be included in several

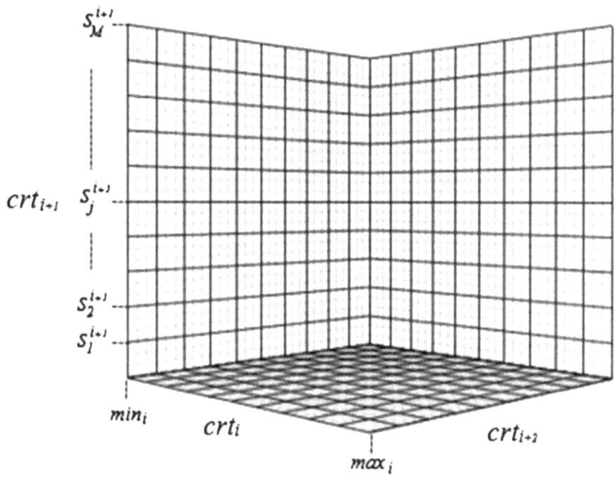

Fig. 4 Resource space definition preset

clusters. However it will be reduced single in one cluster. Concerning this case we don't refer to any specific ontology, allowing users to pick their own.

3.2.2 Case 2: Numeric Criterion

If crt_i is numeric, we consider the whole set of resources to define the range of possible values.

$$crt_i = [min_i, \ max_i] \tag{4}$$

To illustrate the two cases we propose the following space representation i.e. Fig. 4. In this case, crt_i is numeric and crt_{i+1} a semantic criterion.

4 Modified DBSCAN Supporting CMfg Clustering Framework

4.1 Introduction to DBSCAN

The algorithm DBSCAN (Density-Based Spatial Clustering of Applications with Noise), based on the formal notion of density-reachability for k-dimensional points, is designed to discover clusters of arbitrary shape [12]. DBSCAN centers on a random core point and investigate its neighborhood within a given radius that must contains at least a minimum number of points, i.e. the density in the neighborhood has to exceed some threshold. The neighborhood analysis is determined through a distance evaluation, which in our case is based on a Euclidean distance balanced with the weights associated to the criterions.

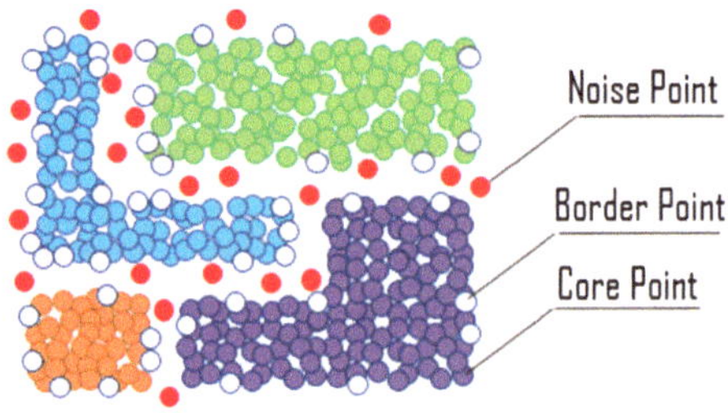

Fig. 5 Resource space definition preset

DBSCAN is based on 2 control parameters i.e. *Eps* and *MinPts*, vectors of the density.

- *Eps*: Maximum radius of the neighborhood.
- *MinPts*: Minimum number of points in an *Eps*-neighborhood of the point *p*.

A point *p* is referenced as a core, border or noise point according to the following definitions (Fig. 5).

- *Core Point*: Point with at least *MinPts* objects within an *Eps* radius.
- *Border Point*: Point on the border of a cluster and warrant of the cluster definition i.e. operating ranges.
- *Noise Point*: Point that can create discordances in the clusterization.

Therefore the *Eps*-neighborhood of a point *p*, denoted by $Neps(p)$, is defined as follow with *q* density reachable or connected and *S* the resource space.

$$Neps(p) = \{q \in S | dist(p, q) \leq Eps\} \quad (5)$$

- *Directly Density-Reachable*: A point *p* is directly density-reachable from a point *q* if *p* belongs to $Neps(q)$.
- *Density-Reachable*: A point *p* is density-reachable from a point *q* if there is a chain of points $p_1, \ldots, p_i, \ldots, p_n$ with $p_1 = q$ and $p_n = p$ such that p_{i+1} is directly density-reachable from p_i.
- *Density-Connected*: A point *p* is density-connected to a point *q* if there is a point *o* such that both, *p* and *q* are density-reachable from *o*.

In our case a noise point will be a resource offering specific features with specific ranges out of the common resource population. DBSCAN can identify them and process clustering with common resources. Furthermore we can reconsider the noise points as core or border points in the same space using a different approach. Then, we can build clusters with more or less specific ranges of operation and unique features for particular demands.

4.2 *Modified DBSCAN Algorithm for CMfg Clustering Framework*

To visualize the algorithm, we developed a virtualization tool powered by the new applications of the canvas feature of HTML5 i.e. CanvasXpress and the modified DBSCAN implemented in Java.

To illustrate the space definition, we consider three fundamental features of hammer drills as criterions e.g. Maximum RPS (rotation per second), Power watts, Drilling material; and obtain the following three dimensional i.e. Fig. 6a.

Algorithm 1. CMfg_DBSCAN()

```
CMfg_DBSCAN(S, eps, MinPts)
   C = 0
   for each unvisited resource R in space S
         mark R as visited;
      N = getNeighbors (R, eps);
      ifsizeof(N) <MinPts
         mark R as NOISE;
      else
         C = next cluster;
         expandCluster(R, N, C, eps, MinPts);
   end
      SC = integer of Count(NOISE)/MinPts;
      for each SC
            select R as NOISE unvisited from Min position vector;
            mark R as visited;
            add R to SC;
            N=createSCluster(N,R);
            add N to SC;
end
forR as NOISE unvisited
mark R as visited;
clusterize R to the nearest SC;
end
endCMfg_DBSCAN
```

CMfg_DBSCAN is the main process creating common and specific clusters. From the first loop it detects whether R is a core or noise point. Then a neighborhood population is gathered through distance evaluation to build and expand common clusters C. The second loop creates specific clusters SC from the resources marked as noise point in order, from the nearest of the origin. The *expandCluster()* function gathers the resources in the neighborhood of the resource R' and evaluate the distance with the cluster centered on R. It merges clusters when R and R' are density reachable (Fig. 6b). Once common clusters created and expanded, DBSCAN call the routine *createSCluster()* for the generation of specific clusters, embedding resources previously marked as noise points. Each specific cluster embeds a number of resources equal to *MinPts*. The clustering phase begins with the nearest resource from the origin and regroups the resources with the minimum distance. The objective is to build clusters of minimum size in space embedding the *MinPts* of resources.

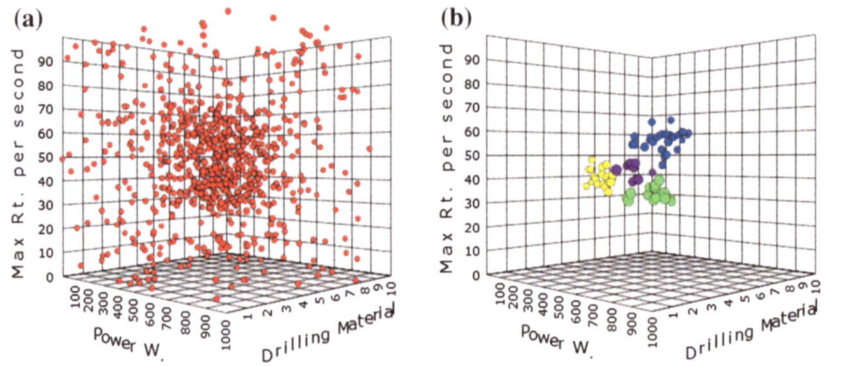

Fig. 6 **a** Resource space definition; **b** cluster generation

Algorithm 2. getNeighbors()

```
getNeighbors (R, eps)
   for each resource R' where R ≠R'
      ifgetDistance(R,R') <eps and ∄R'∈neighbors of R
         Add R'to neighbors of R;
   end
   return neighbors of R;
endgetNeighbors
```

The *getNeighbors()* function gathers the set of resources in the neighborhood of the resource *R* according to the threshold *Eps*. It also avoids duplicated resources to be clusterized more than one time in the same cluster. The distance between the two resources *R* and R' is calculated to determine whether R' is in considered in the neighborhood of *R*. The distance is evaluated through the coordinate of the resource on each axis i.e. the criterion value, and the weight attached to it.

$$dist\left(R, R'\right) = \sqrt{((Rcrt_1 - R'crt_1)\cdot\omega_1)^2 + \cdots + \cdots + ((Rcrt_N - R'crt_N)\cdot\omega_N)^2} \quad (6)$$

Algorithm 3. getDistance()

```
getDistance (R, R')
   for each criterion/axis P
      dP=(R.getP-R'.getP)*P.weight;
      distance=distance+dP*dP;
   end
   distance=sqrt(distance);// Eq. 1.6
   return distance;
endgetDistance
```

Table 1 Clusterization experiments

Case	MinPts	Eps	N. resources	N. common clusters (N. resources)	N. specific clusters (N. resources)
1	10	40	400	9 (327)	7 (73)
2	10	60	400	6 (342)	5 (58)
3	20	40	400	7 (302)	9 (98)
4	20	30	800	16 (721)	7 (79)
5	10	20	1600	25 (1438)	16 (162)
6	10	20	3200	63 (2916)	28 (284)

5 Experiments

We propose a summary table of the experiments including the setup parameters as the result in term of clusters and resource clusterized i.e. Table 1.

To conduct our experiments we create a resource generation function including a notion of probability for the features values to be chosen. As previously we consider a 3D Euclidean space with the three following criterions e.g. Max Rt. Per second, Power W., Drilling material and the three associated weights equal such as $\omega_1 = \omega_2 = \omega_3 = 0.33$. We notice that the population of common and specific clusters vary according to the *Eps* and *MinPts* factors. For the same space definition a larger *Eps* encompass more common resources. It plays a role of precision in the clusterization process, balancing the number of common or specific clusters. A larger *Eps* also means common clusters with larger border, and so larger definition ranges. During these experiments we also notice that a modification of the weights does not necessarily influence a change in the cluster population size, but in the border and definition ranges. For instance a higher weight on the drilling material feature lead to clusters with a short range on this criterion.

6 Conclusion and Future Works

This research paper was mainly motivated by practical issues on manufacturing resource clusterization. Indeed, it has been designed to be integrated on a currently developed Cloud service platform. Our clustering framework has been already proven efficient, matching the desired level of resource search optimization. Meanwhile, several opening can be added to this research. For instance through several searches computation, results concerning the probability of a certain cluster to be selected can lead to re-structure this cluster as to create sub-clusters. The link between the reality of the cluster selection through features search/match process and clusters pre-determined can be an interesting point to ponder. Especially for further optimization of the system-agent responsible for the creation of dedicated manufacturing cloud.

Acknowledgments This work has been partly funded by the MOST of China through the Project Key Technology of Service Platform for CMfg. The authors wish to acknowledge MOST for their support. We also wish to acknowledge our gratitude and appreciation to all the Project partners for their contribution during the development of various ideas and concepts presented in this paper.

References

1. Hongbo, L. (2009). Web-based rapid prototyping and manufacturing systems: A review. *Computers in Industry*, *60*(9), 643–656.
2. Botond Kádár, László Monostori (2001). Approaches to increase the performance of agent-based production systems. In *Engineering of intelligent systems* (Vol. 2070, pp. 612–621). New York: Lecture Notes in Computer Science.
3. Knorr, E., & Gruman, G. What cloud computing really means. *InfoWorld*. www.infoworld.com/d/cloud-computing/what-cloud-computing-really-means-031
4. Bohu, L., Lin, Z., Lei, R., Xudong, C., Fei, T., Yongliang, L., Yongzhi, W., Chao, Y., Gang, H., & Xinpei Z. (2011). Further discussion on Cloud Manufacturing. *Computer Integrated Manufacturing Systems*, *17*(3), 449–457.
5. Estivill-Castro, V., & Yang, J. (2000). A fast and robust general purpose clustering algorithm. In *Pacific Rim International Conference on Artificial Intelligence* (pp. 208–218).
6. Jain, A. K., Murty, M. N., & Flynn, P. J. (1999). *Data clustering: A review.*
7. Rehman, M., & Atif Mehdi, S. *Comparison of density-based clustering algorithms.*
8. Tekbir, M., & Albayrak, S. (2010). *Recursive-Partitioned DBSCAN, Signal Processing and Communications Applications Conference (SIU)*, (Vol. 18, pp. 113–116).
9. Smiti, A., & Elouedi, Z. (2012). DBSCAN-GM: An improved clustering method based on Gaussian means and DBSCAN techniques. *Intelligent Engineering Systems (INES), 16*, 573–578.
10. Dai, B.-R., & Lin, I-C. (2012). Efficient map/reduce-based DBSCAN algorithm with optimized data partition. *Cloud Computing (CLOUD), 5,* 59–66.
11. Tao, F., Hu, Y., Ding, Y., Sheng, B., & Zhou, Z. (2006). Resources publication and discovery in manufacturing grid. *Journal of Zhejiang University SCIENCE A*, *7*(10), 1676–1682.
12. Ester, M., Kriegel, H.-P., Sander, J., & Xu, X. A density-based Algorithm for discovering clusters in large spatial databases with noise. In *Proceedings of 2nd International Conference on Knowledge Discovery and Data Mining (KDD-96).*

Service Systems Modeling and Simulation: The Sergent Distributed Approach

Thècle Alix, Gregory Zacharewicz and Bruno Vallespir

Abstract The economy is dominated by the service sector. Citizens see services as a way to have access to basic or complex commodities, to address environmental problems while manufacturers consider them as a way to differentiate themselves from the competition, to be closer to their customers and to improve the shopper experience. Services are of a huge importance in the national and international economy and are discussed in many domains: human and social science, manufacturing science, business domain, IT domain, etc. Several concepts related to service have merged as well as new scientific disciplines. General issues linked to service design, service implementation, service operation management, service quality, service modeling and simulation, product-service system design are still under consideration and the multiplicity of the domains concerned failed to come up with unanimous answer. This paper proposes a contribution to service modeling and simulation that can potentially be used in any area. The proposed model is based on the most relevant concepts coming from a specialized literature review on services. A distributed simulation model of service is then proposed.

Keywords Service system · Conceptual modelling · Distributed simulation · G-DEVS

T. Alix (✉) · G. Zacharewicz · B. Vallespir
University of Bordeaux, IMS, UMR 5218, 351 cours de la Libération 33400 Talence, France
e-mail: thecle.alix@ims-bordeaux.fr

G. Zacharewicz
e-mail: gregory.zacharewicz@ims-bordeaux.fr

B. Vallespir
e-mail: bruno.vallespir@ims-bordeaux.fr

K. Mertins et al. (eds.), *Enterprise Interoperability VI*,
Proceedings of the I-ESA Conferences 7, DOI: 10.1007/978-3-319-04948-9_30,

1 Introduction

The service sector is particularly important in the current economy and is now between 70 and 80 % of the gross national product of the developed and developing countries. The tertiary sector drove the recent job creation, contributing to more than 4 million of "full-time equivalents" between 1990 and 2009 in France while the other business sectors lost 1.4 millions over the same period. In 2009, it represented about three quarters of the value added and employed 89 % of the working population.

Today, the proportion of unsatisfied services demand lead to the creation of new companies belonging to the service sector and de facto to the increase in the number of employed persons in this sector. As a result is an increase of the competition in both market and non-market-services which rises many questions such as the management of the service production process, the service productivity measurement, the quality measurement of the services supplied and of the activities contributing to provide them, as well as the measurement of the real service value-added for customers. Accordingly, it becomes imperative for those companies producing services to rationalize and to industrialize their practices and to position themselves clearly in the competition regarding their value-added.

To prepare this shift, obviously approaches from the secondary sector of production could be reuse. Unfortunately, optimization models stemming from industry do not allow to analyse the service value chains because of its specific characteristics: a fuzzy border between production and distribution, a context dependent value, a customer participation to the production even to the design of the service, only few physical movement of material and finally the impossibility to reduce the hand of work.

Before focusing on the rationalisation of service production activities, it is of vital importance to define the production ins and outs and to clarify upstream what a service is. On the later point, it must be noted that each domain interested in services proposes its own definition and characterization of the concept. Accordingly, the paper proposes a conceptual model of services as broad as possible, based on a multidisciplinary literature review on the service concept and on the service production as well as some approaches that directly or indirectly address service production. At the end, the model is simulated to verify dynamically the correctnesss of its behavior regarding a temporal sequence chaining.

2 A Multidisciplinary State of the Art of the Service Concepts

Some disciplines such as the Management Sciences are being interested for a long time in services. Its contribution is multiple. Regarding definition, a service is defined at the same time by opposition to goods [1, 2], as an activity [3, 4] and as a result [5, 6]. Each point of view leads to introduce specific notions: some works aims at characterizing the system producing the service often called service delivery system [7]; while others propose specific service characteristics called IHIP characteristics standing for immateriality, heterogeneity, inseparability, and perishability [8–11].

Computer sciences address the concept of service through the service oriented architectures (SOA) or the IT services. A functionality is decomposed into a set of functions or of services supplied by components. A business service is a company functionality that seems to be atomic from the service consumer point of view. In this frame, a service is a connection to an application offering an access to some of its functionalities [12]. It appoints an action executed by a "service provider" component to the attention of a "service consumer" component, possibly based on another system. A Service Broker enables to choose the service that fits the best the requirements among the available ones.

The functional analysis or value engineering method used in product design does not evoke directly the concept of service but focuses on the concept of function that is similar to a certain extent to the concept of service. The functional analysis was firstly used to identify the needs within the frame of new products design [13]. The most known approach that joins the functional analysis into this frame is the value engineering method. It is characterized by a functional approach proposing to formulate the problem in terms of objectives rather than in terms of solutions. For that concern, the value engineering method rests on several fundamental concepts: the value [14], the need [15] and the function [16]. The functions of a product define what the product (or product subset) is doing or what it is going to do. That perspective is very close to the service concept.

The service science, management and engineering (SSME) addresses several concepts relative to services. It aims at a better understanding of design, evolution, and emergent properties of service systems and also how the innovation leads to productivity gains in this sector [17]. The SSME consider the services as processes, performances or else results that a person (company, organization) realizes for the profit of another company/organization. Several approaches such as "the triptych of technical criteria" [18] or the "triangle of services" are proposed [19].

3 Service System Modelling

This section presents a conceptual model of service based on the above-presented contributions. The objective is to embrace the main part of these contributions to be as generic as possible. The result is a conceptual model based on the fundamental concepts related to services [20–22].

Fig. 1 Basic principle and representation of a sergent

3.1 Basic Principles: The Sergent

The basic principle of the proposed approach is that a service is an interaction within a service provider/consumer couple oriented from the provider towards the consumer. The intensity of the service delivered grows in the same way than a parameter that characterises the provider/consumer coupling. A provider is identified with regard to its function and is noted P. A consumer is characterized by its need and is noted C (Fig. 1, left part).

Of course, an object cannot be limited to be a provider or a consumer. That is why objects are generally considered to be able to behave simultaneously as both.

The initial principle presents an object as a one service provider or consumer. Obviously, a complex object is able to provide several services and/or to consume several ones. Accordingly, an object can be a provider/consumer of several services (Fig. 1, right part). Service delivery relations become then more complex i.e. each relation is defined for a given service. An object is then part of a service delivery series, linked to another upstream object as service consumer and a downstream object as service provider.

The "object" talked about up to now will be called an "*agent of service*" or a "*service agent*" or, in short, a "*sergent*".

3.2 Service Delivery Process Dynamics

For realizing the service delivery process, a consumer and a provider require to be coupled. The coupling steps are illustrated in Fig. 2.

The abovementioned situations suppose that the service delivery process can only be led during the coupling. Obviously, the interaction between the service supplier/consumer is the main part of the service delivery process. However, in more complex cases both actors can require to be prepared in an upstream phase (pre-process) and to get free in a downstream phase (post-process) (Fig. 3).

The corresponding phases are the following ones:

- **Initialization**: this phase does not require the coupling to be established but requires to know that the service must be provided. Information on the service need is necessary to activate the phase.

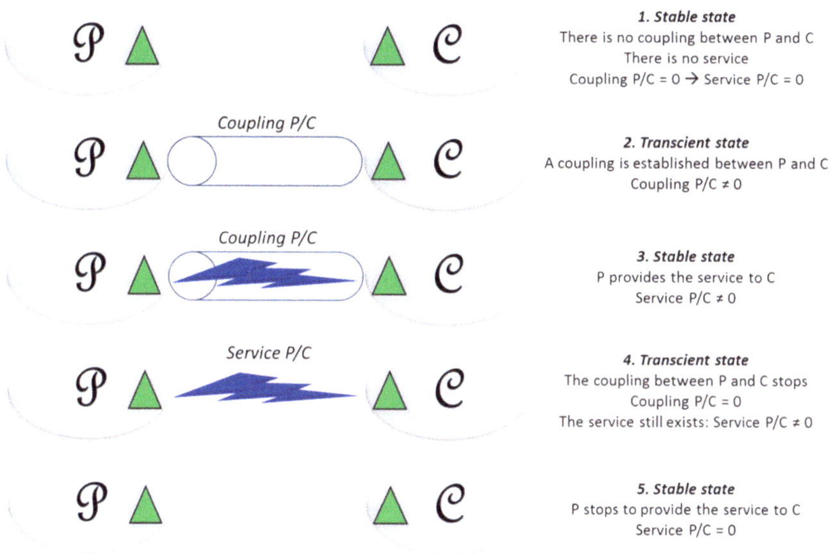

Fig. 2 Coupling dynamics

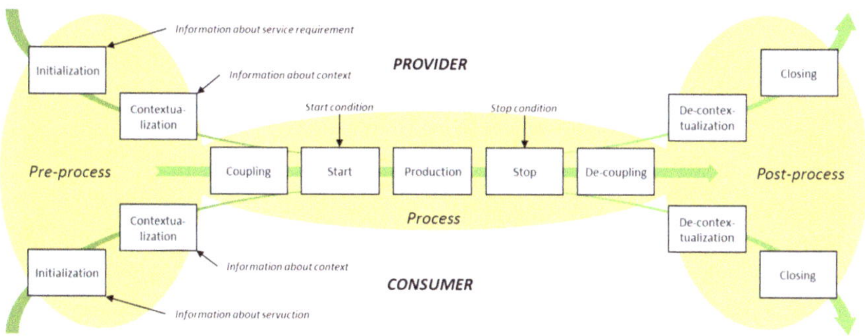

Fig. 3 The whole service delivery process

- **Customization and contextualisation**: in case the service is not standard, a phase of customization based on information coming from the consumer is to be envisaged. The contextualisation focuses on the adaptation to the context (consumer, surrounding conditions, etc.) of the service to be provided and of the service delivery process.
- **Closing and de-contextualisation**: phases exist when both actors require a process to close the activity. This process is similar to the initialization phase one but occurs after the service delivery.

3.3 Function Capacity and Need Load

A sergent may use its function in different ways: possibility to implement a function one or several times, in a successive or simultaneous way. These operational situations differ according to the sergent capacity to implement a function. Conversely, a sergent need can lead to different load levels.

Valuation of capacity and load: the capacity is a variable belonging to an interval [0, Max_c]. When the capacity is lower than Max_c, the sergent is able to implement the function but not with its maximum potential. The function capacity and need load are time variables. In particular, the provider capacity can be low or non-existent because the sergent is occupied or not operational for example.

Capacity variability: several phenomena can lead to a temporary or long-lasting capacity variation of a sergent to provide a service. When a sergent provides a service, its capacity decreases. Once the service provided and the service delivery process ended, two situations can occur: either the capacity provides to its initial value (long-lasting function) or the capacity maintains its new value (consumable function). The first case corresponds to non-perishable sergents while the second one corresponds to the consumption of not renewable resources.

In the case of long-lasting functions, several phenomena can appear: wear (capacity decreases each time the process is launched), learning (capacity increases each time the process is launched) and unlearning phenomenon (capacity decreases because the process is not launched as often as it should).

3.4 Provider/Consumer Matching

The provider/consumer matching is based on the following mechanism.

Function statement. A provider declares his functions in a "service directory" that indicates the precise nature of the proposed functions. The function statement is to be made according to a standard expressed in a "service repository". The nature of the function is the static part of the statement while the capacity that can be used at a moment corresponds to the dynamic part (Fig. 4, left part).

Matching. The dynamic is the following (Fig. 4, right part): (1) expression of needs by the consumer, (2) comparison to functions reported in the service repository, (request), (3) choice among providers able to fulfil the need, (4) provider selection.

4 Service System Simulation

Indeed, the service modelling requires also the modelling of the interactions between multiple services; this process can lead very quickly to a significant level of complexity from a static point of view and even more in a dynamic view. We

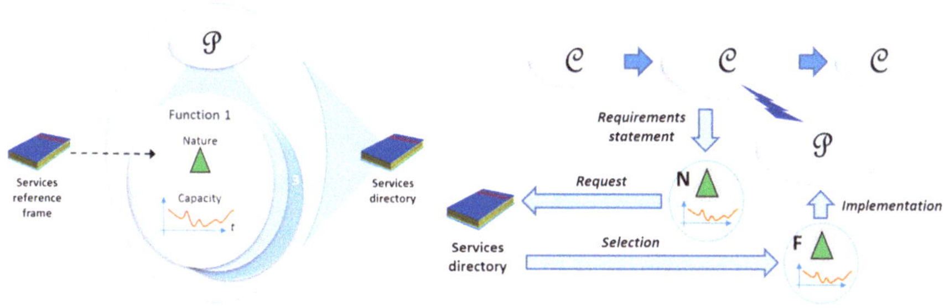

Fig. 4 Provider/consumer matching

proposed in previous works [23, 24] a first tentative to represent service in simulation. We therefore focus in this paper on the establishing operations for a single service coupling in the G-DEVS formalism [25]. This discrete formalism is selected because of its formal property and its time management capability. We propose to model each service component through a G-DEVS model based on attributes. The model attributes are described from a qualitative and quantitative point of view and all elements (actors and material) that interact within its environment are required. Once the description is complete, the described component can integrate a G-DEVS based library of service components: the service repository. The prospect of a break in service into four subsets then seemed obvious. These four models are:

- The sergent requesting the service (here sergent B is the consumer),
- The service provider (here sergent A),
- The coupling as the association between two sergents that will achieve the service delivery process, and
- The service delivery process 'service providing').

4.1 Coupled model

The coupled model (Fig. 5 shows the global architecture) presents the G-DEVS component required for the global simulation of the process. The components 1 and 2 are the A and B models. The component 3 is a coupling model used to connect the models paired for the service. The component 4 is used to orchestrate the process steps defined Sect. 3.2.

4.2 The Service Requester: Sergent B

The G-DEVS atomic model corresponding to the sergent B is an applicant for a service and its operations are detailed. This model describes the sergent behavior during the process of service delivery coupling.

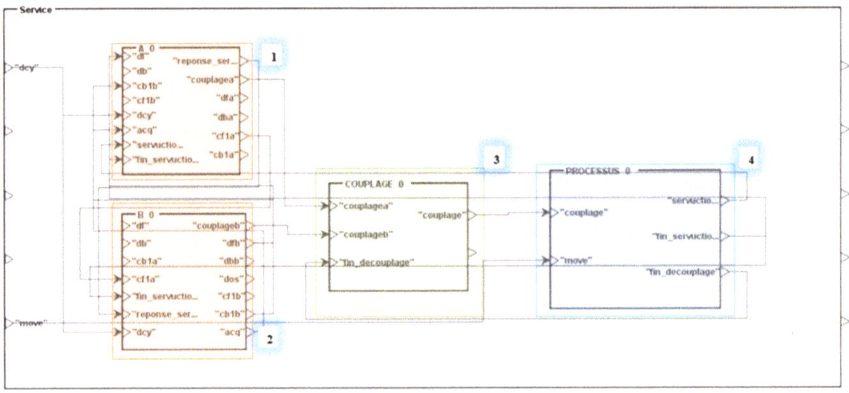

Fig. 5 Architecture of the coupled model of service delivery

The G-DEVS model follows the coupling steps described in §3. It communicates with the service delivery model. To assume the simulation execution, the states, event and temporal information have been added. These data are not related in this example to any information coming from a real system. The goal of the simulation is the verification of different correct communication sequences. Operations are as follow:

1. Every five time units (again: arbitrary chosen), the model B launches the comparison of its P/S potential intensity over a threshold value. This comparison is expressed as a condition on the internal transition.
2. The model function sends a request (DFb1) to the service provider (sergent A).
3. Positive response is received when sergent A is able to achieve this service.
4. Negative response is received when A is not available or not competent for this service achievement, a request will be send to another sergent.
5. Sending an acknowledgment to sergent A to tell them that the service can be achieved and to lock between them a delivery process.
6. Sending "ok_coupling" to external produce model.
7. Pending the external event "end_SD" meaning that the production of the service is completed.
8. Back to the waiting phase, the service was rendered, the intensity of sergent B to decrease the capacitance value function of sergent A demand. We note that some strategy can keep the value of the sergent B since some service potential is infinite.

4.3 The service provider: Sergent A

This section is introducing the G-DEVS atomic model of the service provider sergent A. An explanation of the model and its operations is detailed here.

1. The model A is expecting a request from sergent B.
2. A tests the ability to get the function (FC1a) and the load required by sergent B (NL1b). If the ability and capacity of A is greater than or equal to the need and load of B then the service is feasible.
3. Same as step 2; if the capacity is lower than the load of B then the service will not be feasible.
4. The service is not feasible, the supplier is informed the consumer via a message object "NOK".
5. The service can be done, A sends to B "OK".
6. The model waits for the acquittal of sergent B to produce the service.
7. The model awaits a response from an external "process" models and indicating that the service is in progress.
8. The model pends an external event "end_SD_OK" from the "process" model.
9. The process is over and the ability to provide the function of the particular service A can be reduced (consumption) or increased (experience).
10. The service is completed. The ability of A can stay decreased or can recover its initial value with a gain of experience that increases its ability (the hypothesis can be to gain 10 % capacity acquired for each service delivery).

4.4 Coupling and Decoupling

This atomic model enables the coupling orchestration of two sergents before the service delivery process and manages the notion of decoupling at the end of the service. This model is labelled 3 in Fig. 5. Its main operations are the followings. It starts by waiting an event from the sergent A informing that it is ready and looking for a delivery process coupling. Then it is waiting for an event from the sergent B answering that it is available and capable regarding the load and competence required by A. Then the coupling can be realized. Gathering the information, the model is informing both participants. Then it is waiting for an event informing of the end of the service delivery to return to the standby state. This last step is producing the end of the coupling.

4.5 The Service Delivery Process

A last G-DEVS atomic model is required to define the characteristics and simulate the service delivery process behaviour (service delivery model 4 in Fig. 5).

This model starts by waiting an event to be in-formed that a coupling is ok. At this time, it computes the characteristics of this service delivery including coupling, duration, quantity of load and experience acquired at the end. It informs by output sending the "SD" settings to the model participants and set them in progress. When the service is ended, it informs the participants by sending to output "end_SD".

The simulation concludes and illustrates the choreography between the supplier sergent and the receiver sergent. It shows clearly the communication steps during the service delivery. Nevertheless, the question is still on quantities and values used to define the potential of the sergent and the duration of the steps.

5 Conclusion

The paper presents a conceptual model of service and service delivery which takes into account notions and concepts stemming from disciplines initially remote but which become integrated rather naturally into a coherent set. The proposed concepts contribute to a first level of model that requires nevertheless to be completed. Others concepts need to be explored and some questions need to find response. For example, how to select a provider when the function capacity is similar from one to another? or else, in the case of service composition, how to decide the composition? and how to ensure its coherency?

Considering perspectives to that work, three concepts need to be explored.

At first, considering the fact that a way to approach the concept of service is to make the difference between buying a service and buying a product providing a service, it is necessary to take into account into the model all the elements focusing the transactional aspect around the sergent or around its functions.

Then, the sergent life cycle was not envisaged. Nevertheless, knowing in which phases of its life cycle a sergent is really able to implement its functions, or knowing what it becomes in the other phases or identifying the phases when it will consumes a service are so many questions which have to find an answer within the framework of a model of life cycle.

On the simulation aspect, the paper has provided a preliminary work to define the root concepts that can be modelled and simulated. It has focused on splitting the service delivery into different steps used to couple a service provider and a consumer. It opens the research in this domain where few works considering quantitative and dynamic aspects already exist. As perspective the authors are considering the definition of a simulation platform running an open space populated by autonomous sergents entities. Once a sergent needs a service, it will try to connect to a service supplier in its neighbourhood (that needs to be defined).

Finally, the measure of a service remains a huge problem. In equal function, can the service of a provider be considered as upper to that of another one? This problem of the measure is certainly the most difficult to be settled. Nevertheless, it would enable to clear up some other aspects presented here.

Finally, this model has started to be tested on academic cases [20]. The tests on real cases have to be processed I the near future.

References

1. Rathmell, J. M. (1966). What is meant by services? *Journal of Marketing, 30*, 32–36.
2. Judd, V. C. (1964). The case for redefining services. *Journal of Marketing, 28*, 58–59.
3. Berry, L. L. (1980). Services marketing is different. *Business Journal, 30*, 24–29.
4. Lovelock, C., Wirtz, J., & Lapert, D. (2004). *Marketing des services*. London: Pearson Education.
5. Hill, T. P. (1977). On goods and services. *Review of Income and Wealth, 23*(4), 315–338.
6. Zarifian, P. (2001). Valeur de service et compétence, in: *La relation de service: regards croisés*, 28, L'Harmattan.
7. Eiglier, P., & Langeard, E. (1987). *Servuction, le marketing des services*. Mc Graw Hill.
8. Rosengren, W. R. (1968). Organisational age structure and orientation toward clients. *Social Forces, 47*(1), 1–11.
9. Issac, J. (1996). *Les codes de déontologie: outils de gestion de la qualité dans les activités de services*. PhD thesis, univ. Paris Dauphine.
10. Dumoulin, C., & Flipo, J. P. (1991). *Entreprises de services: 7 facteurs clés de réussite*. Ed. d'organisation.
11. Shostack, G. L. (1977). Breaking free from product marketing. *Journal of Marketing, 41*, 73–80.
12. Raymond, G. (2007). *SOA: Architecture logique, principes, structures et bonnes pratiques*.
13. Miles, L. D. (1961). *Techniques of Value Analysis and Engineering*. McGraw Hill
14. Grandhaye, J. P., & Soltan, E. (2002). Value management project: Processing unforeseen events in design and production activities; a methodology and a case study. In *5th European Project Management Conference*.
15. Perrin, J. (2004). *Valeurs et développement durable : questionnement sur la valeur économique*. L'harmattan.
16. Grandhaye, J. P., & Poisson, P. (2001). Le Management par Valeur pour Concevoir et Rédiger un Projet d'Etablissement. International Congress AFAV.
17. Zhao, G., Perros, H. G., & Xin, Z. (2009). How service science management and engineering (SSME) can evolve to an academic discipline? *International Journal of Web Engineering and Technology, 5*(4), 459–467.
18. Spohrer, J., Maglio, P., Bailey, J., & Gruhl, D. (2007). Steps toward a science of service systems. *IEEE Computer, 40*(1), 71–77.
19. Tabas, L. (2007). *Designing for service systems*. ISD Symposium, UCB I School Report.
20. Touzi, W. (2001). *Conceptualisation et modélisation de la production de service: application aux domaines de la santé et de l'enseignement*. PhD thesis, Univ. Bordeaux 1.
21. Touzi, W., Alix, T., & Vallespir, B. (2012). *Contribution au développement d'un modèle conceptuel du service et de la production de service*. MOSIM, Bordeaux, France, 6-8 June 2012.
22. Touzi, W., Alix, T., & Vallespir, B. (2012). Contribution to the development of a conceptual model of service and service delivery. *IFIP APMS conference* (pp 24–26). Rhodes, Greece, September 2012.
23. Zacharewicz, G., Alix, T., & Vallespir, B. (2009). Services modeling and distributed simulation DEVS/HLA supported. *Winter Simulation conference* (pp 13–16). Austin, Texas, December 2009.
24. Alix, T., & Zacharewicz, G. (2012). Product-service systems scenarios simulation based on G-DEVS/HLA: Generalized discrete event specification/high level architecture. *Computers in Industry, 63*(4), 370–378.
25. Giambiasi, N., Escude, B., & Ghosh, S. (2000). GDEVS: A generalized discrete event specification for accurate modeling of dynamic systems. *Transactions of the Society for Computer Simulation International, 17*(3), 120–134.

Impacts Assessment of Collaboration on Carrier Performance

Samir Amrioui, Esther Leleu and Nicolas Malhéné

Abstract Freight transport is an essential element of our economy. It allows the goods to travel from their production place to customers. In France this activity implies many companies. Most of them are very little ones and are threatened by the economic crisis. This article studies how collaborative approach allow those companies finding solution to improve their performances. We simulate different scenarios which highlight benefits in particular in dynamic context when a last minute unforeseen change occurs. As interoperability is essential for collaboration implementation, companies must integrate its dimension in their development policy.

Keywords Interoperability · Collaboration · Transport · Performance · Last minute unforeseen change

1 Introduction

Road transport became an essential part of any logistics system. The entire hegemonic place of road transport operators leads to consider them as essential actors in providing a formal assurance of fulfilling the condition of the best performance in supply chains (SC). In France there is more than 36,000 road transport companies and logistics which are employing more than five hundred thousand

S. Amrioui (✉)
IMS/LAPS, University Bordeaux1, Talence, France
e-mail: samir.amrioui@eigsi.fr

S. Amrioui · E. Leleu · N. Malhéné
EIGSI, La Rochelle, France
e-mail: leleu@eigsi.fr

N. Malhéné
e-mail: nicolas.malhene@eigsi.fr

K. Mertins et al. (eds.), *Enterprise Interoperability VI*,
Proceedings of the I-ESA Conferences 7, DOI: 10.1007/978-3-319-04948-9_31,

wage-earners. 95 % of these companies have less than 50 employees and 65 % have less than 10 employees. This is one of the domain's weaknesses. The actual economic crisis is dramatic for smallest companies. The statistical study made by FNTR (French National Federation of Road Transport) states the bankruptcy of 900 companies the first semester of 2013. This situation points out the lack of resources for smallest companies. Collaborative approach is an interesting solution in order to reinforce the position of those enterprises.

This paper aims at evaluating how interoperability can facilitate the implementation of collaborative approach in order to develop their potential and reinforce their position.

First, we present a performance indicators framework dedicated to transport activities. Then, based on selected indicators, we expose the benefits of collaboration. We have found that collaboration would enable an efficient approach to the market's challenges, to keep enterprises' self-reliance, to customize the services provided, to make an offer of services, and enhance the rate of the use of resources, etc. It could also decrease the environmental damages linked to freight transport. In France transport activities contribute to 28 % of the greenhouse gas and 10 % of particle emissions. Finally we describe a collaborative information system architecture.

2 Transport and Logistics Performance

2.1 The Concept of Industrial Performance

Logistics performance is viewed under different axes. Zhou and Benton [1] and Roth et al. [2] consider the performance in terms of technology, visibility and information sharing. Ritchie and Brindley [3] specifies that performance analysis must be linked to the management of risks and resilience, especially in uncertain environments. For example, the SCOR model for the Supply Chain Council (SCC) is organized around four dimensions: reliability of business performance, flexibility/responsiveness, cost of the supply chain and rotation of capital employed [4].

The cost remains one of the most mentioned factors in the supply chain performance [5, 6]. The customer satisfaction rate is also an important factor [6]. Other studies focused at sharing, exchange and information technologies and on the practices of supply chain and inter-organizational communication [7]. Some authors retrace the relationship between the use of technology, information sharing, collaboration and business performance [8].

2.2 Performance Framework for Road Freight Transport

The scopes of frameworks for integrated supply-chain management such as SCOR [4], Global MMOG/LE [9] are primarily related on the industrial point of view. In this section we presented a guide to good practices or a performance framework mainly dedicated to companies from road freight transport domain.

The framework is represented throught a set of indicators classified according four levers for logistics performance. We defined it throught interviews with several managers of the road freight industries realized in the frame of PRODIGE [10, 11], a French project dedicated to product-oriented routing problems. We also contacted the French Federation of Transport Logistics and different transport companies. The objective was to identify how companies of the road freight sector evaluate their performance and how they control their activity.

2.3 Performance Framework for Road Freight Transport

The framework is based on four levers logistics performance (see Fig. 1):

- Lever of productivity which includes productivity indicators that aim to measure the return on the transport activity;
- Lever of charge which is linked to the difficulties of operating optimally the vehicle capacity. The competitiveness in transport is rationally based on the vehicle capacity which offers better cost controls;
- Lever of quality of service which aims to measure customer satisfaction according to the job performance;
- Lever of eco-logistic which relates to the environmental policy and sustainable development.

We choose specific indicators from the performance framework to compare the performance of a group of carriers working in a collaborative way against the performance carriers working in an independent way. We chose the most relevant indicators in the freight transport sector.

For **load lever**, two indicators have been selected:

- **The mass loading rate** represents how much the truck is close to its maximum mass load;
- **The volumetric loading rate** represents the extent to which the truck is close to its maximum volumetric load. In this study, it is based on the maximum number of pallets which can be handled by the vehicle.

For **productivity lever**, two indicators have been selected:

- **The empty kilometers rate**, which measures the kilometers covered by the trucks when they are moving without carrying any pallet in comparison of the total distance travelled;
- **The tkm (or ton-kilometer) per delivery tour rate**, a tkm is a unit of freight which represents the transport of one ton of goods over one kilometer. When a vehicle totals more tkm than another it can be because it handled more goods or carried the goods further than the other. So it's more productive.

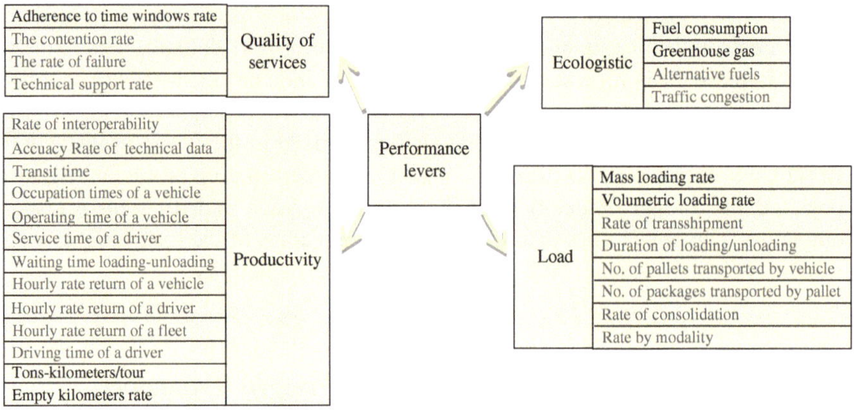

Fig. 1 Detail of indicators of referentiel of performance

For **ecologistic lever**, two indicators have been selected:

- **The greenhouse gas emissions**, which are expressed in a kilogram equivalent of CO_2 and commonly used to measure the impacts of one activity on climate change;
- **The fuel consumption**, closely linked to the previous indicator, which concerns goods traffic. It is largely used by carriers as a cost indicator.

For **quality of services lever**, one indicator has been selected. The **adherence to a time window rate** measures the capacity of carriers to deliver the goods in the window of time indicated by the consignee. However, in our model we considered the respect of this time window as a major constraint, i.e. no solution is accepted if the delivery does not occur in the defined window. As a result this rate is always at 100 % and not monitored in the following sections.

3 Analysis of Collaboration Case Studies

The impact of collaboration on carriers performance have been evaluated through modelling two theoretical but classical cases:

- The first case studies the interest of collaboration in a "static context". Carriers work in a collaborative way or in a none collaborative way in order to optimise their delivery tours to fulfill established orders. The collaboration is studied in a proactive way in order to increase productivity, decrease costs and enhance the visibility in the market of the carriers as a group of transport companies;
- The second case aims at evaluating the interest of collaboration in a crisis context characterized by an event: an implaned urgent order arrives when the tours have already begun.

3.1 Simulation Model

The simulation model represents a VRP (Vehicle Routing Program) with pickups and deliveries. Each customer is identified by their pickup and delivery nodes and can be visited by several vehicles during a tour (see Table 1).

The mathematical model is represented by the following equations:

Notation:

N	The set of all customers;
K	The set of all vehicles;
L	The set of orders (pickups and deliveries);
T_{ij}	The travel time between two customers i and j;
D_{ij}	The distance between two customers i and j;
i_{demand}	The number of pallets for a customer i;
l_{pickup}	The pickup node for the order l;
l_{delivery}	The delivery node for the order l;
k_{capacity}	The capacity of the vehicle k;
$[t_{\min,i}, t_{\max,i}]$	The time window for the customer i.

Decision Variables:

x_{ij}^k	Whether or not vehicle k takes the path between i and j;
y_i^k	Whether or not vehicle k visits the customer i;
z_l^k	Number of pallets treated by the vehicle k for the order l;
c_j^k	Number of pallets transported between the path i and j;
t_i^k	The arrival time at the customer i;
a_i^k	The order of visiting customers by vehicle k;
s_i^k	The service time of the vehicle k at the node i.

3.2 Collaboration and Planning of Delivery Tours

In order to study the performances of collaboration and quantify the benefits which can be expected the number of carriers and the characteristics of their fleet were fixed. Only delivery tours are modeled. All the trucks start in the same city where a warehouse is used to dispatch goods on a regional level.

Three carriers are working in the same region. In order to simplify the analysis, all carriers have similar fleet composed of diesel vehicles:

- Two 7, 5-ton trucks carried 12 pallets using 22 L/100 km of fuel;
- Two 12-ton trucks carried 15 pallets using 24 L/100 km of fuel;
- Two 19-ton trucks carried 20 pallets using 27 L/100 km of fuel.

Table 1 Description of the mathematical model used

	Equations	Function
Objective function	$Z = \min \sum_{i=0}^{N} \sum_{j=0, i \neq j}^{N} \sum_{k=0}^{K} x_{ij}^{k} * D_{ij}$	To minimize all costs (distances) of all selected elements of the tour
Constraints	$\sum_{i}^{N} x_{ij}^{k} = y_{j}^{k} \quad \forall k \in K, \forall j \in N, i \neq j$	To ensure each vehicle can visit one customer only one time.
	$\sum_{i}^{N} x_{ji}^{k} = y_{j}^{k} \quad \forall k \in K, \forall j \in N, i \neq j$	
	$(z_{l}^{k} \geq 1) \Rightarrow \left(y_{l_{pickup}}^{k} + y_{l_{delivery}}^{k} = 2\right) \quad \forall k \in K, \forall l \in L$	To guarantee the same vehicle must visit the pickup and delivery nodes for defined orders
	$\sum_{i}^{N} x_{ip}^{k} - \sum_{j}^{N} x_{pj}^{k} = 0 \quad \forall k \in K, \forall p \in N, i \neq p, j \neq p$	To ensure the tour continuity
	$x_{ij}^{k} \Rightarrow \left(a_{i}^{k} + 1 = a_{j}^{k}\right) \quad \forall k \in K, \forall i \in N, \forall j \in N, i \neq j$	
	$\sum_{l}^{L} z_{l}^{k} = l_{paletes} \quad \forall l \in L$	Capacity constraints
	$\sum_{l}^{L} \sum_{k}^{K} z_{l}^{k} = i_{demand} \quad \forall i \in N, l_{delivery} = i$	
	$y_{i}^{k} \Rightarrow (c_{i}^{k} \leq k_{capacity}) \quad \forall k \in K, \forall i \in N$	
	$x_{ij}^{k} \Rightarrow (t_{i}^{k} + T_{ij} + s_{i}^{k}) \leq t_{j}^{k} \quad \forall k \in K, \forall i \in N, \forall j \in N, i \neq j, i \neq 0$	To respect the total travel time for vehicles and the time window for each customer
	$x_{0j}^{k} \Rightarrow \left(T_{ij} + t_{min,0}\right) \leq t_{j}^{k} \quad \forall k \in K, \forall j \in N, j \neq 0$	
	$y_{i}^{k} \Rightarrow t_{min,i} \leq t_{i}^{k} \leq t_{max,i} \quad \forall k \in K, \forall i \in N$	
	$k_{maxtime} \geq \left[\sum_{i,j}^{N} x_{ij}^{k} * (T_{ij} + s_{i}^{k})\right] \quad \forall k \in K, i \neq j$	
	$z_{l}^{k} = 1 \Rightarrow (a_{l_{pickup}}^{k} < a_{l_{delivery}}^{k}) \quad \forall l \in L, \forall k \in K$	To guarantee each order is picked up before being delivered

Table 2 Benefits and disadvantages of collaboration in comparison of carriers alone

Number of consignees	Mass loading rate (%)	Volumetric loading rate (%)	Empty kilometers rate (%)	Tkm per tour (%)	Fuel consumption (%)	Greenhouse gas emissions (%)
21	−8.7	−2.2	−4	+11	−7	−7
24	−3.7	+1	−2	+3	−6	−6
27	+2	+8.6	+1	+21	−10	−10
30	−6	+0.2	−5	+10	−3	−3

They also have an identical number of consignees. Various markets are tested by varying the total number of consignees in the market between 21 and 30.

A consignee can receive 1 to 11 pallets with an average to 6 pallets. The average weight of a pallet is 200 kg varying between 100 and 300 kg.

The planning of delivery tours is done for each carrier working separately. Another calculation is made in a perfect collaboration context where all vehicles are shared and all information about the consignees' orders is pooled through interoperable information systems.

The following table presents the evolution of the performance of the global system (see Table 2).

The carriers who work in a collaborative way save 3 to 10 % fuel in comparison to carriers who work in an independent way with their own fleet. In addition to the associated environmental benefits these results also prove that considerable savings can be done on energy bills. A virtuous cycle can be created if the money is reused to buy more environmental-friendly trucks like hybrids or electrics ones.

The collaborative approach improves tkm per delivery tour and can increase productivity up to 20 %. The sharing of vehicles allowed choosing the most suitable vehicle for the delivery in terms of capacity or fuel consumption.

In all the modeled markets these productivity gains allowed the carriers to use one truck less to deliver the orders than in the case with no collaboration. With the same fleet the companies which collaborate are able to manage more consignees and deliveries than if they don't work together.

Comparatively speaking, the second indicator of productivity, the empty kilometers rate, is generally lowered. That can be explained by the fact that the sharing of vehicles allowed reducing the total kilometers to a greater extent compared to empty kilometers. Indeed, as all the journeys are delivery tours, all vehicles go back empty to the storehouse. So the empty kilometers can't be completely avoided. This characteristic makes the empty kilometers more difficult to reduce than the others.

Similarly, loading rates are also not suited to measure the performance of collaboration transport systems. As explained above, the empty kilometer rate tends to be lowered, and as a result, the loading rates decrease. But the gains in tkm tend to improve them. This means the loading rates don't give a clear picture of the collaborative system.

Table 3 Cost of the urgent order in interoperability system and classical system

Number of consignees	Kilometers added		Tkm per tour added		Fuel consumption added	
	A (%)	C (%)	A (%)	C (%)	A (%)	C (%)
21	56	10	33	4	60	10
24	20	10	7	7	21	10
27	5	5	2	4	3	5
30	24	8	10	3	22	8

3.3 Collaboration in Managing Last Minute Unforeseen Changes

In order to test the potential of collaboration to improve the management of a last minute unforeseen change, we propose the following scenario: an urgent client order is imposed upon one of the carriers after the tours started and trucks left warehouses with the planning established before. An entire new planning is not possible considering the fact that the trucks have begun the deliveries and are loaded with the pallets of defined consignees. A solution has to be found with the trucks on the road by making a few changes in their itineraries to allow one of them to handle the added pallets.

Table 3 presents costs associated to the last minute unforeseen change, evaluated on the three most relevant indicators. These costs depend upon the way this change is managed: the carrier answers by himself (A) or the answer is done through a collaborative way (C). These costs are compared in percentage to the initial costs.

When a carrier delivers an additional order all alone he must often make greater efforts to manage it. However, a group of carriers does not have the same problem because it has more available trucks. Therefore, it is more probable there is one of those available trucks near the pickup point of the new order than a carrier all alone. A group of carriers collaborating their deliveries covers on the average 15 % less kilometres than a carrier who delivers all alone to deliver the urgent order.

Even in the cases where there is no truck in the neighbourhood the cost in terms of kilometres or fuel of the new order is always reasonable in comparison to the cost of daily activities of the group. This cost is also more foreseeable because it's around 5 to 10 % of the cost of previous tours.

Like the 27 clients case in our study, there are also some cases where the collaboration does not add any benefit to crisis management. It can be explained by the fact that the first planning, before the urgent order, allows a truck of a single carrier delivering all alone to be nearer the point of pickup than in the planning of a collaborating group of carriers. However, the performance of groups planning in a crisis is still really much better than the performance of crisis planning by any of the three carriers who were delivering all alone. This is due to the gains on general planning, not completely offset by the urgent order.

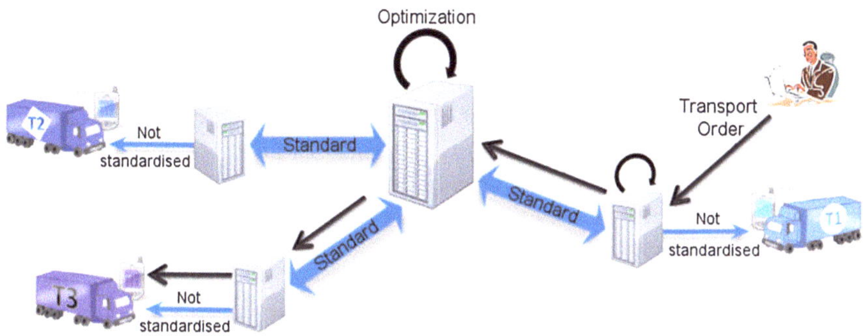

Fig. 2 Collaborative information system architecture

4 The Use of Information and Communication Technologies

In order to facilitate collaboration, the logistics providers must take the necessary steps to make their organizations and systems interoperable.

The use of Information and Communication Technologies (ICT) and Decision Support Systems to support collaboration continually evolves [12]. Currently, some technologies and standards facilitate interoperability between logistics providers (see Fig. 2):

- **EDIFACT** (Electronic Data Interchange For Administration, Commerce and Transport) is the international standard for EDI messages. It describes a number of messages (transactions) that occur frequently, the syntax (structure) of these messages and the protocol past which these messages are exchanged.
- **GS1 eCom** is standard from the GS1 organization. It provides a universal structure for electronic business messages, which enables to communicate business data between trading partners rapidly, efficiently, accurately and cost-effectively, irrespective of internal hardware or software differences.

5 Conclusion

Today, performance indicators are used by big transport companies. Nevertheless, in France this sector is composed of numerous very small companies which do not use performance indicators to manage their activities. We propose a global framework based on four levers which allow to manage these activities.

We also studied how collaborative approach has a positive influence on these levers through theoretical cases. Simulation demonstrates that productivity can be increased (+20 %) and fuel consumption can be decrease (−10 %), without any change in the carriers' fleets. The benefits on "crisis management" is always positive highlighted. Moreover as additional costs are supported by the group, they never rise dramatically in comparison to the daily costs of the tours and they can be more and more foreseeable.

However the implementation of a collaborative approach supposes the deployment of interoperable information systems. An important financial effort is required for companies. This effort can be considered as a real lock for very small companies, often threatened by the crisis. They don't have the financial possibility to invest in new information systems.

Two main solutions can help overcoming this economic barrier. In the first one, big companies integrate the small ones in collaborative networks. It implies that small companies are more or less constrained by biggest ones and their policy. In the second solution, some tools such as interoperable platforms are developed in order to allow the smallest companies to join collaborative networks. Different services are supported by these platforms. This allows small companies to improve their performance while they keep their independency.

References

1. Zhou, H., & Benton, W. (2007). Supply chain practice and information sharing. *Journal of Operations Management, 25*(6), 1348–1365.
2. Roth, A., Cattani, K., & Froehle, C. (2008). Antecedents and performance outcomes of global competence: An ampirical investigation. *Journal of Engineering and Technology Management, 25*(1–2), 75–92.
3. Ritchie, B., & Brindley, C. (2007). Supply chain risk management and performance: A guiding framework for future development. *International Journal of Operations & Production Management, 26*(3), 303–322.
4. SCOR. Home. Retrieved Nov 02, 2013, from http://www.supply-chain.org/scor
5. Kaplan, R., & Johnson, H. T. (1987). *Relevance lost: The rise and fall of management accounting*. Boston: Harvard Business School Press.
6. Kaplan, R., & Norton, D. (1996). Linking the balanced scorecard to strategy. *California Management Review, 39*(1), 53–79.
7. Sakka, O., Botta-Genoulaz, V., & Trilling, L. (2009). Modélisation des facteurs influençant la performance de la chaine logistique. [Hal 00389129, version 1], pp. 1–7.
8. Sanders, N. (2007). An empirical study of the impact of e-business technologies on organizational collaboration and performance. *Journal of Operations Management, 25*, 1332–1347.
9. ODETTE. Global MMOG/LE. Retrieved Nov 02, 2013, from http://www.odette.org/services/mmog
10. Deshamps, J. C., Amrioui, S., & Malhéné, N. (2011). Indicateurs de performance. Livrable Prodige 1.3, Bordeaux.
11. Malhéné, N., & Deschamps, J. C. (2010). De la traçabilité au routage des produits en transport routier. *8èmes Rencontres Internationales de Recherche en Logistique, RIRL'10*, vol. 29, 30 Septembre et 1er Octobre 2010. Bordeaux (France).
12. Amrioui, S., Malhéné, N., & Deschamps, J. C. (2012). Treacability in collaborative logistics : How to use EPCGlobal solution in transport reconfiguration. *6th IEEE Conference on Digital Ecosystem Technologies—Complex Environment Engineering* (*IEEE DEST-CEE 2012*), June 18–20, 2012. Campione d'Italia (Italy).

Part IX
Enterprise Interoperability Science-Based

Decision Tree and Agent-Based Approach to Specify Inter-Company Cooperation Using Offline Services Composition

Meriem Kermani, Mahmoud Boufaida and Emmanuel Paviot-Adet

Abstract With the deep development of economic globalization, companies tend now to collaborate closely with each others to improve their competitiveness. The problems of interoperability have recently been the subject of considerable amount of studies. In this paper, we propose a mediation based approach, which allows to a set of heterogeneous companies to cooperate. So, they can form a company network called SoS "system of systems". The purpose of this solution is to keep the company architecture and to ask the mediator that is a software-based agent to play an intermediary role between companies, and to make the transformation between companies as well. We define a dynamic and cooperative inter companies model. The dynamic aspect gives the possibility to change the collection of services involved in the collaboration. Our model combines the agent technology and the decision trees paradigm. This last facilitates making decision by selecting the services that best meet customer needs, in order to create a composite service. The realization of the offline composition process by mediator saves the response time. Once the service is executable it will be published to permit its reuse.

Keywords Multi agent systems · Decision tree · Interoperability · Cooperative process

M. Kermani (✉) · M. Boufaida
LIRE Laboratory, Constantine 2 University, 11 rue des cousins Hamrouch, Constantine 25000, Algéria
e-mail: kermanimeriem@hotmail.fr

M. Boufaida
e-mail: mboufaida@umc.edu.dz

E. Paviot-Adet
LIP6 Laboratory, Pierre and Marie Curie University, Paris, France
e-mail: emmanuel.paviot-adet@lip6.fr

K. Mertins et al. (eds.), *Enterprise Interoperability VI*,
Proceedings of the I-ESA Conferences 7, DOI: 10.1007/978-3-319-04948-9_32,

1 Introduction

Today the definition of new organization models of companies is driven by the increased competition, the variability of customer demands and the communications performance. In this new context, companies have understood the importance of inter-cooperation and they implement models based on cooperation (virtual companies, companies network..). These forms of organization have a significant impact on information systems.

Our research work concerns with the question of the on demand cooperation considering that the information system of company is a central element of this problem. Being given the nature of the on demand cooperation, we consider two principles. The first one is related to the dynamism and the lifetime of the cooperation. The dynamism aims to respond quickly to a business opportunity. Concerning the lifetime, the objective is to reduce the implementation time of the collaborative process by composing services while improving the quality of the obtained composite service. The second principle concerns the openness. In fact, openness has an immediate consequence on increasing the number of potential partners. Therefore, a strong need of an entity that manages firstly the various services offered by partners and secondly, that allows the setting of the security level required by companies. Though it is important that every company can communicate and report what it can offer in terms of services. This does not deny the fact that they can be rivals. Thereafter, the presence of a third entity that provides these features is of paramount importance. To answer to the limitations and issues described above we propose mediation architecture between different companies which participate in the collaboration process. The use of mediator ensures the interoperability between companies, and a certain level of security and allows the interconnection processes of different companies.

The dynamic and cooperative autonomous processes required for this interaction must then integrate a representation of the user knowledge and behaviors, and have a real ability to communicate. The multi-agent approach provides an abstraction level suitable for this problem. In fact, multi-agent systems (MAS) allow a coordination of the interacting agents' behavior in a company in order to perform tasks or solve problems [1–3]. We focus on the communication capabilities of agents to allow them to be used as basic elements for the design of the complex collaborative systems.

In order to facilitate the decision taking and the choice of services that best meet the customer needs in the shortest time, we use the paradigm of the decision trees in the software agents inside the mediator component.

This paper is structured as following. Section 2 is dedicated to a comparative study between multiple research works. Sections 3, 4 and 5 are consecrated to the study of our proposal which will be illustrated by an example in Sects. 6 and 7. In Sect. 8, we will give a discussion for positioning our solution compared to the existing work. Finally we will make some conclusions about our work and we will talk about the extension of it.

2 Some Related Works

Several authors and approaches have focused on resolving the issue of inter-company cooperation offering tools and approaches for the design of information systems. The proposed works have the information system at the center of their concerns and qualify it as an indispensable support for the implementation of the company strategy. Research works such as [4–6] were attached to the analysis and design domain of process-oriented information systems. Several methods and tools of information systems engineering have emerged such as component-based approaches, (MDE) Model Driven Engineering. These methods have been proposed to meet the evolution of the current information system context: evolution in terms of the information system architecture (heterogeneous and distributed), and evolution in its use (open and accessible for different users). Since a few years the concept of (SOA) Service Oriented architecture has been rapidly spread and widely accepted as a supporting architecture of the company information system.

Although many project such as shape [12], that have studied the aspects of companies cooperation and have proposed solutions to the different heterogeneities that may exist, but the problem still remains an open domain of research. During our bibliographical study we note that researchers in this field have different views. We will present a panorama of solutions. We studied the overall work according to the used approaches and technologies, the nature of collaborative processes if it is static or dynamic, and the field possible for the use of the proposed solution.

Touzi et al. [7] adopted a non-standardization approach, which proposes the design of a mediation system between information systems. A mediator is considered as a vector for the interoperability of heterogeneous information systems. She decomposed the proposed approach on its business, logical and technical projections. The proposed practices are similar to the MDA and allow one to anticipate the tooling of a model transformation workshop. This method was also adopted in [8] by Truptil et al. to solve a crisis management problem. The approach used in [11] is a standardization approach which consists on the proposition of a framework where all the participating company must respect it.

Among the works that have used the agent paradigm for solving the inter-company cooperation problems, the authors of [3, 9] have proposed global architectures, integrating the technology of web services. The difference that Namin et al. [3], proposed, is the integration of web services and software agents within the internal structure of the company and by adopting software agents within the UDDI registry, they introduced some agent components to help service requesters to select the most appropriate service provider. Few studies have treated the notion of dynamic cooperation or the 'on demand cooperation'. For instance, the work of Boukadi et al. [2] offers such a mechanism of cooperation based on important concepts that are: description, publication, discovery, and the composition of services and the adaption of the services to the context.

Mallek et al. [10] have a different vision. They proposed an approach to detect problems before a real collaboration, then analyzing and finding solutions to each partner.

After studying the achieved research of the inter-company cooperation context, we note that these works are still unable to provide a model of cooperation that can mask the heterogeneity of ISs (information systems) of companies. In addition the few studies that have used software agents have not addressed the decision-making of these agents. We also note that these studies have not taken into account the dynamic aspect, so the process is fixed from the beginning and the addition or deletion of a company disrupts the entire system. In this paper, we propose a mediation architecture, which allows the creation of a dynamic process between companies. We adopted the mediation in order to involve companies that offer interesting services without using standard tools. To allow the reuse of the composite service, we propose its publication in a network.

3 Our Proposal

In this section, we propose a mediation architecture between services of companies that are participating to the cooperation. According to Lascoux [13] the mediator "provides a framework with its own benchmarks, consisting of operating and communication rules, and steps process. It begins with the acknowledgement of the parts position in terms of legitimacy, to the formalization of an agreement as satisfactory as possible for the parts". In this context, the source and the need of mediator between information systems is their heterogeneity, even making their interoperability complicated. Mediation must successfully make them interoperate following a well defined steps process and operating rules. Specificities acknowledgement of each information system in the network is essential. It is on the basis of the specificities that the mediator of the information systems will offer satisfying solutions. In our architecture the mediator is used to perform a dynamic intercompany process, offering the possibility to change the collection of services involved in the collaboration. The intercompany process will be published to be used as a web service, to permit the visibility as well and to be reused by other customers. Figure 1 shows an overview of the architecture.

The different parts of this architecture are:

- Companies or Partners of collaboration: A set of companies participating in the collaboration. These companies offer accessible services via the public part of their information system. The informations about each service is stored in a services directory.
- Mediator: is the core of the architecture, it is acting as an intermediate between the requester and the partners. It is composed of:
  - Interface agent: it receives and makes the processing of the applicant request to build a decision tree, which will be sent to the coordinator agent.

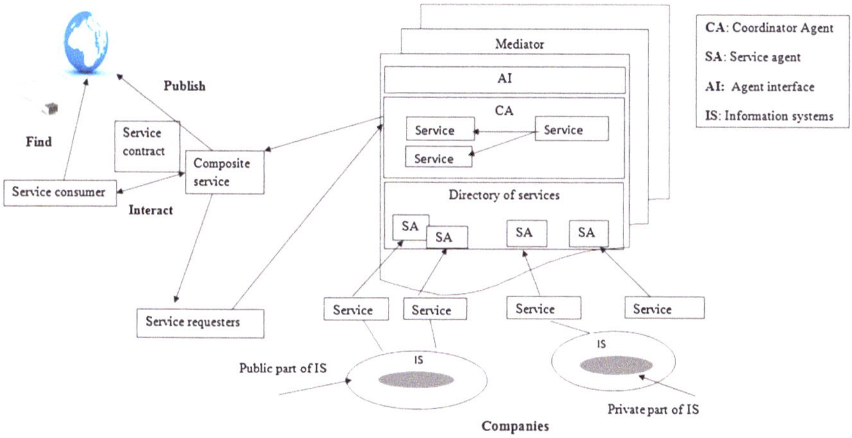

Fig. 1 Overview of the our proposed architecture

- Services directory: in this directory we have services agents; each of these agent is related to a service proposed by a company. These agents represent information about each service, the information is organized as: a local decision tree in each agent service, and service address, communication protocol, and the details of the service purpose.
- Coordination Agent: it allows creating a composite service to meet the client request. It has a global decision tree that is created from the set of local decision trees. This tree is used to make a decision which allows choosing the best services to meet customer demand.

• Applicant service: it represents the launch process.

We have proposed the general architecture of the coordination agent, respecting the three properties of independence, communication and intelligence. This architecture is inspired from the modular theory of Fodor [15] (Fig. 2).

• Communication Module is messages support between the agent and its coordinator. All interactions between the coordination agent and the services agent or the interface agent pass through this module.
• Mailbox is used to submit messages. This queue box type FIFO (**F**irst **I**n **F**irst **O**ut) is used for storing messages in order to process them asynchronously.
• Representation module receives the request sent by the agent service as a tree, and then reformulates the global tree (the tree that includes all local trees published by the different service agents) following the order of criteria sent by interface agent.
• Information Management Module contains information related to the service published by partners (protocol, address, message structure, description of the

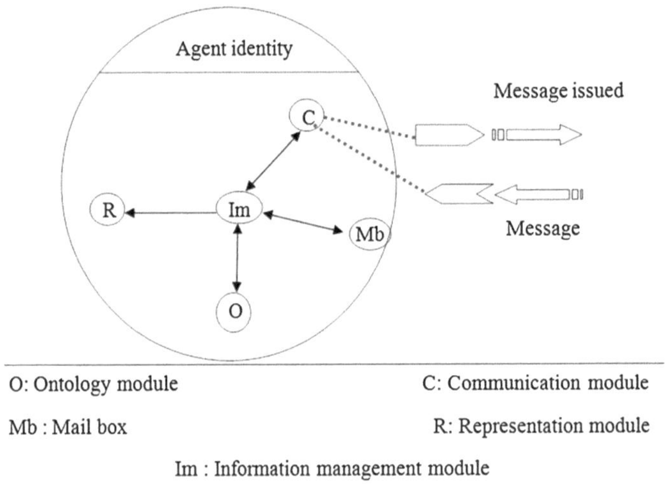

Fig. 2 Structure of the coordinator agent

roles of services). Each service agent uses a local ontology to represent vocabulary and concepts in its application domain.
- Ontology module this module deals with the search for correspondences between agents to perform a cooperative task.

4 Cooperation Mechanism Based on Decision Trees

To mask the heterogeneity of the participating companies in the cooperation, and to facilitate decision making we propose the use of decision trees as a learning mechanism in our agent. A decision tree can be used to clarify and find an answer to a complex problem. The structure allows our coordinator agent to take a problem with multiple possible solutions and display it in a simple and easily understandable format that shows the relationship between different events or decisions [14]. We propose the use of two types of trees, with the following format: Model $_{\text{decision tree}}$ *(Initial node, Branches, Intermediate node, Leaf)*

- Local tree: shows the details of the services offered by the participating companies to cooperation. This tree is used by the service agents.
- Global Tree: is created from the fusion of local trees, and used by the coordinator to quickly select the services that best meet customer needs. Each service in the composition is defined by a path to the global tree of the coordinator agent.

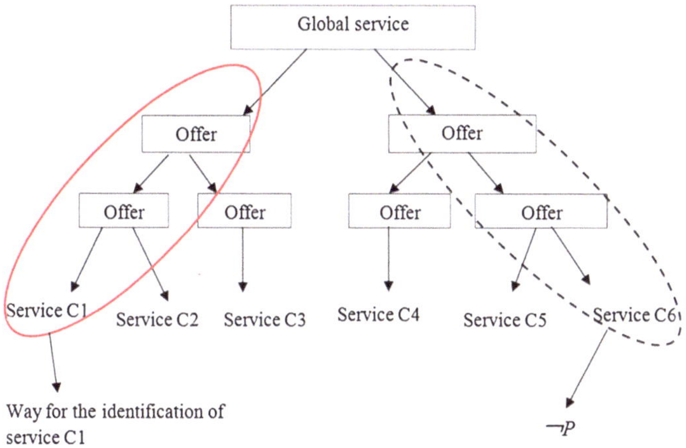

Fig. 3 Internal structure of the global tree

The Fig. 3 shows the structure of the global tree.

- *Initial node*: each initial node of a global tree located in coordinator agent represents the global service which includes the services representation of different companies.
- *Branches:* they connect parent internal node to a child node. They represent the variable value tested in the intermediate parent node to the child node. Each branch corresponding to a question, allows choosing the path to follow.
- *Intermediate node:* represents the deals offered by services, with a sort of category, it means that each path or branche belongs to the same service of the same category.
- *Leaf*: each leaf refers to a class; this class identifies a service company which means that the path from the root to a leaf of this tree gives details of one or more service.

Rule: the conjunction of all tests that have been taken during the tree paths, from the root to the leaf identifies the service C1 or multiple services [C1…C n]; each test is either the predicate P associated with intermediate node if we go on the left after the node or the complement of the predicate ¬ P if we go on the right.

5 Offline Services Composition Process

In this section we will explain the construction of dynamic intercompany process. Its creation is based primarily on the selection of multiple services and then specifies the sequence flow between them. The Fig. 4 shows the selection steps and the offline services composition. We show in this figure the entities responsible for each action, and the operations that we propose for the passage from an action to another. Our process is divided into three principal parts.

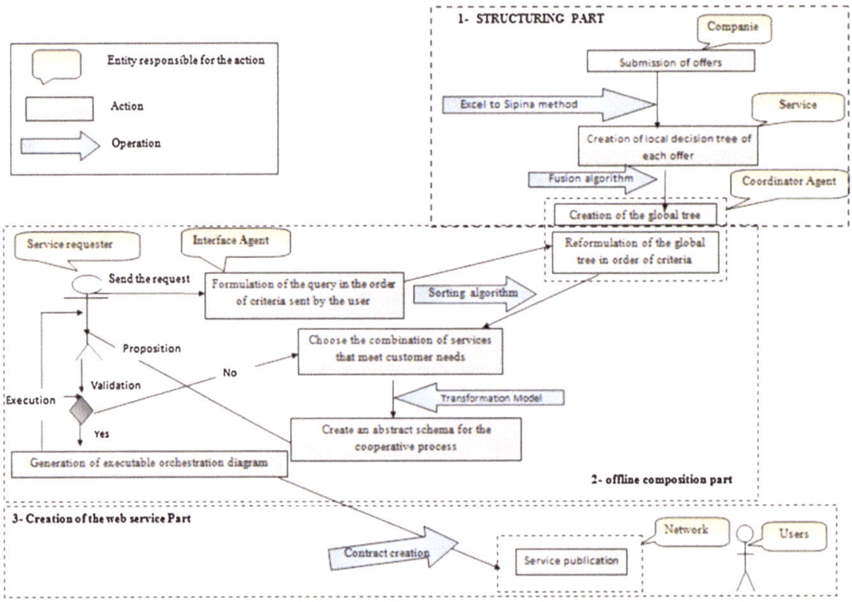

Fig. 4 Process of the offline services composition

- Structuring part: It includes the submission stages of the offers, by companies participating to the cooperation, the creation of the local decision tree for each offer by Service agents, and the creation of the global tree by the coordinator agent.
- Offline Composition part: this part of the process is triggered by customer demand, we follow each stage of this part from the process to arrive to the generation of an executable orchestration diagram.
- Web service creation part: After creating the composite service we follow the publishing step of executable service to permit its reuse.

6 Example

We will treat in this part the realization of a response process of an emergency case which is an accident, using three existing entities named: Hospital, SAMU and Police.

- The Hospital service: each hospital proposes a description including its position, the free operating rooms and the laboratory, thus all the available tests, in our case we need the GPS data of the hospital. The description of each hospital is published in a service agent and represented by a local decision tree. All hospitals will be formalized by a global decision tree stored in the coordination agent.

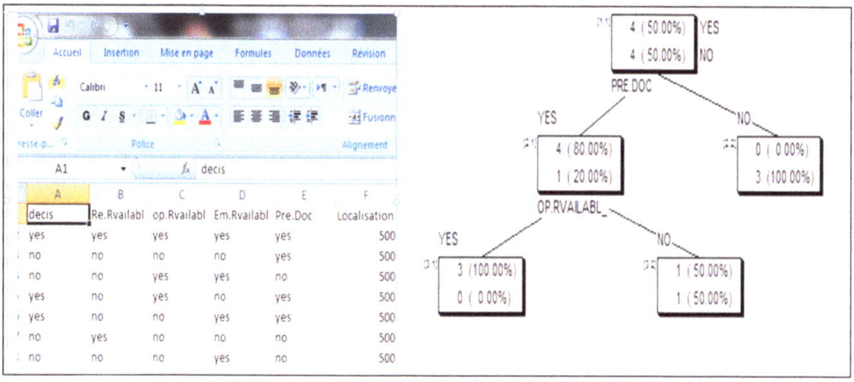

Fig. 5 Generalization of the local decision tree from Excel using Sipina Version 3.2

- The SAMU service: It provides a description of its services such as the emergency care team, the possible route and the expected duration of each route. We have service agents representing the SAMU by a local decision trees and the coordination agent representing the global tree.
- The Police service: It provides the position of the police stations and available patrols, which will be represented by decision trees in the local services agent. All the police stations are represented by the global decision trees stored in the coordination agent.

Using these three entities, we will create a composite service to respond to an emergency which is called 333. This service will allow users to report an accident at a given position. The service will take the necessary steps and return to the user the expected intervention delay in seconds. The internal scenario which will lead the interactions between services is the following:

The coordination agent will firstly ask the accident base to determine whether the reported accident is known, if so a negative value is simply returned to the service requester. Otherwise, we will first find the nearest hospital to the accident position, we use the hospital GPS data and we determine both the SAMU and the police to be sent. We then add this to the accidents base before returning to the user the intervention time that was given by the SAMU. The scenario here is voluntarily simplified, staying inspired by a real-life case. The aim is indeed to facilitate understanding and to demonstrate the feasibility of our development proposal through a simple but a realistic example.

We have chosen to use the SIPINA method for the generalization of our local trees, SIPINA is software but it is also a training method. It generalizes trees by introducing an additional operation which is the fusion, during the induction of the prediction model. The Fig. 5 shows the generalization of the local tree from Excel files. After following all the method steps we obtain the following tree.

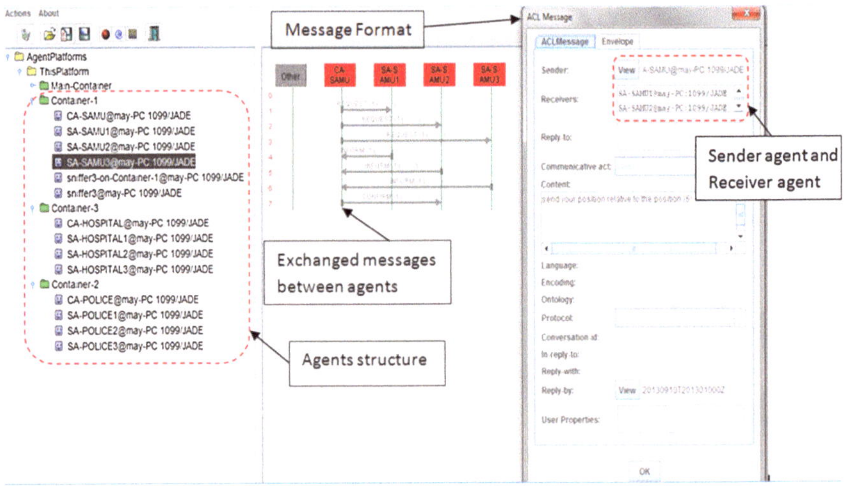

Fig. 6 Communication between agents JADE

7 Some Implementation Aspects

Currently we are implementing a model of our dynamic inter-company process to examine the feasibility of the offline services composition. The behavior of agents is created with the jade platform, Fig. 6 shows the structure of agents, and some exchanged messages between our agents, for example a coordinator agent of SAMU service named "CA-SAMU" sends a request to the service agents "SA-SAMU" the representatives of the different SAMU, then this services respond with a messages "inform" given their position, the coordinator agent select the service that best meets the needs and sends a message "confirm" to the service agent concerned, the traces of exchanged messages between agents are made with the help of sniffer agent. The third part of the figure shows the structure of a message "request" sent by the agent "CA-SAMU" to the "CA-SAMU" agent.

8 Discussion

Our solution is a distributed system evolves in a dynamic environment where the partners of cooperation change, and when it comes to designing this type of system, agent technology is suitable, because MAS not only allow the sharing or distribution of knowledge, but also the achievement of a common goal. Agents in our work learn the decision from the user in certain situations to support him in other situations. The decision problem in the case of the cooperative distributed systems is a complex problem. Also we must divide our decision problem into several problems. In our case, we choose to use decision trees, because they can model simply graphically and quickly a complex measured phenomenon. Their readability, time processing and the lack of a priori necessary assumptions explain their current popularity. The aim of using training agents with the paradigm of decision trees is to select the best services that meet customer needs.

Our contribution has four originalities that distinguish it from the existing works, firstly it takes into account the dynamic aspect, in this case our partners change and so do the service composition according to the request, then we can have a composed service for each request. Secondly the offline achievement of a large part of the composition computing saves the response time by defining an execution order of services. The third thing is the learning of agents by the decision tree. Finally the non-standardization approach or the use of mediator allows a better cooperation between the heterogeneous companies.

9 Conclusion

In this paper, we discussed the problems of interoperability of the intercompany collaboration process. Trying to respond to these problems, we studied and analyzed the existing approaches and mechanisms in the literature. Then we discussed the creation problem of dynamic process. We have proposed an offline services interconnection of different companies and to make it, we used a mediator based on software agents that learn decision making by the decision trees. Our future work includes refinement of our model by definition of an algorithm to merge the local trees according to a test on each arc. We will define a transformation model to create matches between decision trees created with SIPINA and agents implemented in Java.

References

1. Hoeta, S., & Saboureta, N. (2010) Simplification de la mémoire des agents en utilisant des FMDP. In *18th Journées Francophones sur les Systèmes Multi-Agents (JFSMA'10)* (pp. 13–22).
2. Boukadi, K., Vincent, L., Ghedira, C., & Maamar, Z. (2010). CSMA: Context-based, service-oriented modeling and analysis method for modern enterprise applications. *International Journal of Organizational & Collective Intelligence*, *1*(2), 1–28
3. Namin, A. S., Shen, W., & Ghenniwa, H. (2006). Implementing enterprise collaboration using web services and software agents. *International Journal of Networking and Virtual Organizations: Special Issue on Organizing Collaborative Supply Chains and Networks*, *3*(2), 185–201.
4. Indulska, M., Green, P., Recker, J., & Rosemann, M. (2009). *Business process modeling: Perceived benefits*. ER2009 (458–471) Gramado, Brazil.
5. Henderson, S., Gonzalez, S. P., & Ralyté, J. (2007). *Situational method engineering: Chunks or fragments*. The CAiSE Forum, Trondheim.

6. Rolland, C. (2009). Method engineering: Towards methods as services. *Software Process: Improvement and Practice, 14*, 143–164.
7. Touzi, J., Bénaben, F., & Pingaud, H. (2008). Prototype to support morphism between BPMN collaborative process model and collaborative SOA architecture model. In *4th International Conference on Interoperability for Enterprise Software and Applications.*
8. Truptil, S., Bénaben, F., & Pingaud, P. (2011). On-the-fly adaptation of crisis response information system. In *Conference on Management of Emergent Digital EcoSystems.*
9. Brahimi, M., Boufaida, M., & Seinturier, L. (2009). Multi-agent architecture for developing cooperative e-business applications. *International Journal of Information Systems and Supply Chain Management, 2*(4), 43–62.
10. Mallek, S., Daclin, N., & Chapurlat, N. (2011). An approach for interoperability requirements specification and verification. In *Third International IFIP Working Conference.*
11. Hui, L., Anne, F., Cutting, D., & Bourey, J. P. (2010). Use of ontology for solving interoperability problems between enterprises. In *11th IFIP Working Conference on Virtual Enterprises.*
12. http://www.shape-project.eu.
13. Lascoux, J. L. (2007). *Pratique de la médiation*, Editeur ESF, ISBN-10: 2710118599.
14. http://www.investopedia.com/terms/d/decision-tree.asp.
15. Fodor, J. (1983). *The modularity of the mind.* Cambridge, MA: MIT Press.

Extending the Ontology of Enterprise Interoperability (OoEI) Using Enterprise-as-System Concepts

Wided Guédria and Yannick Naudet

Abstract In this paper, we discuss the use of the General Systems Theory, for Enterprise Interoperability (EI). We review the main systemic concepts and models, highlighting the systemic concepts related to the enterprise domain, that are important for interoperability. In particular we survey the Ontology of Enterprise Interoperability (OoEI) based on the Framework for Enterprise Interoperability (CEN/ISO 11354) and grounded in systemics. We then extend the OoEI by a systemic model of enterprise. This will help locate problems and solutions for interoperability with better granularity.

Keywords Enterprise-as-system · Enterprise interoperability · Framework · Modeling · System theory

1 Introduction

Research has significantly advanced in the field of Enterprise Interoperability (EI) over the past 10 years, and EI has become an important area of research, ensuring the competitiveness and growth of European enterprises. It studies the problems related to the lack of interoperability in organizations, and proposes novel methods and frameworks to contribute with innovative solutions to these problems.

In our research work, we focus on applying a general systemic approach to Interoperability, with the aim of building a domain-independent framework for

W. Guédria · Y. Naudet (✉)
Henri Tudor Public Research Centre, 29 Avenue John F. Kennedy, Luxembourg-Kirchberg, Luxembourg
e-mail: yannick.naudet@tudor.lu

W. Guédria
e-mail: wided.guedria@tudor.lu

K. Mertins et al. (eds.), *Enterprise Interoperability VI*, Proceedings of the I-ESA Conferences 7, DOI: 10.1007/978-3-319-04948-9_33,

Interoperability modelling and decision-aid based on model-driven problem detection. Systemic means for us mainly the General System Theory, based on seminal works of Bertalanffy [1], Le Moigne [2] and Walliser [3]. Interoperability is considered from a problem-solving perspective, where associated issues appear when two or more incompatible systems are put in relation. The ontology of interoperability (OoI) [4] is the first component of this framework and constitutes a meta-model which is the first systemic formalisation of Interoperability [5]. Its application to EI requires building a dedicated meta-model, specialising the concepts of the OoI to the enterprise domain. This is the purpose of the Ontology of Enterprise Interoperability (OoEI), which is proposed in Naudet et al. [6] as a combination of OoI with the meta-model of the Framework of Enterprise Interoperability (FEI) [7], which goal was to offer a classification framework for problems in EI. From the FEI, OoEI inherits three specific dimensions, namely interoperability barriers, interoperability concerns and interoperability approaches, where concerns target four enterprises levels (Business, Processes, Services and Data). This first proposal of OoEI is however not sufficient to precisely locate interoperability problems inside an enterprise, and it is necessary to complete it with an enterprise model. Finally, this model needs also to follow a systemic approach, so that its concepts can be related to the systemic concepts in OoEI.

In this paper, we propose a generic meta-model that extends the OoEI with concepts specific to the enterprise domain, following a systemic approach. The paper is structured as follows. In Sect. 2, we recall the main reasons in favor of a systemic approach for interoperability and detail systemic solutions to solve interoperability problems, extending from previous work of Naudet et al. [5]. Section 3 provides an outline of the Enterprise Interoperability domain by surveying the OoEI, including the EI-specific parts from the FEI. Section 4 proposes a systemic view of the enterprise and gives a formalization that forms an extension to enhance the OoEI. An illustrative example is then presented in Sect. 5. Finally, conclusion and future work are given in Sect. 6.

2 General System Theory for Interoperability

2.1 Systemic Core of Interoperability

The General System Theory (GST) comes from cybernetics and was first introduced by Bertalanffy [1]. It has a holistic view and can exceed the limits of classical theories in tackling complex problems. The idea of using this theory, for our research work, comes from the need to have a general approach, which can be applied indifferently in different domains concerned by interoperability: IT, human organizations, enterprise, information systems, etc. This common base seems necessary to avoid reinventing the wheel for each new domain addressed. So far, as it was shown in Naudet et al. [5], interoperability is a systemic concept by nature: it is about *systems that interact*. A system can be characterized by the elements composing it, which can themselves be organized in sub-systems potentially interconnected [1].

2.2 Exclusion and Domination

As a complement to the systemic modelling of Interoperability presented in Naudet et al. [5], we introduce here generic systemic solutions that can solve interoperability problems, all being in particular applicable to an enterprise. According to Walliser [3], there are three classical ways that a system can use to adapt its organization or coordinate its sub-systems, to realize its objective:

- *Exclusion*: defined by the rejection of a problematic subsystem. This solution is used for an interoperability problem due to a system failure or a non-adjustable system. A non-adjustable system is a system that cannot use the adjustment solution.
- *Adjustment*: defined by the modification of the system's structure while keeping its original objectives. The adjustment solution can be used for an interoperability problem caused by the system's structure, when the system is not adaptable. An adaptable system is an adjustable system that can react to changes and adapt its structure or behaviour [8].
- *Domination*: defined by the limitation of the action field of a sub-system. This solution is applied when systems need to have a common reference or strategy for the enterprise and wish to avoid the adjustment costs. The difference between this solution and adjustment is that in the domination solution, the system does not keep its original objective instead it adopts the objective of the dominant system during the interoperation. If the same example is used with the adjustment solution, one of the systems would adopt the model of the other one.

3 Enterprise Interoperability Domain

Dealing with EI requires considering the enterprise from a general perspective, taking into account not only its different components and their interactions but also the environment in which it evolves and the interfaces through which it communicates with its environment. An enterprise is considered as a complex system in the sense that it has both a large number of parts and the parts are related in ways that make it difficult to understand how the enterprise operates and to predict its behaviour [9].

The Ontology of Enterprise Interoperability (OoEI) [6] provides a meta-model for Interoperability that takes this general perspective, as given by a systemic approach. It comprises a systemic model inherited from the generic OoI, centred on the concept of *System*, where a system can be composed of *System Elemen*ts

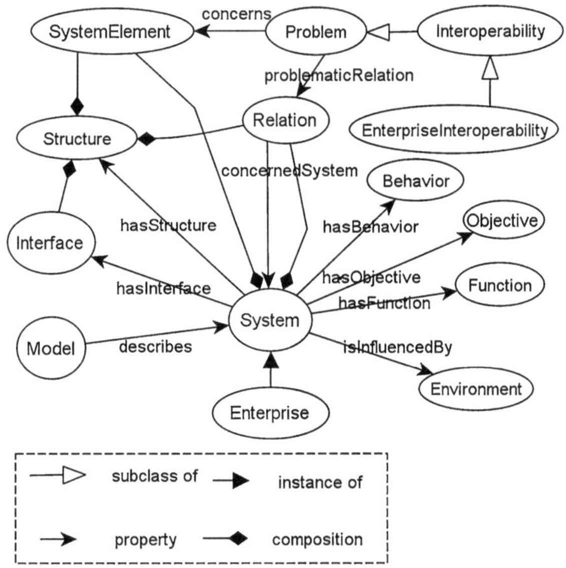

Fig. 1 Extract from the OoEI meta-model [10]

and *Relations*. Interoperability is implemented as a subclass of the *Problem* concept. Problems of interoperability exist when there is a relation, of any kind, between incompatible systems in a super- system they belong to or system they will form. Incompatibilities can concern different parts of the system: objectives, behaviours, functions, interfaces, structure or models. Figure 1 shows an extract of the OoEI model. An exhaustive description can be found in Guédria [10].

EI problems and solutions can be related to three Interoperability dimensions, as defined in the Framework for Enterprise Interoperability (FEI), which has been integrated in the OoEI. These are: Interoperability aspects (conceptual, organizational and technical), Interoperability concerns (business, process, service and data) and Interoperability approaches (integrated, unified and federated).

As illustrated by Fig. 2, these dimensions are modelled by dedicated concepts, all subclasses of the wider concept *EnterpriseInteroperabilityDimension*: *InteroperabilityAspect*, *InteroperabilityApproach*, and *InteroperabilityConcern*.

Interoperability barriers (represented by the *InteroperabilityBarrier* concept) are defined as incompatibilities preventing interoperability to take place: they are the source of interoperability problems and concern a specific aspect that is handled by some dedicated solution.

4 Systemic Approach of Enterprise

Extending the definition given by Bertalanffy [1], we define a system as "*a bounded set of inter-connected elements forming a whole that functions for a specific finality in an environment, from which it is dissociable and with which it*

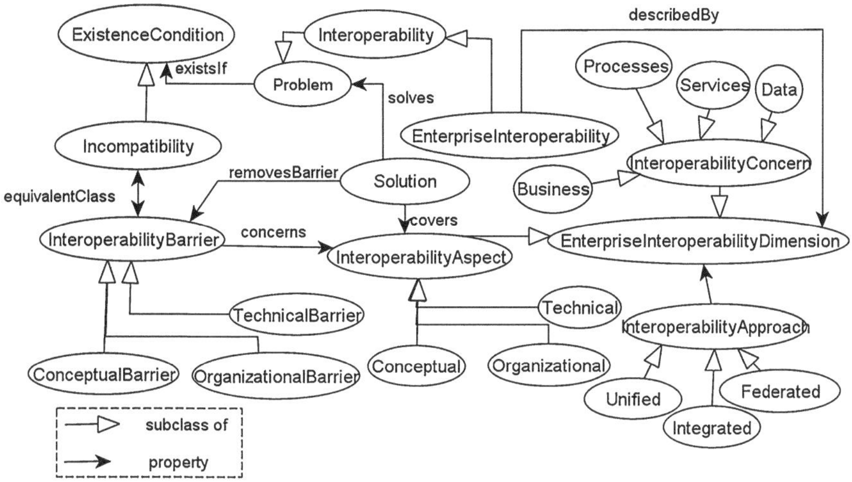

Fig. 2 FEI dimensions of enterprise interoperability in OoEI [10]

exchanges through interfaces" [6]. This definition of a system is the basis of the systemic core of the OoEI as illustrated in Fig. 1.

As shown earlier by Bertalanffy, enterprises are simply specific kinds of organisations and have to deal with the same issues. They are systems whose goal is to produce goods or services. We detail in the following the systemic model of an enterprise, starting from recalling some backgrounds on complexity, to better understand why enterprise is a complex system.

4.1 Systems and Complexity

Systems can be classified in two categories: complicated systems and complex systems [11, 12]. Complicated systems are characterized by a behaviour that can be predicted by analysing the interactions between components. Complex systems are systems for which the behaviour cannot be predicted by such an analysis.

At a very basic level, a system is defined as being complex because it is difficult to predict its behaviour and a system is complicated because it is difficult to understand its behaviour but it is understandable, especially by the system-maker [9]. A widely held viewpoint is that complexity is due to the large number of interacting parts [13]. However, complexity arises from not only the number of parts in the system, but also from the interrelationships of the system parts and the emergent behaviour that cannot be predicted from the individual system parts alone [14]. Thus, as identified by Giachetti [9], two characteristics make systems complex: the number of parts and the network of relationships between the parts. A system with many parts is at least complicated to understand and may also be complex. To be complex, the relationships between the parts must be such that system behaviour becomes difficult to understand and predict.

Within this context an enterprise is considered as a complex system in the sense that it has both a large number of parts and the parts are related in ways that make it difficult to understand how the enterprise operates and to predict its behaviour. According to Giachetti [9], an enterprise is a complex, socio technical system that comprises interdependent resources of people, information, and technology that must interact with each other and their environment in support of a common mission. The term enterprise is used because it encompasses all types of organizations: companies, government, not-for-profit, supply chains, virtual enterprises, as well as parts of a company such as a division or program [9].

4.2 Structure of an Enterprise-as-System

An enterprise can be defined as *a complex, open, heterogeneous and adaptive socio-technical system*. As an open system, it exchanges with its environment through its inputs and outputs (its interface), applying some transformation to inputs to produce the outputs. Both inputs and outputs can take on many forms: energy, material information, knowledge etc. An enterprise is obviously a heterogeneous system since it involves many components of different kinds (both human, machines and material resources). Finally, it needs to be an adaptive system: reaction and adaptation to its environment (e.g. market, technology advances, regulations and laws) is vital.

From a systemic point of view, an enterprise can be modelled as three main sub-systems [2]: *an operating or physical system* that transforms inputs flows into outputs, according to a defined goal; *a decisional or pilot system* that takes decisions; and *an information system* responsible for transmitting information in the enterprise and in particular linking the two other systems. Today, this decomposition constitutes a basis for most systemic enterprise modelling techniques, and in particular for the GRAI model [15]. The GRAI Integrated Methodology (GIM) [16] describes how the components of an enterprise are linked. It identifies the importance of decision support in the enterprise reference architecture, and describes a two stage process to design decision roles in the enterprise system.

The decisional system provides a structure of decision centres, which can be office bearers or autonomous units at different levels in the organization. The physical system comprises the machines, components and resources that the enterprise has at its disposal for generating profits and wealth. The enterprise does business through its physical system. Hence, the activities of the physical system are affected by the decisions made in the decision centres, e.g. for a supply chain, decisions drive goods movement. The information system is critical for the propagation of decisions to the lower levels of the decisional system and to the physical system. If the right information about the decision is transmitted to

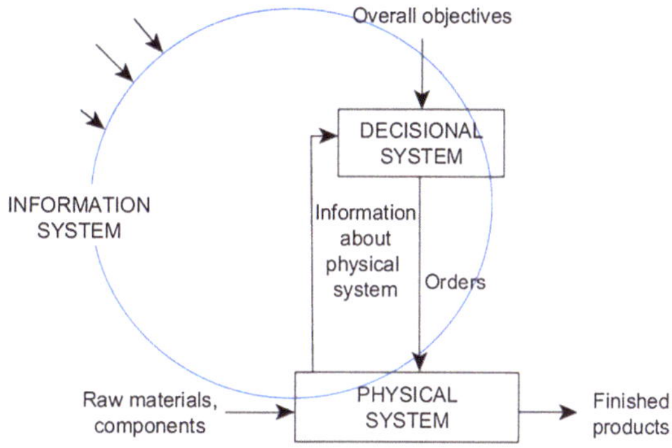

Fig. 3 The enterprise subsystems [16]

the right decision centre at the right time, the physical system will act correctly. Otherwise, it will not take any action or even worse, perform a wrong action.

The Enterprise-as-system can be illustrated by the model on Fig. 3. The objectives of the enterprise are ensured by the decisional system, which take them as inputs to send "orders" to the physical system [17]. Moreover the pilot system communicates with the environment relating to the system's aims, accepting orders, making commitments, and exchanging any other information with the environment that is necessary to make decisions about how to control the physical system to successfully achieve overall system aims and objectives [17]. The decisional system relies on models of the physical system to act (i.e. to make its decisions). However, for these models to reflect reality to a sufficient degree, the decisional system must receive information, or feed-back, from the physical system.

4.3 Interoperability Model for Enterprise-as-System

In the OoEI, the Enterprise-as system can thus be modelled as a composition of the three subsystems or *SystemElements*: *Decisional System*, *Physical System* and *Information System*. All three are concerned by interoperability issues and can be related to interoperability concerns and enterprise components to help locating problems with a better granularity using a systemic model of an enterprise as illustrated in Fig. 4, where for the sake of clarity only the enterprise system extension of OoEI is shown. We introduce in this new model the concept of *Enterprise Level*, which represents the layers of enterprise in general. The four interoperability concerns are thus also subclasses of this new concept.

The physical system is concerned with the interoperation of physical facilities such as: IT infrastructure, tools, products or human resources. For the decisional system, taking into account interoperability implies making decisions so that the impact on the enterprise systems can be minimized. This concerns mainly

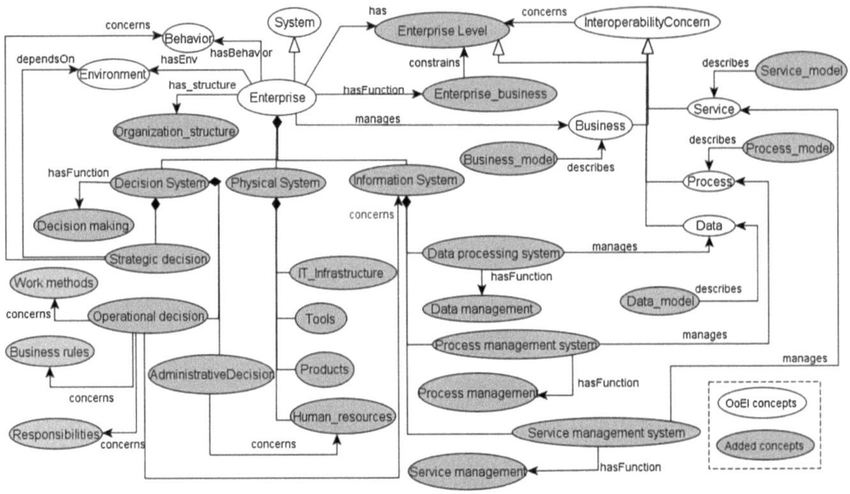

Fig. 4 Interoperability concerns in enterprise-as-system context

operational, administrative and strategic decisions. The information systems interoperability concerns the exchange of information between two systems; it is thus composed by the data processing system, the process management system and the service management system. The *Enterprise business* concept is used to denote the enterprise function such as: delivery of products and services to customers.

5 Illustrative Example

To better understand the application and use of the above defined conceptual framework, i.e. the OoEI extended with the systemic enterprise model, we develop here an illustrative case study. Our case concerns a company EA1, a construction company in Europe. They have a reputation for delivering high-end homes and providing excellent customer service and quality construction. Fully licensed and insured, they are known to be professional builders that guarantee the construction and the remodeling of a house that is unique for customers (kind of dream homes). EA1 Company faces an interoperability issue with the chimneys' company EA2 for the installation of the chimney in the house of a customer.

EA2 does not give authorization to the construction company EA1 for the installation of the chimney, due to the non-respect of the electric sheaths (ES) measure defined by their business rules and by the standard NFC 15-100. Indeed, EA1 does not respect the 8 cm from the chimney flue (CF), in the attic and the electrician ensures that there is no problem. The prime contractor supports its company and says that EA2 has to accept the current installation which is compliant to their internal business rules. The chimneys' company however refuses to

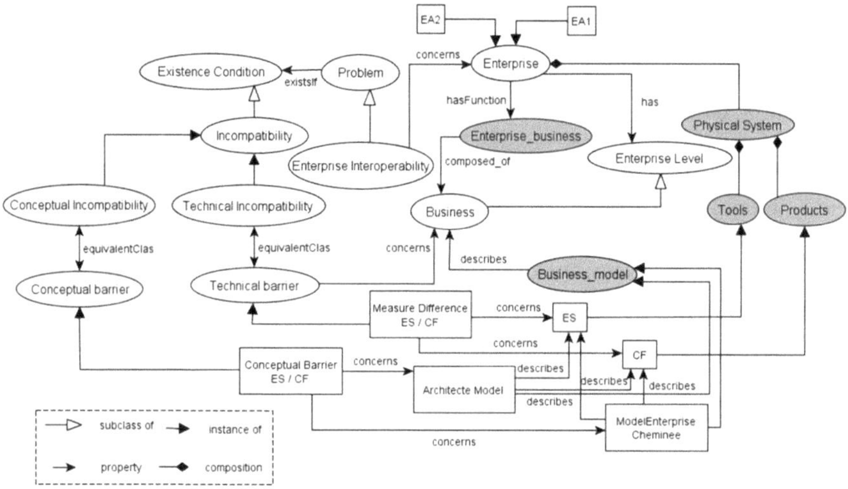

Fig. 5 Illustrative example with two enterprises EA1 and EA2

Fig. 6 Solving interoperability problem between EA1 and EA2

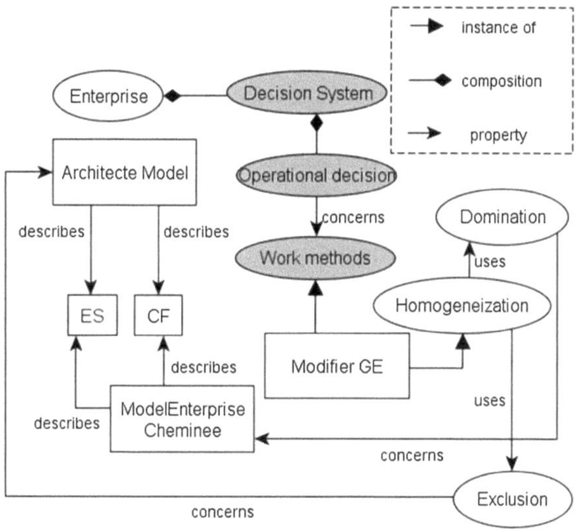

accept any other solution: as long as the sheaths are not 8 cm minimum from the chimney, the installation won't be possible for EA1. An application of our model to this example can be given by Fig. 5.

A possible solution to the above described problem would be the alignment of the business model of EA1, following the one of EA2, this solution is called: homogenization (more details can be found in Chen et al. [7]. In our case, the homogenization uses two of the systemic solutions described in Sect. 2.2: *Exclusion* and *Domination*. The *Exclusion* solution consists in rejecting the current electric sheaths (ES) model used by EA1. Then, the associated *Domination* solution consists in modifying the current ES by adopting the measure of 8 cm from the CF defined by the business rules of EA2. This solution won't be easy for EA1 but is essential to continue the building process of the concerned house. An *Adjustment* solution is not possible for this case, as EA1 does not respect the standard (NFC 15-100) and the EA2 is the company that has the authority to block the building process of the customer' house and a part of the business of EA1 (the installation of chimneys).

To solve the problem, the decision system of EA1 is responsible to choose the adequate solution for the diagnosed problem as shown by Fig. 6.

6 Conclusion and Future Work

In this paper, we have proposed a systemic meta-model of enterprise as an extension of the OoEI and a detailed description of the enterprise-as-system. This model is based on a systemic approach where an enterprise is considered as a complex system. Adopting this approach allows in particular to have a general view of the enterprise and to have a model that stays valid whatever the kind of the studied enterprise. The proposed model allows having a detailed description about the structure of the enterprise and its elements and then to point a particular element within the enterprise that may be the origin of an interoperability problem. Its main goal is to provide a framework to facilitate the diagnosis of the EI problems and their resolution process. This was illustrated through a case study in the construction domain.

The work reported here still needs to be refined and assessed on other concrete cases. The proposed model is a first version that needs to be refined and completely integrated into the OoEI. This will, in particular, allow the automatic reasoning on EI problems and solutions through ontologies and logic rules [18].

References

1. Bertalanffy, L. V. (1968). *General system theory: foundations, development, application*. New York, USA: Georges Braziller, Inc.
2. Le Moigne, J. L. (1990). La modélisation des systèmes complexes, Dunod.
3. Walliser, B. (1977). Systemes et modeles, introduction critique a l'analyse des systemes. Editions du Seuil.
4. Naudet, Y., Latour, T., Haussmann, K., Abels, S., Hahn, A. & Johannesonn, P. (2006). Describing interoperability: the OoI Ontology. *In workshop on enterprise modelling and* Ontologies *for interoperability (EMOI-INTEROP)*.
5. Naudet, Y., Latour, T., Guedria, W., & Chen, D. (2010). Towards a systemic formalisation of interoperability. *Computers in Industry Integration and Information in Networked Enterprises 61*(2), 176–185.

6. Naudet, Y., Guedria, W., & Chen, D. (2009). Systems science for enterprise interoperability. In *I-ESA '09 Proceedings of the International Conference on Interoperability for Enterprise Software and Applications China.*
7. Chen D, D., Dassisti, M., & Elvester, B. (2007). *Di.3: Enterprise interoperability framework and knowledge corpus—final report.* Technical report, INTEROP Network of Excellence, IST—Contract no.: IST-508 011.
8. Guedria, W., Naudet, Y., & Chen, D. (2009). Contribution of System Theory to Develop Enterprise Interoperability In 8th Congrès International de Génie Industriel (CIGI), Bagnères de Bigorre (France).
9. Giachetti, R. E. (2010). *Design of enterprise systems: theory, architecture, and methods* (1st ed.). Boca Raton: CRC Press Inc.
10. Guédria, W. (2012). *A contribution to enterprise interoperability maturity assessment* (Ph.D. Thesis, University Bordeaux 1).
11. Morin, E., & Le Moigne, J. L. (1999). L'intelligence de la complexité. Editions L'Harmattan.
12. Clergue, G. (1997). *L'apprentissage de la complexité.* Paris: Editions Hermes.
13. Rouse, W. B. (2003). Engineering complex systems: implication for research in systems engineering. *IEEE transactions on Systems, Man, and cybernetics—Part C: Applications and Reviews, 33*(2), 154–157.
14. Sutherland, J. W., & Van Den Heuvel, W. J. (2000). Enterprise application integration and complex adaptive systems. *Communications of the ACM, 45*(10), 59–94.
15. Doumeingts, G., Vallespir, B. & Chen, D. (1998). Chapter decision modeling GRAI grid. In *Handbook on architectures of information systems* (313–337). New York: Springer.
16. Chen, D., Vallespir, B., & Doumeingts, G. (1997). GRAI integrated methodology and its mapping onto generic enterprise reference architecture and methodology. *Computers in Industry, 33*, 387–394.
17. Doumeingts, G., Vallespir, B., & Chen, D. (2006). GRAI griddecisional modeling. In P. Bernus, K. Mertins, & G. Schmidt, (Eds.), *Handbook on architectures of information systems, International handbooks on information systems* (pp. 321–346). Berlin: Springer.
18. Eiter, T., Ianni, G., Polleres, A., Schindlauer, R., & Tompits, H. (2006). Reasoning with rules and ontologies. In P. Barahona, F. Bry, E. Franconi, N. Henze & U. Sattler (Eds.), *Reasoning web* (pp. 93–127). Berlin, Heidelberg: Springer.

System of Systems Design Verification: Problematic, Trends and Opportunities

Mustapha Bilal, Nicolas Daclin and Vincent Chapurlat

Abstract System of Systems (SoS) Engineering (SoSE) requires to be able to model and to argue the quality of the modeled solution, thanks to various objectives prior to any other efforts. This paper presents and discusses the development of an approach to support SoSE activities and particularly to achieve SoS modeling and verification. First, requested models are identified and illustrated here on Virtual Enterprise domain (VE). Second, it is proposed to merge two complementary verification approaches, formal proof and simulation. This allows us to ensure particularly the stability, integrity and control expectations of the proposed SoS solution, and must encompass particularly three main SoS characteristics chosen here that can impact SoS stability, integraty and controllability. These characteristics are connectivity, particularly subsystems' interoperability abilities, evolution and emergence of behaviors and properties which are due to the subsystems' interactions when fulfilling the SoS operational mission. For this, a formal properties specification and proof approach allow the verification of the adequacy and coherence of SoS models with regard to these characteristic and to stakeholders' requirements. Then, simulation based on Multi Agents Systems (MAS) allows the execution of the architectural model of SoS. This allows to detect potential emergent operational scenarios and then to obtain an approached behavioral model of the SoS. This MAS is enriched by concepts and mechanisms allowing to evaluate some criteria to facilitate and guide the identification of such operational scenarios.

M. Bilal (✉) · N. Daclin · V. Chapurlat
Laboratoire de Génie Informatique et d'Ingénierie de Production, ENS Mines Alès, Parc scientifique G. Besse, 30035 Nîmes Cedex 1, France
e-mail: Mustapha.Bilal@mines-ales.fr

N. Daclin
e-mail: Nicolas.Daclin@mines-ales.fr

V. Chapurlat
e-mail: Vincent.Chapurlat@mines-ales.fr

K. Mertins et al. (eds.), *Enterprise Interoperability VI*,
Proceedings of the I-ESA Conferences 7, DOI: 10.1007/978-3-319-04948-9_34,

Keywords System of systems · Interoperability · System engineering · Verifications · Formalization · Emergence · Properties proof · Virtual enterprise · Multi-agent system

1 Introduction

The concept of complex system is defined but also often seen as a little limited by several authors when considering large and heterogeneous systems involving other called complex, technical as sociotechnical and interacting systems. The notion of System of Systems (SoS) has been then introduced. Indeed, these ones, such as Virtual Enterprise (VE), military coalition forces or even crisis management system, have particular characteristics. Moreover, SoS design is distinguished from classical system design [1]. In SoS design it is required to focus designers' attention on interfaces to design, and that to allow subsystems to improve interoperability.Therefore, SoS Engineering (SoSE) remains today an open issue.

In this paper, we define and focus first on such characteristics. Second, we propose a conceptual and tooled approach allowing designers to model a SoS and to analyze the resulting models taking into consideration some of these particular but expected characteristics of a SoS. Presented here, the first phases of an ongoing research project mixing modeling concepts coming from System Engineering, Enterprise modeling, formal verification and advanced simulation techniques.

The paper is structured as follow. The problematic, related to the modeling and verification issues for SoS Engineering (SoSE) and expected results are presented in Sect. 2. The proposed SoSE approach is introduced in Sect. 3. Finally, Sect. 4 concludes this paper, drawing the orientation of future works.

2 Problematic and Expected Results

2.1 SoS: Concept and Definition

The short literature review presented below allows us to fix the most relevant characteristics of a SoS and its subsystems to study, in order to help designers building SoS models and assuming qualities of the modelled solution. Therefore, a SoS is seen as a combination of systems (subsystems) together to fulfill some kind of capability that a system alone cannot fulfill. It can be considered as a complex system [2]. Furthermore, [3] and [4] mention the following characteristics allowing to distinguish SoS from large and complex but monolithic systems:

- *Operational Independence of the Elements: SoS is composed of subsystems which are independent and useful in their own right.*

- *Managerial Independence of the Elements: The subsystems are separately acquired and assembled but maintain a continuing operational existence independent of the SoS.*
- *Evolutionary Development: The SoS does not appear fully formed. Its development and existence is evolutionary with functions and purposes added, removed, and modified with experience.*
- *Emergent Behavior: The SoS performs functions and carries out purposes that do not reside in any subsystem taken isolated but reside in the various interactions between these subsystems. The principal purposes of the SoS are fulfilled by these behaviors considered here then as emergent behaviors.*
- *Geographic Distribution: The geographic extent of the subsystems is large. Large is a nebulous and relative concept as communication capabilities increase, but at a minimum it means that the subsystems can readily exchange only information and not substantial quantities of mass or energy.*
- *Connectivity: To enable the SoS, subsystems are capable of building links among their interfaces and destroying them dynamically. The SoS places a huge reliance on effective connectivity in dynamic theatres of operations.*
- *Diversity: The SoS can only achieve its higher purpose(s) by leveraging the diversity of its constituent systems.*

Some SoS characteristics (autonomy, belonging, diversity and geographic distribution) are well defined and several works are developed in terms of methodology and tools about these ones [5]. Therefore, the here proposed work takes an interest in three characteristics: Connectivity, Evolution and Emergence.

2.2 SoS Engineering Problematic

It is admitted that SoS Engineering (SoSE) can be distinguished from System Engineering (SE) [1, 6]. Indeed, a SoS results essentially from assembling and interfacing of, in most cases, existing systems in order to fulfill a specific mission (to provide goods and services in agreements with stakeholders' requirements). However, these subsystems must remain independent and have to remain capable of achieving their own mission while SoS is existing. Therefore, they are selected and involved under various conditions and constraints, particularly their interoperability and performances, that have to be characterized prior the assembling. Indeed this assembly establishes various interactions between the subsystems. In this context, interoperability takes on its full meaning when considering these interactions that make these subsystems able to work together. On the one hand, the interactions between subsystems are expected in order to allow to the SoS to fulfill its mission.On the other hand, these interactions imposes to have interfaces of various types: technical (e.g. software), organizational (e.g. communication rules), human/machine (e.g. touchscreens) or logical at a high level of abstraction (e.g. resource utilization). Therefore, designers' attention has to be then concentrated on interfaces-to-design in order to ensure the connectivity. Furthermore, SoS should be able to evolute, and by consequence, this evolution, in parallel with the various interactions between the subsystems, can be at the origin of emergent behaviors and properties that remain not easy to identify and can be considered eventually as beneficial or damaging.

Various properties, such as proposed in [7], characterizing the SoS cannot be directly deduced and linked to the set of the properties which characterize separately all the subsystems. In the same way emergent behaviors remain, by definition, not easy to detect in a simple and efficient way. The SoS complexity, the determination of potential interactions and the large number of behavioral scenarios, cannot be totally explored and analyzed. Therefore, several design verification techniques can be used with more or less good results from informal to formal ones in order to keep the SoS characteristics maintained all over its life cycle. Moreover, these approaches and methods are largely used to help detecting errors or mistakes during design activities. Hence, avoiding drawbacks in case of SoS design can be solved by defining an efficient SoS model and adopting model verification techniques of various types.

To synthesize, SoSE process evokes a decision when assembling subsystems. It requires modeling and verification techniques and tools. Therefore, a first barrier is selecting subsystems considering and checking their capabilities and abilities to be/to stay interoperable, and to optimize the impact of a set of properties (functional, behavioral, ilities,[1] constraints[2] and performance expectations), for a more or less long time while SoS has to fulfill its mission. Defining the requested interactions, the needed interfaces and, finally, the global architecture of the SoS is a second barrier. Defining metrics and verification or validation techniques allowing proving and checking the same set of properties mentioned previously, that affect the so-called *analysis perspectives* is a third barrier. The Analysis perspectives are defined as:

- *Stability*: is the quality that reflects the ability of a system to maintain its viability (it characterizes the relationship between system's structure and its cohesion).
- *Integrity*: characterizes the relationship between system's behavior and its consistency.
- *Controllability (performance)*: is the quality that reflects the system's ability to achieve its mission (it characterizes the relationship between the functions to be performed by the system and the given service's compliance).

These three barriers have to be treated keeping in mind (1) that a design process involves various disciplines, (2) the size and expected characteristics

[1] Refers to the non-functional requirements such as maintainability, safety, security etc.

[2] Legal, deployment, implementation, etc.

(connectivity, evolution and emergence) that might impacts the set of properties, and (3) the sociotechnical nature of the requested SoS. The next section presents why SE is not sufficient for SoS to respond properly to our problematic.

2.3 SoSE Needs

Some authors assume that SE proposes sufficient principles and processes, suitable for the SoSE [8]. However, according to the specificity of a SoS (size, complexity, characteristics) and the *analysis perspectives* we chose, SoSE induces stronger effort for designers.

First, **requirements engineering** activities for a SoS (functional as non-functional) are more complex. In addition to the classical "-ilities" such as reliability, maintainability, availability etc., new "-ilities" such as interoperability, flexibility, adaptability and composeability are imposed during SoSE. Therfore, designers need an enriched requirement model which includes these new "-ilities".

Second, choosing and assembling the subsystems, which are able to provide requested capabilities/capacities and they respect **model based system engineering** principles, requires having an adapted modelling languages in order to achieve SoS modelling and verification expectations. Due to the specificity of a SoS, some current available modelling languages (e.g. behavioral) remain not sufficient for embedding or representing chosen SoS characteristics. Moreover, existing modelling languages do not consider emergent phenomena. Hence, a new behavioral modelling language has to be defined; here based on the enrichment of existing ones. Furthermore, an architectural model (functional and organic) must be proposed allowing rendering SoS architecture characteristics. The challenge is then to formalize the modelling languages (ML) that can be used in order to allow to model and to assess the interactions between subsystems. ML must permit to design requested interfaces allowing managing these interactions without inducing huge modifications or dysfunction of each subsystem. These interfaces can be of various types: technical (respecting general standards of physical interconnections of technical systems), physical (hardware), informational (knowledge, information and data exchange protocols), organizational (separation process public/private, protocols and rules of organization, control, taking responsibility, delegation, etc.) or HMI (human machine interface). These interfaces allow designers to ensure the necessary *interoperability* of subsystems [9].

Third, SoS model must allow to check if the modelled solution respects the modelling systems and stakeholders requirements [10]. Indeed, ML must also allow designers to attest that the SoS model is well constructed, well-formed and coherent with these requirements. This has to be done by **verifying**, interpreting and analyzing the obtained SoS models through various methods (formal and semi-formal ones). Formal methods [11] mathematically reasons (proves) the correctness of a given design and the systems specification as well. Formal model verification will allow designers to establish and justify that the models represents accurately the SoS system. As a result, when stating that a system has been formally verified, there should be a detailed explanation of what was formalized and what properties where verified. However, it has been many years since the formal methods are being used but their integration into the industry is still limited, mainly due to state-space explosion problem and the need of a significant knowledge in advanced mathematics. Model Checker is a formal method used for verification of systems. For instance, on the one hand, applying Model Checking techniques [12] needs to describe the system with a formalized modeling language, to formalize the properties by a specification language and to apply a deductive algorithm or calculus for the verification. On the other hand, simulation can be also used simultaneously with formal proof techniques. It is an easy technique that can be automated and is very scalable. It helps to study designs in its early stages. Furthermore, the simulation can ensure a partial validation of a model.

The effort to be made in order to deal with the various issues related to the SoSE process has been presented in this section. The following section shows that to perform SoSE process, we need (1) a set of ML to be identified (2) a set of concepts, an architecture framework—that can be handled by using these ML and (3), to attend to activities in order to ensure and verify the quality of the design, its adequacy and feasibility by merging, in a complementary way, the formal properties specification and proof approach with the simulation approach whatever the size of the SoS and the emergent behavior that might be produced .

3 Proposal of SoS Engineering (SoSE) Approach

To perform SoSE process, SoS modeling and verification concepts have to be identified. Starting with the **modelling phase**, two packages of concepts are requested: *SoS modeling concepts* package and *SoS model management* package. First, from SE point of view, *SoS modeling concepts* package is decomposed into environmental, functional, organic, behavioral and requirement concepts. Second, *model*, *refinement principle* and *view* are categorized into *SoS model management* package. This package allows us to manage the *SoS modeling concepts* package.

Environmental concepts group the *context, enabling system, resource, flow* and *stakeholder*. Functional concepts group the *function, construct (parallelism, sequence, iteration,* etc.*), flow (data, energy* and *material), item, interaction* (characterized by *attributes—Time, Shape* or *Space), resource, effect* and *risk*. The organic concepts group *the interaction, component, link, interface, resource* and *technical indicator (performance, constraint, "-ility"*). The behavioral concept groups at least *component, configuration, transition, capacity, capability, operational scenario, interaction, risk* and *effect*. The requirement concepts group the *role, need, requirement, capability, activity, process, capacity, operational scenario, functioning mode, interaction, life cycle* and *system/subsystem configuration*.

After fixing the most relevant concepts in the **modelling phase**, we present in the following paragraph the relationships phase or what we call the **bonding phase**. In this phase, a definition of some concepts is given with the relationships between each other.

Each stakeholder has concerns and *needs* (functional and non-functional) that are expected to be met by the SoS. For example, in a virtual enterprise (VE) [13], seen as a SoS, the company's stakeholders can be: employee, supplier, community, owner, investor, government, etc. Each one has its own *needs* and concerns (e.g. owners'/companies' needs profitability, longevity, market share, market standing, succession planning, raising capital, growth, social goal, etc.). These *needs* are transformed into a set of *requirements* to be satisfied by the SoS. A requirement modelling language is used in order to formalize the needs into requirements (e.g. SBVR, natural language, formal language). Moreover, the requirements are classified into various categories (functional, operational, performance, human factors, "-ilites", constraint, interfaces etc.).

Once the *Requirements* are well defined, they are evaluated by some *technical indicators*: criticality, safety, security, interoperability, maintainability, availability, adaptability, flexibility, ilities, performance, constraints and many others. The *stability*, *integrity* and *control* are the **analysis perspective** (basic principles) which we consider in our research. The *technical indicators* are verified in order to determine how they affect the analysis perspectives of the SoS. This will allow to optimise their impact. For example and in the context of VE, the performance is strongly related to three main types: delay, quality and cost. The performance indicator is high: (1) when the VE is capable of respecting the time constraints (accomplishing a mission in a given interval of time), (2) when the quality of the accomplished mission is high and (3) when the cost is low. Moreover, the stability of a VE is the capability of executing the mission/objective whatever the internal/external changes. The integrity is a concept of expectations' and outcomes' consistency in order to keep unambiguous position in the mind of various enterprises forming the VE. Therefore, the integrity here is described as the state of being whole, complete and always in perfect condition.

The *model* is a concept on which the *technical indicators* are described. A SoS is modeled throughout subsystems models. These models have unknown (or partially known) *capacities/capabilities* and *performances*. Their *interactions*, that allow them to fulfill their mission and to get connected to other subsystems or with the environment, remain not clearly identified and modeled. Therefore, this imposes to have a clear architectural model. This architectural model will allow to the subsystems models to be represented with the same formalism and a same level of detail.

As stated in the previous section, some *modeling languages* already exist, each with their strengths and weaknesses, in order to meet SoS design Verification and Validation (V&V). In this way, three approaches can be envisaged for our purpose: (I) the modeling language (ML) is fully adapted and can be directly used, (II) the ML partially covers the concern and it is required to extend it and (III) no ML related to the concern exists and it is necessary to develop new model.

As far as requirements models are concerned, existing ones are adapted to model stakeholders' and subsystems' requirements. They enable concurrent engineering processes to work more efficiently through models and they give a concise picture of the boundaries and constraints that it is expected to operate within a large and complex systems like the SoS.

An environmental model has to be proposed. It should contain the stakeholders, the context and the subsystems. However, a global behavioral model is difficult to build due to the connectivity and interoperability which are a major reason behind the appearance of emergence. Thus, a behavioral model based on interaction and effects models has to be proposed.

A behavioral model reproduces the behavior of the SoS. Due to the dynamicity in the SoS, the behavioral model will never be a global model but it will be able to cover a wide range of behaviors (including the emergent ones) through a simulation technique. The approached behavioral model to build will be a description of how the subsystems will interact together, with the actors and with any entity which is out of the SoS's boundary and from here comes our proposal to build this model based on interactions and effects models, no more described in this paper.

An interaction model is proposed to describe how subsystems have to exchange flows. Various interfaces are then defined in order to ensure the necessary connectivity (interoperability) of the subsystems respecting or, if needed, managing reverse effects due to these interactions. These interfaces establish the *links* (e.g. protocol, synchronization, collaboration, delegation rules, etc.) which exist between SoS subsystems. These links transport a *flow* (continuous, discrete or hybrid). Moreover, this interaction model should include the *effects,* which induce some kind of *risks* (e.g. technical, managerial, human, financial etc.).

In our case, the following table (see Table 1) presents all models used to ensure SoS modeling. These models are consistently related to the concerned views. One of the main differences between the architecture of a complex system and the SoS architecture is its dynamic in its reconfiguration [14]. Therefore, an architecture for the SoS has to be proposed. Proposing an architecture is to define the fundamental organization, its subsystems, the interactions between these subsystems, the environment, and the principles which guide SoS design and evolution.

Once the concepts and the models are well identified, a choice of the modeling languages has to be taken. In that sense, SysML seems allowing requirements description through its Requirement Diagram and interaction description (interoperability) through its Activity Diagram, Sequence Diagram and State Machine Diagram. However, SysML is considered as semi-formal modeling language and it remains too limited for building other models. Therefore, it has to be enriched and formalized in order to permit formal checking and to facilitate simulation. For example, the Requirement Diagram of SysML allows us to collect and organize all the textual requirements of the subsystems. However, a SoS has its own characteristics/requirements to be considered (connectivity, evolution and emergence). These new requirements evoke some new concepts that cannot be modeled by SysML (coordination of communication between the subsystems, adaptability, confidence etc.).

Table 1 SoS models and views

views	SoS models
Functional, logical and physical views	Architectural model (including physical, functionnal, interface and interaction models)
Requirements view	Requirement model
External view	Environmental model
Behavioral view	Behavioral approached model

Verifying the SoS model, whatever may be its size and the complexity of its subsystems, is not being yet fully discovered by the research. All the verifications that have been done concern only specific application domains [15]. We consider that formal verification is not efficient since: (1) it is not sufficient for the challenge to establish SoS Integrity, stability and performance such that they fulfill valid requirements of their users with expected quality and are constructed in cost effective way and (2), pure formalization and verification can only prove a correct relationship between formal specifications and implementations but cannot prove that the SoS meet valid requirements. Therefore, it is important to have a mathematical formalization of engineering concepts. Moreover, engineering concepts in systems are mostly complex and abstract and they are difficult to define properly, to understand and to justify.

The verification methodology presented in this paper is based on the use, in a complementary way, of formal proofs (Model Checker) and simulation techniques through the Multi-Agent Systems (MAS). However, using Model Checker for verification requests to have a global and deterministic behavioral model of SoS, but this model cannot be described when facing emergence phenomenon. Therefore, the use of Model Checker should be at a definite instant of the simulation. The methodology to follow is summarized in the following steps:

1. A scenario is initialized (parameter models),
2. Some scenarios are emerged,
3. The simulation is stopped at a time “t”,
4. Information (the set of properties: functional, behavioral, ilities, constraints and performance expectations) are retrieved from agents,
5. An agent responsible of using Model Checker will verify this set of properties and their impact on the analysis perspective.

The choice of MAS refers to the fact that it is natural and an effective solution to deal with complex situation in distributed environments [16], it allows modeling and simulating the parallel evolution and the interactions of various complex subsystems independently and it can answer to the individual failure of one of the elements without degrading the whole system. Moreover, it is widely used in various domains (transportation/parking [17], biomedical science, crime analysis, environment evaluation etc.).

In our case, agents represent the active entities of the SoS (subsystems). However, the subsystems can be of various natures which raises the importance of defining various kinds of Agents, such as: Intentional Agents, Rational Agents and Situated Agents. A BDI technology (Beliefs, Desire, Intention), used in some MAS, allows to model more accurately, the knowledge and rules of behavior to be exhibited by the agents which model each subsystem.

To model the SoS architecture, Multi-Agent models structure is defined according to six dimensions:

- Agent model represent the active entities (subsystems)
- Environment model is where the SoS exists and with which they interact
- Interaction model manages the interaction between the subsystems
- Model Checker model uses formal proof techniques applied to a set of properties
- Evaluation model uses evaluation techniques applied to properties translating analysis perspective expectations (stability, integrity and controllability)
- Organization model defines constraints and rules on the interaction model.

As stated previously, the simulation will allow emerging some scenarios. However, an emergence can be beneficial, harmful, or neutral in its effect. It is the primary mechanism for both success and failure in SoS. Therefore, we need to determine what the harmful and beneficial ones are.

In our research, we consider two types of emergence out of four proposed by [18]. Detecting and filtering the emergent behaviors is achieved through some criteria: the emergence need to be observable at some level, novelty, coherence irreducibility (a complete account of an entity will not be possible at lower levels of explanation and which has novel properties beyond prediction and explanation), interdependency between levels, non-linearity, plausibility and credibility. Further details about each criterion will not be shown in this paper due to the lack of space.

4 Conclusion and Prospects

This paper has introduced the importance of SoS design model verification through the complementarity between the formal proof techniques and simulation in order to verify SoS design model and to detect errors and emergent behaviors (which are due to interaction/connectivity and interoperability) in early stages of architectural and interfaces design. Moreover, we have seen how a set of properties can impact the analysis perspectives (stability, integrity and control—performance) of a SoS.

The SoSE approach presented consists on identifying the models to verify inside a SoS, identify the concepts, choose the modeling language and enriched it, and then start to verify these models by a mathematical formalization and by an adequate verification tool simultaneously with the simulation.

We aim to develop a Meta model for SoSE covering architectural model, users' and systems' requirements model, behavioral approached model (including interaction and effect based models) and environmental model, then to propose a mathematical formalization of elements from this Meta model. We are willing to propose as well a repository of expected SoS properties and a verification techniques in order to simultaneously check properties and simulate the approached behavioral model.

References

1. Blanchard, B.S., & Fabrycky, W.J. (2011). *Systems engineering and analysis* (5th ed.).
2. Chapman, W.L., & Rozenblit, J. (1995). Complexity of the system design problem. In *Proceedings of the 1995 IEEE International Symposium and Workshop on Systems Engineering of Computer Based Systems*, (pp. 51–57).
3. Maier, M. W. (1998). Architecting principles for systems-of-systems. *Systems Engineering, 1*(4), 267–284.
4. Stevens Institute Of Technology, Castle Point On Hudson, Hoboken, NJ 07030: Report On System Of Systems Engineering. August 2006.
5. DeLaurentis, D. (2007). *Research foundations, School of Aeronautics and Astronautics*. West Lafayette: Purdue University.
6. Sheard, S. (2006). *Is systems engineering for "systems of systems" really any different? INCOSE insight* (Vol. 9, Issue 1).
7. de Weck, O. L., Ross, A. M., & Rhodes, D. H. (2012). Investigating relationships and semantic sets amongst system lifecycle properties (-ilities). In *Third International Engineering Systems Symposium CESUN 2012*, June 18–20, 2012, Delft University of Technology.
8. Clark, J. O. (2009). System of systems engineering and family of systems engineering from a standards, V-model, and Dual-V model perspective. In *Systems Conference, 2009 3rd Annual IEEE*, March 23–26, 2009 (pp. 381–387).
9. Mallek, S., Daclin, N., & Chapurlat, V. (2012). The application of interoperability requirement specification and verification to collaborative processes in industry. *Computers in Industry, 63*(7), 643–658.
10. Chapurlat, V. (2013). UPSL-SE: A model verification framework for systems engineering. *Computers in Industry, 64*(5), 581–597, ISSN 0166-3615.
11. Woodcock, J., Larsen, P. G., Bicarregui, J., & Fitzgerald, J. (2009). Formal methods: Practice and experience. *ACM Computing Surveys (CSUR), 41*(4), 19.
12. Clarke, E., & Schlingloff, H. (2000). *Model checking. Handbook of automated reasoning*. Amsterdam: Elsevier.
13. Camarinha-Matos, L. M., & Afsarmanesh, H. (1999). *The virtual enterprise concept*. Portugal: Kluwer Academic Publishers.
14. Saunders, T. et al. (2005, July). In United States Air Force Scientific Advisory Board Report on System of Systems Engineering for Air Force Capability Development SAB-TR-05-04, July 2005.
15. Drusinsky, D., Michael, J. B., & Shing, M. T. (2004). Behavioral modeling and run-time verification of system-of-systems architectural requirements. In *Proc. CCCT* (Vol. 4).
16. Khosla, R., Dillon, T. (1997). Intelligent hybrid multi-agent architecture for engineering complex systems. In *Proceedings of the 1997 IEEE international Conference on Neural Networks* (Vol. 4, pp. 2449–2454).
17. Bilal, M., Persson, C., Ramparany, F., Picard, G., & Boissier, O. (2012, June). Multi-Agent based governance model for Machine-to-Machine networks in a smart parking management system. In *Communications (ICC), 2012 IEEE International Conference on* (pp. 6468–6472). Ottawa, Canada: IEEE.
18. Jochen, F. (2005) *Types and forms of emergence*. Kassel: Kassel University Press.

Enterprise Interoperability Science Base Structure

Keith Popplewell

Abstract A science base for enterprise interoperability was first proposed in 2006, as a mechanism to formalize knowledge being generated by researchers and applied by industry to facilitate collaboration between enterprises through mutual interoperability of their enterprise systems. Subsequently the community of researchers and exploiters of Enterprise Interoperability research addressed this issue as a collaborative group, culminating in a project funded by the European Commission FP7 programme. In this paper we explore the structure for an Enterprise Interoperability Science Base defined in this project, based on analysis of its purposes, the knowledge already available from pragmatic research, and the lessons learned, both on interoperability and the theoretical structure of a science base. The resulting science base is now evolving from the body of knowledge used for its initial population to embrace new research results and issues.

Keywords Enterprise interoperability · Science base

1 Introduction

Having grown out of purely pragmatic requirements of industry to achieve enterprise systems interoperability in collaborative projects, the need was progressively recognised first in the research roadmap developed by the IDEAS project, funded by the European Commission in 2002 [1] and subsequent revisions of this. Version 4 [2] identified 4 grand challenges, including the need to establish a Science Base for Enterprise Interoperability: a structured and evolving repository

K. Popplewell (✉)
Faculty of Engineering and Computing, Coventry University, Priory Street, Coventry CV1 5FB, Coventry, UK
e-mail: k.popplewell@coventry.ac.uk

K. Mertins et al. (eds.), *Enterprise Interoperability VI*,
Proceedings of the I-ESA Conferences 7, DOI: 10.1007/978-3-319-04948-9_35,

of domain knowledge providing a source of application knowledge to industry, and a statement of current state of the art to researchers and industry alike.

The research domain acquired the title Enterprise Interoperability (EI), and as a result of the IDEAS roadmap a network of excellence, INTEROP-NoE was funded by the European Commission's Framework Programme 6, becoming one of a cluster of EU projects working in the domain, and entitled the Enterprise Interoperability Cluster (now re-named the Future Internet Enterprise Systems Cluster) [3]. The Enterprise Interoperability Science Base grand challenge, developed over several years by the EI Cluster, was then addressed by the ENSEMBLE [4] project, funded by the EU from 2010 to 2012. This paper outlines the resulting structure for an Enterprise Interoperability Science Base (EISB), building on previous presentations at I-ESA 2012 [5].

2 Defining the Structure of an Enterprise Interoperability Science Base

2.1 Methodology

Extensive review of other more or less neighbouring scientific domains revealed that there is no view of the definition of a science base common to all, or even a related set of, scientific domains, although good examples exist, including for example that for software engineering science [6]. We therefore submit that the definition of a science base is to a degree dependent on the nature of the domain and the purpose for which it is designed and maintained, and indeed the definition for a particular domain will evolve as the needs of the domain evolve with its maturity.

This implies, in turn, that the definition and therefore the structure of a science base must be related to the purposes of its creation and application. In the context of a pure science, this may perhaps be primarily concerned with maintenance of scientific rigor and consistency, demonstration of experimental repeatability, and documentation of domain knowledge to support re-use and reduce duplication. We might also expect such a science base to recognize the uncertainties and indeed conflicts present in current domain knowledge, and to identify the general lines of research needed to resolve such issues: this after all is the conventional scientific approach to knowledge discovery and validation.

However in the context of applied science domains, where the science is advanced so as to meet the needs of application outside the scientific community, the purposes of a science base are extended to enable efficient support of application. The requirements of a pure science base remain essential but the additional layer of application related knowledge is necessarily added. The process of defining a structure for such an applied science base therefore must begin from the perspectives of its purposes as both a repository of scientific knowledge, and as a reference for application of domain knowledge.

From these objectives we derived a high-level structure for the EISB, whose components can be related to technologies which may support their eventual implementation. The approach adopted was not to go directly to a final structure but to take a crowd sourcing approach. The initial structure was presented to the Enterprise Interoperability research and application communities, hereafter referred to as the Enterprise Interoperability constituency, to interested parties from neighbouring and other scientific domains, and to as wide an audience as chose to participate. The structure and supporting documentation [7] was made available through an online consultation tool enabling contributors to attach comments and especially contributions, which could be associated with content, with a granularity ranging from overall comment on the document to individual paragraph or illustration. This process, whose results are summarized below, had a major and positive impact on the specification of EISB components, as well as on defining the context in which it might be populated and applied.

The process of structure definition was completed through definition of the tools and technologies needed to support the components. The process of population is also described in below.

2.2 Defining Objectives

The overall objective in creation of an EISB is to formulate and structure the knowledge gained through pragmatic research in the domain over the last decades and more. This establishes that there is indeed a coherent and specific body of scientific knowledge and understanding attributable to that research, and renders it accessible to both future researchers aiming to build on existing knowledge, and to those wishing to use it in industrial applications. Without such an EISB there is danger of repeating research, and missing opportunities for application.

In particular, the EISB aims:

1. To document and catalogue domain knowledge. In this context knowledge may embrace factual knowledge, and methodologies for application.
2. To identify application areas for domain knowledge items. This will include a taxonomy of problems addressed by the Enterprise Interoperability domain, and the domain solutions to these problems.
3. To identify approaches for application, which may combine methodologies to achieve integrated solutions for complex problems. These should, if possible, characterise problems in sufficient detail to eliminate inappropriate methods and prioritise those which are applicable.
4. To identify domain related problems which are currently not resolved or addressed in the knowledge base, and which should be prioritised for research.
5. To identify related problems addressed in other sciences, directing attention to the appropriate knowledge in the addressing domain.

6. To support application of Enterprise Interoperability knowledge by clearly documenting the route from domain problems to domain solution approaches, and providing access to the solution methodologies. This may be linked to access to both the knowledge base content and to sources of expertise, consultancy or training to support application.
7. To identify, structure and document fundamental axioms and consequent theorems of interoperability, to form the foundation for establishment of Enterprise Interoperability as a new and self-standing science.

2.3 Knowledge Sources

A Science Base for EI comprises a new set of concepts, theories and principles derived from established and emerging sciences, with a view to long-term problem-solving, as opposed to short-term solution provision. In order to effectively capture the work already undertaken in the domain, the state of the art as reported in [7] can be classified in the following categories:

- Concepts and Positions.
- Methods, indicatively consisting of the following sub-categories:
  - Mathematical models with mathematical statements (especially equations) of facts, rules, principles, or other logical relations;
  - Econometric models specifying the relationship that is believed to hold between the various economic quantities pertaining a particular phenomenon under study. An econometric model can be derived from a deterministic economic model by allowing for uncertainty or from an economic model which itself is stochastic;
  - Optimization models applying mathematical and computer programming techniques to the construction of deterministic models;
  - Descriptive models defining systems and operations in order to gain insight into their functioning.
  - Simulation models;
  - Frameworks and conceptual models based on well-established formal methods;
  - Proof-of-concept;
  - Tools;
  - Experiments;
  - Case Studies;
  - Surveys (Empirical Data);
  - Standards.

Popplewell et al. [7] also identifies a set of 12 scientific areas related to Enterprise Interoperability. Extensive and deep "knowledge mining" performed in relation to the requirements of interoperability within these 12 scientific areas leads to the identification of a list of macro-issues that can be considered coincident with the sub-areas. Even a rapid analysis of the documents referenced there testifies to a large number of initiatives and studies in each sub-area with a heavy prevalence of "local" analyses, hypotheses, and experiments, while attempts at holistic views and syntheses are less popular. In no sub-area can we say that all the specific issues have been adequately satisfied by existing methods, tools, or standards, for two main reasons:

- *The accelerated pace of technology*, in terms of digitalized management of an ever increasing number of aspects of our every-day life, requires a continuous adjustment of standards and tools. The landscape is evolving.
- *The increasing "globalization of everything"* highlights ever newer requirements for transparent and effective interoperability, at all levels of the "research chain" from conceptualization to the provision of operating tools. The landscape is expanding.

3 A First Proposal for an Enterprise Interoperability Science Base Structure

Initial work based on the objectives, requirements and pre-existing knowledge discussed above led to the proposal of the initial structure for an EISB, destined to be the basis of further consultation within the Enterprise Interoperability constituency.

We considered that the content of a base for a science like Enterprise Interoperability may therefore consist of the following categories of knowledge:

- Formalization of the Problem space: A taxonomy of the spectrum of application and theoretical problems addressed by the domain, organized so as to be used to characterize real applications and to link these to elements of the solution space.
- Formalization of the Solution space: The converse of the problem space, as it provides a taxonomy of knowledge available for the solution of domain application problems. In turn this links to methodologies and tools in the domain knowledge base.
- Enterprise Interoperability Knowledge Base: The domain knowledge base contains both structuring and methodological knowledge. The former defines the structure of the domain as perceived by the Enterprise Interoperability constituency:

  - A taxonomy of topics within the domain knowledge;
  - The scientific principles which provide the foundation of knowledge in the domain, and of both future research and application;
  - Relationships between these topics, the problem space and the solution space;
  - Relationships between domain knowledge and knowledge embedded in related scientific domains.

The main components of this structure for Enterprise Interoperability are discussed below. In the context of Enterprise Interoperability, structural knowledge is likely to consist of appropriate taxonomies and frameworks supporting understanding of the overall content and the relationships between Problem and Solution Spaces.

Methodological knowledge maintains understanding of how problems (both in research and application) may be addressed in the domain, and is based on both formal frameworks and processes, and on experience of domain stakeholders both individually and as a constituency. Typically this might contain, in this context, formal models, solution algorithms, simulation tools and assessment tools. However structural knowledge must also provide understanding of how to combine methodologies to solve more complex problems, or inter-related problems.

In order to support the enterprise application of domain knowledge, the knowledge base should also embrace the knowledge of how application will impact on the enterprise. This, in the case of Enterprise Interoperability, includes for example value scenarios and business models surrounding applications, based on both assessment and analysis (using the above mentioned methodologies and tools), and on experience of real implementations.

Enterprise Interoperability is universally acknowledged to be an interdisciplinary domain, and as such interacts closely with other domains, as described in relation to neighbouring scientific domains below. Understanding of the relationships with other domains, the contributions to be drawn from their respective science bases, and the necessary references to content, expertise and training, is therefore also a significant part of the Enterprise Interoperability Knowledge Base.

Figure 1 shows the summary of proposed content of the Enterprise Interoperability Science Base. The main components described above are shown in the central area, whilst inputs to the Problem Space arise from application requirements. Knowledge and experience derive from application, which provides knowledge in all three components. Another major contribution to the Knowledge Base comes from the research community. The objectives of the Science Base are summarized on the right of the figure as outputs derived from the Science Base, thus providing the rationale for its development and maintenance.

4 The Enterprise Interoperability Science Base Knowledge Base

4.1 Explicit Knowledge

For the EISB formulation, it is proposed that explicit knowledge be handled by a reference ontology. This establishes a common language for sharing and reusing knowledge about phenomena in a particular domain, i.e. Enterprise Interoperability in this case.

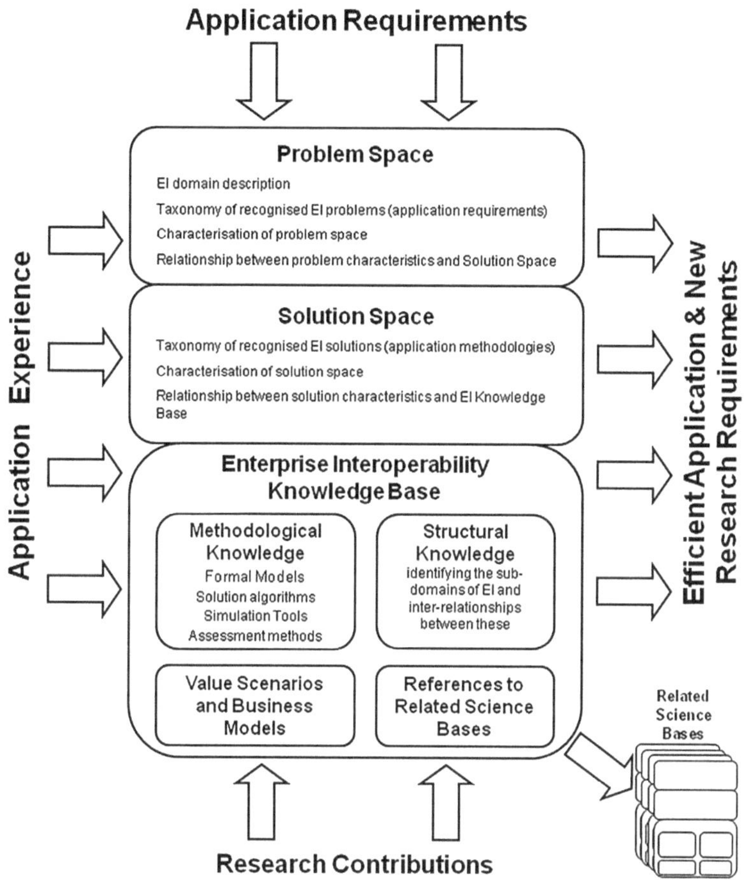

Fig. 1 Initial view of enterprise science base structure and content

The reference ontology enables terminology sharing, and cross-referencing interoperability terms, problems, solutions and expertise. Since normally ontology building is a long process, and involves gathering human knowledge from many experts, a four-step methodology was proposed, as described in Fig. 2.

4.2 Enterprise Interoperability Structural Knowledge

The European Interoperability Framework (EIF) adopted four layers of interoperability, (Technical, Semantic, Organizational, and Legal & Political Context Interoperability), and in this context, in order to identify a proper structure for Enterprise Interoperability, which can at a second stage be mapped to these four layers it is necessary to focus on the real object of observation, which is the

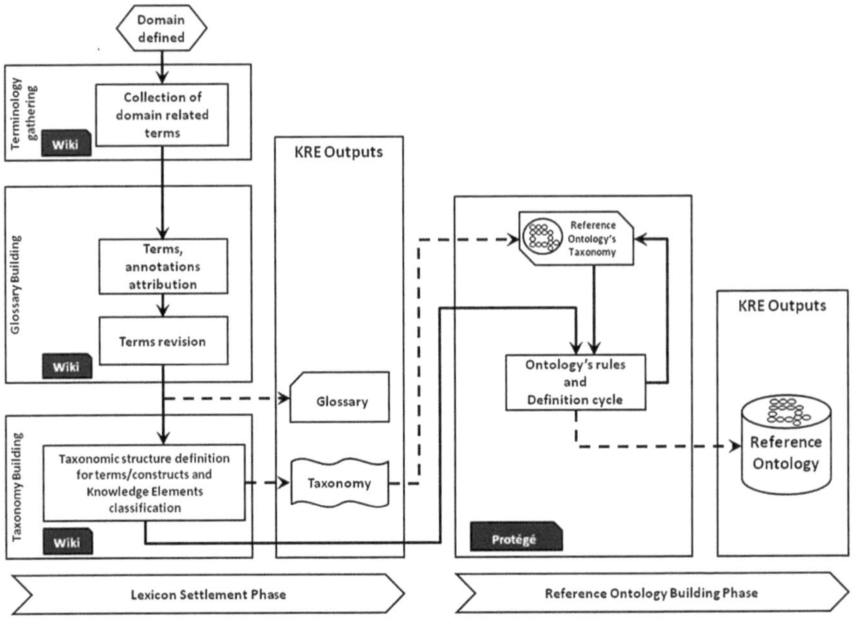

Fig. 2 Methodology for EISB ontology building

"Enterprise", and analyzing its core components to identify the implied interoperability needs.

An Enterprise, as defined in [8], is "…an organization designed to provide goods, services, or both to consumers." The main ingredients of such a system are the following:

- Infrastructures referring to all the facilities and non-human assets possessed by an enterprise, which are used for their operation. Under infrastructures, software platforms, hardware systems, building facilities, automobiles, etc. can be classified.
- Data used for the business transactions within and outside the boundaries of the enterprise. This includes the documents, application forms, transactional data exchanged by the enterprise.
- Processes including all the related, structured activities or tasks that produce a specific service or product.
- Policies embracing the different rules that are applied either due to external (e.g. legislation, business association rules, etc.) or internal (e.g. working hours, dress code, etc.) factors.
- People with all the human resources that are part of an enterprise system.

Starting from these core ingredients of an Enterprise, and by analyzing the current technological trends and the background knowledge of the domain of

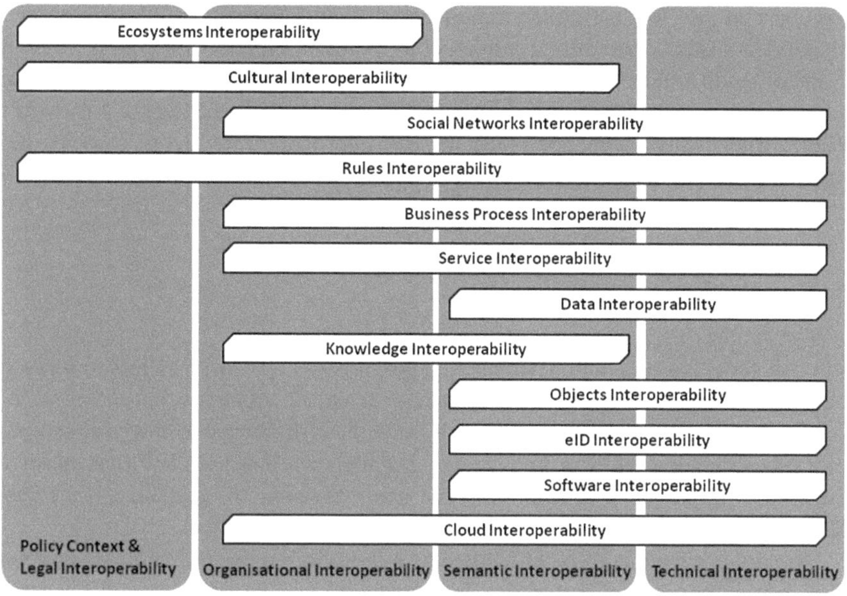

Fig. 3 EI scientific areas mapping to interoperability layers

Enterprise Interoperability, the first Enterprise Interoperability Scientific Areas (SA) are formulated, these being the fundamental areas and constituting the 1st granularity level of EI, as following:

Data Interoperability.
Process Interoperability.
Rules Interoperability.
Objects Interoperability.
Software Interoperability.
Cultural Interoperability.

These scientific areas, whilst being the core components and as such the most important areas of Enterprise Interoperability, are however incapable of solving all interoperability related problems, as enterprises are constantly becoming more complex, with disappearing boundaries, loosely coupled architectures and virtual resources. Those changes are very well reflected in the Qualities of Being (QoB) for future enterprises that are pursued by the Grand Objectives of the FInES Research Roadmap [9], namely Inventive Enterprise, Cloud Enterprise, Cognizant Enterprise, Community-oriented Enterprise, Green Enterprise, Glocal Enterprise. Enterprises are restructuring themselves and try to align with new trends they regard as important for their viability, such as the Future Internet. In this respect, the issue of interoperability becomes even more complex, as not only are new technologies such as social networks or e-ID ae constantly being taken up by enterprises (having an impact on their overall operation), but also as there is a need for constant and flexible collaboration between all enterprise systems in order to respond in a timely and effective manner to the requirements of the global market.

In order to reflect this deeper complexity the first granularity level scientific areas were expanded through progressive levels, to include the twelve presented in Fig. 3, where they are mapped onto the EIF interoperability layers, showing how they span the layers.

5 Conclusion

We conclude that it is both desirable and possible to construct an EISB to meet the objectives discussed above. Whilst many scientific domains have followed a similar path from pragmatic solution of immediate problems, to the establishment of a recognizable corpus of knowledge and interests, it is not clear that in any of these cases a deliberate attempt has been made to define the domain science base. Rather this has become accepted through custom and practice.

The drivers for this to be attempted as a specific intellectual and philosophical exercise in the case of Enterprise Interoperability arose from several pressures, discussed above, and have resulted in the definition of an accepted (within the domain) structure for a science base. This opportunity to take advantage of the crowd-sourcing and consultation capabilities available to the Enterprise Interoperability constituency through recent developments in internet technology has been exploited, and the need of research funding bodies, as well as industry, to be able to access and assess available domain knowledge has provided some of the motivation.

The result is an EISB structure together with definition of the tools needed to populate and drive its evolution. The sustainability of the science base is therefore technically achievable, but it remains to be seen whether policy and funding pressures support the Enterprise Interoperability constituency in making full use of the opportunity so presented.

Acknowledgments This work has been partly funded by the European Commission through the Project ENSEMBLE Envisioning, Supporting and Promoting Future Internet Enterprise Systems Research through Scientific Collaboration (Grant Agreement No. 257548). The authors wish to acknowledge the Commission for their support. We also wish to acknowledge our gratitude and appreciation to all the ENSEMBLE Project partners for their contribution during the development of various ideas and concepts presented in this paper.

References

1. Doumeingts G., & Chen, D. (2013). Interoperability development for enterprise applications and software. In P. Cunningham, M. Cunningham, & P. Fatelnig (Eds.), *Building the knowledge economy: Issues, applications, case studies*. Amsterdam: IOS Press.

2. Li M. -S., Cabral, R., Doumeingts, G., & Popplewell, K. (2006). *Enterprise interoperability: A concerted research roadmap for shaping business networking in the knowledge-based economy*. Brussels: Commission for the European Communities.
3. FInES Cluster, from FInES Cluster. Retrieved March 27, 2013, from http://www.fines-cluster.eu
4. ENSEMBLE, from FInES Cluster. Retrieved March 27, 2013 from http://www.fines-cluster.eu/fines/jm/FInES-Private-Information/ensemble.html
5. Popplewell, K. (2012). Towards the definition of a Science Base for Enterprise Interoperability: A progress report. In M. Zelm, R. Sanchis, R. Poler, & G. Doumeingts (Eds.), *Enterprise interoperability*. Hoboken: Wiley.
6. Shaw, M., & Clements, P. (2006). The golden age of software architecture. *IEEE Software*, *23*, 31–39.
7. Popplewell, K., Lampathaki, F., Koussiris, S., Mouzakatis, S., Charalabidis, Y., Goncalves, R., & Agostino, C. (2012). EISB state of play. European Commission
8. Sullivan, A., & Sheffrin, S. (2003). *Economics: Principles in action*. Upper Saddle River: Pearson Prentice Hall.
9. Missikoff, M., Charabilidis, Y., Goncalves, R., Popplewell, K., & FInES Research Roadmap 2025-v3.0, from FInES Portal. Retrieved March 22, 2013, from http://www.fines-cluster.eu/fines/jm/Publications/Download-document/323-FInES-Research-Roadmap-2025-v3.0.html

Part X
Standards for Interoperability

A Contribution to Software Interoperability Verification of Manufacturing Application

Henri Basson, Michiko Matsuda and Dukki Chung

Abstract With an increasing number of distributed heterogeneous manufacturing systems, interoperability is becoming more and more critical for the reliability, performance, and security of working applications. In order to control the development and evolution of applications interoperability, crucial decisions should be made at the earliest stages of development. These decisions need to be incorporated in the detailed design of the components that make up the application in order to accomplish a reliable and efficient solution at the implementation level. For a complete solution the application architecture must be designed using a methodology that can be verified. The components used in the implementation must be verified against the evolving application requirements. These components need to be checked for coherency of the design as the application evolves to meet new application requirements. The two checking types, in conjunction with an integrated approach, constitute the main goal of interoperability verification. This paper presents elements of an integrated modeling of interoperability verification. The implemented model is designed to develop automatic interactive tools which facilitate the tasks of interoperability verification.

Keywords Interoperability design · ADL · UML diagrams · Software artifacts modeling · Software interoperability verification

---

H. Basson (✉)
LISIC, EILCO, University of Lille North France, (Littoral),
56 Rue Ferdinand Buisson, Calais, France
e-mail: basson@lisic.univ-littoral.fr

M. Matsuda
Kanagawa Institute of Technology, 1030, Shimo-Ogino,
Atsugi-shi 243-0292 Kanagawa, Japan
e-mail: matsuda@ic.kanagawa-it.ac.jp

D. Chung
Rockwell Automation, Cleveland, 1 Allen-Bradley Drive Mayfield Hts,
Cleveland, Ohio 44124, USA
e-mail: dchung@ra.rockwell.com

K. Mertins et al. (eds.), *Enterprise Interoperability VI*,
Proceedings of the I-ESA Conferences 7, DOI: 10.1007/978-3-319-04948-9_36,

1 Introduction

With a continuously increasing number of distributed and heterogeneous systems being developed, their interoperability is becoming more and more critical for the reliability, performance, and security of working applications. In order to assure a satisfactory level of quality and to control the evolution of applications interoperability, major decisions should be made in the earlier stages of the systems development cycle, and more specifically, in the design phase.

In order to match the evolution of functional and quality [5] requirements, manufacturing applications are often subject to changes aiming at meeting updated system requirements. However, any application evolution requires changes which may impact the interoperability of components or systems which already exist or are in development. In an evolution context, an interoperability verification of changes requires:

- the identification of interoperability elements and components which are affected by the applied change,
- the analysis of the impacts generated on affected components and their interoperability,
- an estimate of the cost of validating versus not validating the applied change.

Considering the increasing size and complexity of heterogeneous manufacturing applications, in this paper a contribution towards an integrated model is proposed to develop tools to better achievement of interoperability verification. It is assumed that the development of software units for the manufacturing application under consideration is supported by an architectural description at the highest design level and adequat UML diagram for the detailed design level.

An important question at this level is how to best to specify the application interoperability requirements and design. How can the implementation realizing the design be checked, the goal being to verify whether the adopted interoperability solutions were implemented according to the adopted specification and design [3, 4].

For architectural description of application artifacts, several languages have been proposed [1] and compared to specify, at the top level, application components, interconnections, configuration, and behavioral constraints. On other side, depending on the languages used, interoperability mechanims such as service invocation, data exchange, data sharing, and message communications between components are more or less explicitly described. Our goal of interoperability verification requires a language providing an explicit enough description and enough detail to perform interoperability verification.

In our work we propose to use an ADL (Architecture Description Language) to emphasize the interdependencies among software components. To identify the parts of the application requiring the implementation of interoperability mechanisms at the level of each individual component, a UML activity diagram is used to designate the part of code requiring the use of interoperability mechanisms.

This paper is structured as following: In Sect. 2 is presented the context of our contribution where Sect. 3 defines what is meant by interoperability verification to identify the needed artifacts permitting to realize it. A short presentation of ADL concepts is given before presenting our modeling elments in Sect. 4. A structural description of prototype intended to validate the proposed contribution is given in Sect. 5. The last section mention the conclusions and perspectives of our work.

2 Present Contribution Context

In the context of ISO standards development addressing industrial automation systems and integration, ISO 15745 has been developed to provide a framework to define the elements and rules that describe the integration models and application interoperability profiles for the interoperability of manufacturing applications [6]. ISO 16100 specifies a framework for the interoperability of a set of software products used in the manufacturing domain and to facilitate integration into manufacturing applications. As a part of complementary extension, the series 16300 is devoted to specifying a framework for verifying and validating the interoperability of MSUs (Manufacturing Software Units) having a set of capabilities that meet the functional requirements of a target manufacturing application solution [7]. Considering the corresponding volume of work, this paper focuses on only the interoperability verification addressed by ISO 16300-3 [8]. In this paper, a contribution in terms of modeling features and corresponding prototype are proposed to investigate the feasibility of guidelines provided by 16300-3 series.

3 Interoperability Verification

In the context of the development and revision of manufacturing software systems, interoperability verification aims at checking the degree of matching between the software interoperability requirements specification as described at the design level and its coding at the implementation level. The verification requires a model which provides detailed knowledge of the designed components, interoperability, and corresponding implementation in the developed application.

Nevertheless, in the context of the evolution of manufacturing software systems, interoperability verification aims at checking whether the chosen refinement to be applied to the current state of the software system has been implemented according to the updated application design specification. This is the case of what can be called top-down development. In the other case of reverse or bottom-up development, the validation concerns the coherence between the updated implementation and the revised application specification updated according to the implemented interoperability design (Fig. 1).

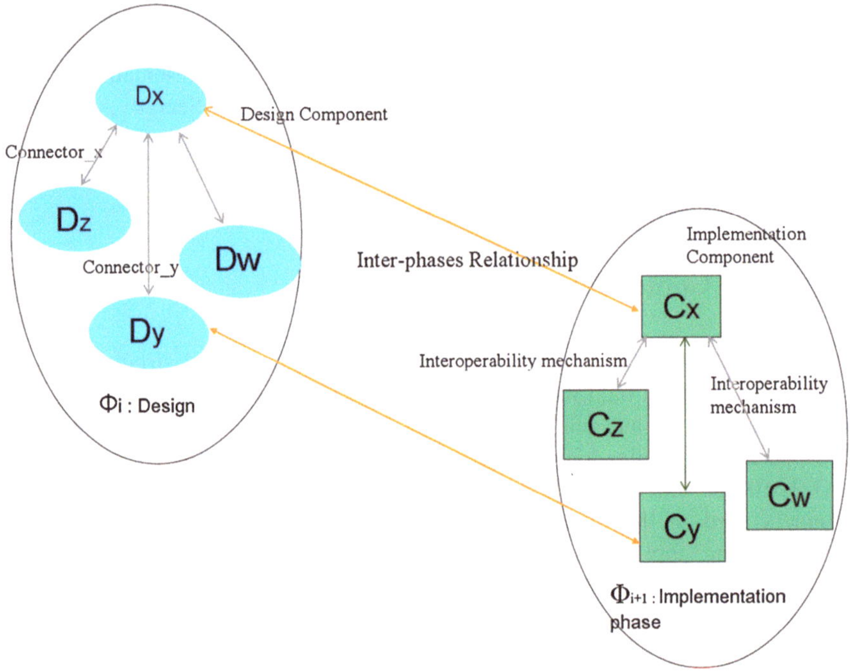

Fig. 1 The verification process target in the context of application evolution

In this context the term "evolution" is defined to encompass the integration of new components, or their revision, replacement, deletion, or update.

3.1 Required Description for Interoperability Verification

As already mentioned, in developed or evolving applications, not only a detailed design specification is required, but also a detailed implementation description must be available. It is necessary to elaborate a model integrating descriptive knowledge of the design level artifacts, the implementation level components and the mapping between the two levels.

3.1.1 Architecture Description Languages

With the increasing complexity and heterogeneity of manufacturing software applications, system architecture description languages have become necessary for better understanding of the frequently changing systems and the control of their evolution. Software architecture is defined [1] as "a high level program model that describes a system's major pieces (its components) and how they interact (its connectors)". Between manufacturing software units, the interactions can be realized using a wide range of interoperability mechanisms providing capabilities for information exchange, service invocation, communication messaging, and other service exchange between software units.

Several approaches have been proposed aiming for architectural specification and analysis via several candidate Architectural Description Languages (ADLs) [1, 2]. To build a generic representation of an architecture description, independently of any particular ADLs, we propose to define common concepts shared by the major ADLs in order to integrate any description.

3.1.2 Interoperability Specification Inside ADL Diagram

The interoperability types, explicitly represented by an ADL diagram are one or more of the following:

- a required access to a file shared by two or more software manufacturing units;
- a required access to a database shared by two or more software manufacturing units;
- a message sent to one or more software manufacturing units;
- a message reception by a specific software manufacturing unit. Message transmission and reception are performed using a communication protocol provided by the manufacturing software environment.
- a direct invocation of function or service provided by the other software manufacturing unit;
- an indirect call of a function or service provided by another software manufacturing unit. This last type of call needs the availability of an appropriate middleware. One or more middlewares such as OMG CORBA, RMI JAVA, Microsoft DOM, or OPC-UA shall be used according to the programming languages and operating systems of the hosting environment of the manufacturing software units. Above an example of an ADL architectural description aiming at specifying various types of interconnection between system components [Fig. 2].

4 ADL and Knowledge Modelling for Interoperability Verification

The proposed model ASAM (Architecture Software Artifact Model) allows the representation of the Manufacturing Software System Architecture using an ADL where are used the major concepts shared by the main ADLs [3]. These concepts include the component concept [1] designating a software unit composed of interface and body, the components interoperate throughout connectors to be deployed in respect to an adopted configuration.

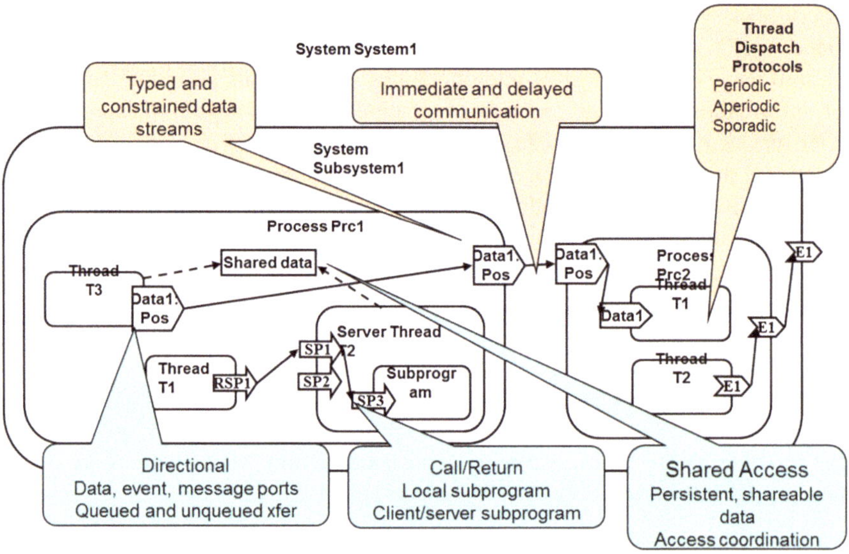

Fig. 2 An illustration example of an ADL description [2]

4.1 Common ADL Concepts

At the highest architectural level, four major basic concepts (called the 4C) are concerned and designated by: *Component* which is represented by an interface and an implementation, the *Connector*, *Configuration*, and the *Constraints*.

By considering a wide range of proposed ADLs [1, 2], the *component*, *connector* and *configuration* are refined into detailed elements in order to provide a more precise model. Subsequently, in the proposed model various concepts shared by the considered ADLs are represented (*writing hereafter in italic style*). the *component interface* which includes its provided services that can be invoked by other system components. An *interface* may define *ports* which are destined to be used for communication between *components.* A *port* can be used by one several *connectors*. In all ADLs, an interface is mandatory to describe a distributed *component* in the system architecture. The *implementation* designates the *source codes* realizing the services declared the *component interface* by providing the interfaces and the implementations related to the various ADLs the *connection* provides a link based on some interoperability mechanisms to exchange information between distributed components. The link is usually activated using a *middleware* that allows a component deployed on a subsystem to access programs and data located on other subsystem. A middleware provides a set of services that allows interactions between multiple components running on different sites. The *properties* are destined to specify constraints on the *interface* or the *implementation* of a *component*. These constraints can be functional or qualitative related to system performance, reliability, or security.

4.2 Integrating ASAM Model Notations

To develop a general model to describe, among other aspects, the software units' interoperability, a general model is essential to apply expected reasoning for interoperability validation, the modeling here is limited here for ADL design and Coding, the knowledge about interoperability activations (message sending, message reception, procedure call) inside each software component will be integrated in the same manner as presented for ADL artifacts. In the proposed integrated model:

$\sum\Phi$ designates the set $\{\Phi i : i = 1 \ldots n\}$ of all the development phases of considered manufacturing software application. For the specific needs of the present work, two phases are only considered within $\sum\Phi$:

1. the design phase denoted by Φdsn,
2. the implementation denoted by Φimp.

In each phase a set of artifacts is developed, denoted by:

- $\Phi dsn.\sum A$ as being the whole set of different parts constituting an ADL diagram,
- $\Phi imp.\sum A$ as being the whole set of different parts constituting the code implementing the ADL diagram.

For instance, in a considered ADL diagram, an artifact part Ap_x may designate an ADL component, an ADL connector, or an ADL interface, port, or any other defined part of the considered ADL diagram. An $\Phi_{imp}.\sum A$ element may be an object, a thread or a process, implementing a design constituent or an interoperability mechanism of a connector.

In order to verify the interoperability, we define two necessary mapping relationships between design and implementation artifacts:

- The relationship *describes*(Ap_x, Ap_y) representing the fact that an artifact part $Ap_x \in \Phi_{dsn}.\sum A$ of the design describes an implementation part $Ap_y \in \Phi_{imp}.\sum A$.
- the reverse relationship *implements*(Ap_y, Ap_x) means that coding part $Ap_y \in \Phi_{imp}.\sum A$, is the implementation of design part $Ap_x \in \Phi_{dsn}.\sum A$.

In the context manufacturing application development, the interoperability verification apply the general rule of verification whether each interoperability element described by ADL artifact is implemented in the coding phase.

For the two phases: Φdsn and Φimp designating the Design and implementation phases) the verification rule to be applied is such as:

for all $Ap_x \in \Phi dsn.\sum A$

if $(Ap_y \in \Phi imp.\sum A)$ *and* (*describes*(Ap_x, Ap_y))

then *Check implements*(Ap_y, Ap_x)

endif

endfor

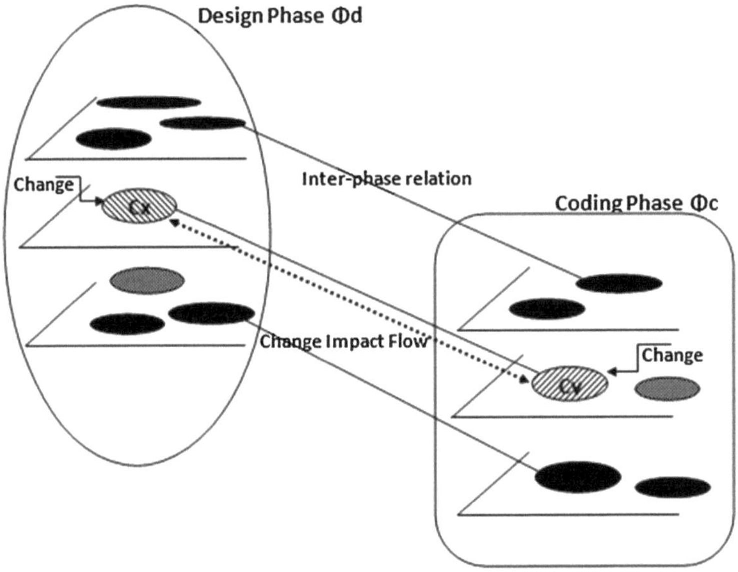

Fig. 3 A change context for interoperability verification

In the context application of manufacturing application change, partially illustrated above [Fig. 3]. As example illustrating the reasoning, which may take place through the interoperability verification process, is the application of the following rule:

if $(Phase(\Phi_{dsn}) \wedge Component(Ap_x) \wedge belongs(Ap_x, \Phi_{dsn}.\sum A) \wedge$
$Phase(\Phi_{imp}) \wedge Component(Ap_y) \wedge belongs(Ap_y, \Phi_{imp}.\sum A) \wedge$
$implements(Ap_x, Ap_y))$ **then**
if $modified(Ap_x, Ap_{x+}) \Rightarrow (modified(Ap_y, Ap_{y+}) \wedge implements\ (Ap_{x+}, Ap_{y+})$
endif

5 Tool Design for Interoperability Verification

In order to develop an interoperability framework, the design of manufacturing activities will be a main reference to be acquired and updated. The system activities are designed using ADL diagrams and other complementary UML diagrams.

Our experimental prototype of a knowledge based framework for interoperability verification is composed of five major software units (Fig. 4):

The experimental prototype is composed of five major software units:

1. A multi-languages parser permitting to analyze the source codes and the architectural description of the current application. The analysis is based on the

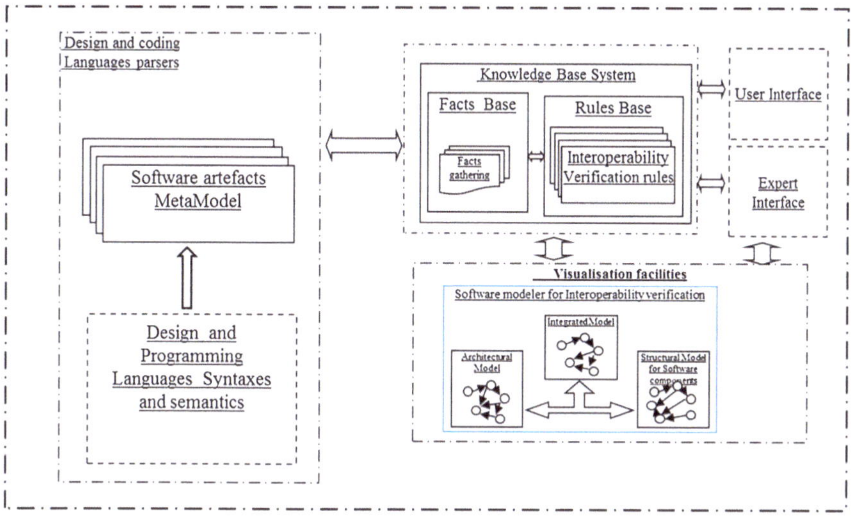

Fig. 4 Tool prototype for Interoperability verification

grammar of each language used in the design and code descriptions. To each application a set of resources is associated, it includes source code files, libraries, architectural and other design descriptions. These resources are analyzed to be represented by the adopted model instantiated into graphs managed by a unit called "Software Modeler".
2. "Software modeler" contains the representation of different graphs representing the facts describing the concerned manufacturing applications instantiated in reference of the proposed modeling. ADL description and corresponding implementing code are instantiated through two graphs where nodes represent the components and edges their various relationships between components. The goal being to check the mapping between the elements of artifacts, representing the adopted interoperability mechanisms at the design level, and their corresponding source code at the implementation level.
3. The Facts Base represents the whole set of facts (tuples) composing different tables of various repertories. It makes available detailed information about any unit or element belonging to manufacturing application structure.
4. The system inference engine contains of interoperability verification rules. Using current evolution events and the Facts Base, It permits according to accomplish a reasoning required by the interoperability verification process. For instance, an intended application change often requires an impact analysis before implementing corresponding updates. Using the adopted Inference engine, the impact analysis is accomplished through the activation (firing) of concerned verification rules.
5. The user interface provides the applications graphs visualization and displays the mapping links between design artifacts and the corresponding software units. It allows to simulate a change operation indicating the concerned

components and associated impact propagation paths. The user interface provides an access restricted to the interoperability verification experts to update the verification Rules Base.

6 Conclusion And Perspectives

The interoperability of distributed heterogeneous software units is becoming increasingly critical for the reliability and performance of manufacturing systems.

In the context of ISO standards development addressed to industrial automation systems and integration, ISO 16300 aims at interoperability verification and validation. We have presented in this paper the basic elements of integrated modelling which will be used to develop tools for performing various interoperability verification tasks. The proposed methodology uses architectural description of capability units of the considered manufacturing application. The proposed modeling approach is independent of any ADL chosen for the system architecture. An attribute graph was proposed to capture an adequate set of knowledge on software unit architectures and their corresponding interoperability. The attributes are chosen to provide knowledge to activate adopted rules of evolution which consitutes a part of a reasoning framework for interoperability verification. We have started the development of a first prototype to validate our proposed distributed unit modeling for interoperability understanding and verification. The current work will continue for protoyping extension to deal with wide range of manufactuting application types and to integrate the process of validation activity as an important part of our work of the ISO 16300-3 development. Other work perspective concerns the link between Interoperability verification and different types of software evolution [9] as well as adapting the proposed modeling for interoperability verification of internet-based applications [10].

References

1. Medvidovic, N., & Taylor, R. N. (2000). A classification and comparison framework for software architecture description languages. *IEEE Transactions on Software Engineering*, 26 (1), 70–93.
2. The Emerging SAE AADL Standard: Tutorial of Real Time Systems Group (RTG), University of Pennsylvania, DASC AADL Tutorial, Part I (p. 33). Pennsylvania, USA.
3. Hassan, M. O., Deruelle, L., & Basson, H. (2009). Towards a change propagation process in software architecture. In *18th International Conference on Software Engineering and Data Engineering* (SEDE-2009, pp. 85–90). Las Vegas,Nevada, USA.
4. Hassan, M. O., Deruelle, L., Basson, H., & Ahmad, A. (2010). A change propagation process for distributed software architecture, ENASE 2010. *In Proceedings of 5th International Conference on Evaluation of Novel Approaches to Software Engineering.* Athens, Greece.

5. Melab, N., Basson, H., Deruelle, L., & Bouneffa, M. (1999). Performance of object-oriented code: Profiling and instrumentation. *Proceedings of the IEEE International Conference on Software Maintenance* (pp. 982–911), Oxford, UK.
6. ISO 15745-1 (2003). *Industrial automation systems and integration—open systems application integration framework—Part I: Generic reference description*. ISO, Geneva, Switzerland.
7. Matsuda, M. (2012). *Manufacturing software interoperability services which ISO 16100 brings about* (pp. 60–70, Vol. 0122). Lecture notes in business information processing series,New York: IFIP, Springer.
8. ISO/TC 184/SC 5/WG 4 N464, PWI 16300-3 (2013). *Interoperability of capability units for manufacturing application solutions—Part 3: Verification and validation of interoperability among capability units.*
9. Wilde, N. et al. (2013). Interoperable Systems and Software Evolution: Issues and Approaches. In *Digital enterprise design and management* (pp. 45–56). New York: Springer Berlin Heidelberg.
10. Bandyopadhyay, S., Sengupta, M., Maiti, S., & Dutta, S. (2011). Role of middleware for internet of things: A study. *International Journal of Computer Science and Engineering Survey (IJCSES)*, *2*(3).

Interoperability Frameworks for Health Systems: Survey and Comparison

Wided Guédria, Elyes Lamine and Hervé Pingaud

Abstract Several research work and initiatives have been proposed in the literature to identify the dimensions of interoperability and to define a framework that provides organizing mechanism and knowledge of this field in a structured way. However, the lack of a common understanding and a consensus on these dimensions is one of the biggest barriers to true interoperability. In this paper, we aim to identify the required dimensions that have to be taken into account to facilitate interoperability between health systems by (i) a survey of the main research works and initiatives dealing with interoperability in the health domain and (ii) the investigation of a comparative analysis of the interoperability reference frameworks based on specified criteria.

Keywords Health system · e-health · Interoperability · Framework · Dimensions

W. Guédria
CRP Henri Tudor, 29 Avenue John F. Kennedy, 1855 Kirchberg, Luxembourg
e-mail: wided.guedria@tudor.lu

E. Lamine (✉) · H. Pingaud
ISIS Department, University of Toulouse, CUFR J.F. Champollion, Avenue Georges Pompidou, 81104 Castres, France
e-mail: Elyes.lamine@mines-albi.fr

H. Pingaud
e-mail: herve.pingaud@univ-jfc.fr

E. Lamine · H. Pingaud
University of Toulouse—Mines d'Albi, CGI Campus Jarlard, Route de Teillet, 81013 Albi Cedex 09, France

K. Mertins et al. (eds.), *Enterprise Interoperability VI*,
Proceedings of the I-ESA Conferences 7, DOI: 10.1007/978-3-319-04948-9_37,

1 Introduction

eHealth interoperability is a fundamental prerequisite to further improve individual health care and well-being and ensure high quality and safe services.

The continuous improvement of the patients' health care, in terms of accessibility, coordination and continuity of care, requires better interchange and management of information flow; and cooperation between stakeholders involved in the health care management process of the patient. This requires having interoperable socio-technical systems.

Although this requirement for interoperability is recognized as a cornerstone for improving the quality of healthcare and the efficiency of the overall health system organization, it is however very difficult to achieve.

Indeed, developing an interoperable system or conduct system interoperation should be considered with multiple perspectives, various dimensions and with different types of approaches.

Several research work and initiatives have been proposed in the literature to identify the dimensions of interoperability and to define a framework that provides organizing mechanism and knowledge of the domain in a structured way (e.g. Nehta [1], HIS-IF [2], eHealth EIF [3], PHS [4], etc.). However, the lack of a common understanding and a consensus on these dimensions remains a challenge.

In this paper, we survey and compare the main research works and initiatives dealing with interoperability in the health domain in order to identify the limitations and contribute to fostering the interoperability of health systems.

The paper is structured as follows. In Sect. 2, we survey the main interoperability frameworks in the health domain. In Sect. 3 we compare the presented frameworks and highlight, respectively, their relevance and coverage to the interoperability domain. Finally, we conclude in Sect. 4 and suggest future research.

2 Survey of Main Interoperability Frameworks

In this section, we display the main interoperability models and frameworks relevant to the health domain.

2.1 *eHealth Europeen Interoperability Framework (eHealth EIF)*

The eHealth EIF framework [3] was developed in the context of a research program funded by the European commission for the interoperability development. It is an application of the generic Europeen Interoperability Framework (EIF) [5] to the domain of eHealth. eHealth EIF aims at providing a set of recommendations and specifications to connect eHealth systems. It identifies four levels of interoperability: legal, organizational, semantic, and technical.

- Legal interoperability aims to "align legislation so that exchanged data is accorded proper legal weight".
- Organizational interoperability aims to "coordinate processes in which different organizations achieve a previously agreed and mutual beneficial goal".
- Semantic interoperability aims to precise "meaning of exchanged information which is preserved and understood by all parties."
- Technical interoperability aims to "discuss technical issues involved in linking computer systems and services".

An overview of these different EIF concepts is given in Fig. 1.

For each interoperability level, the organizations involved should formalize cooperation arrangements in interoperability agreements.

Interoperability governance "covers" the ownership, definition, development, maintenance, monitoring, promoting and implementing of interoperability frameworks in the context of multiple organizations working together to provide (public) services (see Fig. 2).

Six principles were defined based on the generic EIF: security and privacy, transparency, preservation of information, reusability, technological neutrality and adaptability, and openness. Two additional principles have been added to this list: patient centricity and an approach based on use cases. The eHealth EIF proposes a list of ten high-level use cases to the eHealth Network. For more details, please see [3].

2.2 *The E-health Interoperability Framework*

The E-health interoperability framework [1] was developed by the National E-Health Transition Authority (NEHTA) initiatives in Australia. It defines three levels of interoperability across health organizations (see Fig. 3):

- Organizational layer which provide a shared policy and process framework across the E-Health interoperability agenda covering each NEHTA initiative. It includes the Business Processes, standards plan, security policies and Privacy.
- Information layer which provide shared building blocks for semantic (information) interchange including Foundation Components, Value Domains, Structures, common Assemblies, Relationships and Metadata.
- Technical layer is concerned with the connectivity of systems for information exchange and service use. Solutions are based on open standards providing a level playing field for competitive provision of technical solutions.

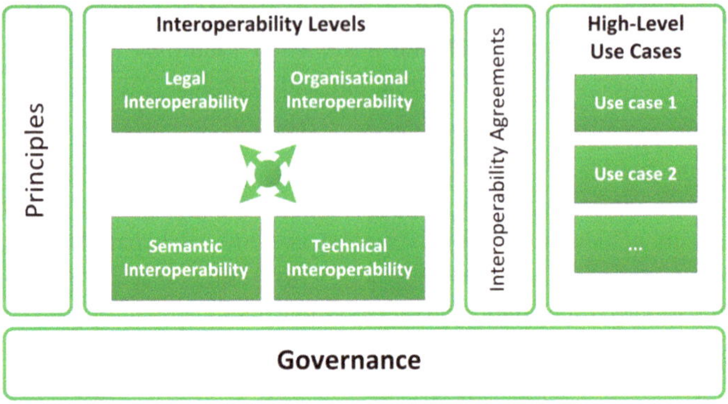

Fig. 1 Structure of the eHealth EIF [3]

Fig. 2 E-health interoperability framework [1]

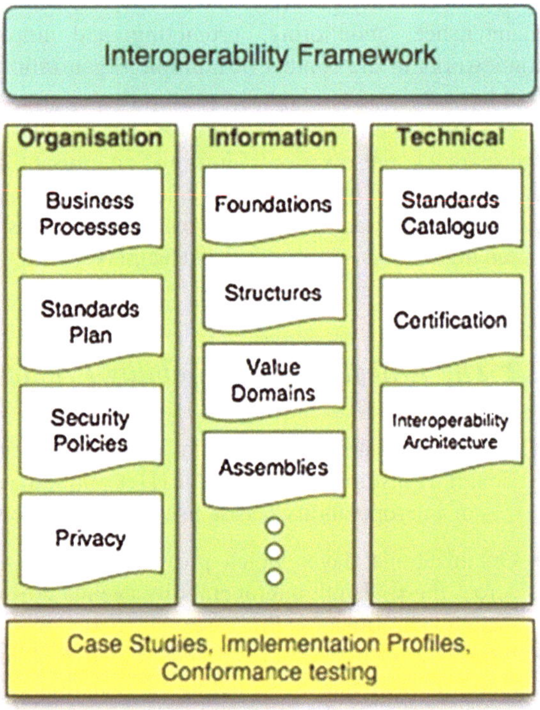

2.3 *Health Information Systems Interoperability Framework*

The Health Information System (HIS) Interoperability Framework is a reference framework created by ASIP Santé (Agence nationale des Systèmes d'Information Partagés de Santé) [6] for the purpose of:

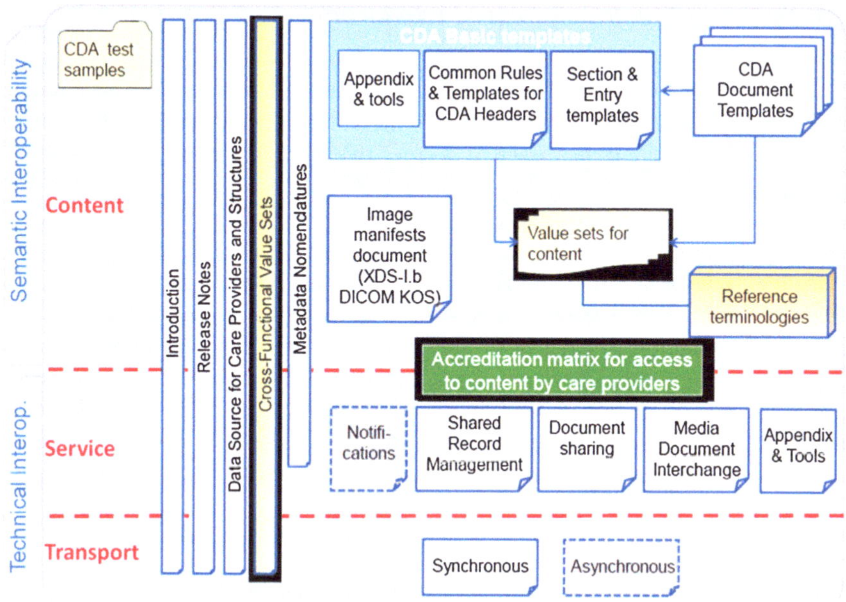

Fig. 3 Organization of the HIS interoperability framework [2]

- Encouraging the development of services for the electronic sharing of personal health information.
- Creating interoperability conditions between HIS systems that meet privacy and security requirements.

This reference framework specifies the standards that must be used for the electronic sharing or transmission of personal health information using HIS systems. The model also specifies how to implement these standards in order to facilitate the deployment of interoperable HIS systems in agreement with privacy and security requirements.

The HIS Interoperability Framework (IF) reference is divided into modules. Modules are distributed across 3 interoperability layers, defined as follows:

- Content layer (semantic and syntactic content): Specification of exchanged or shared content in terms of structure and vocabularies;
- Service layer : Specification of content sharing or exchange services, their rules and usage parameter; and
- Transport layer: Specification of exchange protocols used by services.

The service and transport layers are often referred to as a group as the "Technical Base" of the framework because they are both developed in a technical development track with vendor participation. The content layer is mostly focused on user input and requirements and is developed separately [2].

2.4 Personal Health Systems Framework

Personal Health Systems (PHS) assist in the seamless provision of quality controlled, and personalized health services to individuals regardless of location. They consist of: (1) *Ambient and/or body devices* (wearable, portable or implantable), (2) *Intelligent processing of the acquired information and coupling of it with expert biomedical knowledge* to derive important new insights about an individual's health status, (3) *Active feedback* based on such new insights [4].

As can be seen in Fig. 4, the PHS Interoperability Framework (PHS IF) can be subsumed into two smaller frameworks: (1) *technical and implementation framework*, including standards, profiles and guidelines for their implementation based on elaborated business use cases, identification and authentication mechanisms, security protocols, testing and certification, etc., and (2) an *institutional/organizational framework* encompassing policy issues (e.g., governance, reimbursement), legal and regulatory aspects such as data protection, liability, etc. [4].

3 A Comparison of Interoperability Frameworks

This section develops a comparison of the interoperability frameworks in the health domain. This comparison will be based on the interoperability dimensions as defined by the Framework of Enterprise Interoperability (FEI). Our choice is motivated by the characteristic of the FEI to be a general framework for enterprise interoperability. It is defined within the general perspective of the enterprise-as a system, where a health organization can also be considered as an enterprise.

The FEI [7] was developed within the frame of INTEROP European Network of Excellence (NoE) [8, 9]. It defines a classification scheme for interoperability knowledge according to three dimensions: *interoperability barriers*, *interoperability approaches*, and enterprise *interoperability concerns*.

Before comparing the reviewed health frameworks towards the dimensions of FEI, it is important to give a general overview of the domain covered by these frameworks. This is given by Table 1.

The comparison criteria are based on the three main dimensions of the general framework FEI. The evaluation uses the following notations. The '+++' means there is a strong concern and the model meets better the criteria, '+' denotes that is weak and '++' is in between, '−' means that the model does not meet or address the criteria. The evaluation uses the following notations.

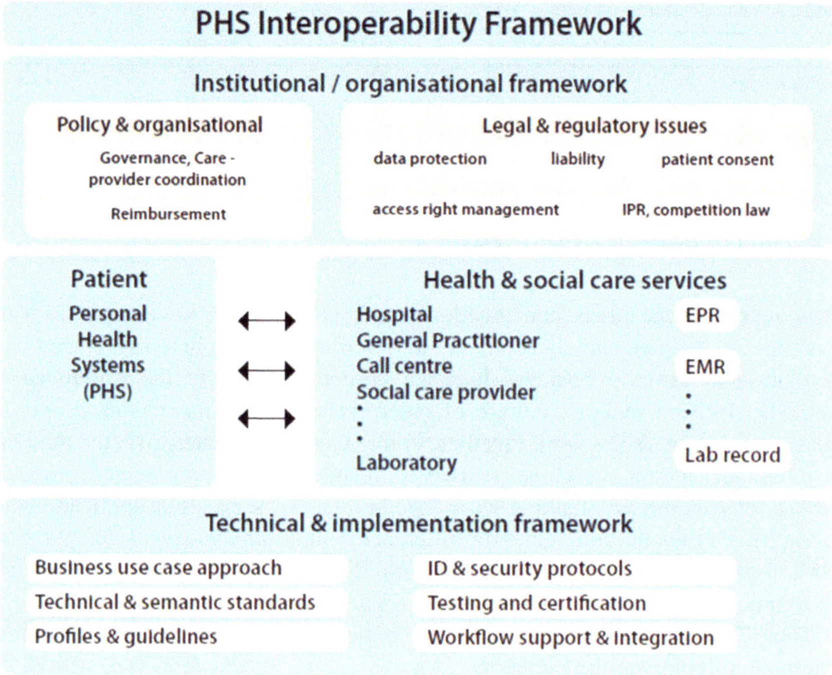

Fig. 4 PHS interoperability framework [4]

Table 1 Overview of health Interoperability framework coverage

	eHealth EIF	HIS IF	eHealth IF	PHS IF
Interoperability dimensions	• Legal • Organizational • Technical • Semantic	• Semantic • Technical	• Organizational • Technical • Informational	Organizational Technical
Case study orientation	yes	–	Yes	yes

3.1 Interoperability Barriers

According to FEI, the establishment of interoperability consists in removing all the identified barriers. Three kinds of barriers are identified: *Conceptual* (syntactic and semantic differences of information to be exchanged), *Technological* (incompatibility of information technologies: architecture and platforms, infrastructure, etc.), and *Organizational* (definition of responsibilities and authorities). Table 2 illustrates the coverage of the reviewed frameworks towards these three barriers.

Table 2 Interoperability barriers coverage

	eHealth EIF	HIS IF	eHealth IF	PHS IF
Technological	+++	+++	+++	+++
Conceptual	+++	+++	++	+
Organizational	+++	–	++	+++

3.2 Interoperability Concerns

They represent the areas concerned by interoperability in an enterprise. Four concerns are defined, namely *business interoperability* (work in a harmonized way to share and develop business between companies despite the difference of methods, decision making, culture of enterprises, etc.), *process interoperability* (make various processes work together. In the interworked enterprise, the aim will be to connect internal processes of two companies to create a common process), *service interoperability* (making work together various services or applications by solving the syntactic and semantic differences) and *data interoperability* (make work together different data models with different query languages to share information coming from heterogeneous systems).

Table 3 illustrates the coverage of the reviewed frameworks towards the four enterprise interoperability concerns.

3.3 Interoperability Approaches

There are three basic ways to relate entities together to establish interoperations: *The integrated approach* (characterized by the existence of a common format for all the constituents systems), the *unified approach*, characterized by the existence of a common format but at a meta-level, the *federated approach*, in which no common format is defined. This approach maintains the identity of interoperating systems; nothing is imposed by one party or another and interoperability is managed in an ad-hoc manner. Table 4 illustrates the coverage of the reviewed frameworks towards these interoperability approaches.

3.4 Discussion

The review of the different aspects and the frameworks coverage with respect to the dimensions defined by the FEI, enables us to identify the main elements in health interoperability that are taken into account by the existing frameworks and the elements that lack to be considered within these frameworks.

Table 3 Interoperability concerns coverage

	eHealth EIF	HIS IF	eHealth IF	PHS IF
Business	++	–	+++	++
Process	+++	++	+++	+++
Service	++	+++	+	+++
Data	+++	+++	++	++

Table 4 Interoperability approaches coverage

	eHealth EIF	HIS IF	eHealth IF	PHS IF
Integrated	–	–	–	–
Unified	–	–	–	–
Federated	–	–	–	–

Organizational, technical and semantic dimensions are the main relevant dimensions that we find in the health interoperability domain. Some of them are missed, as shown by the Tables 1 and 2.

The comparison has clearly shown that no one of the defined frameworks, within the health domain, considers interoperability approaches and thus proposing a way to deal with a specific interoperability problem.

Suggesting a unified framework that takes into account existing ones and integrating the required dimensions of interoperability would allow covering the whole domain of interoperability and going beyond existing approaches.

In order to facilitate the design of this unified framework, we believe that having basis on usage scenarios of the electronic patient record in an extended health pathway, particularly in supporting cross-boarder healthcare, is a very effective way to identify the dimensions of the future framework. These scenarios would be especially needful if considering cases where the patient data are, partially, supplied by information from medical devices. This is the approach that we intend to adopt in our future work to define the dimensions of the unified framework.

4 Conclusion and Future Work

In this paper, we surveyed the main interoperability frameworks in the health domain. A comparison of the considered frameworks was then proposed based on interoperability aspects defined by the general framework: the Framework of Enterprise Interoperability (FEI). This comparison enabled us to identify clearly the missing interoperability aspects in existing frameworks within the health domain. Future work are planned to propose an integrated framework that would allow having a unified approach covering various existing interoperability dimensions and thus going beyond existing ones.

References

1. NEHTA (2007). Interoperability framework. Version 2.0.
2. ASIP Santé-PRAS. (2010). Health information systems interoperability framework (HIS-IF).
3. European Commission. (2013). Directorate-general for communications networks, content and technology. eHealth European Interoperability Framework. ISBN: 978-92-79-30389-0.
4. European Commission. (2011). Enabling smart integrated care: Recommendations for fostering greater interoperability of personal health systems. SmartPersonalHealth publication 2011, Retrieved 2011, from http://sph.continuaalliance.org
5. European Commission. (2010). Towards interoperability for European public services. Annex 2: European Interoperability Framework (EIF) for European public services, COM (2010) 744 final.
6. http://esante.gouv.fr/
7. Chen, D. (2013). Framework for enterprise interoperability and maturity model (CEN/ISO 11354). In *Interoperability for Enterprise Software and Applications: Proceedings of the Workshops and the Doctorial Symposium of the I-ESA International Conference 2010*. Wiley. com.
8. INTEROP. (2007). Enterprise interoperability -framework and knowledge corpus- Final report, INTEROP NOE, Contact no: 508011, Deliverable DI.3.
9. Chen, D., Dassisti, M., & Elvester, B. (2007). *Di.3: Enterprise interoperability framework and knowledge corpus—final report*. Technical report, INTEROP Network of Excellence, IST - Contract no: IST-508 011.

Dynamic Manufacturing Network, PLM Hub and Business Standards Testbed

Nicolas Figay, David Tchoffa, Parisa Ghodous, Ernesto Exposito and Abderrahman El Mhamedi

Abstract Current trends in manufacturing enterprises dealing with complex systems are usage of System Engineering, Product Life Cycle management and systematic utilization of computer aided solutions for all engineering or management activities. In such a context, sustainable and agile infrastructure for emerging digital ecosystems is required, based on open eBusiness PLM standards supporting exchange, sharing and long term archiving of digital models describing behavioural products and implied organizations. This paper will present some new ways to deal with interoperability in such a context, based on Dynamic Manufacturing Networks, constituted by a network of partner enterprises, of enterprise applications involved in the cross-organizational collaboration processes and of the underlying ICT systems. The approach will consist in qualifying such a network to support a portfolio of cross organizational processes supported by a PLM Hub, in particular in terms of interoperabilty and in terms of security, and to map it with available capabilities used by the actual collaboration participants. Then it will be possible to define the efforts required by the partners in order to be able to participate to a given kind of collaboration, making then it possible to participate in any other future collaboration where the partner will play the same role. Doing

N. Figay (✉)
EADS IW, 12 rue Pasteur 92150 Suresnes, France
e-mail: nicolas.figay@eads.net

D. Tchoffa · A. El Mhamedi
University Paris8, 140 rue Nouvelle France 93100 Montreuil, France
e-mail: d.tchoffa@iut.univ-paris8.fr

A. El Mhamedi
e-mail: a.elmhamedi@iut.univ-paris8.fr

P. Ghodous
LIRIS-CNRS, Université Lyon1, Lyon1, France
e-mail: ghodous@liris.cnrs.fr

E. Exposito
LAAS-CNRS, INSA, 7 av. du Colonel Roche, F-31077 Toulouse, France
e-mail: ernesto.exposito@laas.fr

K. Mertins et al. (eds.), *Enterprise Interoperability VI*,
Proceedings of the I-ESA Conferences 7, DOI: 10.1007/978-3-319-04948-9_38,

so, interoperabilty can be established in a continuous and smooth way, allowing all the digital business ecosystem community to build together their interoperability maturity. This approach is completed by the ellaboration of an associated testbed for eBusiness PLM standards allowing the community to accelerate development of required standards as well as their implementation, being by software product solutions or by industrial processes and methods. The principles will be exposed, as well as illustrations coming from research projects such as IMAGINE or IRT-SystemX SIP projects.

Keywords Interoperability · Dynamic manufacturing network · Product lifecycle management · Open standards · Testbed

1 Introduction

1.1 The Business Context

Currently, the new trends within European Aeronautic and Defence are leading to a full reconfiguration of the Supply Chains for various families of Products. If the current industrial programs are reaching a high percentage of subcontracting (e.g. 60 % for Airbus' A380), the target is even higher concerning subcontracting for future programs (e.g. 80 % for future long-range Aircrafts). In addition, if it is aimed reducing the number of tiers-one sub-contractors, there is a global increase of the number of sub-contractors for the whole supply chain from level one to the other levels. This trend is referred as "Virtualization of the enterprise". Another trend is the systematic usage of computers and modelling for a growing set of Engineer disciplines and for collaboration, which comes with the systematic usage of Model Based engineering, being for System Engineering (SE) or for Product Life Cycle Management (PLM). This trend is referred as "Virtualization of the Product". Combining the both trends is leading to border effects in terms of interoperability for collaboration within extended enterprise: ensuring interchange and sharing of digital model of the Product within the extended enterprise and between the used technical applications is today a real challenge, due to the existing heterogeneity between partners concerning the used methods and tools. The non PLM harmonization creates important industrial risks in terms of qualities or delays, and a digital break between huge companies and SME.

In order to respond to such requirements, emerging eBusiness PLM Hubs within large groups (EADS PHC PHUSION) and for the European Aerospace and Defence are being setup, with in one hand the systematic usage of COTS (Commercial Of The Shelves) as components of the collaborative infrastructure and digital engineering chains, and in the other hand the identification of strategic importance of eBusiness PLM standards, as promoted by strategic standardization workgroup such as ASD Strategic Standardization Group.

In such a context, some issues exist with the expected qualities of a cross organizational collaborative platform. Firstly the interoperability, flexibility, robustness, security or other qualities of the platforms are to be adaptative in order to support a continuously changing Supply Chain without endangering the Programs. Current platforms don't support this matter. Second on-boarding process is facing heterogeneous maturity of the members of the Supply Chain when dealing with digital collaboration, making it difficult to constitute Dynamic Manufacturing Networks. Third the appropriate methodologies for setting-up effective and adaptative end to end processes combining internal private processes and cross-organizational collaborative processes through such a platform don't exist yet. Fourth the organizational impacts of the strategy adopted by the Hub initiatives in terms of Architecture, Security and Product Data Management are not always well assessed by involved organizations and stakeholders because of the complexity of these new environments and because of the lack of experience in deploying proposed approaches at this scale. Finally the PLM strategic organizations are facing some difficulties for making Design Office, Production, Customer support and Supply Chain Management communities working together and combining their effort for end to end processes all along the phases of the Product life cycle.

1.2 The Technical Context

In parallel to the evolution of practices of industries, the information and communication technologies are evolving very quickly with a continuously growing rhythm, including those aiming to establish interoperability. Let's consider as example CORBA (Common Object Request Broker Architecture), developed by the community between 1990 and 2002, and with several PLM standard relying on this technology: PDM Enablers or Workflow facilities.

However, new emerging technologies developed for the Internet by W3C and other related organizations became mainstream, such as SOAP (Simple Object Access Protocol) or WSDL (Web Service Description Language). PLM standards followed this evolution, providing new specifications using this formalism (e.g. PDM.net based on SOAP message exchange and PLM Services specifying services with WSDL). Development of CORBA was stopped due to the decreasing involvement of solution providers and to the competition with W3C standards, becoming main stream.

Today, new mainstream technologies are coming, relying on NoSQL databases, Restfull services or Ajax. Once again PLM standards are following, with the emerging OSLC (Open Service for Lifecycle Collaboration). Due to the change of underlying technologies, the provided PLM specifications are not compatible. Due to the time required for implementing a PLM standard, being within software products or industrial processes, there is no time enough to take advantage of such investment before developed interfaces being deprecated due to a new coming technology. Such issues exist not only for distributed services and applications, but also for information models or process models.

The Object Management Group proposed Model Driven Architecture in order to address existence of multiple technological platforms. Business and knowledge domains are captured using UML, as well as applicative services and functions, and the business logic is then "projected" on execution platforms by means of model transformation. If such approach gives a solution for architects for constructing interoperability, it doesn't ensure semantic interoperability—as ontology could do—or prepared interoperability, which required governance of standards by a community of interest or a digital business ecosystem.

2 The State of the Art

According to Ford et al. [1], 6 types of interoperability are mentioned in research papers, demonstrating richness of the interoperability field. Is it needed to agree on a single and precise one? As stated by Morris et al. [2], "We may never have any agreement on a precise definition due to differing expectations that are constantly changing. New capabilities and functions … continue to offer new opportunities for interactions between systems."

Numerous examples are available, such as the definition given by Institute of Electrical and Electronics Engineers [3] or by European eGovernement [4].

One used categorization of interoperability makes clear distinction between information interoperability, service interoperability and process interoperability. Obstr [5] distinguishes syntactic, structural and semantic interoperability. Numerous interoperability frameworks exist, qualifying different type of interoperability, but also systems and organization maturity in relation with interoperability. Figay and Ghodous [6] compares this different interoperability frameworks and characterizes the ideal collaborative system according these different frameworks. It also presents an approach for achieving efficient eBusiness collaboration between enterprises, relying on ATHENA interoperability framework [7, 8] but extending it on the basis of analysis of interoperability brakes leading to non interoperability—which are to be addressed—and providing a set of interoperability enablers which are to be considered. In particular, importance of open source is highlighted by Figay and Ghodous [9]: availability of commodities on the web, i.e. industrial quality software solutions which are open source and which are implementing open standards, is of prime importance for assessing maturity and usability of a standard.

Some of the interoperability brakes are addressed within [10], by proposing a federative framework allowing to prepare and construct operational interoperability for supporting Networked Collaborative Product Development within the extended enterprise. In particular, the proposal of extended hypermodel for interoperability [10] addresses data loss when using different languages involved in a "Model Driven Approach" for projection of business logic on a service oriented execution platform including standardized components delivering horizontal enterprise collaboration services: enterprise service bus, enterprise application server, data transformation, data integration, service composition, enterprise workflow engine and enterprise repositories. Importance for an eBusiness community to govern his standards is also promoted, in order to build the appropriate interoperability maturity level of this community, and in order to prepare the interoperability. A standard is not considered as a technical solution, but as a strategic solution. Finally, iterative approach between research activity and operational implementation is also promoted in order to resolve issue related to silos existing between research and operations. This approach was applied in the Aerospace and Defence context, with creation of the ASD Strategic Standardization Committee [11], which defined the eBusiness PLM standards of European Aerospace and Defence community. In parallel, industries of this community are pushing the Software solution providers to implement these standards, and PLM hubs are being developed, relying partially in principles defined in the thesis. But applying these research results was also the opportunity to discover new brakes and issues which are to be resolved in order to ensure sustainable interoperability at an acceptable price, in order to support secured collaboration within the extended enterprise.

A first issue is related to security, with controlled access to resources provided by a hub. It is partially addressed by the activities undertaken within the Crescendo project [12], where new standard based components were integrated on the original execution platform: enterprise portal based on portal standards (Web Service Remote Portlet, JSR168 [13] and JSR286 [14]), combined with Identity federation component (based on SAML [15]), Single Sign On and Encryption.

A second issue is related to unavailability of enterprise modeling open standard that can be used in order to realize semantic cartography of the PLM Hub and associated network of connected enterprise.

A third issue is related to the important resources required for onboarding process by a partner on a PLM hub, and qualification of interoperability of the applicative and ICT infrastructure which support cross organizational processes in terms of interoperability.

A fourth issue is related to time required for defining a standard and obtaining qualified implementation of this standard, being by software products or by process and methods within the industry.

A fifth issue is related to integration of research results within a single platform with providing ability to create and maintain very fast appropriate skills to support evolution of interoperability framework and associated implementation of reference, a sustainable way.

These issues are addressed through two research initiatives which are presented in the next sections of this paper: the IMAGINE project and the IRT-SystemX SIP (Standard Interoperability PLM) Project.

3 Dynamic Manufacturing Network

The first motivation for defining Dynamic Manufacturing Network is to resolve the third issue related to on-Boarding, qualification of actual resource and identification of required efforts in order to achieve operational and sustainable interoperability at an acceptable price, not only between two enterprise applications, but with all the applications of partners one member of the PLM manufacturing hub want to collaborate with according pre-established and standardized cross organizational collaborative processes. The considered cross organizational processes are "Technical Data package interchange" and "Change and Configuration management". A second motivation is to include Production department of the enterprises within the PLM approach. Today, PLM approach is more used for Design offices and for Integrated Logistic Support. But it remains difficult to have production on board and to identify the relevant manufacturing standards which are to be considered in order to extend the relevant configured set of standards to use in order to improve global efficiency of transversal business processes. Within the framework of the IMAGINE project, we have established new concepts and integrated them in the federative interoperability framework in order to extend PLM eBusiness collaboration scope by including Production and Supply Chain management.

The used execution infrastructure has been defined relying on ATHENA specification, enriched by result of the Crescendo project (operational open source execution platform based on open source components) and extended for the IMAGINE research goals. The result is the cPlaform v2, which is defined a functional way and a technical way. Functional definition provides the high level horizontal functions, which are expected for ensuring efficient collaboration. Technical definitions include definition of all the standards to consider for an overall framework, and a proposal with elected relevant technical solutions implementing the appropriate standards and including those recommended by ASD SSG.

Within the scope of IMAGINE [16], an open standard for enterprise modelling was identified which allows realizing semantic cartography of the considered information system: ArchiMate 2 [17]. In addition, an open source tool was identified which allows visual modelling, on top of Eclipse Modelling Framework, making it possible to use the models in a model based engineering approach. The tool is Archi [18]. ArchiMate makes a clear distinction between Business, Applicative and Technical layers, allowing to qualify and to contextualize the used standards, which can be business, applicative, technical or transversal. Such exercise allows to clarify how the different involved architects—enterprise, processes, information system, software or network—have to consider the standard, and to ensure that the link between the strategic aspect of a standard and the technical aspect of a standard are clearly established: a standard is not only technical solution, it is first a strategic solution.

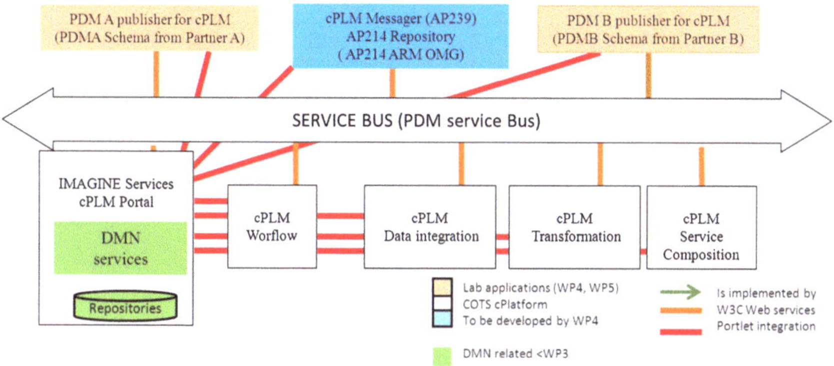

Fig. 1 Functional architecture of the IMAGINE Aeronautic Living Lab

In addition to ASD SSG elected standard, manufacturing standard ISA95 [19], "Enterprise-Control System Integration" was assessed as a way of qualifying expected properties of a production system and assessing actual capabilities of suppliers. IMAGINE partners provided a generic methodology based on this standard for dynamic selection of partners as suppliers. These generic outputs have been customized and integrated within the context of Aeronautic and Defense putting the focus on the virtual digital products, in place of the physical products, and considering the production system of these digital artefacts, qualifying partners and their applicative capabilities for inclusion in the Networked Collaborative Product Development environment. The qualification is made in terms of interoperability and security, and validates that a partner can or not play a given role in the portfolio of cross-organizational processes supported by the collaborative manufacturing PLM Hub. In this context, important ability is the measurement of the efforts to do in order to align capabilities of the partner with required protocols for being able to start the collaboration. Once aligned, the interest is that the partner can not only participate to one single collaboration, but also to any future collaboration where he will have to play the same role for other industrial projects in the future with partners of the hub.

The execution infrastructure, described by Fig. 1, is a service-oriented platform, including enterprise components such as workflow, portal, service bus, data integration, data transformation and service composition. It also includes the IMAGINE services for Dynamic Manufacturing Network.

Following the principle of the federated interoperability framework, all the solutions realizing the functions of the platform are commodities on the Web, implementing relevant configured set of open standards of the framework, for ensuring agility, flexibility, interoperability and security. Component solutions can in addition be replaced by any other solution implementing properly the standards, as integration of the components is based on standardized interfaces and communication protocols. Then technical enterprise applications such as PDM systems

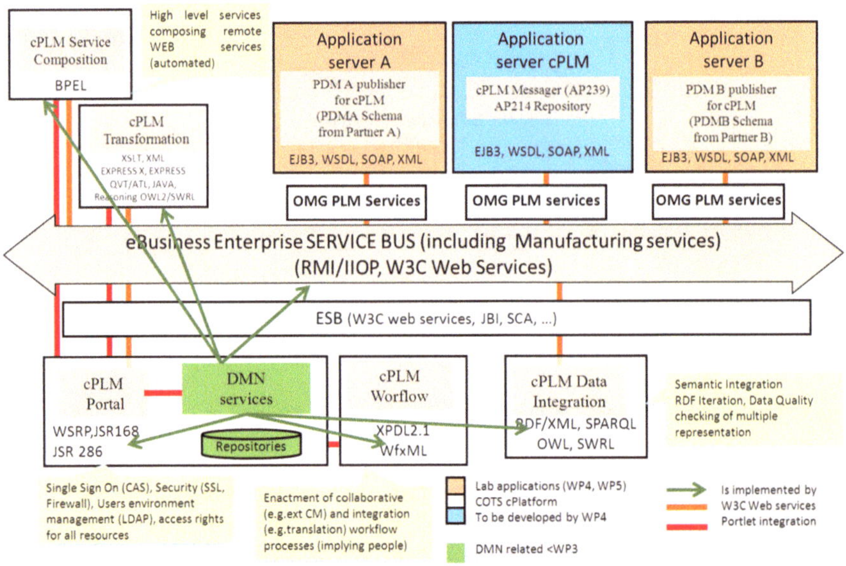

Fig. 2 IMAGINE Aeronautic Living Lab used standards

or ERP systems, can be plugged to the platform by mean of standardized interface and contribute to the collaboration. Shared repositories based on neutral standardized models are also included as temporary storage place for data aggregation and transposition from one partner context to another partner context.

Figure 2 illustrates the set of open standards used in order to ensure expected interoperability for the network of applications plugged on the PLM hub. For each function, open source solutions of industrial quality and implementing properly the standard was identified and assessed in order to build one implementation of the platform.

For Dynamic Manufacturing Network component, the standard being assessed is ISA95. The current efforts are concerning integration of components and standards, with extension of some open source components. The used workflow system is being extended in order to be integrated in the portal, with the access to the DMN repository qualifying the expected and actual participants of cross organizational processes workflow models. Archi extensions [20] are also being developed in order to publish the semantic cartography of the DMN on the hub portal, allowing architects to govern evolution of the DMN infrastructure.

The enterprise service bus will be extended with autonomic self-management and self-protections capabilities. Some innovative capabilities will be integrated for reconfiguration of the network with preservation of the interoperability and automated generation of some collaborative workflow model instances. Integration and implementation of some business cases should be available in the following year, which will demonstrate advantage of the federated interoperability framework and of its extensions related to DMN.

4 Testbed for Interoperability

One important issue is not addressed in IMAGINE, related to the time for developing, assessing and implementing Manufacturing PLM standards, in particular for interconnection of technical enterprise applications dealing with management of product data and process data. It is a stopper for adoption of standards based approach by the industry, even if today no satisfactory approach exists. Being methods for conformance testing defined by ISO 10303, methods adopted by PDM implementer forum or model based framework for recommended practices (OASIS PLCS), development, assessment and deployment of standards and related implementation, being processes or software solutions, are still consuming too much resources and are not aligned with the needs of industrial programs for quick constitution of capabilities needed for dynamic network constitution. For such a reason, a new project was launched, "Standards Interoperability PLM", within the frame of IRT-SystemX [21], with four objectives. The first objective is the development of a generic approach and framework in order to specify and test implementation of standards for multi-disciplinary and multi-sectorial data interchange. The second objective is to develop an experimental research capability for quicker development, assessment and implementation of a configured set of relevant PLM standards covering all the phases of the lifecycle of an industrial product and the different involved disciplines. The third objective is to put in place shared processes, practices and services related to digital business objects which are to be exchanged or shared. The last objective is to create a community approach for informing and implying the different concerned stakeholders and actors who can take benefits of the initiative, taking advantage of collaboration theory for win–win situations. It will involve a first circle of partners, those involved in the project, including industries and universities. A second circle will be constituted for any other company interested by the usage of the test bed. Other stakeholders will be software solutions providers, industrial sectors, standardization organization and Implementer forum, who will be invited to use the results of the project in order to facilitate development, assessment and qualification of standards and their implementation. IMAGINE infrastructure will be reused for the test bed, and extended with complementary functions and capabilities required to support the business cases considered by the project. In particular, services related to management of a repository of uses cases, demonstration scenarios and associated relevant test data sets will be integrated in order to take advantage of DMN facilities. Tests will not be focus on usage of standards one per one, but in the assessment of their simultaneous usage. Interchange or sharing will not be tested between two applications only, but for the set of applications involved in cross organizational processes coupled with local private system engineering processes. Some open source reference components will be automatically generated with minimum set of functionalities required for implementation of standardized services or communication protocols, generated using Model Based Generation approach, as defined by ATHENA [22]. An orchestrator will allow playing scenarios based on execution of instance of cross-organizational collaboration workflow models. Doing so, implementations of standards as processes of reference will be tested simultaneously.

Once the standards based collaboration process is validated, then development and assessment will start for software product or applications implementing the standards. Components of reference will be unplugged and replaced by the implementations to test, and the same testing scenarios will be replay, allowing qualifying the solutions integrated in a dynamic network. Due to the fact the whole collaboration process was assessed first, going till the implementation, workable and shared test scenarios and test data sets will be available as well as the test bed infrastructure in order to accelerate assessment of the implementation scenarios, restricted only to the collaboration processes the partners want to put in place. It should simplify and focus the tests to be performed, and facilitate the work of solution providers in order to implement only but all of what is required in order to support expected collaboration.

5 Conclusion

This paper described emerging requirements for standards based collaboration for industry in order to support system engineering processes and PLM approach, and how the research effort undertaken by the authors contributed to development of new solutions to face the encountered interoperability issues. The federated extension of ATHENA interoperability framework was developed then applied to emerging needs. After industrial usage assessment, the list of interoperability issues was enriched. It led to new initiatives for extending the framework. IMAGINE's Aeronautic Living Lab infrastructure will add Manufacturing Network capabilities applied to network of partners and applications supporting System Engineering processes, responding to the issue of on boarding of partners on a standard based PLM Hub. "Standard Interoperability PLM" project will provide a test bed platform based on the same principles, extending the Lab with use cases, test scenarios and test data set that can be shared and used in order to validate the full set of solutions implied in a set of standardized cross organizational processes, considering not only one standard, but a configured set of relevant standards to be use jointly in order to cover the operational business needs.

Acknowledgments This work has been partly funded by the European Commission through the Project IMAGINE: "Innovative End-to-end Management of Dynamic Manufacturing Networks" (Grant Agreement No. 285132). The authors wish to acknowledge the Commission for their support. We also wish to acknowledge our gratitude and appreciation to the IMAGINE Project partners for their contribution during the development of various ideas and concepts presented in this paper.

References

1. Ford, C., Colombi, J., Graham, S., & Jacques, D. (2008). A survey on interoperability measurement. In *12th ICCRTS "Adapting C2 to the 21st Century"*.
2. Morris, E., Levine, L., Meyers, C., Place, P., & Plakosh, D. (2004). *System of system interoperability*. Technical report CMU/SEI-2004-TR-004.
3. *IEEE standard computer dictionary: Compilation of IEEE standard computer glossaries*. New York, NY. (1990).
4. ISA. (2011). *European interoperability framework (EIF)—towards interoperability for European public services*. ISBN 978-92-79-21515-5.
5. Obstr, L. (2009). *Ontologies and the semantic web for semantic interoperability*. SICOP909.
6. Figay, N., & Ghodous, P. (2009). *Innovative interoperability framework for enterprise applications*. Medes.
7. ATHENA Interoperability Framework. http://athena.modelbased.net/
8. Berre, A. et al. (2007). *The ATHENA interoperability framework*. IESA.
9. Figay, N., & Ghodous, P. (2009). *FLOSS as enterprise applications interoperability enabler*. SITIS.
10. Figay, N. (2009). Interoperability of technical enterprise applications (Ph.D. thesis report, University Lyon 1, 2009).
11. ASD Strategic Standardization Group. http://www.asd-ssg.org
12. EU FP7 Project. (2009–2012). *Collaborative and robust engineering using simulation capability enabling next design optimisation*. http://www.crescendo-fp7.eu/
13. Java Specification Request 168: Portlet specification. Retrieved October 27, 2003, from http://jcp.org
14. Java Specification Request 286: Portlet specification 2.0. Retrieved June 12, 2008, from http://jcp.org/
15. Maler, E. et al. (2003). *Assertions and protocols for the OASIS security assertion markup language (SAML)*. OASIS. http://www.oasis-open.org/committees/security/
16. FP7 IMAGINE project. *Innovative end-to-end management of dynamic manufacturing networks*. http://www.imagine-futurefactory.eu
17. The Open Group. *ArchiMate 2.0*. https://www2.opengroup.org/
18. Archi ArchiMate modeling. http://archi.cetis.ac.uk/
19. ANSI/ISA-95.00.01-2000. *Enterprise-control system integration part 1: models and terminology*. http://www.isa-95.com/
20. ArchiMagine. http://www.eads-iw.net/web/imagine/archimagine
21. IRT SystemX projects. http://www.irt-systemx.fr/projets-rd/
22. ATHENA FP6 IST 507849. *Advanced technologies for interoperability of heterogeneous enterprise networks and their application*. http://interop-vlab.eu/ei_public_deliverables/athena-deliverables/?searchterm=athena

Access Control Framework Within a Collaborative PaaS Platform

Ahmed Bouchami and Olivier Perrin

Abstract Collaboration of organizations in professional context has become ubiquitous. However, the security issues still exist. In this paper, we propose a decentralised hybrid framework for managing identity and access control for collaborative platforms as a service *PaaS*. We propose an approach based on federations that ensure the interoperability within the platform while preserving organizations authentication and authorization mechanisms.

Keywords Authentication · Authorization · Interoperability · Trust · Platform as a service

1 Introduction

Collaboration of organizations in professional context has become ubiquitous. In fact, exchanges are becoming easier due to interactive, dynamic and rich environments offered by the Cloud computing through its various layers (SaaS, PaaS, IaaS). Collaborative platforms allow several companies to host their resources, including processes, data, or services. Thus, the complexity of heterogeneous systems such as Cloud systems has made security issues and enforcement one of the most challenging task. These issues relate to authentication, identity management, and authorization. In this context, every application hosted by the PaaS

A. Bouchami (✉) · O. Perrin
Loria—Inria— Université de Lorraine, UMR 7503, BP 239 54506 Cedex Vandoeuvre-les-Nancy, France
e-mail: ahmed.bouchami@loria.fr

O. Perrin
e-mail: olivier.perrin@loria.fr

K. Mertins et al. (eds.), *Enterprise Interoperability VI*,
Proceedings of the I-ESA Conferences 7, DOI: 10.1007/978-3-319-04948-9_39,

requires to enforce the local security policies of each partner, despite the distributed, collaborative and heterogeneous nature of the applications.

Authentication is the first challenge to address. A first requirement deals with the fact that collaborators have various heterogeneous authentication systems. A second challenge deals the decentralized nature and administration of security mechanisms. Indeed, each partner wants to preserve its autonomy, i.e. the ability to control the access to its shared resources, and each partner will not update authentication mechanism and access control policies.

Another requirement deals with the highly evolutive nature of such environments. It is necessary to take account of the possible partners' fine-grained control policies that can become difficult to manage accurately. Last, the security infrastructure of a PaaS platform must manage interoperability between partners' heterogeneous security policies and authentication mechanisms.

In highly-distributed collaborative environments like PaaS, the pillar of a security architecture is based on the design of three types of collaboration: federated collaboration, loosely coupled collaboration and ad hoc collaboration [1].

In our work, we propose a hybrid collaboration architecture that includes the advantages of the above collaboration types, namely: trust, global meta policy and long term collaboration (federated collaboration), users autonomy (loosely coupled architecture) and isolation of partner data (ad hoc collaboration). We also address the problem of partners' heterogeneous authentication mechanisms in order to have an interoperable identity management system.

The remainder of this paper is organized as follow. We present some related works and our motivation by discussing the limits of classical access control mechanisms in Sect. 2. We then address the identity management (authentication) in Sect. 3. Section 4 discusses the authorization problem and details the proposed architecture. Then, Sect. 5 concludes the paper, summarizes the contribution and outlines future research directions.

2 Motivations and Related Work

Our objective is to provide an architecture supporting access control to shared resources (data, process and services) within a collaborative heterogeneous PaaS platform. Within this PaaS, several organizations share their resources: data, services and process fragments. These organizations have their own authentication and authorization mechanisms, and these mechanisms are heterogeneous. In this context, we must address several challenges.

First, partners prefer to preserve their authentication systems, and don't want to change their application's access control code. Hence, the first challenge is to manage the interoperability between partners' authentication mechanisms. For example, let us suppose that we have three partners within our PaaS, companies A, B and C. These companies have different authentication mechanisms as depicted in Fig. 1. If a user belonging to a company wants to access a remote resource(s), it

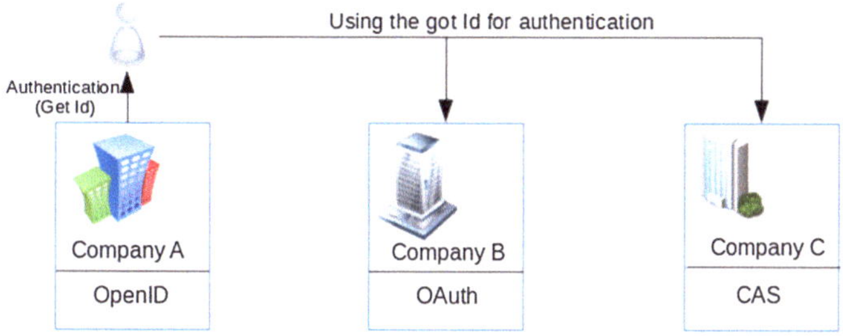

Fig. 1 Federation

should be authenticated by the resource owners' authentication mechanism. However, this mechanism could be different from its own authentication mechanism. For cxamplc, uscr *User1* wants to be authenticated by the authentication mechanism of *company B* (namely, OAuth), while its own is based on OpenID. In such case, the identity of *User1* cannot be recognized by the *B*'s authentication mechanism. So, the problem is *how can we handle this heterogeneity problem*?

The second challenge deals with **authorization**. Assuming that the above authentication problem is solved, all users belonging to a company can be authenticated by the other companies' authentication mechanism. Lets assume that **company A** and **company B** share a set of common resources **R1**. Therefore, **A**'s staff can be authenticated by **B**'s authentication mechanism and thus, allowed to access **R1** with their identity provided by their authentication mechanism (**A**). Furthermore, let us assume the same scenario with **company B**, **company C** and resources set **R2**. Based on this scenario, we conclude that, **A**'s staff are not allowed to access to **R2** and vice versa for **C**'s staff regarding **R1**. Hence, opting only for the authentication mechanisms, we cannot prevent such access. Then, an authorization mechanism must be implemented for each partner. Of cause, we should take into account the fact that partners should be autonomous for their authorization decisions. We can have a different decisions provided by partners related to the same resources. In this context, the federation concept can be useful to lower the complexity of the above problems. Indeed, a federation [2] is a contract between different partners belonging to a same circle of trust for certifying information (especially identity information) exchanged between partners that belong to heterogeneous domains. However, the federation concept can be still limited. This limitation is due to the fact that two or more federations can include some common partner(s), and these partners can be authenticated outside their own federation. For example, *federation 1* in Fig. 2 includes A, B and C companies, meaning that **A**'s staff are recognized by **C**'s and **B**'s authentication mechanisms. On the other hand, *federation 2* includes **B**, **D** and **F** companies. So, we have *company B* is a common partner between both federations. Consequently, *company D* for example is recognized by the authentication mechanism of *company B*, and thus, can access its collaborative **shared** resources protected under *federation 1* that can be owned by **A** and/or **B** and/or **C** companies.

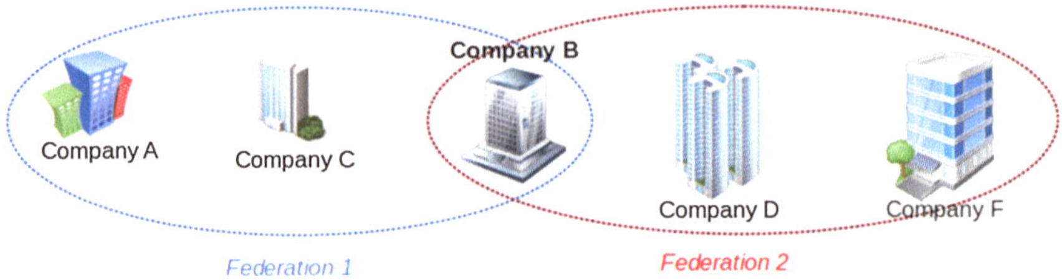

Fig. 2 Federations

Furthermore, partners are free to define their own "circle of trust" by choosing their trusted partners. This depicts the problems to be solved for access control limitation regarding to access rights of each partner.

Several control access systems has been proposed in the literature. The most famous ones are RBAC [3] and XACML [4]. These models do not fit very well with collaborative and/or dynamic distributed environments. The application of the traditional RBAC model may be limited (i.e. static), because it works very well for domains where the roles are previously known, but in our context what will we do if a new user is dynamically included within the federation? Therefore, we must be able to establish a method to correlate roles. XACML overcomes this previous limitation as it based on a Policy Decision Point that uses a set of policies for making decision. However, it still poor regarding our focus in terms of architecture, because it's a centralized approach and as we mentioned before, we need a **decentralized** administration (no unique centralized point of decision mechanism).

3 Identity Access Management

Authentication is the first important issue in our context. The main objective of own security platform is to manage access to resources with respect of digital identities (IAM) [5]. Then, identity access management needs to take into account two aspects, authentication and authorization.

Within a collaborative PaaS, a user *User1* belonging to a *company A* can use a remote protected resource **R** owned by *company B*. This implies that the identity of user A must be recognized and validated by the authentication mechanism of *Organization B*. Various authentication mechanisms like Login/Password, 2 step validation, X.509... can be used. But, the problem occurs when each resource owner don't want to share its identity information and use its proper authentication mechanism. This justifies the need of a decentralized authentication protocol that considers the interoperability between collaborators' heterogeneous authentication mechanisms.

In this context, Single Sign On (SSO) protocols appear to be interesting to handle the problem. They allow maintaining a local authentication for accessing to remote resources. They able to transfer identities information through different PaaS access control layers to shared services, data and resources. OpenID and OAuth [6] are the most known SSO protocols. However, these protocols are generally used for individual authentication while we need the concept of federation. Another limitation is the untrustworthiness of the provided identities, and as we mentioned before, the use of concept of federations is mandatory in our context. Hence, to deal with this limitation, we can opt for the federated identity.

Our first objective is to manage the interoperability between the partners' authentications systems. Several authentication mechanisms exist. However, technically, they are heterogeneous. Each of them is coded in different manner. Hence, it's not easy to implement a method that will handle the mapping and treat all differences between all the existing mechanisms mainly when we want to preserve them as they are. One solution is to opt for a common authentication mechanism that **supports** the majority of the existing ones. Such an authentication mechanism exists under the name of ***LemonLDAP::NG*** owned by the french software vendor *Linagora.*[1]

LemonLDAP::NG is an open source Web *Single Sign On* product (WebSSO) that can handle more than 200000 users used by many organizations [7]. It supports a lot off authentication modules like: LDAP, Database, SSL X509, SAML 2.0/Shibboleth, OpenID, Twitter (OAuth), CAS, etc. LemonLDAP::NG authentication mechanism can delegate its authentication to CAS, OpenID, SAML. It can also provide Identity with the above authentication mechanisms. Users authenticate at the level of LemonLDAP::NG using their identity provided by their organization's authentication mechanism. Actually, *LemonLDAP::NG* can be seen as a bridge between authentication mechanisms [7].

Its main components are:

- **Manager**: used to manage LemonLDAP::NG configuration and to explore sessions. Dedicated to administrators,
- **Portal**: used to authenticate users, display applications list and provide identity provider service (SAML, OpenID, CAS). Portal provides also many other features (see portal for more),
- **Handler**: Apache modules used to protect applications.

LemonLDAP::NG behaves as follows. The user tries to access protected application, and his request is intercepted by Handler. SSO cookies are not detected, so Handler redirects user to Portal. The user authenticates on Portal. The portal checks authentication. If authentication succeeds, Portal collects user data. The portal creates a session to store user data. The portal gets the session key. The portal creates SSO cookies with session key as value. The user is redirected to the

[1] French software vendor and professional services provider focused on Open Source Software company, http://www.linagora.com/.

protected application, with his new cookie. The handler gets session obtained from the cookie. The handler stores user data in its cache. The handler checks access rule and send headers to protected applications. The protected application sends response to Handler. Then, the handler sends the response to the user.

Also, there is another tool that synchronizes a remote companies' users databases with the LDAP format. This tool is ***LDAP Synchronization Connector*** [8].

LL::NG doesn't completely solve the problem of interoperability between **authentication mechanisms**, beacause its technically is very hard to perform such operation, nevertheless, LL::NG decreases the negative impact of this problem. Beacause, LL::NG brings together the majority of effective and existing authentication mechanisms, both, as identity consumer and identity provider. It can also be easily implemented by companies as an upper level layer to control inter-organization IAM.

4 Proposed Architecture

As discussed in Sect. 3, a user who wants to use the platform services must connect through a **web portal** [9]. This web portal is based on LemonLDAP::NG, hence, users can use their own organization's authentication mechanism. After authenticating the user, LemonLDAP::NG gets its credentials information: identity information, role, organization name,.... Then, a user session is created with its identity information.

In order to offer autonomy to the platform (PaaS) users, we propose in our framework (Fig. 3) that organizations will be free to choose their trusted partners to establish their circle of trust and thus their federation partners. Basically, federations are created based on organizations' identities. Then, for each federation we can define the identity of all partners with the associated privileges to be authenticated, and therefore, get access to the federated resources.

However, to handle the aforementioned federation limitations, we propose to set up a security module named *PaaS Federation Security Module* (PFSM) for each federation. We ensure that all *PFSM*s are **not interconnected**. Thus, users who want to request a resource encapsulated within a federation don't includes their organization, will not be recognized by the *PFSM* of the requested resource federation, even in the case of a common federated organization. Furthermore, we propose to create a federation based on resources (Fig. 4). For each federation based on identities, we create one or more sub **logical federation** based on **shared resources**. These federation type aims to allow more fine grained access to some resources regarding some federated organization(s) and/or user(s). For this type of federation we set up another security module named *PaaS logical Federation Security Module* (*PLFSM*) Fig. 5. We detail these two components below.

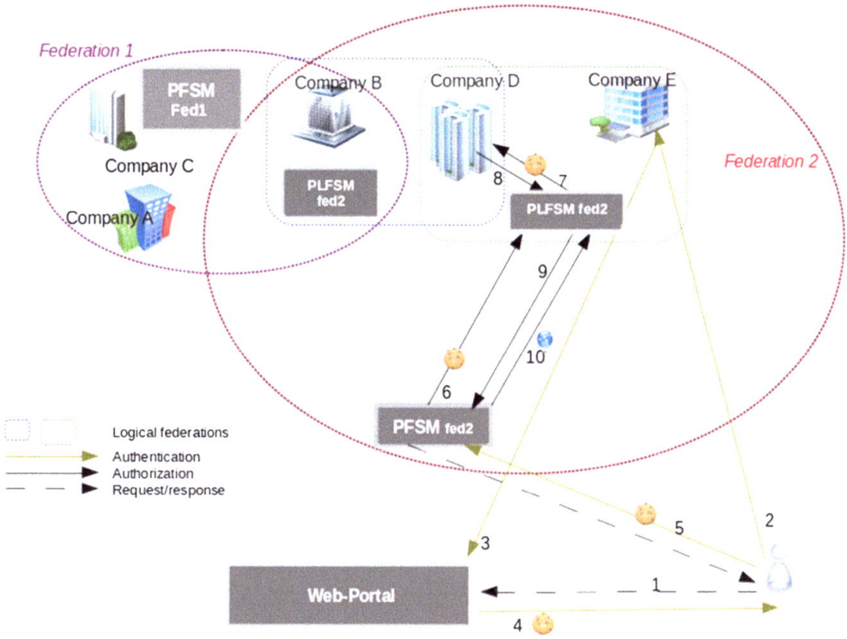

Fig 3 Architecture global overview

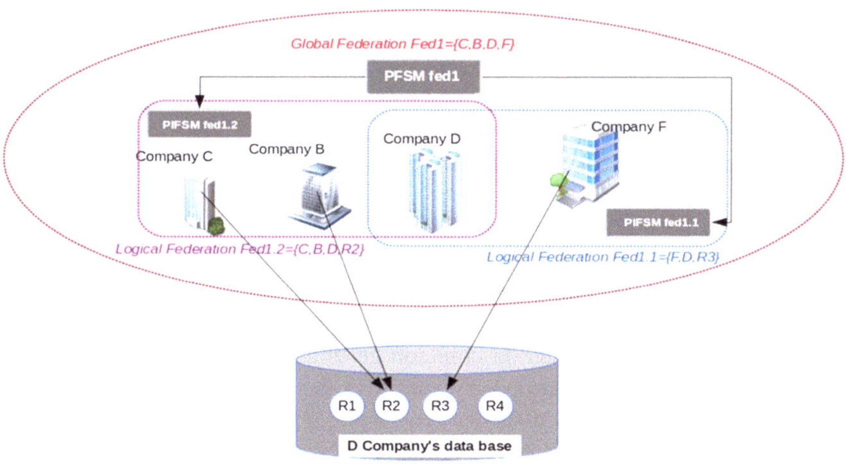

Fig. 4 Logical federation

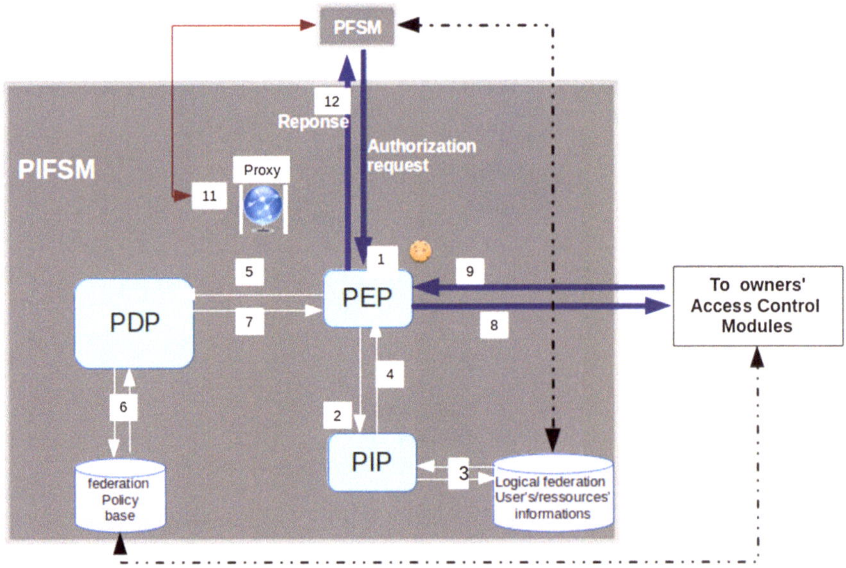

Fig. 5 PaaS logical federation security module overview

4.1 Federation Security Modules (PFSM/PLFSM)

Authorization process in our framework is inspired from the XACML model [4]. It includes four components, the most important one for the authorization protocol is the *policy decision point* **PDP**. To overcome limitations introduced in Sect. 2, we propose a two levels adaptation.

The *PaaS Federation Security Module* is the control access module for each federation within the PaaS. For each identities federation, one *PFSM* is implemented at the level of the PaaS. It contains a global federation users'/resources' information base that contains information about federated partners as: **staff identities' information**, **staff trust values**, **staff roles** of each partner and possibly a black list for some forbidden users. The PFSM also contains global platform policies with the global federation authentication policies managed by the platform administrators. They are based on common general partners' authentication policies. The PFSM is an intermediary unit between the authentication and the authorization processes. It assesses requesters' identities and forwards their requests to the appropriate *PLFSM*s for authorization decision-making. Due to the hight number of received PISFM's authorization decisions, and for the sake of time processing optimization, the PFSM analyzes them with the **Mining log engine,**[2] and then, the PSFM may implements a **proxy module** over the

[2] With the use of data mining algorithms.

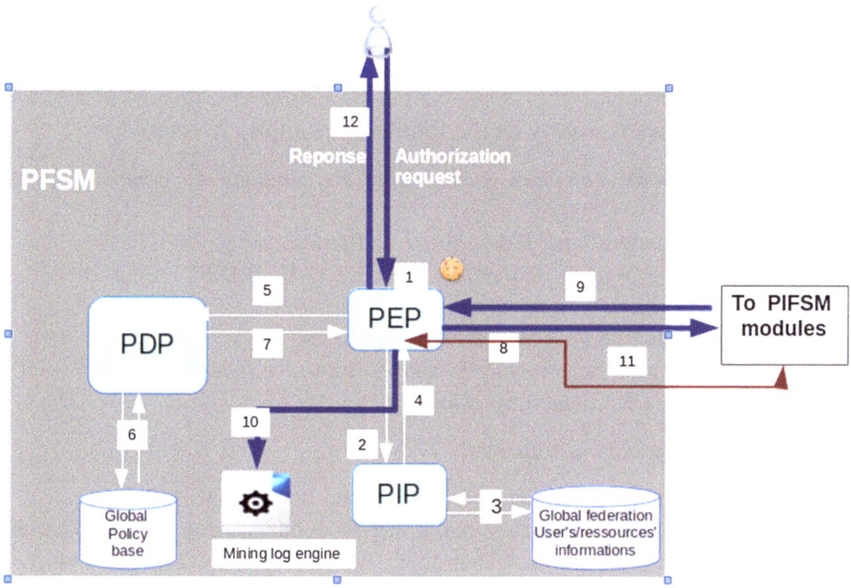

Fig. 6 PaaS federation security module overview

appropriate PLFSM. This **proxy** will contain the owner's defined fine grained authorizations, hence, the requester will be redirected to the proxy for accessing to the requested resource but under the constraints defined by its owner for this resource regarding to the requester profil. As depicted in Fig. 6, the *policy Enforcement Point* **PEP** is the handler of the PSFM. It receives the initial request, forward it to PLFSM for decision-making, receive the initial PLFSM response(s) and finally, forward the PFSM's final decision to the requester. We will explain in details how the process works in the next section.

For each logical federation, the ***PaaS Federation Security Module*** is the control access module to the partners' shared resources within the logical federation. It makes the initial decision based on a role-based approach (that we can not detail here due to the space limitation) defined by resources' owners within the Policy Administration Point. The information base of the PLFSM interacts with the internal *Policy Information Point* **PIP** to look for logical federation users' informations. The information base is connected to the Global federation users'/ resources' information of PFSM module for getting and/or updating these users' information. As the *PFSM*s, the *PLFSM*s are independent and don't communicate even in the same global identity federation.

4.2 Behavior of Our Identity and Access Management in Proposed Framework

We illustrate our framework with the following example. We consider:

- Five organizations: *company A, company B, company C, company D* and *company E,*
- Two separated federations: *federation 1* and *federation 2,*
- Two separated logical federations Fed1.2 = {C,B,D,R2} and Fed1.1 = {F,D,R3} (Fig. 4),
- One company that belongs to the two federations,
- A user belonging to the *company E* that belong to the *federation 2* who want to access a resource R3 owned by *company D.*

The scenario unfolds as follows:

1. A user authenticates through the portal based on LemonLDAP::NG.
2. The web-portal gives the user the ability to connect using its organization's authentication mechanism. Therefore, the web-portal (*LemonLDAP::NG*) redirects the user to its organization's authentication mechanism.
3. The user's organization authenticates the user in question and provides to *LemonLDAP::NG* its identity information.
4. If the user is authenticated by *LemonLDAP::NG*, it gets a session with a cookie containing its identity information.
5. The user uses the provided cookie for requesting a remote resource. In our example, user request a resource(s) owned by *organization D* belonging to the *federation 2* controlled by **PFSM fed2**. Therefore, its identity (contained in the cookie) must be validated as a federation member by the local federation's *PSFM*.
6. After getting user's information, the *PFSM* initially checks if the user's organization belongs to its controlled federation, else, it rejects the request. After this, it performs an assessment based of the global policy contained within the **PAP**. If the initial request assessment succeeds, the *PFSM* forward the user request to the appropriate *PLFSM* for decision-making.
7. After receiving the request, PLFSM evaluates the request, takes a decision based on predefined roles and in the same time forward the request to the resource owner's organization.
8. The resource owner's organization makes a possible additional fine-grained authorization constraints (facets) regarding to the request and/or requester. Then, The resource owner's organization sends back the facets to the PLFSM.
9. PLFSM sends the decision and the possible resource owner's facets to the PFSM as a log files.
10. The PFSM The PFSM performs a log **mining**[3] and makes a final decision based on requested resource owner's decision. Therefore, the PFSM can

[3] Data mining algorithms may be used in the case of a very high information volume.

update the trust values of the requester and set up a proxy that contains the received facets. Then, the PFSM sends the proxy to the PLFSM, and the final decision to the requester.

5 Conclusion

In this paper, we have proposed framework that considers: users' autonomy to make decision (to grant or not the access to their resources), data isolation and federation for managing access control designed for collaborative PaaS platforms. In our framework, both identities and resources are federated: the former are controlled by *PaaS Federated Security Modules*, while the later are by a *PaaS Federated Security Modules*. The framework is based on LemmonLDAP::NG for decreasing the problem of heterogeneous authentication mechanisms, an access control mechanism inspired from XACML and role-based approach with a mining logs engine to ensuring dynamicity for making decision within the platform by estimating the dynamic trust values (e-reputation) of each collaborator within federations.

We plan to focus in our future work on the conception and development of the modules PFSM and PLFSM for a trusted collaborative PaaS environment. Then, we would handle the interoperability between users' heterogeneous access control policies. The platform need a supervision mechanism based on the defined interoperable policy to be sure that the federated collaborative resources will not be compromised by a trusted partner(s). Wich can contribute to compute partners' trust values, with the aim of improving the quality of collaboration and the definition of facets contained in the different Proxies.

Acknowledgments This work is supported by the project OpenPaaS.

References

1. Almutairi, A., Sarfraz, M., Basalamah, S., Aref, W., & Ghafoor, A. (2012). A distributed access control architecture for cloud computing. *Software, IEEE, 29*(2), 36–44.
2. Martin, G., Johannes, M., & Martin, N. (2005). A modeling approach to federated identity and access management. In *Special Interest Tracks and Posters of the 14th International Conference on World Wide Web*, (pp. 1156–1157), ACM.
3. Ferraiolo, D., Cugini, J., & Richard Kuhn, D. (1995) Role-based access control (rbac): Features and motivations. In *Proceedings of 11th Annual Computer Security Application Conference* (pp. 241–48).
4. Markus, L., Seth, P., Rebekah, L., Dennis, K., & Sumit, S. (2003). First experiences using xacml for access control in distributed systems. In *Proceedings of the 2003 ACM Workshop on XML Security* (pp. 25–37), ACM.
5. Prasad, G. (2013). *50 data principles for loosely coupled identity management*. Retrieved May, 2013, from http://fr.slideshare.net/ganeshcprasad/50-data-principles-for-loosely-coupled-identity-management-v1-0/

6. Kaila, P. (2008). Oauth and openid 2.0. From End-to-End to Trust-to-Trust. In S. Tarkoma & J. Heikkinen (Eds.), *Proceedings of the Seminar on Network Security* (pp. 18–22), December 12, 2008.
7. Oudot, C. lemonldap::ng un websso libre. Retrieved October, 2013, from http://fr.slideshare.net/coudot/lemonldapng-un-websso-libre?from_search=19
8. Ldap synchronization connector wiki. Retrieved October, 2013, from http://lsc-project.org/wiki/about/start
9. Prasad, G., & Rajbhandari, U. (2012). Identity management on a shoestring. Retrieved October, 2013, from http://www.infoq.com/minibooks/Identity-Management-Shoestring

Part XI
Interoperability Scenarios and Case Studies

Requirements for Supporting Enterprise Interoperability in Dynamic Environments

Georg Weichhart

Abstract Interoperability in enterprise systems is currently discussed from a system theoretic point of view. In the conceptual work described here, two special instances of systems theory are used as a basis allowing to detail requirements for interoperability in dynamic environments. Chaos Theory and Complex Adaptive Systems Theory focus on the description of properties of dynamic systems where the global system's behavior cannot be determined by summing up behaviors of system parts. First a connection between enterprise systems and the theories are established. The theories are then used as a lens for analyzing and discussing initial requirements for a platform that supports interoperability in a dynamic context.

Keywords E-learning · Chaos theory · Complex adaptive systems theory · Enterprise interoperability

1 Introduction

Recently important steps towards the formulation of a scientific basis for *interoperability* have been made. General System Theory (GST) [1] has been used to discuss interoperability in enterprise systems [2–4]. One important aspect of interoperable systems, in contrast to both *non-interoperable* and *fully integrated* systems, is that interoperable systems are more resilient [5]. Non-interoperable systems do not provide the desired functionality. In fully integrated systems, a failure in one part lets the overall system fail, because of links to other dependent system parts [6].

G. Weichhart (✉)
Department of Communications Engineering—Business Informatics, Johannes Kepler University, Altenberger Straße 69 Science Park 3, 4040 Linz, Austria
e-mail: Georg.Weichhart@jku.at; Georg.Weichhart@metasonic.de

Metasonic AG, Münchner Str. 29–Hettenshausen, 85276 Pfaffenhofen, Germany

K. Mertins et al. (eds.), *Enterprise Interoperability VI*,
Proceedings of the I-ESA Conferences 7, DOI: 10.1007/978-3-319-04948-9_40,

In order to design and engineer a supporting environment for enterprise systems to establish and maintain sustainable interoperability in a dynamic environment, we aim at identifying requirements for a digital platform that facilitates interoperability on different levels. Two special systems theories which place special attention to dynamics are used to support the definition of requirements:

- Chaos Theory
- Complex Adaptive Systems Theory.

The paper is structured as follows. First a brief introduction to systems theory with special attention to interoperability is given. This leads to the combined discussion of the two used system theories. Properties of the theories are discussed and linked to enterprise systems. Existing Interoperability approaches are used to discuss problems identified with respect to dynamic environments.

2 Systems Theories

General System Theory (GST) [1] intends to support the identification of principles that are valid for many systems. GST facilitates the communication between different scientific disciplines. This is done by abstracting concepts of a discipline to form systems. This abstraction allows scientists of other disciplines to use elements and insights described by these systems.

GST builds upon the notion that a system is an organization of connected parts, where each part and the overall system exhibits some behavior. A system is placed in an environment and may have a function and produce some outcome according to a system's objectives [7]. Parts of a system are themselves systems. A system has a state and may be evolving over time, therefore it has a history. The following concept map shows system concepts and links them (Fig. 1).

Naudet et al. [3] have identified a close relationship between GST and the research domain of interoperability.

Due to the dynamics of today's business environment, recognizing organizations and organizational networks as "static" systems is not sufficient [8].

In the following two special sub-theories of GST are used to characterize enterprise systems. Chaos theory and complex adaptive systems theory are theories which have their roots in GST (see above) and put emphasis on dynamic aspects. As there is no general agreement on the properties such systems have [9], selected properties described within these theories are briefly discussed below. These properties are used for establishing a conceptual link between enterprise systems and the theories.

In the following each property is briefly discussed. This is followed by an example that highlights how every aspect is recognized in enterprise systems.

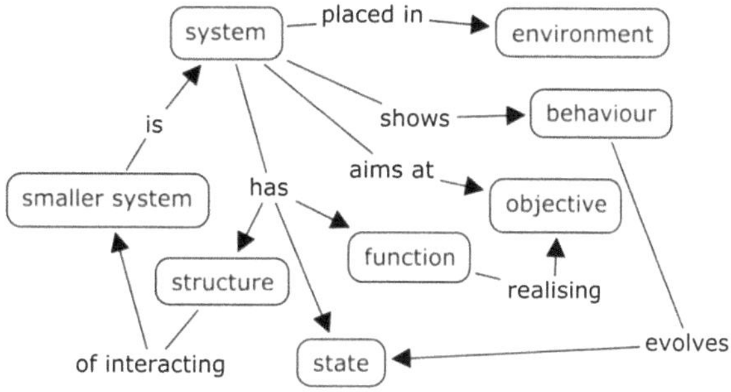

Fig. 1 Concept map describing systems

2.1 Non-linear Interdependence

In a complex system, parts are connected (structure) and show some (individual) behavior. However, due to some non-linearity either in the link or in the part's behavior, the global behavior of the system may not be predicted by summing up the individual part's behaviors.

For supply chains, the example in the following exemplifies links between system parts visible as physical flow of goods, information flows, and links between the supply chains participants' behaviors.

2.2 Path Dependence

The "butterfly effect" (coined by Lorenz [9]) exemplifies that systems are sensitive to initial conditions. Depending on a small change in some part of the overall system, the system's state evolves significantly different in a distant part. In that particular example, a butterfly's wing causes changes in the airflow which amplifies over time and causes a thunderstorm somewhere else.

From a supply chain management point of view, the so called "Beer Game" is used to exemplify how small changes in the demand, lead to a globally observable phenomenon, which is called "bullwhip effect" [10]. The following figure shows small changes of customer orders, which get more and more amplified down the supply chain. The more a company is located at the beginning of the chain, the more amplified order changes get. The (well designed) setting for this example includes communication and transport delays which lead to an amplification of beer bottles ordered. Supply Chain participants place higher orders than required, overcompensating delivery delays (Fig. 2).

2.3 Strange Attractor and Bifurcation

In a dynamic system, a (strange) attractor is a part which attracts other independent parts. The force of attraction is dependent on the distance between these parts.

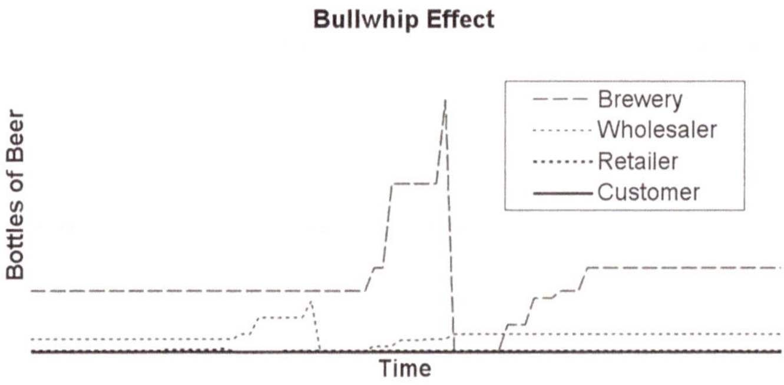

Fig. 2 Bullwhip effect in supply chains

Fig. 3 Bifurcation in a logistic map [11, p. 34]

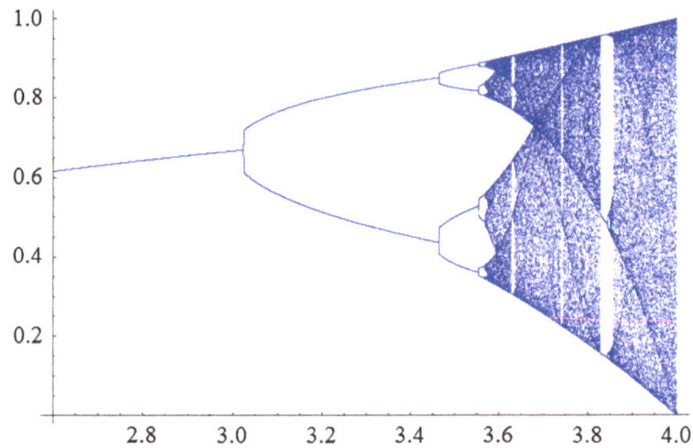

Parts getting close to the attractor remain close. The paths are influenced by attractors and are not fixed or predetermined.

A bifurcation point marks a moment in time where a system's part comes under the influence of another attractor changing its state. The concept of bifurcation is shown in the following figure. It illustrates a system's transition by varying a single parameter (x-axis). On the left the system is in its beginning single steady state. The branches illustrate a period in which the system begins to fluctuate around two, and in the following more states. System parts are influenced by more and more attractors, and the system is getting more complex over time [12] (Fig. 3).

2.4 Active Agents

In a complex adaptive system, "great many independent agents are interacting with each other in a great many ways" [13, p. 11]. The agents follow their individual rules how to interact with other agents. This interaction between agents is a local event. Agents may have sensors and actors to interact with their local environment.

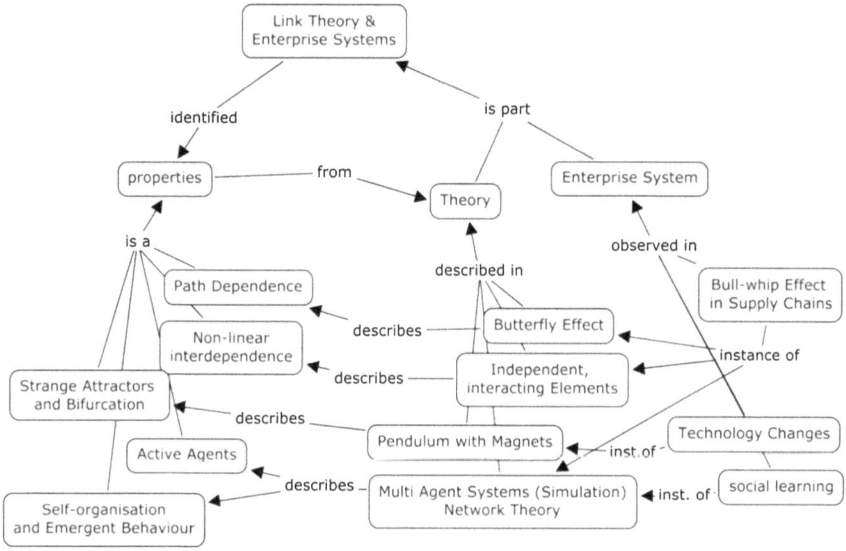

Fig. 4 Properties of the theories in use (cf. [17])

Yet it is important to understand that there is no global control flow, but there are only *local* interactions.

In the view taken, agents in supply networks are enterprises taking part in these organizational networks [8]. Within organizations these agents are human agents [14]. All agents (independent on the observed system scale) act independent and interact with other agents.

2.5 Self-Organization and Emergent Behavior

The lack of global control enables agents to act self-controlled and self-organize as group. As mentioned above, interactions are local to agents taking part in that interaction. Interactions with or without taking the higher system level state into account, facilitates emergent behavior on a higher system level (see above bifurcation).

In a social system, an agent's behavior influences the environment and vice versa [15]. Over time, agents learn from each other, for example through copying successful behavior. However, individual and group learning paths and learning results are not predictable. The performance of a group does not only depend on individuals, but also on the interaction between individuals. Learning and improvement are results of self-organization of individual agents and of groups of agents. Depending on the level observed an agent may be a learning organization taking part in several supply networks [8], or a human agent working in an organization [14].

The following concept map (based on the approach by Novak and Cañas [16]) brings together the above discussed concepts and their relationships (Fig. 4).

3 Existing Interoperability Support

In order to detail and structure the discussion of a dynamic system, we use the above described properties as lenses. This way we conceptualize requirements for supporting enterprise systems to reach and maintain interoperability. However, the theories will not provide direct input to a solution (cf. [17]). This means, chaos theory and complex adaptive systems theory are used to describe *dynamic properties of the environment* in which the enterprise system is placed. The enterprise system has to cope with these dynamic properties. However, it has to be remarked, that the enterprise system might be for example a department within an enterprise, or an enterprise within a supply network.

3.1 Systemic Interoperability

Systemic approaches to enterprise interoperability have been already mentioned above [2–4]. Interoperability problems are discussed within frameworks along the following dimensions and categories (see for example [4, 18]):

Interoperability barriers	Conceptual I Organizational I Technological
Interoperability concerns	Business I Process I Service I Data
Interoperability approaches	Federated I Unified I Integrated
Interoperability solution timing	A-Priori solutions I A-Posteriori solutions

Barriers stop systems to be interoperable. *Concerns* describe the level of the enterprise system on which interoperability problems occur. *Approaches* provide strategies for overcoming the barriers. *Solution timings* refer to the circumstances when interoperability problems may be tackled. A-priori solutions are approaches that allow to anticipate problems and to overcome barriers before systems are build. A-posteriori solutions are approaches that allow to identify and correct problems after they occur in the running system [18].

3.2 Enterprise Architectures and Enterprise Modeling for Integration

In the following we discuss existing interoperability support approaches like Enterprise Modeling (EM)/Enterprise Architecture (EA) and their (potential) role in a platform for supporting interoperability. However, we take a special look on arising challenges when interoperability of sub-systems is the objective for a system in a dynamic environment. This allows the identification of shortcomings of existing approaches and to propose modifications/enhancements of existing approaches to overcome challenges.

To improve organizational interoperability in a complex adaptive system is a challenge. Existing approaches and languages for enterprise modeling (e.g. UEML [19]), and business process modeling (ARIS [20]) support the creation of detailed descriptions. The aim of the models is to support enterprise *integration* (EI). These models may be used as a point of reference to unify the views of *agents* involved in the modeled enterprise system.

But not only enterprise modeling results are of importance. According to Vernadat [21], EI also has the objectives to enable and facilitate communication and coordination in order to allow (independent) actors to collaboratively fulfill the enterprise's goals more efficient and effective. Models facilitate communication through the design of abstractions. The process of modeling facilitates knowledge exchange (i.e. local interactions between agents). For example, Oppl [22] has designed an interactive environment which allows multiple modelers to articulate their views and supports the agents to "negotiate" a unified model.

Enterprise Architectures (EA) aim at providing a uniform representation and supporting the integration of different domains across the enterprise [23]. EA takes an IT centric perspective and supports a unified way to model different parts of an organization including: organizational structures and processes on business level, information systems and applications on IT level, and software and hardware infrastructure on technology layer [23].

3.3 Challenges of Dynamic Environments

While the above approaches support a certain level of integration within enterprise systems, enterprise modeling and enterprise architecture approaches show some shortcomings when being analyzed using the theories discussed above as lenses.

Integrated systems exhibit less resilience than interoperable systems (see also above) [5]. This is of importance if a dynamic environment is assumed. A unified single model is used as point of reference. With larger enterprise systems the modeling process, requires so much effort that when the models are finished, reality has moved on and the models are obsolete. This hampers pursuing enterprise integration goals [24].

3.3.1 Non-linear Interdependence

The interdependence between parts may not be determined a-priori. An approach is needed which enables modeling, monitoring and simulation of flows, links and connections. Due to the non-linearity the global behavior cannot be predicted by observing individual parts. It is necessary to integrate possibilities for monitoring parts and links. The global behavior needs to be simulated based on the monitoring data. Modeling in dynamic systems is only one part of a continuous process.

3.3.2 Path Dependence

Any model or architecture (as any artificial artifact) has a history and at the same time the process of modeling becomes part of history. The model will influence upcoming, future solutions. This influence may have negative or positive consequences for consecutively modeled models.

3.3.3 Strange Attractor and Bifurcation

Any model or architecture hopefully becomes some sort of "strange attractor" in the sense of influencing decisions and decision processes in order to fit to the model. A new created model however might influence decisions taken in enterprise systems to move towards a different direction. This model will trigger a bifurcation point for the system. A different state will be reached by the system after passing this point. Care has to be taken that this process will not suppress the function of the enterprise system, but support the system to better reach its objectives.

3.3.4 Active Agents

Many modeling environments support a single modeler at a time. However, it has to be assumed that multiple agents are concurrently working on a particular model or architecture. A modeling approach has to allow modules which may be modified concurrently by active agents. The modeling language also has to be simple enough to enable agents with diverse backgrounds to participate in the process.

3.3.5 Self-Organization and Emergent Behavior

In order to support self-organization of active agents working concurrently on a model or architecture, a communication infrastructure is needed. This support system should allow working in parallel and the identification and negotiation of (interoperability) solutions (a posteriori and a priori).

4 Supporting Interoperability in Dynamic Environments

There are some challenges which need to be tackled in order to support interoperability in dynamic enterprise environments.

Interoperability seen from this point of view is a process that follows the evolution of the (enterprise) system. It is assumed that the different sub-systems of any enterprise system are in flux.[1] Non-interoperability between two or more subsystems may emerge at any time. Integration is not a solution to this type of system, as any small change in one system part would require actions to be taken at any other part. Interoperability of these parts would assume some level of "autonomy" of each part and provide the means to allow loosely coupled parts. Yet the same point of view requires monitoring and simulation facilities which allow to continuously update the models.

For maintaining the interoperability of enterprise systems (active) agents in the system require support to recognize events that lead to non-interoperability situations. This support may take the form of Key Performance Indicators, or might be hardware sensors which monitor the "state" of the enterprise. Agents are required to communicate with others to self-organize (possible within a small group) and negotiate a course of action for reaching an interoperable state.

In order to understand the paths taken during the discussions, negotiations and decisions by the agents, thc history of the models needs to be recorded. This history will support other agents to understand the paths taken by the team working on interoperability.

The initial work described here is a possible next step following other research on systemic and sustainable interoperability. The major difference to existing approaches is the special point of view which assumes a dynamic, complex and chaotic environment.

Acknowledgments This work has been partly funded by the European Commission through the Project Marie Curie—Industry and Academia Partnerships and Pathways (IAPP) program project: IANES (Grant Agreement No. 286083). The authors wish to acknowledge the Commission for their support. For more information on the IANES project see http://www.ianes.eu.

References

1. von Bertalanffy, L. (1969). *General system theory—foundations, development, applications*. New York: George Braziller.
2. Doumeingts, G., Ducq, Y., & Chen, D. (2009). *System theory to support enterprise interoperability science base. Proceedings of the 15th International Conference on Concurrent Enterprising: ICE2009-Collaborative Innovation: Emerging Technologies, Environments and Communities*.
3. Naudet, Y., Guedria, W., & Chen, D. (2009). Systems science for enterprise interoperability. *Proceedings of I-ESA Conference 2009*, Beijing.
4. Ducq, Y., Chen, D., & Doumeingts, G. (2012). A contribution of system theory to sustainable enterprise interoperability science base. *Computers in Industry, 63*, 844–857.
5. Dassisti, M., Jardim-Goncalves, R., Molina, A., Noran, O., Panetto, H., & Zdravković, M. (2013). Sustainability and interoperability: Two facets of the same gold medal. In Y. Demey, & H. Panetto (Eds.). *On the Move to Meaningful Internet Systems: OTM 2013 Workshops*, (LNCS, vol. 8186), Berlin: Springer.

[1] i.e. in a state of ongoing change.

6. Weichhart, G., Hämmerle, A., & Fessl, K. (2002). *Service-oriented concept of a holonic enterprise—enabling adaptive networks along the value chain. Proceedings of 5th IEEE/IFIP International conference on Information Technology for Balanced Automation Systems in Manufacturing and Services (BASYS)*, Kluwer Academic Publishers.
7. Ackoff, RL. (1971). Towards a system of systems concepts. *Management Science, 17*(11), 661–67.
8. Choi, T. Y., Dooley, K. J., & Rungtusanatham, M. (2001). Supply networks and complex adaptive systems: Control versus emergence. *Journal of Operations Management, 19*, 351–366.
9. Gleick, J. (1987). *Chaos: Making a new science*. New York: Penguin.
10. Senge, P. (1995). *The fifth discipline: The art & practice of the learning organization*. New York: Currency.
11. Weichhart, G. (2013). The learning environment as a chaotic and complex adaptive system. *Connecting Matter, Life, Culture and Technology 1*(1), 36–53. http://www.systems-journal.eu/article/view/130/138
12. Hite, J. A. (1999). *Learning in chaos—improving human performance in today's fast-changing, volatile organizations, houston*. Texas: Gulf Publishing Company.
13. Waldrop, M. (1992). *Complexity: The emerging science at the edge of order and chaos*. New York: Simon and Schuster.
14. Anderson, P. (1999). Complexity theory and organization science. *Organization Science, 10*, 216–232.
15. Bandura, A. (1986). *Social cognitive theory*. In R. Vasta (Ed.), *Annals of Child Development. Six Theories of Child Development.* (vol. 6, pp.1–60). Greenwich, CT: JAI Press.
16. Novak, JD., & Cañas, AJ. (2008). The *theory underlying concept maps and how to construct and use them*. Florida Institute for Human and Machine Cognition (IHMC), 2008. http://cmap.ihmc.us/publications/ResearchPapers/TheoryCmaps/TheoryUnderlyingConceptMaps.htm
17. Weichhart, G. (2013). Supporting interoperability for chaotic and complex adaptive enterprise systems. In YT Demey & H. Panetto. (Eds.) OTM 2013 Workshops, (LNCS, vol. 8186, pp. 86–92), Springer.
18. Naudet, Y., Latour, T., Guedria, W., & Chen, D. (2010). Towards a systemic formalisation of interoperability, *Computers in Industry 61*, 176–185, Elsevier.
19. Vernadat, FB. (2001). UEML: Towards a unified enterprise modelling language. *Proceedings of the 3éConference Francophone de Modelisation et Simulation* (MOSI-M'01).
20. IDS Scheer AG. (2000). Handbuch zur ARIS-Methode, Version 5, Saarbrücken.
21. Vernadat, FB. (2002) Enterprise Modeling and Integration (EMI): Current status and research perspectives. *Annual Reviews in Control 26*, 5–25, Elsevier.
22. Oppl, S., & Stary, C. (2013). Facilitating shared understanding of work situations using a tangible tabletop interface. *Behavior & Information Technology*.
23. Jonkers, H., Lankhorst, M., Quartel, D., Proper, E., & Iacob, M-E. (2011). ArchiMate(R) for integrated modelling throughout the architecture development and implementation cycle. In 13th IEEE Conference on Commerce and Enterprise Computing (CEC), 2011. (pp. 294–301).
24. Weichhart, G., Feiner, T., & Stary, C. (2010). Implementing organisational interoperability—the sudden approach. *Computers in Industry, 61*, 152–160.

An Interoperability Test Bed for Distributed Healthcare Applications

Robert Snelick

Abstract Standards provide the foundation for ensuring interoperability, but if they are not implemented correctly or consistently their value is diminished leading to problematic installations and higher costs. Conformance and Interoperability testing is essential for ensuring standards are implementable and implemented correctly; however, limited budgets often preclude adequate attention to this testing during the product development life cycle. Automated testing can help on both fronts. We propose an Internet-based interoperability test bed that extends a testing infrastructure and conformance testing framework. The operational aspects of the architecture are presented against the backdrop of distributed health information technology applications and a representative case study. Although the concepts and methods are applied to the healthcare domain, they have broad applicability.

Keywords Conformance · Data communication standards · Healthcare · Interoperability · Messaging systems · Testing

1 Introduction

A major challenge for the healthcare industry is achieving interoperability among proprietary applications marketed by different vendors. Each healthcare entity may have to use multiple applications to capture and share administrative and clinical data. Seamless and reliable exchange of information is difficult to attain. Recent mandates in the United States have ignited a renewed push towards interoperable healthcare information systems based on standards. Specification and wide-spread

R. Snelick (✉)
National Institute of Standards and Technology (NIST), 100 Bureau Drive Stop 8970, Gaithersburg, MD 20899, USA
e-mail: robert.snelick@nist.gov

K. Mertins et al. (eds.), *Enterprise Interoperability VI*,
Proceedings of the I-ESA Conferences 7, DOI: 10.1007/978-3-319-04948-9_41,

use of standards provide the foundation for ensuring system capabilities and the ability to exchange information reliably; however, standards alone are not enough. Testing and certification programs are necessary to assess implementations for conformance and interoperability. In prior work [1–3], we presented the testing infrastructure and testing tools we developed for conducting conformance testing. These tools have been used extensively in the US as part of the health information technology (HIT) certification program related to meaningful use of electronic health record (EHR) systems [4–6]; internationally at the Integrating the Healthcare Enterprises (IHE) testing events [7, 8]; and at production sites. We now present an extension of this work by describing the fundamental components of the testing infrastructure and operational facets of an interoperability test bed (ITB). A case study for patient identification cross-referencing is employed to describe a representative use case and to aid in explaining the ITB capabilities and operation. An overview of the applications involved, the roles the applications play, and the interactions between the applications is presented. We then describe how the interactions are tested in the operational environment. Central to the design of the ITB is that it must provide Internet-based, on-line, continual access to conformance and interoperability testing services.

The Health Level Seven (HL7) Version 2 messaging standard [9] is the focus for discussing the aspects of the ITB. HL7 is a data exchange standard for transmitting clinical and administrative information between healthcare applications [9]. Typical HL7 messages include admitting a patient to a hospital or requesting a laboratory order for a blood test. Conformance message profiles are used to constrain the set of data that is exchanged between applications that support the HL7 standard; these message profiles are the source of the conformance requirements that verify correct data exchange [2].

2 Testing Infrastructure

The central design principle behind the construction of a test system is a modular and flexible approach afforded by a testing infrastructure. The testing infrastructure provides reusable components with well-defined interfaces that can be utilized in a test system as needed. The obvious advantage of this approach is that the user can combine the components of the test system into whatever configuration is most suitable for solving the problem at hand. The system is also easily extensible since each test system component runs independently, and adding new services to the test system can be accomplished readily. The test systems described and referenced in this work, including the ITB, adopt this approach.

The testing infrastructure is composed of three types of high-level components: the services, the test harness (services composition), and network functions. Additionally a test management system may be employed, but it is not central to the test infrastructure—it is likely a separate system. Services provide the testing functionality. It is apparent that for this model to be used effectively in conducting testing, it will be necessary to facilitate the interactions between the user and the supported testing services. A test harness is necessary to orchestrate the services to conduct a test. A network is utilized to route messages and may include logging and proxy capabilities. A test management system (e.g., IHE Gazelle [11]) can be used to assign, manage, and track each set of tests. A certification testing lab may utilize such a system in their process.

Services provide a portfolio of testing-related capabilities that are specified to perform unique functions within the testing infrastructure. Each service has well-defined responsibilities and authority, and they work together in collaboration with other services to support the execution of test cases. The generation service creates a message instance based on a message profile (template) and a data set. The validation service evaluates a message instance against requirements stated in the standard. Data content also can be evaluated to support application functional testing. Test agents (also referred to as simulators) are implementations of actors (or applications) that support the functionality of the underlying specification of the actor and need only support the functionality of the actor to support testing of applications. The resource repository contains the artifacts necessary for conducting tests and facilitating test execution processing. Artifacts that support test case execution typically include the test execution script, test case descriptions, test data specifications, test data, and test specifications (e.g., conformance profiles). The testing infrastructure is not bound to a defined set of services; rather, the set is determined by the objectives of the testing goals and system.

3 Test Environments

Recognizing that testing is a complex, multidimensional, and often incremental process leads us to consider the use of multiple environments for conducting testing. In an earlier work [12], we identified three distinct environments and described the testing activities that can be performed within each environment. These environments include the data instance test environment (not discussed here), the isolated system test environment, and the peer-to-peer system test environment. The delineation of environments and their testing capacity is intended to facilitate a more structured approach to testing in which the relationship between test requirements and testing, along with an understanding of the capabilities and limitations of testing tools, is more clearly defined.

In the isolated system testing environment a test is conducted with the SUT and a test tool. Conformance testing, including data exchange and functional behavior, is the main objective in using this model. Functional behavior assessment is achieved with a test scenario in which a sequence of orchestrated transactions is composed to probe certain requirements. The test tool includes functionality of an application (i.e., test agents) that an SUT would typically interact with in an operational environment. Isolated system testing typically accounts for the majority of testing that is conducted. Once a system has successfully undergone conformance testing, interoperability testing usually proceeds more easily.

Peer-to-peer system testing is designed to test interoperability among one or more systems and is the focus of the ITB. The peer-to-peer system testing environment poses different and significant challenges in testing from that of isolated system testing. In this environment, data exchange occurs among a group of systems, and the testing tool no longer has direct interaction with the systems under test. Here an intermediary or a proxy can be employed to intercept, log, and route messages to their intended destination. Peer-to-peer system testing may include some or all of the conformance testing described for isolated system testing. When conformance testing is conducted in advance, peer-to-peer testing specifically targets both syntactic interoperability and semantic interoperability testing. The conformance test cases that were developed for isolated system testing can be leveraged in peer-to-peer testing. The abstract test cases could be identical; however, execution of the test steps, configuration requirements, and assertion assessment will differ. By ascertaining that the conformance requirements are now met in an environment where the SUTs are interacting, we can make a declaration of the interoperability capabilities of the systems.

IHE testing involves a number of incremental testing steps. First, vendors conduct conformance testing of their products. This task involves testing in isolation to determine if the system implements the requirements specified in the standard. These tests are labeled *pre-connectathon* tests and correspond to the isolated system test environment. The NIST IHE Patient Identification Cross-referencing (PIX)/Patient Demographics Query (PDQ) [7, 8] and Patient Care Devices (PCD) pre-connectathon test tools are examples of production implementations of the isolated system test environment. Interoperability testing (peer-to-peer testing) is conducted at an event called a *connectathon* in which scores of vendors bring their products to a central site and live-monitored tests are performed over a period of a week [10]. Such concentrated events are useful as vendors can interact with many other vendors in a short period of time; however, connectathons occur infrequently (once a year in the United States and Europe) and are costly. One objective of the ITB is to provide an intermediate format in the form of an on-line *virtual connectathon* in which interoperability testing is *always* available. In the virtual connectathon environment, participants indicate what they want to test and publish their availability. Once testing partners reach agreement, their systems are configured in the test bed and they can proceed with testing. The ITB leverages the testing infrastructure and implements the peer-to-peer testing environment.

The ITB isn't fundamentally that different from the test system for the pre-connectathon isolated test environment. The main divergent point is testing multiple *real* systems instead of one, i.e., replacing some or all of the test agents in the isolated system environment with vendor products. This difference does, however, present noteworthy technical challenges including the scheduling of participants, sequencing of events and notification to the participants, and capturing/forwarding messages and then mapping the messages to the corresponding test case interaction.

4 Case Study: Patient Identification Cross-Referencing

In this section, we describe an example case study and associated workflow that demonstrates typical transactions among disparate healthcare information technology systems. IHE publishes integration profiles that describe many healthcare workflows [7]. We describe a typical workflow of cooperating patient identifier and document management systems. The data exchange standard involved in this use case is HL7 V2. For the purposes of demonstrating the test execution of the ITB the scenario focuses on the HL7 V2 messaging PIX aspects.

Our example examines a healthcare system made up of a Patient Identifier Cross-referencing (PIX) Domain and a Cross-Enterprise Document Sharing (XDS) system. We examine a typical PIX domain made up of three disparate actors: a PIX Source, a PIX Manager, and a PIX Consumer. A PIX Source is used for adding and modifying patient demographic data; a PIX Manager is used for managing and cross referencing patient identifiers from different domains; and a PIX Consumer is used for querying a PIX Manager for patient identifiers. All communications among the actors are accomplished through the exchange of HL7 V2 messages. An XDS system supports registering and retrieving documents across enterprises within an administrative domain.

Healthcare systems can be divided into various administrative domains, each responsible for managing a set of patient information. Patients may require services provided in different healthcare domains. When this situation occurs, different electronic health records for the same patient may exist in more than one domain. It is clearly desirable to be able to recognize when multiple records exist belonging to the same patient. IHE has addressed this problem by delegating the responsibility for determining when two patient identifiers belong to the same patient, and hence there are two records belonging to the same patient, to the PIX Manager actor. Our case study (Fig. 1) examines some of the data points the PIX Manager must consider; for example, when it is determined that two patients are actually the same patient, how will the information be propagated throughout the healthcare continuum? The resolution is important since a single patient identifier is typically used to retrieve documents about a patient from a repository.

The scenario (workflow) is predicated on a family doctor at Practice Y seeking pertinent medical documents for patient Donna Bee. Practice Y relies on Hospital X's patient management and document repository systems; however, before the doctor can retrieve the documents from the repository he must obtain the patient identifier used by Hospital X's document repository. Practice Y acting as a PIX Consumer queries (via HL7 V2 message) the PIX Manager using its patient identifier for Donna Bee (PY-3322) and requests the master patient identifier in Hospital X's domain. The PIX Manager returns an HL7 response message containing the patient identifier (HX-5924). Once the patient identifier is acquired, it can be used to query for available documents in the registry and retrieve these documents from the repository. This scenario assumes that the relevant documents have already been uploaded into the repository via a document source actor.

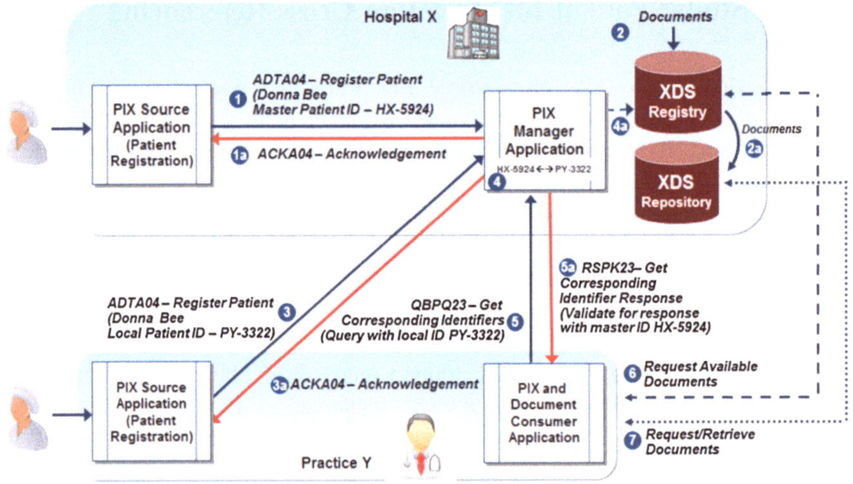

Fig. 1 Patient identification cross-referencing workflow

The three steps below show a test case derived from the above scenario. Although the test case is simple, it provides a representative set of applications and steps that have to be accounted for in testing and supported by the ITB constituent components. One patient (Donna Bee) is registered in different domains. The registration messages are sent to a PIX Manager. A query is sent to resolve a reference to patient Donna Bee who is expected to be found.

Step 1: Hospital X PIX Source sends a registration message (ADT^A04) to register patient Donna Bee in domain HOSP-X. Patient ID is HX-5924. The PIX Manager shall register the patient and send an Acknowledgement (ACK) message back the PIX Source.

Step 2: Practice Y PIX Source sends a registration message (ADT^A04) to register patient Donna Bee in domain PRAC-Y. Patient ID is PY-3322. The PIX Manager shall register the patient and send an ACK message back to the PIX Source.

Step 3: Practice Y PIX Consumer sends a query message (QBP^Q23) to ask for Donna Bee's ID in domain HOSP-X using their ID PY-3322 in domain PRAC-Y. The PIX Manager is expected to respond to the query with Donna Bee's ID HX-5924 in domain HOSP-X.

Successful completion of the case study requires that each application involved in the process correctly performs certain tasks that can be measured based on the application's externally observable behavior. The requirements on the application's external behavior can be formulated as a set of conformance and interoperability testing requirements. Although no explicit interoperability requirements have been defined, successful completion of all of the steps in the workflow provides a prima facie demonstration of interoperability among the systems.

5 Interoperability Test Bed

The interoperability test bed supports the peer-to-peer testing environment that targets multiple systems under-test. Figure 2 depicts a scaled-down design of the architecture that employs a set of fundamental components and operating procedures necessary to conduct testing. A description of each component and its capability is described followed by an abbreviated step-by-step test flow.

5.1 Operational Functions

The interoperability test bed has a number of logical operational divisions. A ***scheduling system*** is required to coordinate the vendors and match up interest in test selection, actors, and time availability. Vendors input their testing interest and capabilities along with time availability, and the scheduler will pair up common requests and will notify the participants. The ***configuration utility*** records the connection and addressing information required for communicating with each participant. This information is necessary for the ITB to prime the proxy to handle intercepting and forwarding messages. ***Test setup*** includes informing each of the participants of the test instructions associated with the test case. These include their roles, actions required, data requirements, and system configuration (e.g., loading test data in the data base). The ***test manager*** controls the overall operation of the ITB. These activities include managing simultaneous test executions and initiating test case instance executions.

The test management system relies on a ***test orchestration*** component to organize and coordinate the collective activities of the vendors and the test system. Information (e.g., instructions and status) is disseminated through a common shared user interface and a participant-perspective user interface. A test script encapsulates the choreographed steps and participant actions. The ***test engine*** is a workflow management tool that directs the execution of defined test cases through the orchestration of testing services. The test execution operates at a level below (within) the test orchestration and coordinates the measurement aspects (e.g., message validation) of the test case instance. The test execution is independent of the test orchestration since its scope is coupled to the test case. Therefore the same test execution script is applicable to isolated system testing. For each task that is performed either by the test system or participant, a status update is ***broadcast*** (served pages) to the participants. This includes a human test manager observing the execution of the test instance from a test management dashboard. As messages progress through the system they are captured, stored, analyzed, and forwarded. The ***proxy***, ***message database***, and ***validation*** provide these capabilities. For each interaction (message), the test execution assembles the various artifacts and submits them to the validation service for evaluation. A machine-processable report is returned and rendered to the participants. A test case instance consists of numerous

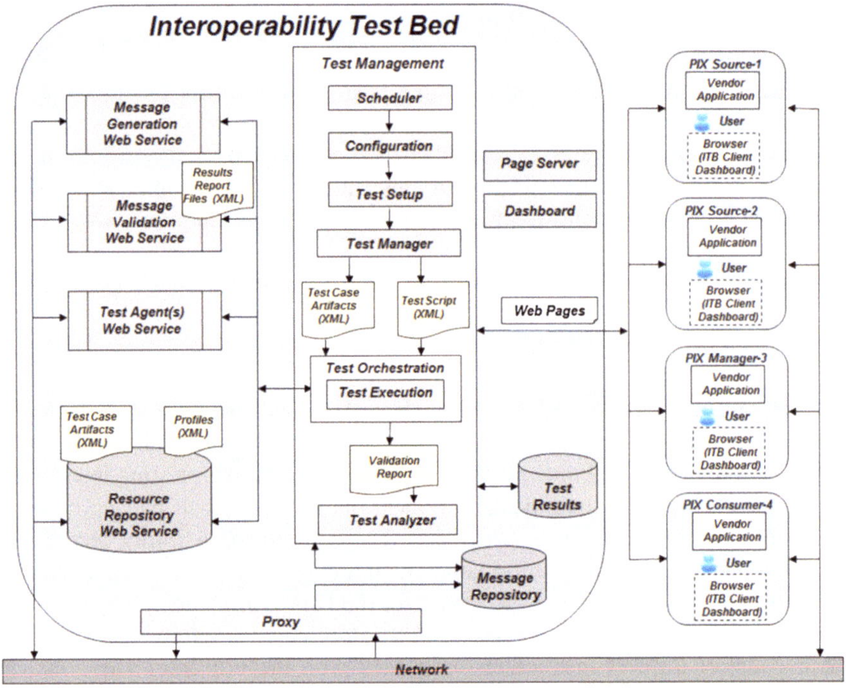

Fig. 2 Interoperability test bed architecture

steps for the applications being tested and creates many individual validation reports. A ***test analyzer*** is necessary to make a collective determination of the test results. The test analyzer can use the test script for navigating, linking, and analyzing the array of validation reports. Upon completion of the test instance execution, the test results are stored for analysis and auditing purposes.

The ITB includes the capability to supply a vendor system(s) when not all the participant systems needed are available or not available at the desired time. These features include message generation and test agent services. Having one SUT and all test agents is essentially equivalent to isolated system testing provided by the pre-connectathon testing tools. The ITB encapsulates this model.

5.2 Test Flow

The test flow describes each of the actions required by the ITB and participants to perform our test case instance. The test case calls for four applications or application modules. There are two PIX sources (PS-1 and PS-2), a PIX Manager (PM-3), and a PIX Consumer (PC-4). Our focus is on the operational aspects of the test instance. Figure 3 illustrates the events as they progress through the ITB for the transaction 1, interaction1 described below.

ITB Actions: PS-1 is provided a test data specification that includes data (e.g., patient name and DOB) associated with the test story and consists of typically available information in the administrative and clinical setting. Together, the test story and the test data specification provide sufficient information to be entered into the SUT for a particular test case such that a message can be generated.

PS-1 Actions: PS-1 loads the test data provided in the test data specification; typically, loading is a manual process performed using the user interface (UI) capabilities of the SUT. Once all relevant test data are loaded into the PS-1, a message is generated and sent. Based on the configuration, the message (intended for the PM-3) is sent to the ITB proxy. PS-1, via the dashboard indicates that it has sent the message; this controlled execution eases the test instance navigation for the test execution engine.

ITB Actions: The ITB proxy captures and stores the message in the message repository along with pertinent message meta-data sufficient for mapping the message to the interaction. The ITB test engine retrieves the message from the message repository and the conformance validation artifacts from the resource repository and invokes the validation service to validate the message. The ITB test analyzer evaluates the interaction along with validation results to determine if any errors detected would prevent the test from continuing. Note that inconsequential conformance issues are not pertinent at this point; the focus is on the delivery of the message and key data elements that might prevent other participants from completing their tasks. If a severe fault is detected, the test orchestration notifies all participants accordingly via the common shared display capability and then ends the test instance. The ITB proxy forwards the message to the PM-3. The test orchestration reports that interaction 1 is completed and the common user interface display is updated accordingly. The validation report for the interaction is made available to the participants. The ITB now instructs PS-2 to begin interaction 2. Note that the ITB also updates the status at intermediate events, for example, when the message is received by the proxy or forwarded.

Upon completion of the test case instance, the test analyzer evaluates the collective results and determines the outcome of the test instance. This information, along with the individual interaction validation reports, is provided to the participants. The vendors can use these test results, as well as logging information and the observed behavior of their system to further assess the performance of their product. A test manager can review the reports to make an assessment of the test instance. All related test instance data are stored in the test management system for auditing purposes if desired.

Although we have presented here a simplified portrayal of the architecture and case study, it is still evident that the ITB provides the capabilities to conduct conformance and interoperability testing in support of distributed healthcare applications for a broad assortment of test cases essential for comprehensive testing. For example, an interoperability test may stipulate incorporation and display of laboratory results into an EHR. In this case, completely automated black

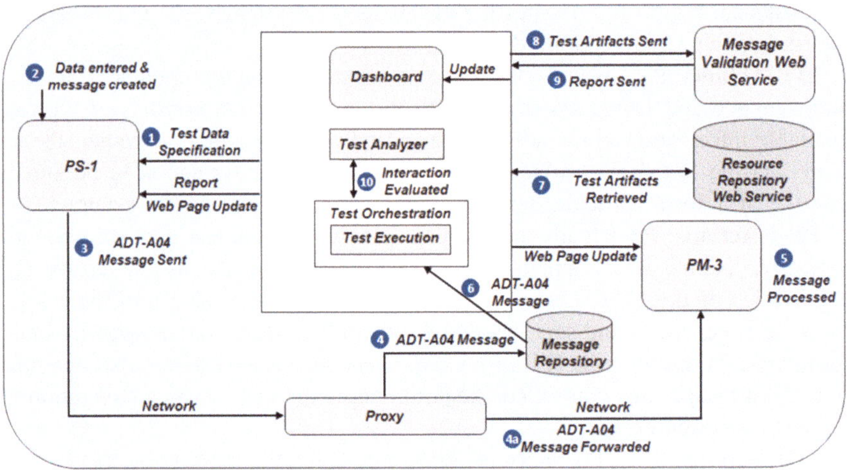

Fig. 3 Test flow for transaction 1, interaction 1

box testing is unattainable; however, an inspection document based on the test data is dynamically created and provided for this test step. Screen shots of the vendor's display and/or data base can be captured and submitted to the test system for visual evaluation by a monitor. A monitor will use the inspection document to verify expected behavior.

6 Conclusion

Interoperability between disparate applications can be achieved better through the use of standardized interfaces. Even if the same standard is implemented in the applications, interoperability is not assured due to two primary reasons: (1) the same sets of options allowed by the standard are not implemented by the developers, a problem that can be addressed with conformance provisions offered by the standard; (2) the standard is implemented incorrectly in the applications. These issues are addressed through conformance testing. Applying conformance processes and successfully conducting conformance testing will not ensure interoperability, but these actions will increase the likelihood of applications interoperating. Beyond conformance testing, interoperability testing is employed. Interoperability testing has been challenging because assembling even a minimum number of application vendors in one place at the same time to conduct this type of testing is difficult. For years, the IHE connectathon has provided an invaluable venue for massive interoperability testing; however, this event occurs only once a year. The proposed *always-available* Internet-based interoperability test bed seeks to fill this gap and complement connectathon events.

References

1. Snelick, R., Gebase, L., & O'Brien, G. (2009). A framework for testing distributed healthcare applications. In Arabina, H. R., & Reza, H (Eds.), *2009 Software Engineering Research and Practice (SERP09), WORLDCOMP'09*. Las Vegas, NV: CSREA Press.
2. Snelick, R., Rontey, P., Gebase, L. & Carnahan, L. (2007). Towards interoperable healthcare information systems: The HL7 conformance profile approach. In Goncalves, R., Mertins, K., & Zelm, M*Enterprise interoperability II: New challenges and approaches* (pp. 659–670). London: Springer.
3. NIST Laboratory Results Interface (LRI). *EHR meaningful use conformance testing tool.* http://hl7v2-lab-testing.nist.gov/mu-lab/
4. NIST Resources and Tools in Support of Health IT Standards and the ONC Meaningful Use certification program. http://healthcare.nist.gov/
5. Centers for Medicare & Medicaid Services (CMS). EHR incentive program. http://www.cms.gov/Regulations-and-Guidance/Legislation/EHRIncentivePrograms/index.html
6. ONC Certification Programs. http://www.healthit.gov/providers-professionals
7. Integrating the Healthcare Enterprises Technical Framework. http://www.ihe.net
8. NIST PIX/PDQ Pre-connectathon Conformance Testing Tool. http://pixpdqtests.nist.gov:8080/
9. Health Level 7 (HL7) Standard Version 2.7. Retrieved January, 2011, from http://www.hl7.org
10. IHE Connectathon. http://www.ihe.net/Connectathon/
11. Gazelle Test Management Framework. http://gazelle.ihe.net/
12. Gebase, L., & Snelick, R. (2010). Testing environments for assessing conformance and interoperability. In Arabnia, H.R., Reza, H., & Deligiannidis, L *2010 Software Engineering Research and Practice (SERP10), WORLDCOMP'10.* Las Vegas, NV: CSREA Press.

Integrated Interoperability Capability Model for Adaptive and Sustainable SMEs

Nader Nada and Zulfiqar Ali

Abstract Interoperability is the ability of making systems and organizations to work together (inter-operate). Integrated interoperability represents the enterprise collective interoperability capability which composed of System interoperability and business interoperability. Business interoperability is further composed of other sub-capabilities: Strategy capability, operational capability, managerial capability, which have indirect impact on the effectiveness of system interoperability. Interoperability is broadly seen as an essential component of SMEs competitiveness and sustainability. Our research primary focus is the 'how' of achieving a state of interoperability that engages technology, people organizational mission, value and culture in collaborations to exchange information, knowledge or services—so that all can advance mission or achieve business success and sustainability. The objective of this research is to develop an effective and integrated interoperability Model for sustainable and adoptable SMEs. The model is empirically validated in Danish SMEs to identify the correlation between the integrated interoperability capability and the adaptive capabilities through an integrated interoperability capability model analysis. The empirical data analysis reveals that there is strong positive correlation between the firms integrated interoperability capability and their adaptive capability including: agility, resilience, and innovation.

Keywords Integrated interoperability · Adaptive capability · SMEs · Resilience · Innovation · Agility

N. Nada (✉) · Z. Ali
Buyukcekmece Campus, Fatih University, 34500 Istanbul, Turkey
e-mail: dr.nader.nada@gmail.com

Z. Ali
e-mail: zulfiqar.engr@gmail.com

K. Mertins et al. (eds.), *Enterprise Interoperability VI*,
Proceedings of the I-ESA Conferences 7, DOI: 10.1007/978-3-319-04948-9_42,

1 Introduction

Interoperability, adaptability and sustainability are becoming indispensable in today's economy: They allow companies to offer innovative products, services and innovative marketing approach by bundling of physical products and electronic services from different partners and to conduct business in a more effective way by redistributing tasks between organizations. Provided the increasing need for external collaboration, firms realize that it is time consuming and expensive to individually negotiate and set up relationships with a larger number of organizations, in particular with SMEs. They are also experiencing that the manual exchange of information and the lacking of interoperability generates significant costs. Brunnermeier estimates that "imperfect interoperability costs the US automotive industry about $1 billion per year and delays the introduction of new vehicles by at least 2 months" [1].

Interoperability is usually discussed in the context of technical integration related to platforms, network devices and communication protocols, as well as syntactic and semantic data formats [2]. This is reflected by the most cited definition of interoperability, which characterizes interoperability as "the ability of two or more systems or components to exchange information and to use the information that has been exchanged" [3]. Over the last decade, internet and web service technologies have significantly fostered interoperability at the transport and communication level [4]. But with the broader use of these technologies, a multitude of interoperability issues have to be solved at higher levels in order to allow for seamlessly integrated collaboration. Whereas many authors have highlighted the need for aligning the semantics, some of them consider interoperability in the broader context of value chain integration. Yang mention Business process compatibility, adaptability of business processes, leveraging legacy assets, support for business transactions and network security services as important factors driving interoperability in the context of integrated value chains [5]. In this broader sense, Legner introduce the term business interoperability which denotes "the organizational and operational ability of an enterprise to cooperate with its business partners and to efficiently establish, conduct and develop IT-supported business relationships with the objective to create value" [6].

In the next sections, we first review prior work in order to define (business) interoperability and establish a baseline for developing the Integrated Interoperability Model (IIM). Second, we will outline a research agenda related to integrated interoperability (interoperability capability) and adaptive capability. Third, we will empirically validate the IIM.

2 Literature Review

Literature revealed that earlier research has mainly focused on the information structures and interfaces or the communication and transport level, whereas recent work introduces a broader perspective on interoperability on the one side, and increasingly addresses semantic aspects and business process compatibility on the other side. Another stream of research investigates product interoperability, in particular in those industries where complex systems are built from a larger number of components. In order to come up with a more holistic view on interoperability, a number of initiatives have tried to systemize and classify the different interoperability aspects into comprehensive interoperability frameworks, among others the e-Government Interoperability Framework (e-GIF), the Levels of Information Systems Interoperability framework (LISI) or the European Interoperability Framework (EIF).

Generally, the initiators of these frameworks have been practitioners or public administrations which are pursuing the goal of standardizing across distributed organizations and avoiding technology vendor lock in. These interoperability frameworks distinguish different layers of interoperability and describe artifacts or guidelines or standards for each of these layers. With the exception of the EIF, they distinguish the infrastructure, data/message and functions/services layer. Only the EIF introduces organizational aspects of interoperability, e.g. the definition of business goals and the modelling of business processes to enable different organizations to work together.

In addition, most frameworks introduce either explicitly or implicitly an evolutionary perspective and suggest a linear advancement from lower to higher levels of interoperability. Peristeras relate existing interoperability frameworks to theoretic concepts from linguistics and semiotics and derive the Connection, Communication, Consolidation, and Collaboration Interoperability Framework (CIF) for information systems interoperability [2]. A more holistic approach is taken by the Business Interoperability Framework which explicitly introduce organizational and management related layers [6]. Today, interoperability is often discussed from a purely technical perspective, focusing on technical standards and IS architectures. While prior research suggests concepts for presenting, collecting, exchanging, processing and transporting data, we think that a systematic analysis of strategic, organizational and operational issues associated with interoperability is currently lacking especially in SMEs.

2.1 *Business Interoperability*

The analysis of many definitions along this spectrum leads to the conclusion that interoperability is a very context-specific concept. Rather than aiming for a single, one-size-fits-all definition, it seems more promising to carefully consider the terms' specific contours in each case up for discussion, but otherwise to operate pragmatically with a rather open working definition. At the definitional core of the multi-layered and multifaceted concept "interoperability" is what we might describe as the ability to communicate and transfer useful data and other information across systems (which include all the departments of an organization), applications, or components.

The term Business Interoperability is defined as "the organizational and operational ability of a company to cooperate with its business partners and to efficiently establish, conduct and develop business relationships with the objective to create value" [7]. Business Interoperability is usually understood as the capability of a company "to exchange information and be able to use it", or simply put: the capability of a company to work together or to collaborate, "working together" of company means that they execute a collaborative business process. Business Interoperability can also be defined as the capability of an organization to execute a collaborative business process among its departments. Business interoperability is a holistic approach which if further dig deep consists and backed up by the business capabilities of a company. Companies are mostly isolated into to information silos which operate individually. Even if company put emphasis on one area will not provide the desired results of integration as interoperability is companywide approach. Business operability is fed by the business capabilities of a company, which includes Strategic, operational and Managerial capabilities.

Business interoperability should be indispensible part of business strategy because every business in order to survive and thrive in a competitive business environment needs to possess a certain level of strategic capability. Operational integration is the need of time especially for SMEs to compete and sustain competitive advantage over their rivals. Interoperability mostly focused on systems by excluding human factors. We have included strategic, operational, and managerial capabilities as subsets of business interoperability to make it a holistic approach for SMEs.

2.1.1 Strategic Capability

Every business in order to survive and thrive in a competitive business environment needs to possess a certain level of strategic capability. The type of strategic capability that the company needs at a specific time is determined by the legitimizing forces and the threats/opportunities in the future business environment [8]. Legitimizing forces are the factors that establish the purpose of the business and the criteria for its success. These forces evolve from the external environment of the international business and involve: The determination of the key attributes measuring successes in the business [9]; the determination of the aggressiveness of behaviour pertaining to each key attribute [10]; the determination of the rules of the game for the business organization; the determination of the driving forces; the power structure of the business organization. Therefore strategic capability can be defined as the set of capacities, resources, and skills that build a long-term competitive advantage for SMEs. It is the ability to work towards a vision built on value adding elements relevant of the firm including interoperability as a strategic priority and a plan for profits that has the right balance between taking advantage of short term opportunities and longer term actions to achieve business sustainability.

2.1.2 Operational Capability

Operational capability is the ability to align critical processes, resources and technologies according to the overall guiding vision and customer focused value propositions coupled with the ability to deliver these processes effectively and efficiently. Indicators of operational capability include process management and performance measurement.

Operational capabilities are firm-specific sets of skills, processes, and routines, developed within the operations management system, that are regularly used in solving the problems faced by a unit and which provide that unit and, ultimately, the firm with the means of configuring the resources of the operations management system to meet the firm's distinctive needs and challenges. The performance of operations is normally associated with competitive criteria (quality, cost, flexibility and delivery) with minor variations have been reported by many authors in previous decades [11, 12].

Operations performance and competitive priorities are repeatedly analyzed based on the logic of trade-offs. Trade-offs can be expressed as a function of two variables that are inversely correlated [13]. The concept of trade-off should guide production decisions on the factory floor and throughout the supply chain [11].

2.1.3 Managerial Capability

Managerial capability is the ability of managers to create a strong workplace and culture which facilitates the employees to grow and engage, and at the same time business goals and objectives are achieved. It includes leadership qualities, collaborative decision making, and nurturing creativity and innovation. Early works in this stream of research have investigated this relation on a Conduct Structure Performance basis by focusing on "traditional" Schumpeterian determinants such as market structure, firm size and company R&D and innovation effort. However, neo-Schumpeterian scholars partly influenced by the management and the Resource-Based-View (RBV) theory of the firm realized that firm idiosyncratic capability in mastering innovation processes have a comparable weight in explaining firm potential to get profit rates higher than competitors when confronted with "traditional" factors.

But measuring firm managerial capacity in producing innovation is far harder than accounting for the role played. This depends on the higher immaterial and fuzzy nature of managerial capabilities that can be approximated by variables that poorly can give an account of the phenomenon. Furthermore, this problem becomes trickier when one wants to separate "general" managerial capability referring to the whole management of firm divisions and activities from the specific entrepreneurial ability of managing innovation processes. Bughin for instance, explored the Schumpeterian links between size, market structure and innovation, by controlling for a series of managerial factors thought of as affecting innovation success rate and efficiency [14].

The problem of this literature is twofold:

1. It does not look explicitly at the Innovative Managerial Capacity (IMC) of companies, but more at the general company capabilities.
2. It uses only "indirect" measures of managerial capacity.

Managerial capabilities are defined as "the capacity of managers to purposefully create, extend, or modify the resource base of an organization". Bititci highlighted the need to better understand what managers perceive and act upon in response to unstable environments [15].

2.2 System Interoperability

System interoperability is deifned as the more technological aspects of collaboration and the interralated components of that system. When a company integrate all the systems by any software or technological means to achieve the flow of data in real time, is called system interoperability. IEEE has defined system interoperability as "the ability of two or more systems or elements to exchange information and to use the information that has been exchanged".

2.3 Adaptive Capability

Adaptive capability can be defined as an organization's "ability to identify and capitalize on emerging market opportunities [16, 17]. This is having the maturity and capacity to adjust the Managerial, Strategic and Operational Capabilities by sensing, preparing and responding to environmental change." Tuominen distinguish between three interrelated aspects of adaptability such as technological aspects (system interoperability), external market aspects, and internal organizational aspects (business interoperability) [18].

The development of adaptive capability is often accompanied by the evolution of organizational forms. Here the Adaptive capabilities consist of three variables, Agility, Resilience, and Innovation which will be described in the following three subsections.

2.3.1 Agility

Agility is the ability of a business to adapt rapidly and cost efficiently in response to changes in the business environment. Business agility can be maintained by maintaining and adapting goods and services to meet customer demands, adjusting to the changes in a business environment and taking advantage of human resources.

Agility is a concept that incorporates the ideas of flexibility, balance, adaptability, and coordination under one umbrella. In a business context, agility typically refers to the ability of an organization to rapidly adapt to market and environmental changes in productive and cost-effective ways. The agile enterprise is an extension of this concept, referring to an organization that utilizes key principles of complex adaptive systems and complexity science to achieve success. One can say that business agility is the outcome of organizational intelligence.

2.3.2 Resilience

The way a business survives and thrives, even when things go wrong. Resilience is the key to sustainability and business continuity.

A resilient SME has three main functions:

- Anticipate, respond and recover from disruptive events effectively.
- Transform when the current business model is no longer feasible.
- Adapt to changing circumstance.

Resilience is important to ensure not only business continuity, but also to protect the bottom line and create advantage over less adaptive competitors. Resilient organizations see an overall increase in their shareholder value and make a healthy recovery, whereas those organizations that fail to plan and deliver an effective response struggle to recover.

2.3.3 Innovation

Innovation is a process that takes ideas and transforms them into something useful. We believe that innovation is a holistic process that involves generating, selecting, developing and implementing ideas. To achieve innovation, the company environment must support the successful management of this innovation process. Innovation is recognised as a source of sustained competitive advantage, survival and growth. Innovation can bring numerous strategic advantages to the business.

These include delivering a unique product or service, changing the nature of competition by anticipating the emergence of a new technology, being first to market, or improving on competitor products or services.

3 Research Problem and Proposed Solution

Based on our literature review on interoperability and adaptability, there does not exist a theoretical Integrated Interoperability Model (IIM) which: (1) introduce the interoperability capability as an integrated and collective interoperability which combines both business interoperability and system interoperability (2) provide the linkage between the interoperability capability and adaptive capability including: agility, resilience, and innovation, and (3) Empirically validating the IIM for SMEs.

3.1 Research Hypothesis

Our research primary focus is the 'how' of achieving a state of interoperability that engages technology, people organizational mission, value and culture in collaborations to exchange information, knowledge or services—so that all can advance mission or achieve business success and sustainability. Our research hypothesis is that higher level of integrated interoperability capability is highly correlated with firm's adaptive capabilities such as agility, resilience, and innovation which have direct impact on SMEs business sustainability.

4 Integrated Interoperability Model (IIM)

The Integrated Interoperability Model is devised to explore the relationship of business capabilities and System interoperability (combined as Interoperability Capability) with Adaptive capability of SMEs. The business capability is divided into three sub-capabilities (Strategic capability, operational capability and Managerial Capability). The interoperability capability consists of two dimensions: (1) business interoperability and (2) system interoperability. The main objective of this study is to find the relationship of the collective Interoperability capability with the firm's Adaptive Capability which consists of agility, resilience, and innovation. Figure 1 depicts the integrated interoperability model in relationship with adaptive capability.

5 Research Method

The empirical data was collected by using an online questionnaire which consists of 116 questions which were designed after formulating the research questions and hypothesis. The questionnaire consists of five sections each dedicated for five

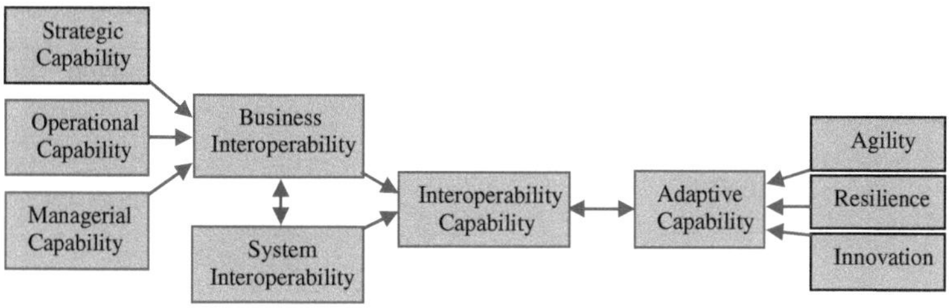

Fig. 1 Integrated interoperability model (IIM)

major variables (Strategic capability, Operational capability, Managerial capability, Adaptive capability and System Interoperability). The questions of Adaptive capability section are further divided into three sub branches which are Agility, Innovation and Resilience. All the questions have equal weight age.

The first section of the questionnaire contained general questions regarding company profile including the number of employees, customers (suppliers), annual turnover. Also how companies rate their business in comparison to your competitors or business sector against three criteria namely market share growth, Productivity growth and Customer satisfaction. The second section is about strategic capabilities, the third section is about operational capabilities, the fourth section is about managerial capabilities, the fifth section is about adaptive capabilities, and the sixth and last section is about system interoperability which assesses if the business has implemented or using an integrated management system (e.g. ERP, CRM, SCM) and how the business activities are being communicated.

The online questionnaire was randomly sent to 100 SMEs in Denmark. Data collection was done in 3 months started from June 2013 till end of August 2013. Thirty companies responded to online questionnaire with a response rate of 30 %. Due the insufficient data and large size of the company 5 responses were excluded from the data analysis. The collected data was analyzed and processed by using a software package JMP 10. For the data analysis we used the Pearson correlation to predict the relationship between variables which are Interoperability capability and adaptive capabilities. The adaptive capability have further three variables which are Agility, Innovation and Resilience. We used Pearson correlation for all variables as they are normally distributed. The Interoperability capability is an independent variable while Adaptive capability along with three sub variables is dependent variable (Table 1).

The table shows the value of Pearson correlation for interoperability capability with adaptive capability and its three sub variables. All the relationships are positive and significant. The relationship between Interoperability capability and adaptive capability is positive and significantly high with coefficient $r = 0.7305$.

The relationship of interoperability capability with three variables of adaptive capability is also positive and significantly high, with highest coefficient value $r = 0.5634$ with innovation and least coefficient value $r = 0.5513$ with resilience.

Table 1 Correlation of interoperability capability and adaptive capability

	Interoperability cap.	Adaptive cap.	Agility	Innovation	Resilience
Interoperability cap.	*				
Adaptive cap.	0.7305	*			
Agility	0.5543	0.5484	*		
Innovation	0.5634	0.9054	0.3409	*	
Resilience	0.5513	0.6523	0.1533	0.3624	*

*Cap. = Capability

All the other correlation coefficients are positive and moderately exhibit strong and linear relationship of adaptive capability with its variables. Adaptive capability have strong relationship with innovation with coefficient value r = 0.9054. The above table also represents the strong and positive relation between the variables of adaptive capability.

6 Conclusion

The objective of this research is to develop an effective and integrated interoperability Model for sustainable and adoptable SMEs. The model is empirically validated in 25 SMEs in Denmark to identify the correlation between the integrated interoperability capability and the adaptive capabilities: agility, resilience, and innovation through an integrated—interoperability capability model analysis. The above analysis validated the IIM model robustness and confirming our research hypothesis that integrated interoperability capability is highly correlated with firm's adaptive capabilities including: agility, resilience, and innovation which have direct impact on SMEs business sustainability.

References

1. Brunnermeier, S. B., & Martin, S. A. (2002). Interoperability costs in the US automotive supply chain. *Supply Chain Management, 7*, 71–82.
2. Peristeras, V., & Tarabanis, K. (2006). The connection, communication, consolidation, collaboration interoperability framework (CIF) for information systems interoperability. *International Journal of Interoperability in Business Information Systems, 1*, 61–72.
3. IEEE. (1990). Standard computer dictionary—A compilation of IEEE standard computer glossaries. New York: IEEE.
4. Alonso, G., Casati, F., Kuno, H., & Machiraju, V. (2003). *Web services: Concepts, architectures and applications*. Berlin: Springer.
5. Yang, J., & Papazoglou, M. P. (2000). Interoperation support for electronic business. *Communications of the ACM, 43*, 39–47.
6. Legner, C., & Wende, K. (2006). Towards an excellence framework for business interoperability. *Proceedings of the 19th Bled eConference eValues*.

7. Urs, G., & John, P. (2007). *Breaking down digital barriers when and How ICT interoperability: Drives innovation*. Cambridge: Berkman Publication Series.
8. Ansoff, H. I. (1984). *Implanting strategic management*. Englewood Cliffs: Prentice-Hall International.
9. Vogel, D. (1991). Business ethics past and present. *Public Interest, 102*(1), 49–64.
10. Grahame, W. K. (1991). Lean, mean and green. *Chief Executive*, *68*, 30–33.
11. Wheelwright, S. C. (1984). Manufacturing strategy: defining the missing link. *Strategic Management Journal, 5*(1), 77–91.
12. Swink, M., & Hegarty, W. H. (1998). Core manufacturing capabilities and their links to product differentiation. *International Journal of Operations and Production Management, 18*(4), 374–396.
13. Hayes, R. H., & Pisano, G. P. (1996). Manufacturing strategy: at the intersection of two paradigm shifts. *Production and Operations Management*, *5*(1), 25–41.
14. Bughin, J., & Jacques, J. M. (1994). Managerial efficiency and the Schumpeterian link between size, market structure and innovation revisited. *Research Policy*, *23*, 653–659.
15. Bititci, U. S., Ackermann, F., Ates, A., Davies, J. D., Gibb, S., MacBryde, J., Mackay, D., Maguire, C., Van der Meer, R., & Shafti, F. (2011). Managerial processes: an operations management perspective towards dynamic capabilities. *Production Planning and Control*, *22*, 157–173.
16. Holley, G. J., Greenley, G. E., Cadogan, J. W., & Fahy, J. (2005). The performance impact of marketing resources. *Journal of Business Research*, *58*, 18–27.
17. Wang, C. L., & Ahmed, P. K. (2007). Dynamic capabilities: A review and research agenda. *International Journal of Management Reviews, 9*(1), 31–51.
18. Tuominen, M., Rajala, A., & Möller, K. (2004). How does adaptability drive firm innovativeness? *Journal of Business Research, 57*(5), 495–506.

Support of Manufacture Enterprises Collaboration Through the Development of a Reference Ontology: A Contribution in Metrological Domain

Carlos A. Costa and João P. Mendonça

Abstract Nowadays, manufacturing environment is populated with computational information files and records with implicit knowledge, which integration has becoming a major problem to seamless computational integration. Moreover, most of these systems are heterogeneous, thus problems of interoperability are frequent and any collaborative environment becomes easily compromised. Ontologies constitute the set of concepts, axioms, and relationships that describes a domain of interest, contributing to harmonize the information flow within computational systems. The distributed and heterogeneous nature of the organizations, in particular networked enterprises, led to the development of different ontologies for the same or overlapping areas, resulting in non-interoperability. This has become the basis for research methodologies to support a reference ontology, contributing to the standardization and development of ontologies within enterprises and virtual network, providing interoperability properties to intelligent systems. This paper exposes how the MENTOR methodology assisted the development and use of a reference ontology in the field of metrology, contributing to manufacturing teams collaboration and systems' integration. The aim is to maintain the different ontologies of each partner, providing networked enterprises with coherent interaction and unambiguous communication. The case study in the field of metrology demonstrates the proposed methodology benefits introduced at collaborative manufacturing level.

Keywords Ontologies · Metrology · Measuring systems · Intelligent manufacturing · Semantic harmonization

C. A. Costa · J. P. Mendonça (✉)
CT2M—DEM, Universidade do Minho, Campus de Azurém 4800-058 Guimarães, Portugal
e-mail: jpmas@dem.uminho.pt

C. A. Costa
e-mail: berto@dem.uminho.pt

K. Mertins et al. (eds.), *Enterprise Interoperability VI*,
Proceedings of the I-ESA Conferences 7, DOI: 10.1007/978-3-319-04948-9_43,

1 Introduction

Nowadays, the global economy is driven by rapid innovation and short development schedules, with consumer's expectative always increasing in terms of performance, quality and products' cost [1]. The management of expectations is the driving force for the strong industrial development, especially the technological development in production processes and manufacturing. In this context, computer support behaves as an aggregator of the multidisciplinary information elements involved, contributing to increased automation and overall efficiency of manufacturing systems.

In parallel to manufacturing automation, and given the heterogeneous nature of the organizations involved, the teams face the challenge of integration and reuse of information, but non-interoperability and heterogeneity are main obstacles to such objective. The involvement of several organizations in nowadays product's life-cycle raises the complexity of the problem in terms of terminology and contexts, adopted product models and respective variables and computational data.

The absence of a common technical vocabulary among different partners is sometimes a source of conflicts that may result in non-interoperability between information systems. In this context, the establishment of the first edition of the VIM (International Metrology Vocabulary) was the first step towards the harmonization of terms and concepts in the field of metrology. Subsequently were presented other proposals and contributions, both from standards organizations and research community. However, given the diversity of the domain ontologies, not all concepts are covered, the issue remaining open therefore.

This led to research in methodologies to support a reference ontology. MENTOR methodology proposed by authors [2], is mainly intended to support an organization to adopt and build or reuse a reference ontology, particularly in this paper in the field of metrology. This reference ontology will improve semantic interoperability among the partners or companies concerning the contents of a standardized model of data representation, while allowing each partner to maintain its own ontology, providing consistent and unambiguous interaction.

The subsequent contents of this paper are organized as follows: Sect. 2 reports the flow of information in the inspection phases; Sect. 3 presents the MENTOR methodology for the development of a reference ontology, showing the benefits of its application; Sect. 4 presents a case study in the field of metrology and demonstrates the proposed methodology and Sect. 5 presents main conclusions.

2 Information Flow in Inspection Phases

During the inspection phases, the processes and the measurement systems need to be flexible and able to verify product compliance automatically, in order to validate the information corresponding to the design and manufacturing stages in a more integrated and adjustable process in time [3]. The variety of systems and software available contribute to such aimed automatic scenario, but a heterogeneous use may result in problems of application and interoperability.

The first edition of the VIM published by ISO in 1984 [4] contributed with the main set of harmonizing concepts and terms in metrology. Some other documents and updates followed, in order to harmonize concepts and terminology within the measurement field. Based on VIM, Abran and Sellami [5], established a model of the concepts and sub-concepts presented therein, aiming in particular to illustrate the various levels of abstraction. However, this modelling does not yet cover all the concepts; it needs to be completed in order to ensure full coherence and consistency of representation.

Another important issue, since it may compromise the automatic planning, inspection and assembly of mechanical components is related to the geometric tolerance. Requicha and Chan [6] proposed a constructive solid geometry (CSG) based scheme, allowing the representation of surface features with tolerances and other associated attributes. Additional developments in the domain may be found in [4–7].

The continuous diffusion of CAD/CAM fostered the Integrated Measurement Process (IMP), in parallel to manufacturing automation, to integrate the control and data sharing, facilitating the flow in good manufacturing systems, and easy, fast and reliable inspections. Feng presented a model for planning the dimensional control that provides the bridge between design and dimensional control of the manufactured products, summarizing some functional requirements and activities of IMP [8]. The additional performance and functional requirements may affect the planning and development of the IMP's activities. Tsai and Cutkosky proposed a model of representation and reasoning for geometric tolerances in project [9]. Hong and Chang [10] reported that, although many efforts have been made towards shaping the representation of geometric tolerances, the field of research is still relevant and active.

Zhao et al. [11] say that the modelling across an enterprise is viable due to the advances in Internet technologies and increasing integration requirements from industry. In this context, these authors present a model representation of geometric tolerances stratified by level of compliance. The model has as its main objective to allow an unambiguous communication among different application domains in an enterprise, and thus, to promote interoperability. The model uses the widely applied ASME Y14.5 M-1994 as its foundation layer, supplemented with additional geometric tolerances information defined by DMIS and STEP to form the corresponding conformance layers that support IMP.

The development and diffusion of CAD/CAM systems and IMP led to the release of ISO/IEC 14598 series [12]. These standards series established the general requirements for measurement methods and quality assessment of software products, and may also be used along the development and maintenance phases. Sharing the same terminology of ISO/IEC 9126 [13], this series is mainly concentrated in setting the concepts in the field of measurement and in the establishment of requirements, recommendations and guidelines for supporting the measurement process. The SQuaRE project [14], based on the standards of

ISO/IEC 14958 and ISO/IEC 9126 series, has its main focus on convergence, consistency and unambiguous concepts to avoid conflicts and promote interoperability. ISO/IEC 25000:2001 [15], was presented as the final result of convergence project between the standards referred and contains an explanation of the transition between them and SQuaRE, also providing information on how to use the ISO/IEC 9126 and 14598 series, in their earlier form.

The automatic assessment of compliance of organizations, products and processes for ISO 9000 led Henry Kim [16] to propose a formal model of ontology for enterprise quality. With a general character, the model also contains a sub-ontology of terms and concepts of measurement.

García et al. [17] identifies similarities, differences and conflicts in terminology and presents a unified approach, with the main objective of contributing to the harmonization of different standards of measurement software. Subsequently, the authors of [18] propose a basic Software Measurement Ontology (SMO) that aims at contributing to the harmonization of the different software measurement proposals and standards, by providing a coherent set of common concepts used in software measurement. This ontology is also aligned with the VIM, although semantic interoperability problems still remain to be solved [19, 20].

Recently, the development of ontologies, as promising techniques with capabilities to solve semantic issues, has been addressed by important companies, research and scientific community. Thus, each company is struggling to develop competencies at this ontological level, but inevitably different perspectives will lead to different final results, achieving different ontologies in the same business domain in the end [21]. One possible solution is to have a reference ontology for a specific domain that all the domain enterprises should share in their business, using a mediator in the interface between the reference and the proprietary.

3 MENTOR: Methodology for Reference Ontology Development

A reference ontology development may follow the MENTOR methodology, which is adopted in this work. Its main objective is to help an organization to adopt, reuse and/or build, a domain reference ontology, through several main steps as semantic comparisons, basic lexicon establishment, mappings among ontologies and other operations on knowledge based representations [22]. This methodology considered the state-of-the-art in terms of ontologies merging and concatenation (and applications and tools [22–25]) plus trends in the research field [26, 27] as well). In a quick overview, Fig. 1 (upper part) depicts the state diagram of the lexicon settlement phase.

The terminology gathering step concerns to the process of collecting all relevant terms in a previously defined specific domain. All the participants in the process should give their inputs. There is no rule from where the terms should

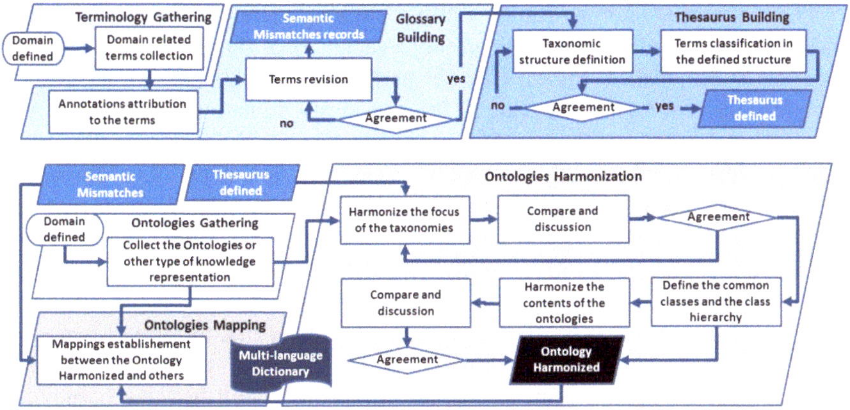

Fig. 1 MENTOR phases and steps

come: they are related with the domain established. Tools for automatic extraction of domain related terms can be found, but there is always need of a human checking before closing the terms list to avoid missing any domain terms. All the terms provided from the contributors are acceptable in this step. Nobody has authority to erase other's participant term. The term should be collected with reference to the contributor in order that each contributor provides term's annotation in the next step.

The Glossary Building step intends to build a glossary in the domain defined. It starts with annotations attribution to the terms collected. Each contributor should provide the annotations for his own terms. After having all the terms provided with annotations, it proceeds to the terms revision cycle. In this cycle it would be useful to use a multi-language dictionary in case of the organization members don't use the same natural language. The dictionary would help translations to the language agreed for the reference ontology. The terms revision process can have semantic and syntactic cases of mismatches, recorded as a semantic mismatch for future mappings using the proposed Mediator Ontology (MO). After a careful revision of all the terms with a successful agreement in their meaning consolidation, the glossary is defined from the terminology list in the domain specified. Another output from this process is the semantic mismatch records: this is made using the Mediator Ontology.

The Thesaurus Building step is composed by a cycle where firstly, the knowledge engineers define a taxonomic structure from the glossary terms, establishing some as thesaurus node terms. Secondly, the other terms are classified to the right paths in the existent taxonomic structure, being the thesaurus leafs. If there is an agreement in the structure and in the terms classified, the thesaurus is defined. If not, the cycle starts again from the taxonomic structure definition. The thesaurus defined will enhance the ontology harmonization process in the next phase.

The Reference Ontology Building phase is where the reference ontology is built and the semantic mappings between the organizational ontologies and the reference one is established. Figure 1 (lower part) describes this phase. The first step comprehends ontologies gathering in the previously domain defined. Other types of knowledge representation could be used as input for the harmonization ontologies process together with the thesaurus defined in the previous phase. The harmonization method for building ontologies, proposes the development of a single harmonized Ontology in two cycles where first the structure is discussed until having agreement on it, resulting in the definition of the common classes and the class hierarchy, and then the same process for the ontology contents definition is followed. From this process new semantic conflicts risk to arise. After agreement, the resolution could be recorded in the Mediator Ontology for further mapping establishments. With all the agreements accomplished, the harmonized ontology is finalized together with the mapping tables, describing the ontological relationships between the harmonized ontology and each one of the individual ontologies through the use of the semantic mismatches records. Semantic difficulties related to the natural language of the potential users of the harmonized ontology are likely to happen. To assist on it, the ontology is complemented with a multi-language dictionary where a set of normalized tokens gives the reference to the corresponding concepts and definitions in different native languages.

4 A Metrological Case Study

The competitive and demanding digital world of manufacturing business has led SMEs to consider the search for products in electronic format as an important method for parts selection and supply. Within networked enterprises, this is a major achievement, since news possibilities arise, like information retrieval, tasks automation and knowledge capture and re-use. Metrological stage is commonly an intermediary task, evaluating design conformance and manufacturing performance.

Considering the specific example of measuring a mechanical component, many product data models may be provided from different CAD applications, each one usually representing their specifications in different formats, with heterogeneous contents and classification. In common, most of them diverge from available advisory ISO standard designations or VIM vocabulary as described in previous section. Thus, the need to align applications and semantics, to exchange products data emerged as a priority to solve the dilemma. Figure 2 describes the validating scenario, where a set of enterprises agreed to work together to supply a big common client with various mechanical parts which are built collaboratively.

The first step is to follow a methodology which will guide the applications evaluation activities [2] determining interoperability level.

In the second step, to establish a common semantic level it is developed a reference ontology to the endeavours that are working together (right part of Fig. 2). The MENTOR methodology is used to develop such reference ontology.

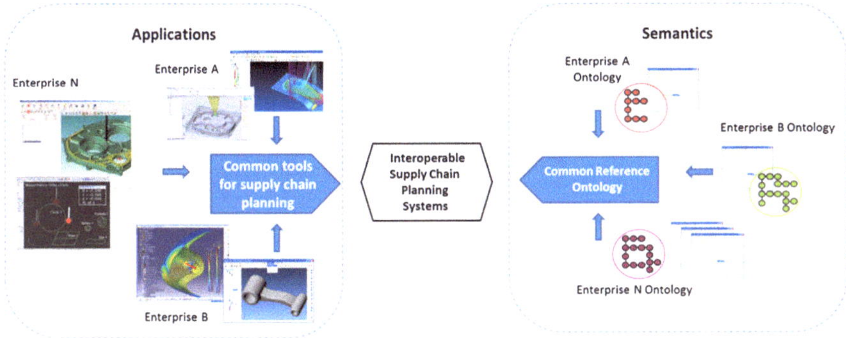

Fig. 2 Case study and validating scenario

During the reference ontology building phase, it is produced a mediator ontology which records, and if necessary translates, all the semantic operations performed in this process. Figure 3a shows a subset of the terms used by a Reference Ontology concerning the mechanical domain. In this example, when enterprise B receives a request of a "Hexagonal Bolt" with a "M16 diameter", the message is translated to a "hexagonal head Bolt" with a "Thread = Metric" with a "nominal diameter = 16" assisted by terms included in the reference ontology. Although these mappings are relatively smooth to operate at Classes level, the complexity increases when going deeper in the established hierarchic semantics. Now, it was found that Enterprise B has two terms that Reference Ontology doesn't consider, which are Maximum Distance and Minimum Distance. It was found too that an Enterprise C, acting in the same domain of interest, has also the term Tolerance not considered in the reference ontology.

During the harmonization phase, domain engineers from both enterprises decided to use them as reference. Thus, the concepts Maximum Distance and Minimum Distance, used as a property by enterprise B, may result in non-interoperability relatively to enterprise C. The same may occur due the term Tolerance used by the last one. Hence, three new additional concepts were created in reference ontology: Maximum Diameter, Minimum Diameter and Circularity. The first two defined in ISO 1101 Annex B as the maximum and minimum allowable variability in the process or characteristic in analysis; and the last concept defined by the difference between the radius of two concentric circles, whose value must be less or equal than the tolerance. MO logs all this operations, keeping the consistency between the ontologies. The example of Fig. 3b explains what happens in Enterprise B side when using the MO to translate messages according to reference ontology. The *maximum diameter* and *minimum diameter* reference concepts, which appear in the communication content, is replaced by MO to *maximum distance* and *minimum distance*. This way, Enterprise B gets and understands the messages with its own terminology and semantics, while the communication with external partners is under a common interoperable framework to all the endeavours.

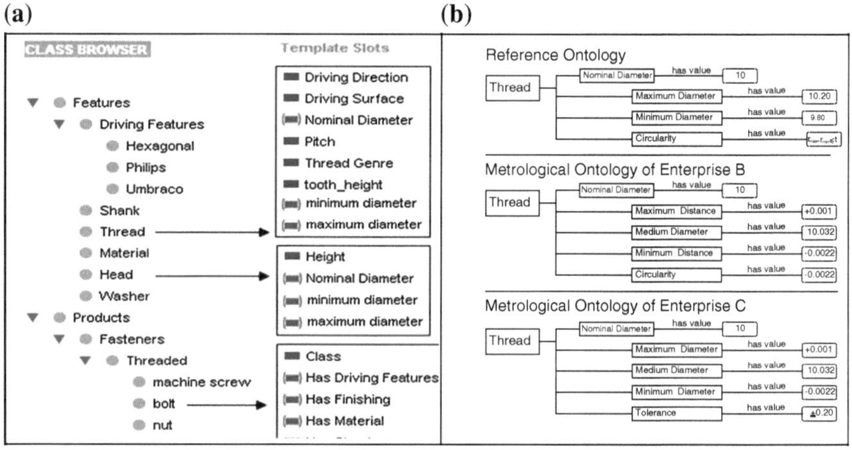

Fig. 3 **a** Reference ontology overview. **b** Mediator's translation example

One immediate advantage of the use of MO, is that the method enables the computational systems of any enterprise to smoothly communicate with external parties as it was using the Reference Ontology. This is also the main motivation that enterprises may consider to adopt the Reference Ontology building process, independently of their domain expertise.

5 Conclusions

Within networked enterprises, agreement on a reference ontology is presented as a contribution to promote interoperability, herein focused in the field of metrology. The reference ontology appears to meet the needs of the harmonization of manufacturing systems, enabling communication with each partner operating in the same domain in a consistent and unequivocal way. The reference ontology aims to improve interoperability between the companies involved, respecting proprietary models for standard data representation.

Those models, in order to ensure full coherence and consistency of data representation, granted contributions by several researchers, as well as major international organizations for standardization in the area.

The paper proposes the use of the MENTOR methodology to reinforce computational systems communication in what Inspection terminology respects, within a set of manufacturing enterprises working together, to seamless communicate between each other. Through this methodology, collaborative effectiveness may be experienced since they are enabled to understand each other using its own syntax and semantics present in its own data representation when supported by a metrological reference ontology. This achievement is assessed by software components evaluation and post conformance testing procedures. In this context, a subset of terms used by a reference ontology for the mechanical domain and related metrology are described as a case study, demonstrating the contribution to increase interoperability level of networked enterprises and collaborative assistance to manufacturing teams.

References

1. VDI 2206. (2004). *Design methodology for mechatronic systems*.Berlin: Beuth Verlag.
2. Sarraipa, J., Silva, J., Jardim-Goncalves, R., & Monteiro, A. (2008). MENTOR—a methodology for enterprise reference ontology development. In *2008 4th International IEEE Conference on Intelligent Systems*.
3. Pereira, P. H. & Hocken, R. J. (2006). Characterization and compensation of dynamic errors of a scanning coordinate measuring machine. *Precision Engineering*, *30*, 85 95.
4. http://www.inmetro.gov.br/metcientifica/vim/vimGum.asp
5. Abran, A., & Sellami, A. (2002). Initial modelling of the measurement concepts in the ISO vocabulary of terms in metrology. In *Proceedings of IWSM2002*, Magdeburg, Germany.
6. Requicha, A. A. G., & Chan, S. C. (1986). Representation of geometric features, tolerances, and attributes in solid models based on constructive geometry. *IEEE Journal of Robotics and Automation RA-2*, *3*, 156–166.
7. Turner, J. U. (1990). Relative positioning of parts in assembly using mathematical programming. *Computer-Aided Design*, *22*(7), 394–400.
8. Feng, S. C. (1996). A dimensional inspection planning activity model. *Journal of Engineering Design and Automation*, *2*(4), 253–267 (Special issue on Tolerance and Metrology for Precision Manufacturing).
9. Tsai, J. C., Cutkosky, & M. R. (1997). Representation and reasoning of geometric tolerances in design. *Cambridge Journals*, *11*(04), 325–341. http://journals.cambridge.org/action/displayAbstract?fromPage=online&aid=4203944
10. Hong, Y. S., & Chang, T. -C. (2002). A comprehensive review of tolerancing research. *International Journal of Production Research*, *40*(11), 2425–2460.
11. Zhao, X., Pasupathy, T. K., & Wilhelm R. (2006). Modeling and representation of geometric tolerances information in integrated measurement processes. *Journal Computers in Industry*, *57*(4), 319–330.
12. ISO/IEC, ISO/IEC 14598:2001. (2001). *Software engineering—product evaluation*. Genova, Switzerland: International Standards Organization.
13. ISO/IEC, ISO/IEC 9126:2001. (2001). *Software engineering—product quality*. Genova, Switzerland: International Standards Organization.
14. Azuma, M. (2001). SQuaRE: The next generation of the ISO/IEC 9126 and 14598 International Standards series on software product quality. In *Proceedings of European Software Control and Metrics (ESCOM)*, London, England.
15. ISO/IEC, ISO/IEC 25000:2005. (2001). *Software engineering—software product quality requirements and evaluation (SQuaRE)—guide to SQuaRE*. Genova, Switzerland: International Standards Organization.
16. Kim, H. M. (1999). Representing and reasoning about quality using enterprise models, PhD thesis, Department of Mechanical and Industrial Engineering. Canada: University of Toronto.
17. García, F., Bertoa, M. F., Calero, C., Vallecillo, A., Ruiz, F., Piatinni, M., & Genero, M. (2005). Towards a consistent terminology for software measurement. *Information and Software Technology*, *48*, 631–644.

18. García, F., Bertoa, M. F., Calero, C., Vallecillo, A., Ruiz, F., Piatinni, M., & Genero, M. (2008). *Efective use of ontologies in software measurement. The knowledge engineering review*, Vol. 00:0, Cambridge: Cambridge University Press.
19. Drejer, A. (2003). Gudmundsson, Agnar Exploring the concept of multiple product development via an action research project. *Integrated Manufacturing Systems, 14*(3), 208–220.
20. Silva, João P. M. A., Jardim-Goncalves, R., Monteiro, A. C., & Steiger-Garcao, A. (2004). Make the most of interoperability along PLC stages: A framework based on multilevel integration. DETC2004-57733, Salt Lake City, Utah, USA.
21. Jardim-Goncalves, R., Sarraipa, J., Agostinho, C., & Panetto, H. (2009). Knowledge framework for intelligent manufacturing systems. *International Journal of Intelligent Manufacturing, 12*(3), 281–293.
22. Fernandez, M., Gomez-Perez, A., & Juristo, N. (1997). METHONTOLOGY: From ontological art towards ontological engineering. In *Symposium on Ontological Engineering of AAAI* (pp. 33–40). Stanford, California.
23. Stumme, G., & Madche, A. (2001). FCA-Merge: Bottom-up merging of ontologies. In *7th International Conference on Artificial Intelligence* (*IJCAI '01*) (pp. 225–230). Seattle, WA.
24. Noy, N. F., & Musen, M. A. (2003). The PROMPT suite: Interactive tools for ontology merging and mapping. *International Journal of Human-Computer Studies*, *59*(6), 983-1024.
25. Gangemi, A., Pisanelli, D. M., & Steve, G. (1999). An overview of the ONIONS project: Applying ontologies to the integration of medical terminologies. *Data & Knowledge Engineering, 31*, 183–220.
26. Uschold, M., & King, M. (1995). Towards a methodology for building ontologies. In *Proceedings IJCAI-95 Workshop on Basic Ontological Issues in Knowledge Sharing*, Montreal, Canada.
27. Nicola, A., et al. (2005). *Proceedings of 16th International Conference on Database and Expert Systems Applications* (*DEXA*) (pp. 655–664). Denmark, LNCS no. 3588, Berlin: Springer.

Author Index

K. Mertins et al. (eds.), *Enterprise Interoperability VI*,
Proceedings of the I-ESA Conferences 7, DOI: 10.1007/978-3-319-04948-9,

MIX
Papier aus verantwortungsvollen Quellen
Paper from responsible sources
FSC® C105338

If you have any concerns about our products,
you can contact us on
ProductSafety@springernature.com

In case Publisher is established outside the EU,
the EU authorized representative is:
Springer Nature Customer Service Center GmbH
Europaplatz 3, 69115 Heidelberg, Germany

Printed by Libri Plureos GmbH
in Hamburg, Germany